TO

THOMAS G. SAVAGE, S.J.

1926–1975

VIVID IN MEMORY

༺༻

CONTENTS

THIRTY-SIX

PERSIAN THROW RUGS

LET'S START indoors. Let's start by imagining a fine Persian carpet and a hunting knife. The carpet is twelve feet by eighteen, say. That gives us 216 square feet of continuous woven material. Is the knife razor-sharp? If not, we hone it. We set about cutting the carpet into thirty-six equal pieces, each one a rectangle, two feet by three. Never mind the hardwood floor. The severing fibers release small tweaky noises, like the muted yelps of outraged Persian weavers. Never mind the weavers. When we're finished cutting, we measure the individual pieces, total them up—and find that, lo, there's still nearly 216 square feet of recognizably carpetlike stuff. But what does it amount to? Have we got thirty-six nice Persian throw rugs? No. All we're left with is three dozen ragged fragments, each one worthless and commencing to come apart.

Now take the same logic outdoors and it begins to explain why the tiger, *Panthera tigris*, has disappeared from the island of Bali. It casts light on the fact that the red fox, *Vulpes vulpes*, is missing from Bryce Canyon National Park. It suggests why the jaguar, the puma, and forty-five species of birds have been extirpated from a place called Barro Colorado Island—and why myriad other creatures are mysteriously absent from myriad other sites. An ecosystem is a tapestry of species and relationships. Chop away a section, isolate that section, and there arises the problem of unraveling.

For the past thirty years, professional ecologists have been murmuring about the phenomenon of unraveling ecosystems. Many of these scientists have become mesmerized by the phenomenon and, increasingly with time, worried. They have tried to study it in the field, using mist nets and bird bands, box traps and radio collars, ketamine, methyl bromide, formalin, tweezers. They have tried to predict its course, using elaborate abstractions played out on their computers. A few have blanched at what they saw—or thought they saw—coming. They have disagreed with their colleagues about particulars, arguing fiercely in the scientific journals. Some have issued alarms, directed at governments or the general public, but those alarms have been broadly worded to spare nonscientific audiences the intricate, persuasive details. Others have rebutted the alarmism or, in some cases, issued converse alarms. Mainly these scientists have been talking to one another.

They have invented terms for this phenomenon of unraveling ecosystems. *Relaxation to equilibrium* is one, probably the most euphemistic. In a similar sense your body, with its complicated organization, its apparent defiance of entropy, will relax toward equilibrium in the grave. *Faunal collapse* is another. But that one fails to encompass the category of *floral* collapse, which is also at issue. Thomas E. Lovejoy, a tropical ecologist at the Smithsonian Institution, has earned the right to coin his own term. Lovejoy's is *ecosystem decay*.

His metaphor is more scientific in tone than mine of the sliced-apart Persian carpet. What he means is that an ecosystem—under certain specifiable conditions—loses diversity the way a mass of uranium sheds neutrons. Plink, plink, plink, extinctions occur, steadily but without any evident cause. Species disappear. Whole categories of plants and animals vanish. What are the specifiable conditions? I'll describe them in the course of this book. I'll also lay siege to the illusion that ecosystem decay happens without cause.

Lovejoy's term is loaded with historical resonance. Think of radioactive decay back in the innocent early years of this century, before Hiroshima, before Alamogordo, before Hahn and Strassmann discovered nuclear fission. Radioactive decay, in those years, was just an intriguing phenomenon known to a handful of physicists—the young Robert Oppenheimer, for one. Likewise, until recently, with ecosystem decay. While the scientists have murmured, the general public has heard almost nothing. Faunal collapse? Relaxation to equilibrium? Even well-informed people with some fondness for the natural world have remained unaware that any such dark new idea is forcing itself on the world.

What about you? Maybe you have read something, and maybe cared, about the extinction of species. Passenger pigeon, great auk, Steller's sea cow, Schomburgk's deer, sea mink, Antarctic wolf, Carolina parakeet: all gone. Maybe you know that human proliferation on this planet, and our voracious consumption of resources, and our large-scale transformations of landscape, are causing a cataclysm of extinctions that bodes to be the worst such event since the fall of the dinosaurs. Maybe you are aware, with distant but genuine regret, of the destruction of tropical forests. Maybe you know that the mountain gorilla, the California condor, and the Florida panther are tottering on the threshold of extinction. Maybe you even know that the grizzly bear population of Yellowstone National Park faces a tenuous future. Maybe you stand among those well-informed people for whom the notion of catastrophic worldwide losses of biological diversity is a serious concern. Chances are, still, that you lack a few crucial pieces of the full picture.

Chances are that you haven't caught wind of these scientific murmurs about ecosystem decay. Chances are that you know little or nothing about a seemingly marginal field called island biogeography.

❦

THE MAN

WHO KNEW ISLANDS

❦

BIOGEOGRAPHY is the study of the facts and the patterns of species distribution. It's the science concerned with where animals are, where plants are, and where they are not. On the island of Madagascar, for instance, there once lived an ostrichlike creature that stood ten feet tall, weighed half a ton, and thumped across the landscape on a pair of elephantine legs. Yes, it was a bird. One thousand pounds of bone, flesh, feathers. This is no hypothetical monster, no implausible fantasy of Herodotus or Marco Polo. In a ramshackle museum in Antananarivo, I've seen its skeleton; I've seen its two-gallon egg. Paleontologists know it as *Aepyornis maximus*. The species survived until Europeans reached Madagascar in the sixteenth century and began hunting it, harrying it, transforming the ecosystem it was part of, scrambling those bounteous eggs. A millennium ago, *Aepyornis maximus* existed only on that single island; now it exists nowhere. To say so is the business of biogeography.

As practiced by thoughtful scientists, biogeography does more than ask *Which species?* and *Where?* It also asks *Why?* and, what is sometimes even more crucial, *Why not?*

Another example. The island of Bali, a small mound of volcanic rock and elaborate rice terraces just off the eastern tip of Java, in Indonesia, once supported a unique subspecies of tiger. *Panthera tigris balica*, it was called. Java itself had a different subspecies, *Panthera tigris sondaica*. Meanwhile the island of Lombok, just east of Bali across a twenty-mile stretch of ocean, had no resident tigers at all. Today the Balinese tiger exists nowhere, not even in zoos. It was driven extinct by an intricate combination of the usual factors. The Javanese tiger is probably also extinct, though some people cherish a feeble hope. Sumatra still has a few tigers, and again those belong to a distinct subspecies. Tigers can also be found in certain regions of mainland Asia, but not in the northwestern part of the continent, nor in Africa, nor in Europe. Once they existed as far west as Turkey. Not anymore. And Lombok, no smaller than Bali, with forests no less inviting, is still as tigerless as ever.

Why, why not, why? These facts, and their explanations, represent biogeography. When the same sort of attention is focused specifically on islands, it becomes island biogeography.

Island biogeography, I'm happy to report, is full of cheap thrills.

Many of the world's gaudiest life forms, both plant and animal, occur on islands. There are giants, dwarfs, crossover artists, nonconformists of every sort. These improbable creatures inhabit the outlands, the detached and remote zones of landscape and imaginability; in fact, they give vivid biological definition to the very word "outlandish." On Madagascar lives a species of chameleon barely more than an inch long, the smallest chameleon (among the smallest land vertebrates of any sort, actually) on the planet. Madagascar was also the home of a pygmy hippopotamus, now extinct. On the island of Komodo lurks that gigantized lizard we've all heard of, hungry for flesh and plausibly nicknamed a dragon. In the Galápagos an ocean-going iguana grazes underwater on seaweed, flouting the usual limits of reptilian physiology and behavior. In the central highlands of New Guinea you can catch sight of the ribbon-tailed bird of paradise, no bigger than a crow but dragging a pair of grossly elongated tail feathers, white streamers like the tail on a kite, as it swims heavily through the air of a clearing. On a tiny coral island called Aldabra, in the Indian Ocean, lives a species of giant tortoise less famous though no less imposing than the Galápagos tortoises. On Saint Helena there existed, at least until recently, a species of giant earwig—the world's largest and no doubt most repulsive dermapteran insect. Java has its own species of rhinoceros, another pygmy. Hawaii has its honeycreepers, a whole group of bizarre birds seen nowhere else. Australia has its kangaroos and other marsupials, of course, while the island state of Tasmania has its devil, its bettong, its pademelon, and its spotted-tailed quoll, marsupials too peculiar even for mainland Australia. The island of Santa Catalina, in the Gulf of California, has a rattleless rattlesnake. New Zealand has the tuatara, the last surviving species of an order of beaky-faced reptiles that flourished in the Triassic period, before the peak of the dinosaurs. Mauritius, until Europe invaded, had the dodo. The list could go on, with no diminution of weirdness. The point is that islands are havens and breeding grounds for the unique and anomalous. They are natural laboratories of extravagant evolutionary experimentation.

That's why island biogeography is a catalogue of quirks and superlatives. And that's why islands, those outlands, have played a central role in the study of evolution. Charles Darwin himself was an island biogeographer before he was a Darwinist.

Some of the other great pioneers of evolutionary biology—notably Alfred Russel Wallace and Joseph Hooker—also gathered their best insights from fieldwork on remote islands. Wallace spent eight years collecting specimens in the Malay Archipelago, the empire of islands (and therefore of biological diversity) that now goes by the name Indonesia. Hooker, like Darwin, was lucky and well connected enough to get a place on board one of Her Majesty's ships. It was the *Erebus*, sent off (like Darwin's *Beagle*) on a round-the-world charting expedition, from which Hooker went ashore on Tasmania, New Zealand, and an interesting little nub called Kerguelen Island, halfway between Antarctica and nowhere. Decades later, Hooker was still publishing studies of the plants of New Zealand and his other island stops.

The trend started by Darwin, Wallace, and Hooker has continued throughout this century, with important studies done in New Guinea, in the southwestern Pacific, in Hawaii, in the West Indies, and on Krakatau after the big eruption. It came to a culmination of sorts—or at least to a turning point—in a dense little volume titled *The Theory of Island Biogeography*, published in 1967. *The Theory of Island Biogeography* was a daring, fruitful, and provocative attempt by two young men to merge biogeography with ecology and transform them into a mathematical science. Wherever we wander in this book, you and I, we'll never be far from that one. The two young men were Robert MacArthur and Edward O. Wilson. They derived their theory in part from the patterns of distribution of ant species that Wilson had found among the islands of Melanesia.

Islands have been especially instructive because their limited area and their inherent isolation combine to make patterns of evolution stand out starkly. It's such an important truth I'll repeat it: Islands give clarity to evolution. On an island you might find those enormous tortoises, those flightless birds weighing half a ton, those pygmy chameleons and hippos. Generally you will also find fewer species, and therefore fewer relationships among species, as well as more cases of species extinction. All these factors result in a simplified ecosystem, almost a caricature of nature's full complexity. Consequently islands have served as the Dick-and-Jane primers of evolutionary biology, helping scientists master enough vocabulary and grammar to begin to comprehend the more complex prose of the mainlands. *The Origin of Species* and *The Theory of Island Biogeography* are only two among many landmarks in biological thought that owe their existence to the contemplation of islands. Another is *Island Life*, the first major compendium of island biogeography, published in 1880 by Wallace.

Alfred Russel Wallace was a modest man from a humble English background, the eighth child of amiable, unfocused parents. His father was trained as a solicitor but never practiced, preferring to work as a librarian, dabble disastrously at business ventures, or grow vegetables. After the senior Wallace lost his inheritance through bad investments, the family as a whole lost its foothold in the middle class. Alfred had to leave school at age fourteen and go to work. He learned the surveying trade. It was an early start toward a life of hard, honest, narrow toil, but Alfred broke clear of it. He educated himself during stolen evening moments at workingmen's institutes and public libraries, then escaped from his niche and from England altogether on a wild youthful adventure, and eventually turned himself into the greatest field biologist of the nineteenth century. Most people know of him, if at all, as the fellow who stumbled upon Darwin's most famous idea just before Darwin himself got around to publishing it.

Charles Darwin was a generation older, and had come back from his eye-opening journey aboard the *Beagle* twenty years earlier. Soon after returning, he had conceived his great theory. But the theory was a heretical notion in its early Victorian context, and Darwin was a cautious man, so he had spent those twenty years nurturing it in secret. The heretical notion, specifically, was this: that species had evolved, one from another, along lineages that were continuous but continuously changing, by a process of competitive struggle and differential survival, which Darwin labeled natural selection. Others before him (including Jean Baptiste Lamarck, Georges Buffon, and his own grandfather, Erasmus Darwin) had entertained the view that species were shaped by some form of evolution. None of those earlier evolutionists, though, had offered a persuasive explanation of *how* species evolved, because none had hit on the concept of natural selection. That concept remained Darwin's secret insight, his carefully guarded intellectual treasure, while he devoted much of two decades to gathering evidence and framing arguments that would support it. Then one day Darwin received a manuscript in the mail from a young, obscure naturalist named Wallace—and the Wallace manuscript, to Darwin's horror, contained his own precious concept. Wallace had found his way to it independently. For a brief heartsick period, Darwin believed that the younger man had eclipsed him and preempted his life's work by staking a just claim to priority. As things developed, however, with Joseph Hooker's collusion, Wallace and Darwin announced the concept simultaneously. For a variety of reasons, some good and some shabby, Darwin received most of the recognition; and Wallace, in consequence, is famous for being obscured.

But that's just the cartoon version of a complicated, troubling story. The cartoon neglects much, including Alfred Russel Wallace's role as the patriarch of biogeography.

Darwin was a more ingenious theorist, admittedly. Hooker became the preeminent botanist of his time. Wallace, much later in life, involved himself in certain crankish interests (including the Land Nationalization Society, an anti-vaccination crusade, and spiritualism) that have made it easier for historians to treat him unfairly. Still, Wallace remains the most heroically appealing, at least to my crankish taste. No doubt I'm biased by the fact that he, unlike Darwin and Hooker, was an impecunious freelancer.

On June 13, 1856, after a twenty-day passage down from Singapore aboard a schooner called the *Rose of Japan*, Wallace stepped ashore on the island of Bali. He was just passing through, en route east toward Celebes. The layover in Bali lasted two days. Following his normal routine, Wallace collected some birds and insects. He scouted the landscape. Then he rejoined the *Rose of Japan* and crossed the strait to Lombok. In physical distance, it's a short trip.

3

THE GATEWAY to the island of Lombok is Padangbai, a ferry port on the east coast of Bali. It's a scruffy little town, far off track from the glittery destinations of international beach tourism—Sanur, Kuta Beach, Nusa Dua—for which Bali has become lamentably famous. At Padangbai, there's no notable surf. No upscale hotel. From the pier, Lombok shows itself as a blue gray silhouette on the eastern horizon, like a cold morning sun, offering only the dimmest promise. Of all the Western and Japanese travelers who jet into Bali each year, not one in a thousand crosses eastward to Lombok. Not one in a thousand has any reason to. Nobody would claim that the less famous island is dramatic, except in a rarefied biogeographical sense.

The ferry departs at two o'clock, I learn, but it's already full. This unwelcome information comes from Nyoman Wirata, my translator, driver, and friend. I can see that there's more information lurking shyly behind Nyoman's eyes. I accept the bad news, and I wait.

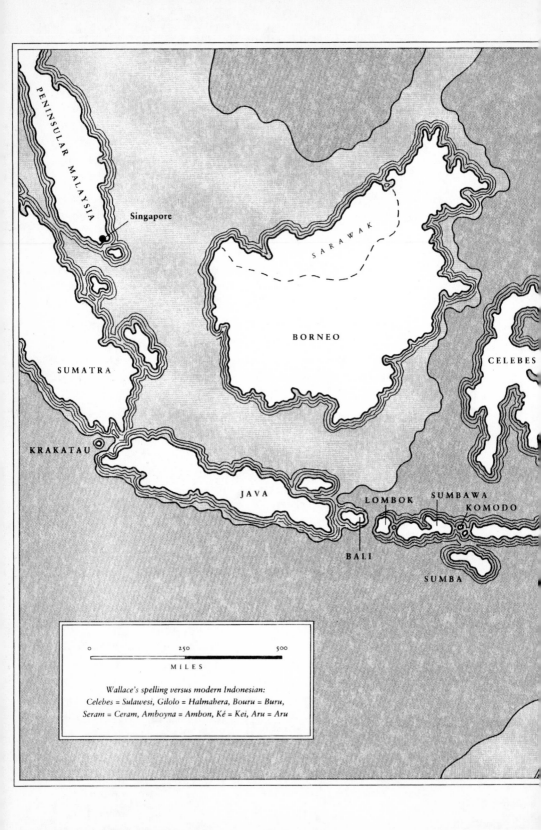

PENINSULAR MALAYSIA

Singapore

SARAWAK

BORNEO

CELEBES

SUMATRA

KRAKATAU

JAVA

LOMBOK

SUMBAWA

KOMODO

BALI

SUMBA

0   250   500
MILES

Wallace's spelling versus modern Indonesian:
Celebes = Sulawesi, Gilolo = Halmahera, Bouru = Buru,
Seram = Ceram, Amboyna = Ambon, Ké = Kei, Aru = Aru

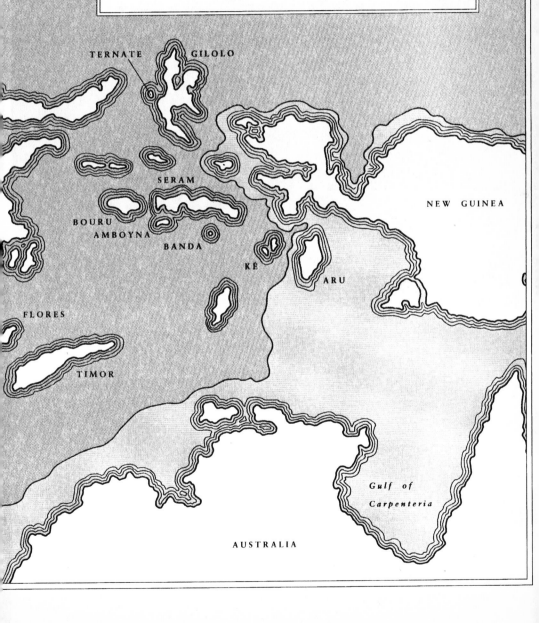

THE
MALAY
ARCHIPELAGO

TERNATE   GILOLO

SERAM

BOURU
AMBOYNA
BANDA

KÉ

ARU

NEW   GUINEA

FLORES

TIMOR

Gulf of
Carpenteria

AUSTRALIA

Then again, maybe the ferry he's *not* full, Nyoman says. A five-thousand-rupiah bribe to the ticket agent might open up some space.

Nyoman is a sweet-natured Balinese tailor unaccustomed to brokering waterfront bribes. He's a sterling young man whose real trade is shirtmaking and whom for this expedition I've shanghaied away from his shop. His tailoring business and his home are in Ubud, a little wonderland of a town in the uplands, full of temples and art studios and clever mask carvers and clanging *gamelan* orchestras. Despite the tourism boom that has tarted up its face in recent years, Ubud remains graceful, pious, more spiritual than most towns on the island—damn well more spiritual than Padangbai—and my trusted pal Nyoman is a true Ubudian. He's so excruciatingly polite and ingenuous that if bribing a ticket agent seems to him faintly advisable, in the mind of the agent it's no doubt mandatory. Having made that extrapolation, I give Nyoman a fistful of rupiahs and full discretion, then I go off to find food.

Padangbai has the character of any border town, though the border crossing here happens to be a boat trip. The streets are full of predatory cab drivers and part-time masseuses and aggressive toddlers who offer to change money. There's none of the artistic ethos nor the architectural charm found in the highlands around Ubud, and the Hindu-Animist ceremonial element is absent too. Lombok, just across the strait, is predominantly Moslem, and the Moslem influence seems to begin right here at the point of embarkation. Beer is suddenly harder to come by, for instance. Still, Padangbai has a certain roguish forthrightness. "Hey. Hello. Where are you going?" the sidewalk hustlers holler demandingly at me. "What you want?" Whatever it is, they can deliver. But all I want is to sit down, out of the sun and preferably in the presence of a cold drink. I catch up with my other traveling partner, a Dutch biologist named Bas van Balen, and we investigate the offerings of Padangbai.

Skinny as a Quaker, with a wiry black Smith Brothers beard and an imperturbable digestive tract, Bas has been ten years in the country, studying birds. He's fluent in the Indonesian language and in sixty different patterns of avian song, but this will be his first trip across to Lombok. Bas and I find a *warung* for some *makanan* and inquire whether *ada nasi campur*. In English: We find a café and order the rice-with-today's-leftover-meat. The first night I fell in with him, Bas led me off to a street stall for *rajak cingur*, which turned out to be cow snout and bean sprouts in muddy gravy. The cow snout was acceptable, but the thick orange drink he recommended for washing it down—something called *jamu*, resembling a barium cocktail flavored with buttermilk and garlic—was a onetime experience, thanks. Today, in retreat from Padangbai's midday heat, we quench our thirst with bottles of tepid Temulawak, a peculiar but inoffensive soft drink made from the pulp of a tuber. And then, with time to kill and Lombok on my mind, I reread a certain passage by Wallace.

It comes from *The Malay Archipelago*, the narrative account of his eight-year expedition among the islands.

> During the few days which I stayed on the north coast of Bali on my way to Lombock, I saw several birds highly characteristic of Javan ornithology.

This was in 1856, the third year of his Malay travels, as he was working his way eastward from an earlier base at the town of Malacca on the Malaysian peninsula. Java in that era was a familiar outpost to English and Dutch colonialists, and its resident bird life was well known. Wallace had noted the unsurprising fact that at least some of those Javanese birds occurred also on Bali, to the east across a narrow strait. He mentioned a half-dozen species, including the yellow-headed weaver, the rosy barbet, and the Javanese three-toed woodpecker. Then he added a noteworthy observation:

> On crossing over to Lombock, separated from Bali by a strait less than twenty miles wide, I naturally expected to meet with some of these birds again; but during a stay there of three months I never saw one of them, but found a totally different set of species, most of which were utterly unknown not only in Java, but also in Borneo, Sumatra, and Malacca.

Conspicuous among these other species was the white cockatoo, absent from Bali, absent also from Java, more characteristic of Australia. A mysterious line of some sort had been crossed.

It came to be known as Wallace's Line. As drawn in those days, it split the gap between Borneo and Celebes. Continuing southward, it split also that narrower gap between Bali and Lombok. As far as the pioneer biogeographers were concerned, these two sibling islands, Bali and Lombok, belonged to two different realms. West of the dividing line lived tigers and monkeys, bears and orangutans, barbets and trogons; east of the line were friarbirds and cockatoos, birds of paradise and paradise kingfishers, cuscuses and other marsupials including (farther east in New Guinea and tropical Australia) the ineffable tree kangaroos, doing their clumsy best to fill niches left vacant by missing monkeys. Bali and Lombok were the two little islands—similar in size, similar in topography, similar in climate, nestled shoulder to shoulder—where the zoological differences were most sharply delineated. Wallace himself wrote that "these islands differ far more from each other in their birds and quadrupeds than do England and Japan." And from the earliest moment he had been wondering why.

The secret of Wallace's Line, as eventually discovered and clarified by later biogeographers, is deep water. The sea gap between Bali and Lombok is narrow but deep, because Lombok lies just *off* the continental shelf while Bali, so close, lies just *on* the shelf's southeast edge. Having once been connected by dry land, Bali shares much of its flora and fauna with Java, Borneo, Sumatra, and the Malaysian peninsula, whereas Lombok, and all the other islands to the east in deep water, are truly at sea. Although Alfred Wallace didn't offer this insight himself, he elucidated the pattern that led to it. That pattern leads to much more as well. The narrow strait between Bali and Lombok is one of the sites where biogeography first came into focus.

I leave Bas drinking coffee in the *warung* and go off to buy water and bananas. On a lark I buy also some unfamiliar fruit, scaly brown globes the size of golf balls, and later ask Bas to identify them. Ah yes, *salak*, he tells me. They're more commonly known as lizard-skin fruit; they grow on a tree that resembles a date palm. Peeling one, I find the flesh cream-colored and squeaky firm. It tastes like a cross between pineapple and garlic. In passing I wonder whether this might be the secret ingredient of *jamu*, the nasty orange drink. If not, then how many different Balinese foods taste like a cross between garlic and something? I'm still puzzling over that one when Nyoman appears, his innocent Balinese face split with a guilty grin, to announce that he has bribed us onto the ferry.

As the boat pulls away I get a last glance at Padangbai. The shining silver turret of a mosque stands above the *warungs*, the shops, the beached praus, and a big cheesy billboard for Bintang beer. Bintang is the Budweiser of Indonesia, cheap, democratic, ubiquitous, the main difference being that (thanks to the Dutch influence and a Heineken recipe, I gather) Bintang tastes good. On this island, even Moslem restaurants will serve it to you, provided that you're obviously an infidel. On the next island, things will be different. *Selamat datang*, says the Bintang sign. Welcome, it tells us, as we leave.

The crossing takes three hours. The upper deck is crowded with Indonesians on holiday, on pilgrimage, on business, on journeys to see family. A traveling salesman is working the crowd, trying to sell false teeth from a cardboard display box. He carries molars, incisors, canines, a half-dozen different specimens of each type, shrink-wrapped individually like antihistamine tablets—a nice selection, though no one is buying. Like any good salesman, he appears undaunted by failure and rejection. Probably he's just amusing himself on his way to a promising circuit of the dentists of Lombok.

Two miles out of Padangbai I see a small white butterfly, beating its way eastward on the wind. The butterfly is a suggestive datum. Here in the sea gap between Bali and Lombok, making its reckless attempt to cross Wallace's Line, it stands as an emblem for larger topics—isolation, speciation, dispersal, all connected within the still larger subject of island biogeography. Isolation is the most obvious of those three, since of course islands are inherently isolated. Their isolation fosters speciation, the process in which one species splits into several. But the isolation of islands isn't absolute, and some creatures always manage to breach it—that's dispersal. Maybe the white butterfly from Bali will make landfall in Lombok and play some genetic role there among a population of similar butterflies . . . though most likely, no, it will fall short. The arduousness of dispersal across stretches of ocean is part of what gives island biogeography its special clarity. Many creatures set out, few arrive. The failure and death of potential dispersers enforces isolation long enough for speciation to occur.

Speciation, let's remember, is a modern word. Within the Victorian world from which Darwin and Wallace went off on their journeys, it had never been spoken or written. Most people hadn't even thought it. Most people believed in a God who had created all species from scratch and put them on Earth in elaborately illogical arrangements— tigers here but not there, cockatoos only as far as this or that point, rhinos on this or that tropical island but not on the neighboring one, white butterflies in some places and blue butterflies in others. Those arrangements, it was assumed, somehow suited divine whimsy. Biogeography until the mid–nineteenth century served to prove nothing except that the deity had a disjointed and inscrutable sense of pattern.

I lose sight of the sea-crossing butterfly. I gape at the horizon. Halfway to Lombok, with folks lying asleep on the benches and on the rusty iron floor, with even the tooth salesman at rest, stewards appear smartly on deck to administer our lifejacket demonstration. In crisp Indonesian the stewards recite a program of safety procedures. I listen attentively, enjoying the lilt of the language, comprehending zilch. Later I discover a bilingual sign; the English half says:

> IN AN EMERGENCY, KEEP YOUR HEAD AND FOLLOW THE CREWS IN-
> DUCEMENTS CONFUSION MAKES THE SITUATION WORSE. ON
> LEAVING SHIP, GIVE PREFERENCE TO LADIES, CHILDREN AND
> OLDS AND TIDY UP YOURSELF TO HAVE YOUR HANDS FREE WITH
> ONLY VALUABLES AS POSSIBLE. THE LIFE JACKETS ARE STOWED IN
> EACH CABIN KINDLY CONFIRM THE STORED PLACE AND HOW TO
> WEAR IT BY "HOW TO PUT ON THE LIFE JACKET" AND CHECK THE
> LEAVING ROUTE BY THE MAP OF LEAVING ROUTE.

No doubt it sounds more reassuring in Indonesian.

At sunset, as we glide into a well-sheltered harbor on the northwestern coast of Lombok, the ship's loudspeaker croons out a Moslem chant. It's the evening prayer, nasal and soothing. The volcanic cone of Gunung Rinjani, Lombok's highest mountain, rises over us to darken the twilight. Menacing gray clouds are coming in from the east, from Komodo and Timor, but the water of this bay is preternaturally slick. Offshore, a small prau with a blue sail cruises weightlessly. A pair of beefy Australian tourists appear on deck, ruddy-faced from Bintang and armed with video cameras to film their arrival for later experiencing. And I think of Alfred Russel Wallace during his years of lonely travel.

4

IN February of 1855, the year before he reached Bali and Lombok, Wallace was holed up in Sarawak, one of the states on the island of Borneo. He had the use of a little house at the mouth of a river. It was the wet season. "I was quite alone," he would recount fifty years later in his autobiography, "with one Malay boy as cook, and during the evenings and wet days I had nothing to do but to look over my books and ponder over the problem which was rarely absent from my thoughts." The problem to which he alluded was: What is the origin of species? Unlike Charles Darwin two decades earlier, Wallace had started off on his travels with that question already in mind. The interlude in Sarawak is important because there, for the first time, he attempted an answer.

His answer derived from biogeography. "Having always been interested in the geographical distribution of animals and plants," he would recall—and having boned up particularly on Malayan insects and reptiles, at the British Museum, before he embarked on this trip—he realized that "these facts had never been properly utilized as indications of the way in which species had come into existence." The data of plant and animal distribution were simply accepted as mysterious givens. Charles Darwin himself had offered many of those data for sheer entertainment value in his narrative account of the *Beagle* voyage, first published in 1839. (In this original edition the book was catchily titled *Journal of Researches into the Geology and Natural History of the Various Countries Visited by H.M.S. Beagle, under the Command of Captain Fitzroy, R.N. from 1832 to 1836.* To us it's more familiar in later editions as *The Voyage of the "Beagle."*) But even Darwin's *Journal* was just a scientific travelogue, a pageant of colorful creatures and places, propounding no evolutionary theory. The theory would come later. Wallace was correct in his recollection, looking back on the 1855 circumstances, that the biogeographical data had never been properly utilized.

Charles Darwin in 1855 was a patient, sedulous forty-six-year-old. He knew that the notion of evolution by natural selection was going to incite a shitstorm of resentment in Victoria's England, and not just from rock-headed clergy. Most of Darwin's own friends among the scientific elite and the comfortable gentry were also disposed toward loathing that notion. For Alfred Wallace, the situation was slightly different. He knew, as Darwin did, that a cogent theory of evolution would rattle tea carts in the rectories, raise eyebrows in the country mansions, and put solid chaps off their sherry at the better clubs. But Wallace, unlike Darwin, didn't care. He was a hungry young man with nothing to lose.

5

THOSE UNUTILIZED facts of biogeography were a byproduct of the great age of discovery. They had piled up in libraries and museums during the previous four centuries as the more remote regions of sea and land had surrendered to exploration by adventurous Europeans who kept their eyes open and later told tales. From all over the planet, travelers had been bringing home reports of wondrous, unexpected beasts. In 1444, for instance, a Venetian named Nicolo de Conti wrote an account of his travels, in which he mentioned the white cockatoo. De Conti had just returned from the Malay Archipelago.

The first voyage of Christopher Columbus yielded eyewitness accounts of macaws, caimans, manatees, and iguanas, none of which had a place in the belief structure of an orthodox European. The later Columbian voyages added South American monkeys, peccaries, and the Hispaniolan hutias, a group of tree-climbing rodents that were ratlike without quite being rats. Antonio Pigafetta, who sailed with Magellan, published a journal that mentioned the golden lion marmosets of eastern Brazil, dainty monkeys that do look like lions, as Pigafetta claimed, "only yellow and very beautiful." Pigafetta also described rheas, which he may have mistaken for ostriches, and penguins, which he could hardly have mistaken for anything. About that time, also from the Americas, the first report of a hummingbird reached Europe; the first accurate report of an armadillo; the first report of a toucan; the first reports of giant anteaters and sloths and vicunas and one useful bird that was quickly welcomed to the Old World as a domestic, the turkey; the first report of the American bison, a giant-headed animal almost too majestic to be real. Imagine how these prodigies must have unnerved a complacently pious sixteenth-century European. Imagine it this way: You think you know every bird in the world, and then someone shows you a curl-crested aracari.

God had supposedly stopped creating after the sixth day. But now, as the wider world opened to the taking of a more thorough biological inventory, it seemed that God had stayed busier than anybody had dreamed.

One theological tenet threatened by the more thorough inventory was the Genesis account of Noah's ark. A boat that was literal rather than symbolic, and therefore finite rather than infinite, could have carried only so many passengers—even if it did measure three hundred cubits by fifty cubits. If the ark had previously been full, with the roster of new animals it became hopelessly crowded.

Was there room aboard that boat for a pair of capybaras? Around the year 1540 Juan de Ayolas sighted this giant South American rodent, "a sort of waterboar, half hog, half hare." Was there room for a pair of sloths? About the same time Gonzalez Fernando de Oviedo described this creature as "one of the slowest beastes in the worlde, and so heavy and dull in mowynge that it cannot scarcely go fyftie pases in a hole day." Was there room for possums? From a forest in Venezuela, Vicente Pinzón brought back a live one to Spain. It was a furred animal like none seen before. It carried its young in a pouch.

The American voyages, the American fauna, were just a beginning. The African continent had its own share of novelties. Elephants, rhinos, lions, and hippopotamuses had been known since the years of the Roman empire, but slightly less spectacular animals, such as warthogs and civets, still seemed remarkable to travelers in the sixteenth century. Toward the end of that century, during a visit to eastern Africa, Friar Joanno dos Sanctos caught sight of an aardvark. He was impressed by the extensible tongue, so useful for slurping up ants, and by certain other incongruous features, such as "the snout very long and slender, long eares like a mule, without haire, the taile thick and straight of a spanne long, fashioned at the end lyke a dystaffe." In India, about the same time, a Dutchman named Jan Huygen van Linschoten saw a pangolin. It was the size of a dog, he reported, with the snout of a hog, small eyes, two holes where its ears should have been, and its whole body "covered with scales of a thumb breadth," which Linschoten took to be "harder than iron or Steele."

From the exotic East came reports of exotic birds. There had been de Conti's account of the cockatoo and of several other parrotlike species that he had seen on one of the islands of the Banda cluster, in the far southeastern corner of the Malay Archipelago. There was the testimony of a Portuguese diplomat named Tomé Pires, who around 1512 stopped at the Aru Islands, still farther east, and laid eyes on the gaudy skins of another unimagined group of birds. The Aru Islands, considerably more remote even than Banda, were richly endowed with biological oddities. The specimens that Pires saw, presumably in a small trading settlement where they had been offered by native hunters, were "birds which they bring over dead, called birds of paradise, and they say they come from heaven and that they do not know how they are bred." This seems to have been the first European report of an especially decorative and significant avian family. Besides the two species native to Aru, another forty-some species of bird of paradise remained to be found in the forests of New Guinea, northeastern Australia, and other small offshore islands of the vicinity.

Actual specimens reached Europe slowly and in mangled form—as dried skins, with the feet (sometimes also the wings and the head) chopped away in the process of preservation. Van Linschoten, the Dutch traveler who had seen a live pangolin but never a live bird of paradise, offered his readers the unreliable assurance that *no one* had ever seen these birds alive, because they spent all their lifetimes in the upper atmosphere. They were footless and wingless, according to him, ethereal creatures of unearthly beauty and unearthly habits that fell to the ground only at death. That misapprehension remained strong through the mid–eighteenth century, when the systematizing Swedish biologist Carl von Linné (known more commonly, through his writings, as Carolus Linnaeus) gave a formal scientific description and a Latin name to the larger of the two Aru species. Linnaeus called it *Paradisea apoda*, the footless. We shouldn't judge Linnaeus's credulousness too harshly, since such a bird wasn't much more implausible than other aspects of the new zoological reality that people were forced to assimilate. If the giant bird *Aepyornis maximus* could exist without wings, why not a modest-sized *Paradisea apoda* without feet?

Fruit bats, dingoes, and cassowaries were also reported. Gradually the European surge of discovery and conquest was extending itself toward the Antipodes. In 1629 a Dutchman shipwrecked on an island off western Australia caught sight of a wallaby and lived to tell. Some people found it too much to believe. A large pouched mammal in upright posture, bouncing around on two oversized feet, with a face like a deer and ears like a rabbit? Right. Skepticism about Australian marsupials lasted more than a century. Then in 1770 the British ship *Endeavour*, commanded by Captain James Cook, landed along the northeastern coast in what is now tropical Queensland. Aboard the *Endeavour* was Joseph Banks, a wealthy amateur naturalist who had bought himself a place on the trip in order to add to his collections.

Joseph Banks was arguably the prototype of the shipboard naturalist on the voyage of exploration, a role that became more important when it was later filled by Darwin, Hooker, and T. H. Huxley, among others. Banks himself, unlike the best of those later shipboard naturalists, was no theorist. But he played his part. One day a whomping big beast was sighted, tracked with greyhounds, and shot. Having examined the carcass, Banks wrote: "To compare it to any European animal would be impossible, as it had not the least resemblance to any one that I have seen." The native people thereabouts, dark-skinned and peaceable folk who hadn't yet been given reason to want to murder Cook's party and feed the bodies to currawongs, explained that their name for the leaping animal was *kanguru*. Banks took the specimen back to England and commissioned George Stubbs to do a painting. From Stubbs's oil someone did an engraving, eventually published with Cook's journal. The kangaroo became famous.

So did *Raphus cucullatus*, the giant flightless pigeon from a little island called Mauritius in the Indian Ocean. This bird reached Europe during the early seventeenth century, initially in the tales of returned travelers and then probably in the form of several live specimens. It was such a ridiculous creature—round and ungainly, gawkish and risible—that its renown spread by hearsay and cartoon. It seemed to burlesque the very essence of birdness. Its anatomical components appeared badly matched—short legs, big body, tiny wing stubs, huge beak, with no harmony to the whole. Who could resist being amused by such an ornithological joke? From the memoirs of an English theologian, Sir Hamon L'Estrange:

> About 1638, as I walked London streets, I saw the picture of a strange fowle hung out upon a clothe and myselfe with one or two more then in company went in to see it. It was kept in a chamber, and was a great fowle somewhat bigger than the largest Turky Cock, and so legged and footed, but stouter and thicker and of a more erect shape, coloured before like the breast of a young cock fesan, and on the back of dunn or dearc colour. The keeper called it a Dodo.

Some historians suspect that the captive animal seen by L'Estrange was actually a Réunion solitaire, another largish bird that might have been casually mistaken for a dodo. The solitaire was slightly more gracile but also flightless, and found only on the island of Réunion, not far from Mauritius. Dodo or solitaire, it was a novelty to Europe.

An interesting sidelight to this wave of biological discovery is that some of the more vivid animals were adopted for map illustrations. As early as 1502 a Portuguese mapmaker drew red and blue macaws into place in Brazil, counterpointing the green and gray parrots with which he decorated Africa. Monkeys also showed up on maps of South America almost as soon as the roughest outlines of that continent could be drawn. A pangolin appeared on the world map done by Pierre Descelliers in 1546. In 1551 Sancho Guttierrez portrayed a bison in North America on his map of the world. Toward the end of the century, another artist drew a bird of paradise—footless—on a map of southeastern Asia. Offering a few splashy pictures to enliven their charts, whether accurately or misleadingly, these Renaissance mapmakers were committing biogeography.

As the inventory increased, so did the difficulty for biblical literalists. Noah's ark was getting too full. Some of the pious scholars haggled desperately about how to interpret the word "cubit." No no, not the *little* cubit, they said—those measurements in Genesis 6 must refer to *big* cubits, six times as large. It was a way of stretching the ark and of buying time, but meanwhile there came ever more species of monkey and pigeon and kangaroo. Other scholars offered ingenious suggestions for retroactive improvement of Noachian boat design. A German Jesuit named Athanasius Kircher published *Arca Noë in Tres Libros Digesta*, containing much learned argument from historical and philological sources as well as an engraving of the ark as he fancied it. Kircher's ship is a boxy, three-story structure resembling a Super 8 motel, beneath which appears no trace of a hull. Unquestionably it would have allowed efficient division of space into many stalls and cages, but it doesn't look seaworthy. Maybe seaworthiness, like creation itself, was the province of miracles. In the Kircher engraving, a pair of lions, a pair of snakes, a pair of spoonbills, a pair of ostriches, a pair of horses, a pair of donkeys, a pair of camels, a pair of dogs, a pair of pigs, a pair of peafowl, a pair of rabbits (Noah, watch out!), and a pair of turtles are waiting their turns to climb the gangplank. It's a crowded scene, though the level of biological diversity is relatively minuscule.

As described by Janet Browne in *The Secular Ark*, her excellent history of the foundations of biogeography, Father Kircher's ark included just 150 kinds of birds and about the same number of other vertebrate animals. Fish, naturally, didn't need a boat. Neither did those reptiles and insects that arose by spontaneous generation. With such cleverly rigorous exclusions, Kircher pared down the total of species requiring rescue and got them all aboard, thereby keeping the good ship of biblical literalism afloat just a while longer.

But his passenger manifest was too small to accord with reality. By the end of the seventeenth century, naturalists were aware of 500 bird species, 150 quadruped species, and about 10,000 species of invertebrates. Fifty years later, when Linnaeus began putting things in order, those numbers were still growing quickly. Linnaeus himself named and catalogued almost 6,000 species. The ark was overbooked.

6

THE ARK as a physical vessel was only one tenet of scriptural orthodoxy under threat. Another was that mountaintop where the boat had
made landfall and unloaded. Conventional belief had fixed on a
mountain in eastern Turkey as the site—Mount Ararat, as we know
it—though Genesis itself gives no geographic coordinates. Grounding Noah's menagerie in eastern Turkey left the distributional anomalies of the world's fauna unexplained. If all the surviving animals
disembarked at Ararat after the deluge, how did various species end
up where they did?

At the simplest level, this was a problem of dispersal. How did the
rheas, being flightless, get from Turkey to South America? Did they
walk? How did the polar bears get to the Arctic, given their need for
cold temperatures, frozen oceans, and toothsome seals? It must have
been a long, hungry, hot trudge across the Caucasus. How did those
troublesome kangaroos get to Australia? Exactly how far could the
bloody things jump?

At a more intricate level, the problem also encompassed disconti-
nuities in distributional patterns. If the kangaroos hopped from
Ararat to Australia, why had none lingered in Asia? If the cockatoos
had flown out from the ark (as the raven and the dove had done earlier, on their scouting missions for dry land) and headed southeast,
why did they fly all the way to Lombok before choosing to stop and
establish a presence? Why had they spurned all the intermediate terrain? What had been so off-putting about, say, Bali?

Linnaeus himself was among the first to see the story of Ararat and
the ark as irreconcilable with biological reality. He doubted that a pair
from each species of land animal could have fit on a single boat, no
matter how Noah had calibrated cubits. Linnaeus also doubted the vision of those animals disembarking onto a mountaintop. Instead he
offered a symbolic interpretation, differing slightly from the literalistic one. He suggested that Ararat must represent a primordial island,
from which all created species had dispersed: "If we therefore enquire
into the original appearance of the earth, we shall find reason to conclude, that instead of the present wide-extended regions, one small island only was in the beginning raised above the surface of the waters."
This island, as Linnaeus saw it, was the site of Eden and the original
home of all God's creatures. It was "a kind of living museum, so fur-

nished with plants and animals, that nothing was wanting of all the present produce of the earth." The island was mountain-shaped, encompassing a stratified spectrum of climatic zones—a tropical zone near the base, a temperate zone farther up, a polar zone surrounding the summit. Each type of animal was matched to its "proper soil" and its "proper climate" within one of the stratified zones. As the primeval waters subsided, allowing steadily more land surface to emerge, the different zones became enlarged across the world's landscape and the distribution of animals expanded accordingly. Each type of animal was dependent upon its own zone, and each type had a physical form suiting it snugly—and immutably—to that zone.

Linnaeus had hit upon a richly ambivalent line of thought, leading in one direction toward insight, in another direction toward a new kind of confusion. Toward insight: The talk about "proper soil" and "proper climate" sounds like the ecological concept of a species adapted to its niche. But that concept wouldn't be developed until much later. Toward confusion: The immutable linkage that Linnaeus imagined between the form of an animal and its "proper" place enticed some eighteenth-century thinkers to embrace the idea of *special creation*.

This idea implied divine involvement in every picayune aspect of floral and faunal origination. It swept aside all those quibbles (big cubits versus little cubits, for instance) that afflicted literalistic interpretations of the ark story, allowing Noah's flood to be accepted again as a true but nonliteral parable. What did not relax into parable, in light of special creation, was the deity's direct role. That bit was still literal and historic. And now He appeared busier than ever.

According to the idea of special creation, God had functioned both as almighty creator and as chief zoological usher, designing particular beasts for particular places and seeing to it that they got to those places. He had made the polar bears in just such a way as to suit them for the Arctic—and so that's where He had put them. He had shaped the kangaroos and set them down in precisely the location where they belonged—Australia. He had personally adapted the capybara to its life in South American swamps. He had divided His attention among a number of distinct geographical realms and flattered each of those realms as a center of creation. He had invented whole floras and faunas, integral communities, every one appropriate to its physiographic context. He had decreed into existence mature ecosystems. Let there be neotropical moist forest, He had said, and lo. Let there be sclerophyllous chaparral, He had said, and it was so. Let there be taiga, and let it be graced with great herds of antlered ungulates, with a few snow-hardy predators, with wildflowers that bloom hurriedly in mid-summer, with fast-reproducing small mammals whose population sizes fluctuate drastically, and with more mosquitoes than even I can count, He had said, et cetera. This newly imagined God of the late eighteenth century was a hands-on, follow-through sort of guy who committed himself to details and showed no knack for delegating power. He made himself immanent and manifest the laborious way, with one little intervention after another. It was a thorough ontological vision, the theory of special creation. It solved everything and nothing. Misconceived but beguiling, it became the new orthodoxy. It reigned long enough for Darwin and Wallace to confront it.

Linnaeus had meanwhile exerted still another form of influence. As a teacher in Uppsala, he had nurtured and energized students. Many of those students became scientific travelers, wandering out across the planet to fulfill their mentor's intellectual program by studying groups of living species in varied places. Janet Browne mentions some of these far-flung Linnaean protégés: Osbeck sailed to China, Solander was on board the *Endeavour* with Captain Cook and Joseph Banks, Gmelin went to Siberia, Koenig to Tranquebar (wherever that is), and Carl Peter Thunberg got into imperial Japan, after which he published his *Flora Japonica*, by definition a work of island biogeography. What sets this school of biological chroniclers apart from the earlier voices—such as Pigafetta, de Conti, and Columbus—is that Carl Peter Thunberg and his peers were traveling for the express purpose of collecting biogeographical observations, not just blundering upon them in the course of imperial exploration.

Flora Japonica is only one item in the bounteous bibliography that resulted. A friend of Linnaeus named J. F. Gronovius had already published *Flora Virginica*, presumably based on fieldwork in Virginia, and Linnaeus had launched his own career with *Flora Lapponica*, derived from his adventurous youthful trek across Lapland. Gmelin's botanical journeying across Asia yielded *Flora Sibirica* and later *Flora Orientalis*. Whether from the example set by Linnaeus (who had held the chair in botany at Uppsala) or for other reasons, most eighteenth-century biologists devoted themselves to the identification and mapping of plants, in preference to animals. Von Jacquin produced *Flora Austriacae*, Lightfoot produced *Flora Scotica*, Oder produced *Flora Danicae*, and Richard Weston, with no fetish for scientific Latin, bless his soul, published *English Flora* in 1775. That was just four years after a notable fellow named Johann Reinhold Forster had offered his *Flora Americae Septentrionalis* (the title means "North American Plants"), undeterred by the disadvantage of never having seen North America. Eighty years later, as Alfred Wallace sat in Sarawak waiting for the wet season to end and pondering the great mass of facts that had "never been properly utilized," these books were among what he had in mind.

Johann Reinhold Forster, of the North American flora, comes forward through time more vividly than the rest. Born in Prussia, he trained as a minister in order to placate his father, though his own real interest was natural history. For some years he held a pastorate near Danzig, but ultimately the pastorate didn't hold him. In 1765 he visited Russia, and the following year he emigrated to England, hoping to establish himself as a scholar. He was a scrambler, like Wallace himself, living on whatever odd employments he could find, whatever opportunities he could manufacture. He published a *Catalogue of British Insects* in 1770. For a while he taught languages and natural history at an institution wonderfully named the Dissenters' Academy, in Lancashire. He became known for translating and editing some of the works by Kalm, Osbeck, and other scientific travelers, as well as for his own writings, and in 1772 he was elected to the Royal Society. Later that year, by a stroke of luck, Forster was picked to replace Joseph Banks as naturalist on James Cook's second voyage.

The voyage would eventually last three years, taking Forster to some of the more spectacular biological provinces of the southern hemisphere, but it was a troubled journey for him in personal terms. He didn't get along with the sailors, and Captain Cook seems to have found him insufferable. Forster lacked Joseph Banks's easygoing, sybaritic charm. He was part Prussian and part taxonomist, after all. One historian has called him "a tiresome, rather mean-spirited pedant and a prude." Another has derided, wrongly, his "miscellaneous low-powered scientific writing." The prudishness must have stood out badly on a voyage that alternated grim weeks of exploration along the coast of Antarctica with rest-and-recuperation layovers back up in Tahiti, where Cook's men needed only red parrot feathers to buy sex with Tahitian girls. Forster disliked what he witnessed in Tahiti, and he said so. His impolitic moralism had consequences. James Cook took umbrage at Forster's sour company and, when they arrived back in England, urged the British Admiralty to forbid

Forster to publish. Cook published his own account of the voyage. Forster's son, who had gone along as an illustrator, slid around the ban on his father and also published. Forster himself finally beat the ban too, publishing *Observations Made During a Voyage Round the World, on Physical Geography, Natural History, and Ethic Philosophy*. Whatever he meant by "ethic philosophy," it certainly didn't encompass the Tahitian trade in red parrot feathers.

A round-the-world journey wasn't possible in those years without stops for provisions at some interesting islands. Cook had seen New Zealand and Tahiti, among others, and the same logistical imperatives had made the Azores, the Madeiras, the Canaries, and the Mascarenes familiar to European mariners (therefore also to European science) at an early stage. Although each of those island groups was itself in the middle of nowhere, each sat on a sailing route between somewhere and somewhere. The faunas of Tasmania, Kerguelen Island, and Hawaii also received attention. Nautical explorers and the naturalists who sailed with them had occasion to see more of the world's biological marvels than did landlubbers—even well-traveled landlubbers—in Europe and Asia, because so many of those biological marvels were unique to remote islands. But Forster was more percipient than most seagoing naturalists of his time. He noticed something that the others had missed, seemingly obvious but vastly significant. He wrote:

> Islands only produce a greater or less number of species, as their circumference is more or less extensive.

Big islands harbor greater diversity than little ones. Johann Reinhold Forster had discerned the relationship between species and area. It was just two logical steps, and two hundred years, from that insight to ecosystem decay.

7

DURING THOSE wet days in Sarawak at the start of 1855, Alfred Wallace wrote a scientific paper, "On the Law Which Has Regulated the Introduction of New Species." Little known now and little read in his own time, it merits recognition as one of the historical landmarks of evolutionary biology. The word "Introduction" in the title, sounding bland and noncommittal, was actually packed full of subversive meanings.

He sent the manuscript to London on a mail steamer, and by September it had been published in a good journal, *The Annals and Magazine of Natural History*. Wallace was proud of this paper, his first major attempt at conceptual biology. As months passed he waited expectantly, as a young writer or scientist will do, for reactions. From one friend he got a letter congratulating him lefthandedly on how "simple and obvious" the central idea seemed now that Wallace had stated it, and misquoting the paper's title. Otherwise, silence. Back in England, Charles Darwin read the paper in his own copy of the *Annals*, made some marginal scratches, took some notes. But Darwin was a clerkish reader by habit, making scratches and notes on virtually everything; despite his programmatic attentiveness, he seems to have missed the paper's point. Wallace had used the word "creation" as well as "introduction" in discussing the origin of new species, and Darwin evidently dismissed him as a believer in special creation. Several years later, Darwin would reread the paper with keener perception and a jealous sense of panic.

A few other scientists had taken notice of Wallace's paper, and two even said so to Darwin. One of these was Sir Charles Lyell, England's preeminent geologist. The other was Edward Blyth, an Englishman stationed in India with whom Darwin corresponded. But neither Lyell nor Blyth communicated appreciation to the author. To them, this fellow Wallace was an unknown, a commercial collector and trader, a purveyor of stuffed birds and mounted insects for the curio market. Collection was crucial to the advancement of biology, but if the collector sold specimens (as Wallace did, in order to live), he was liable to be looked on as déclassé. Furthermore, Wallace was still tramping his way from one island to the next all-hell-and-gone out in the Malay Archipelago. He had no social standing and no fixed address. Did anyone know how to reach him? Did anyone care? Having attended some meetings of the Entomological Society during his last sojourn in London, he was not quite a total outsider, but he stood more out than in. And though the discipline of biology had not yet in those years become a profession, regulated by academic credentials and protocols, it was at least a semi-exclusive club. The club's membership consisted largely of gentlemen with inherited wealth, such as Darwin, and of country clergy, who performed their clerical chores mainly on Sunday and were free to watch birds or collect beetles during the week. The club met in London, in Paris, in Edinburgh, in Cambridge, in Berlin, in a few other places, while Alfred Wallace on the far side of the world coped with malaria and rotting feet.

Wallace's sales agent, back in London, heard mutterings from some naturalists that young Mr. Wallace ought to quit theorizing and stick to gathering facts. Besides expressing their condescension toward him in particular, that criticism also reflected a common attitude that fact-gathering, not theory, was the proper business of *all* naturalists. Those few clubbable scientists who did find merit in Wallace's theorizing acknowledged it quietly, if at all. Two months after the Sarawak paper appeared, Charles Lyell himself started a notebook on the subject of species transmutation, putting Wallace's name at the top of the first page. Lyell recognized what Darwin still didn't: that Alfred Wallace, whoever he was, stood on the threshold of something big.

The punchline of Wallace's paper was:

> Every species has come into existence coincident both in space and time with a pre-existing closely allied species.

Though he called it a law, it was really a description—but a provocative description, with a cargo of radical implications. He stated it near the start of his manuscript and repeated it near the end, each time in italics so no one could miss it. *Every species has come into existence coincident both in space and time with a pre-existing closely allied species.* Description or law, it challenged the theory of special creation and bruited the idea of evolution in a tone of thunderous innuendo.

8

THE TENRECS, for example, are closely allied species, coincident both in space and time.

The tenrecs are a family of extraordinary mammals—about thirty living species, spanning roughly the same size range as mice and rats but showing their own remarkable spectrum of physiological adaptations. Although a related subfamily of mammals (known as the otter shrews) occurs in western and central Africa, all other tenrecs are unique to the island of Madagascar. Wallace in his years of travel never reached Madagascar. Luckier, less intrepid, I do. At the Parc

Botanique et Zoologique de Tsimbazaza, in a suburb of Antananarivo, I find a young Englishman who is happy to share his tenrecs.

The Englishman's name is P. J. Stephenson. He is a doctoral student from the University of Aberdeen, transplanted to Madagascar for his research. He wears a white lab coat, which is no longer white, and a deceptively sleepy grin. He has just been building terraria so his hands are glopped with glue, P.J. explains. He rubs one hand through his hair. It's long, feral blond hair, suitable for a drummer in a band. P.J. blinks. Maybe he's dazed to be speaking English again suddenly; maybe I've interrupted a chain of tenrecoid thoughts. Maybe, glue or no, he was just catching a nap before I arrived. His lab area is a pair of stony gray chambers in a garage behind one of the Parc buildings, where not many visitors intrude. Tenrecs, he says, you're interested in tenrecs? Come.

I follow him to the back.

"Look here." P.J. lifts the lid off a terrarium. "This little guy is really freaky." We gaze down. The terrarium contains a substrate of wood shavings and some items of furniture. Otherwise it seems to be empty.

P.J. turns over a small piece of wood. No tenrec beneath that. He snatches up a cardboard tube and peeks down its bore, seeing only daylight. Then another tube, and another. Nothing. He checks the wood again, still nothing there, and the tubes again, his hand moving fast. "Hold on. He's hiding," says P.J. "Or else we've just had an escape. I hope not. Wait a minute, here he is." P.J. scoops up a handful of wood shavings. He cradles them toward his nose. He spreads his fingers. Wood shavings fall. "No. Wait. Here now." He scoops again. Finally, success and relief: In his palm is a tiny mammal.

Face like a carrot, dark little eyes, gray fur; it resembles a shrew. It isn't a shrew. It's a member of the species *Geogale aurita*, otherwise known as the large-eared tenrec. "He's a termite specialist," P.J. says fondly. "Hides in rotting wood. That's how I found him, actually— busting apart pieces of rotting wood."

P.J. intends to spend two years in Madagascar, busting apart more pieces of rotting wood, collecting more tenrecs, studying the peculiarities of their physiology. They embody some interesting questions. Why, for instance, do some tenrecs mature so quickly? One species reaches sexual maturity at the age of about forty days. Then again, certain other species mature rather slowly. And why do some of them produce such large litters? One species can carry as many as thirty-two fetuses, which is unusual among mammals. Then again, certain other species produce litters as small as one or two. Some of them have exceptionally low metabolic rates, and some don't. Some are capable of going into a state of torpor—turning their metabolism way down, saving energy—and some aren't. They're very diverse. What sort of mechanism controls the shift to torpor? And how can they have such a high reproductive rate, some of them, yet such a low metabolic rate? And what's the relationship between metabolic rate and body temperature? Between metabolic rate and diet? Why is this little termite eater so similar to a shrew in its anatomy and its ecology, yet so drastically different in how its body burns fuel? Why?

Having heard P.J. rattle off these questions, I'm ready to hear answers. I'll gladly surrender myself to the premise—for an hour or two, anyway—that tenrec physiology is the most intriguing of all branches of biological inquiry. What then, exactly, does it teach us? But P.J. doesn't have answers, not yet. What he has is a head full of curiosity, a lab full of animals, sticky hands, and two years to think. He sets the little termite eater back into its cage, where it instantly disappears.

P.J. is maintaining quite a number of different species, but even his menagerie is just a modest sample of tenrec diversity. That diversity is part of what makes the group notable—they aren't only peculiar, they're peculiar in a profusion of different ways. Experts on tenrec taxonomy note that a single genus, *Microgale*, encompasses at least sixteen species: *M. cowani, M. dobsoni, M. gracilis, M. principula, M. brevicaudata*, and so on. The experts sort all these microgales into four categories, based on anatomical and ecological factors. There are burrowing, short-tailed microgales without much talent for leaping; there are surface-foraging microgales, also short-tailed, also nonleaping, that have some marginal talent for climbing; there are long-tailed surface-foragers that are good climbers; and there are very long tailed microgales that merit description as "climbers and ricochettors among branches," which sounds fantastic even for Madagascar. There's a *Microgale longicaudata*, whose name signals its membership among the very long tailed. None of this will be on the quiz at the end of the book.

The other genera include tenrecs that resemble moles, tenrecs that resemble hedgehogs (or, to an American eye, miniature porcupines), and an elusive aquatic tenrec named *Limnogale mergulus*, which lives the life of a small river otter. P. J. Stephenson himself is one of very few scientists who have ever glimpsed *Limnogale mergulus* in the wild.

He and another bloke sat on a riverbank all of one night for that privilege, he tells me, being savaged by mosquitoes and holding their foolish damn flashlights.

P.J. shows me more tenrecs in more terraria. He shows me tenrecs in cardboard boxes, housed there temporarily while he cobbles together still more terraria. In another shed, he says, he's got larger tenrecs in wooden cages. And here's a tenrec in its own little submarine—that is, a sealed gas-monitor chamber suspended inside a tank of water—where P.J. can measure its oxygen use. Alongside the tank sit an electronic console and a needle-graph printer. Suddenly a thunderclap jars the sky above Antananarivo and the lights flicker in P.J.'s lab. The needle-graph printer emits a disconsolate click. It has turned itself off.

"Weather. Hold on." He resets the machine. "That could be a problem. We get a storm and everything's liable to shut down. Not good for the flow of data." The flow of data is lifeblood to a doctoral student. Breathing deeply, P.J. stirs a hand through his unruly yellow hair.

Tomorrow he will go off to collect more tenrecs from the wild. Having mentioned that, he remembers: "I've got to send a fellow out to get me some food." He moves for the door.

What kind of food, I wonder aloud, thinking I might cadge an invitation to join his field trip.

"Crickets," says P.J.

The tenrec family belongs to the order Insectivora—nominally the insect eaters, though not every species confines its diet to true insects such as crickets and termites. The streaked tenrec *Hemicentetes nigriceps* eats earthworms. The aquatic *Limnogale mergulus* eats frogs and crustaceans.

The Insectivora are sometimes considered to be among the most primitive mammals surviving today, and the tenrecs to be the most primitive of the Insectivora. Although "primitive" is an invidious word that raises objections from some biologists, the intended point is that tenrecs preserve certain traits that have been dispensed with by other mammal groups in the course of evolution. A tenrec commonly has bad eyesight. Its body temperature fluctuates. It possesses a cloaca, as birds do, instead of two separate openings for the reproductive and the digestive tracts, like other mammals. The male tenrec has no scrotum, instead retaining its testes inside its abdomen. The female gives birth to helpless young with their eyes and ears closed.

The slow development of those young, in some species, prolongs their dependence on the mother and their vulnerability to enemies. Bad eyesight, unsteady temperature, cloacal plumbing, internal testes, blind and deaf newborns—each of these represents a disadvantage in the struggle against competitor species and predators. The question arises, then, how the tenrecs have survived at all.

The answer is simple. They have survived by getting to Madagascar, where competition and predation are not nearly so intense as on the mainland.

Every species of tenrec (except for those anomalous African otter-shrew forms) is endemic to Madagascar—meaning, native to this island and nowhere else. Their ancestors probably arrived sixty or seventy million years ago, near the end of the age of dinosaurs, at a time when mammalian evolution was still in its earliest stages. As the island became isolated from Africa (by geological splitting that widened and deepened what we now call the Mozambique Channel), only a few other mammal lineages got aboard: lemurs, rodents, viverrids, and the ancestors of that pygmy hippopotamus you've already heard about. An African aardvark reached Madagascar within the last few million years, somehow, but it didn't last. And an occasional small flock of wayward bats has come in on the winds. Compared with what exists on mainland Africa, Madagascar's list is dramatically short. The mammalian fauna is meager. Among the African mammal groups *not* present on the island (at least until humans began arriving by boat, bringing livestock and pets, causing extinctions, and otherwise muddling the biogeographical record) were these: the cat family, the dog family, elephants, zebras, rhinos, buffalo, antelopes, camels, and rabbits. Also, there was no Madagascan giraffe. There was no bear. There were no monkeys or apes, no otters or hyraxes, no porcupines. Within this gentle world of reduced competition and reduced predation, during the long eons of unbreached isolation, the tenrecs not only survived but prospered.

They multiplied. They spread all over the island. They took hold in the rainforests on the eastern slope of Madagascar, in the drier forests of the central plateau and the west, in the thorny desert of the south. They adapted to fill ecological niches that were otherwise empty. They diverged variously from their original type and were variously transmogrified into those thirty-some distinct species. They mitigated competition between one species and another by diverging still further. The scientific term for this process is *adaptive radiation*.

P.J. remembers a moment, earlier in his studies, when he picked up a volume of symposium papers and "there was this talk about adaptive radiation. I was reading it, and I just thought, 'Tenrecs. You're waiting to say tenrecs, that's what you're trying to say.' As an example. And it cited half a dozen things, but it didn't mention tenrecs. Whereas I think it's probably the most classic adaptive radiation I've ever seen in my life. Where you've got ancestral stock, it arrives on an island, finds all the niches available. Evolution takes its course. And it adapts to fill all the niches. I mean, you can't *get* something better." Blessed is the doctoral student who so loves his beasts. "You've got shrews, you've got hedgehogs, you've got moles. You've got things you can't possibly even describe. That have just evolved. It's absolutely unbelievable." Again he rubs at his hair, ever vigilant against the possibility that it has somehow gotten itself combed.

In the scientific literature, the tenrecs of Madagascar have been divided into two subfamilies, the Oryzorictinae and the Tenrecinae. Included among the Tenrecinae are all the larger, prickly-haired species that resemble hedgehogs. The Oryzorictinae include all the smaller species resembling moles and shrews. Somewhere between these two groups stands a missing link, a species called *Cryptogale australis*—in English, "the secretive southern tenrec." It has a skull like the Tenrecinae, body size and teeth like the Oryzorictinae. It's missing in the sense that it's apparently extinct, known only from remnant bones.

Now let's recall again what Wallace wrote in Sarawak: *Every species has come into existence coincident both in space and time with a pre-existing closely allied species*. No matter that he had never reached Madagascar or studied tenrecs. He was finding the same pattern among the Borneo butterflies, among the cockatoos and the macaws, as well as in published reports of plant and animal distribution gathered from all over the world. The unstated point behind Wallace's "law" was that closely allied species come into existence not only *near* one another but *from* one another. The theory of special creation could not account for such patterns persuasively. A theory of evolution could.

Wallace's Sarawak paper was an overture. It hinted toward evolution but stopped short of explaining how the process might work. At that point in time, he had no theory to offer. Charles Darwin did have a theory but wasn't yet ready to offer it.

"Let me show you something," says P. J. Stephenson. "Here. This fellow's *really* freaky."

He lifts another lid to reveal a tenrec as hefty as a muskrat. It's a garish thing with black-and-white stripes like a skunk, sharp quills like a porcupine, a nose like an anteater. This is *Hemicentetes nigriceps*, P.J. says, the famous earthworm specialist. It lives at the edge of the central plateau and in the rainforest of the eastern slope, and it does pretty well around rice paddies also. The quills are barbed and detachable so that they can come away in the muzzle of an enemy. Under attack, it erects those quills and makes bucking motions that threaten the attacker with a muzzle-load of painful regret. The black-and-white stripes, as with a skunk, offer a conspicuous form of visual warning: Mess with me and you'll wish you hadn't. Predators may be scarce on Madagascar, but they aren't nonexistent. The viverrids in particular, those mongoosey carnivores, are capable of forcing a tenrec to defend itself. Another nice little adaptive feature of *H. nigriceps* is that some of the quills have become modified into an organ of stridulation. They can be scraped together, like a bow on the strings of a fiddle, to produce a ratchety squeak by which a female signals her young. Having advanced so far along its own odd evolutionary path, *H. nigriceps* strains the meaning of the word "primitive."

P.J. reaches down dotingly and gives the tenrec a gentle rub.

So I reach down too. P.J.'s appreciation is contagious. The quills are benign when laid flat to the body. I pay my respects to *H. nigriceps* with a cordial little nudge.

"They're rather nasty biters, actually," P.J. says.

9

You MUST READ the paper by this fellow Wallace, Charles Lyell told his friend Darwin. Well, Darwin already *had* read it. He had even written his few notes on a piece of blue paper and pinned that into the back of the journal: "Wallace's paper: Laws of Geograph. Distrib. Nothing very new." Lyell disagreed.

Although his own work in geology was revolutionary and had been a strong influence on both Darwin and Wallace, Charles Lyell in 1855 held no revolutionary view of the origin of species. He still subscribed to the theory of special creation. Wallace's paper unsettled him. Lyell began his species notebook in November of that year under the heading "Wallace, Index Book 1," and the early entries look like an energized private rebuttal to the Sarawak paper. Focusing particularly on the subject of islands, Lyell made more than a hundred pages of notes about the distribution of animals and plants on the Azores, the Madeiras, the Canaries, Saint Helena, New Zealand, and other places, including the Galápagos. He asked himself how closely the species of land snails found in Britain might resemble the species in mainland Europe. And what about the land snails of the Madeiras, four hundred miles off the coast of Morocco? How many of those species were found also in Europe? How many were uniquely Madeiran? How many turned up on one of the two Madeiran islands but not on the other, just twenty-five miles away? Did the larger Madeiran island support more endemic species than the smaller island? What did it mean, what did it mean?

Around the same time, Lyell started discussing island patterns with Darwin. They shared facts. To a lesser degree they shared thoughts. But it was Alfred Wallace who had set Lyell going.

Wallace's paper had made the interesting point that if God indeed performed special creation, producing custom-designed species to occupy each zone of landscape, then God had shown a strong bias toward geologically *old* islands. That is, old islands had received far more endemic species than young islands. Old islands had even received more endemic species—in proportion to area, anyway—than had continents. Of course, some of the more obdurate of special creationists still didn't believe that any island was significantly older than any other, since by their biblical chronology the Earth itself had been created just six thousand years earlier. But Charles Lyell, whose whole geological vision was premised on the immensity of past time, certainly did distinguish young from old. The old-island pattern caught his interest.

Wallace had cited Saint Helena as one case of "a very ancient island having obtained an entirely peculiar, though limited, flora." He had added: "On the other hand, no example is known of an island which can be proved geologically to be of very recent origin (late in the Tertiary, for instance), and yet possesses generic or family groups, or even many species peculiar to itself." This pattern could be explained two ways. The first possible explanation was that in the very distant past, God so loved islands that He graced them inordinately with wondrous forms of plant and animal, whereas in more recent epochs, either God's fondness for islands has flagged or else His creativity has. If He's not a diminished God of inconstant powers, then, He's a capricious God of inconstant tastes. The second possible explanation was less blasphemous: Species evolve.

The necessary conditions for evolution, as proposed in the second explanation, are isolation and long stretches of time. Since old islands have offered prolonged isolation, they have produced a larger share of evolutionary novelties.

Sir Charles Lyell was an honest scientist as well as an orthodox Victorian creationist. For months, within the privacy of his notebook, he sorted data on the distribution of land snails and the correlations with geologic history. This species appeared here but not there. This island was old, that one young. Lyell could see dimly where it all pointed, but he wrote: "The origin of the organic world, like that of the planet, is beyond the reach of human ken. The changes of both may perhaps eventually be within our ken but ages of accumulation of facts & of speculation may be required." Behind the scientific calm, behind the reserved judgment, behind the litany of snails, he was desperate.

10

MADAGASCAR, home of the tenrecs and so many other bizarre species, is an old island. It's so old that even geologists can't agree how old, not even to the nearest thirty million years.

Once it was part of Gondwanaland, the huge southern continent that also included Africa, India, Australia, South America, and Antarctica before tectonic forces split them and they drifted apart. As the splitting progressed, Madagascar may have remained temporarily connected to the east coast of Africa—possibly snugged in beside Mozambique, or else farther north, adjacent to Kenya. Or maybe it separated early from the African mainland but remained connected to India. By that second view, India and Madagascar eventually came un-linked, India rumbled off like a runaway beer truck toward its colli-sion with southern Asia, and Madagascar stayed parked where it was. Meanwhile the sea floor between Madagascar and Africa subsided. At first the sea was still shallow enough that a moderate lowering of its level (caused by global cooling and glaciation near the poles, sucking away water from the oceans) would have exposed a land bridge be-tween Madagascar and the mainland. Episodes of connection and separation, reflecting climate and sea level, may have alternated throughout many millions of years. But as the sea bottom continued subsiding—sinking deeper and deeper—the channel finally became a permanent ocean gap. Severed irretrievably from the mainland, Madagascar was launched on its voyage of insularity.

Evidence for the various hypotheses comes from deep-sea drilling cores, bathymetry, sediment comparisons, paleomagnetic data. The bones of lemurs are also suggestive. Regardless of uncertainties about Gondwanaland, regardless of India, and regardless of glaciology, we can safely assume that by about sixty million years ago Madagascar was isolated. From then on, it stayed isolated.

Bali, by contrast, is a young island, with a young island's geological and biological characteristics.

The strait between Bali and Java is only a few hundred feet deep. The strait between Java and Sumatra is roughly the same. Two hundred feet of sea depth might sound considerable if you're a bad swimmer aboard an old wooden boat, but in the great movements of land, water, and time it's trifling and transient. The Strait of Malacca, between Sumatra and peninsular Malaysia, is also shallow. These three little straits are the only water gaps separating Bali from the Malaysian mainland. Lower the sea level by a few hundred feet and Bali becomes a province of Asia.

It has happened. Probably it has happened many times, once for each ice age since the landmasses of the Indonesian archipelago took their present shape. The most recent occurrence seems to have ended about twelve thousand years ago, with the last major freeze of the Pleistocene epoch. What we now call Bali is just one promontory in a zone of shallow ocean and volcanic rock known as the Sunda Shelf, which includes also Java, Sumatra, Borneo, the Java Sea, and the Gulf of Thailand. During that last episode of low sea level, it would have been the Sunda Peninsula, a vast region of tropical forests and swamps. *Panthera tigris*, the tiger, was among the animals that roamed the region.

Elephants roamed there too. Orangutans. Rhinos and tapirs and wild pigs and leopards, sun bears and moon rats and pangolins. Many of those animals presumably extended their distributional ranges into the southeasternmost reach of the peninsula—that is, down to and including Bali. After the sea rose again, some remained. We know that *P. tigris*, for one, persisted on Bali, because it was present there at the start of the twentieth century. The Balinese population of tigers had diverged just enough, as I've mentioned, to be considered a distinct subspecies, *Panthera tigris balica*.

If Bali were an older island, its tiger population might have diverged more, enough to constitute a fully distinct species. Let's imagine that species with its own imaginary name, *Panthera balica*. Given still more time, there might have been still more divergence—to the genus level, say, making Bali's beast as different from the original tiger as a tiger is different from a cheetah. Maybe the Balinese species would have changed color, or acquired a different pattern of striping, or developed a special aptitude for stalking wild banteng cattle on the island's volcanic slopes. And probably, owing to certain island-specific factors I'll describe later, the Balinese tiger would have been smaller, maybe as small as a leopard or a cougar. Let's call it *Micropanthera balica*, the Balinese dwarf tigeroid. A hypothetical beast, but quite plausible within the context of evolution on islands.

We can push this hypothetical scenario still farther. During a subsequent glaciation, when the sea level again fell, some individuals of *Micropanthera balica* might have roamed back across the muddy isthmus to Java. By this time, let's say, those Balinese tigeroids would have been incapable of interbreeding with Javanese tigers. They would have feared and avoided the bigger cats. But maybe they would have found ways to survive in Javanese forests, and then multiplied, coexisting discreetly with the native species. Now again we bring the sea level up, renewing the disjunction between Java and Bali. Imagine another million years of isolation and divergence, during which the Javanese population of *Micropanthera* becomes a new species, distinct from its ancestral population on Bali as well as from the more distant relatives with which it shares Java. Are you with me? Call that new species *Micropanthera javanica*.

Let's take stock. In place of a single species represented by two subspecies, there would now be three distinct species: *Panthera tigris*, *Micropanthera balica*, and *Micropanthera javanica*. Isolation plus time yields divergence. This is how insularity contributes to the origin of species.

Evolution happens slowly. Twelve thousand years is just a hiccup in the evolutionary scale of time. Extinction happens faster, and insularity contributes to that process too. Twelve thousand years ago Bali was the big toe of the Sunda Peninsula, an outpost of fertile land that may have supported small populations of pangolins, rhinoceroses, sun bears, orangutans, and elephants. None of those animals, nor the tiger, is native on Bali today. Why not?

The answer is complicated. But we're getting there.

I I

OLD VERSUS young is just one crucial dichotomy that affects the biological richness of an island. Small versus large is another. Continental versus oceanic, in the sense that biogeographers use the terms, is a third. These three dichotomies give pattern to a world of spectacular confusion.

Madagascar is a large island, fourth largest on the planet, behind only Greenland, New Guinea, and Borneo. It measures 230,000 square miles, roughly the size of Montana and Wyoming together. Like Montana and Wyoming, Madagascar is psychologically distant from the rest of the world and burdened with a culture that venerates cows. Bali is a small island. It measures 2,100 square miles, just a hundredth the size of Madagascar or Montana-Wyoming. Bali is roughly the size of Yellowstone National Park, though in Bali you see fewer Winnebagos.

Large islands harbor more species than small islands, as a general rule. To be specific, Madagascar harbors more species than Bali—many more birds, many more plants, many more reptiles, many more insects, and (though Madagascar itself isn't rich in this category) significantly more mammals. Among primates, for instance, Madagascar has about thirty living species of lemur and a dozen others recently extinct. Bali has a leaf monkey and a long-tailed macaque.

Most of the Madagascan species are endemic; they evolved there and occur nowhere else. Eighty percent of the plant species are unique to the island. Among trees alone, more than ninety percent. More than ninety percent of Madagascar's reptiles, nearly all of the amphibians, all of the tenrecs, and (if several small offshore islands are counted as part of the Madagascan region) all of the lemurs are endemic. Even among birds, with their greater powers of dispersal, half are special to Madagascar.

Most of the Balinese species, on the other hand, aren't endemic to Bali. Many of those species occur also on Java, and some on Sumatra and mainland Asia. Bali does have *Leucopsar rothschildi*, informally known as the Bali starling, a gorgeous and nearly extinct bird with gleaming white plumage and a mask of turquoise. But aside from the starling and the extinct subspecies of tiger, there are few other creatures that Bali can claim exclusively.

This difference in endemism derives from several causes. The size of the two islands has played a role. So has the difference in elapsed time since each became insular. One other factor is the difference in distance of isolation. Madagascar is much more remote, sitting 250 miles off the African coast. Bali is just a long downwind spit from Java. The remoteness of Madagascar, along with its size and its ancientness, has made it more conducive to speciation and the maintenance of endemism.

But these two very different islands, Bali and Madagascar, do share one essential trait: They are both *continental* islands, as distinct from *oceanic* islands. That is, each was formerly connected to its neighboring continent.

Continental islands tend to be close to the mainland, whereas oceanic islands are more remote. A continental island generally lies on a continental shelf, surrounded by shallow water and therefore subject to reconnection with the mainland by a land bridge during episodes of lowered sea level. Continental islands, for that reason, are known also as land-bridge islands.

An oceanic island is one that never has been and never will be connected to a mainland. It comes into existence as a rising welt off the deep ocean floor, elevated into daylight by some geological process—most commonly, volcanic eruption. After a relatively short lifetime it gets eroded by waves and disappears again below the surface of the sea. The Galápagos are volcanic islands in midocean. The Hawaiian Islands are volcanic. Mauritius and Réunion are volcanic. Among the planet's newest volcanic islands is one called Surtsey, which came steaming up near the southeastern coast of Iceland in 1963. The land-building action of coral is sometimes also involved in erecting oceanic islands. Coralline limestone is laid down just beneath the water's surface and then elevated by volcanic or tectonic pressures. Guam, part limestone and part lava, is a case in point.

Whether built from elevated coral or from extruded lava, every oceanic island comes up from below, like a gasping whale. It starts its terrestrial existence, therefore, completely devoid of terrestrial forms of life. This is the most fundamental distinction between the oceanic and the continental categories. Every terrestrial animal on an oceanic island, and every plant, is descended from an animal or plant that arrived there by cross-water dispersal after the island was formed. A continental island like either Bali or Madagascar, in contrast, already contains a full community of terrestrial species at the moment of its isolation.

A continental island begins with everything, and everything to lose. An oceanic island begins with nothing, and everything to gain. Island biogeography, over the past century and a half, has been the scientific record of those gains and losses.

12

"FROGS ARE NOT found in volcanic islands," Charles Lyell told his notebook in April of 1856. What did it mean, what did it mean?

He added that "Darwin finds frogs' spawn to be very easily killed by salt water." Darwin for years had been conducting quiet little experiments on the question of whether various animals and plants could disperse across a wide stretch of sea. Would the seeds of a given terrestrial plant survive weeks of immersion in salt water? In some cases the answer was yes. Would the eggs of a frog survive that treatment? No. Would an adult frog survive? No. Was it a coincidence that the creator, whoever He was, however He worked, had neglected to put frogs on remote islands? Maybe not. Lyell's faith was weakening.

The subject of islands had begun to occupy Lyell's attention several years earlier, when he and Lady Lyell took a cruise to the Canaries and the Madeiras. These two island groups, standing not far apart in the eastern Atlantic, were regular stops on the oceanic trade route of those years, salubrious and civilized getaways where a Victorian gentleman and his wife could check into a decent hotel. Lyell's main purpose there was to study the volcanic geology. But he couldn't help noticing some remarkable species and some remarkable patterns among the fauna and flora. Many of the Madeiran beetles were endemic; a good number of those endemic species occurred on one Madeiran island but were absent from the other. The land snails of the Canaries caught his eye too. And at least one island of the Canary group, Grand Canary, was strangely empty of wild mammals. Also with reference to Grand Canary, he noted: "I never was in a country where the vegetation was so exclusively . . . unEuropean & so peculiar."

When he got back to England, Lyell sorted through his biological collections and notes, trying to make sense. Late in the autumn of 1855 he confided to his sister in a letter: "It seems to me that many species have been created, as it were expressly for each island since they were disconnected & isolated in the sea. But I can show that the origin of the islands, which are of volcanic formation, dates back to a time when"—and then, instead of dilating on the evidence, he cut himself short. Although in a letter to Darwin, say, he might follow that line out relentlessly, in a letter to his sister he wouldn't. "But I must not run on as it would take me too long to point out how all these bear on one & the same theory—of the mode of the first coming in of species."

A week later he picked up *The Annals and Magazine of Natural History* and read Alfred Wallace's paper, the one written in Sarawak, about "closely allied species." Immediately he started his new species notebook. He began filling it with data from the Canaries, from the Madeiras, from his readings and correspondence about island biogeography; packing it with insular snails, insular beetles, insular plants, and other closely allied species coincident in space and time. Lyell was just realizing what Wallace had lately grasped and what Darwin had known for two decades: that the answer to the riddle of evolution was best sought by a study of islands.

13

IN the Sarawak paper, published that autumn in London for all to see and most to ignore, Wallace wrote: "Such phenomena as are exhibited by the Galapagos Islands, which contain little groups of plants and animals peculiar to themselves, but most nearly allied to those of South America, have not hitherto received any, even a conjectural explanation." It was a polite way of saying that the one man in England who had famously described the Galápagos flora and fauna, namely Charles Darwin, had apparently failed to appreciate their significance.

At the time, 1855, Darwin was still chiefly known for his *Journal of Researches* from the *Beagle* expedition, which despite the ungainliness of the full title had done rather well as a travel book. The Galápagos stopover had occupied only a very small fraction of the *Beagle*'s five-year voyage, but Darwin's literary treatment of the islands was vivid, forming one of the *Journal*'s most memorable sections. And besides bringing home a written account, he had also brought a great trove of Galápagos specimens—bird skins and pickled reptiles and insects and plants—which various specialists subsequently worked on for years. His *Journal* had been reissued in a revised and less expensive edition in 1845, with some additional description of the Galápagos fauna, based on what the specialists had done in the meantime with his specimens. "The natural history of these islands is eminently curious, and well deserves attention," he had written. "Most of the organic productions are aboriginal creations, found nowhere else; there is even a difference between the inhabitants of the different islands; yet all show a marked relationship with those of America, though separated from that continent by an open space of ocean, between 500 and 600 miles in width." Darwin had celebrated the archipelago as "a little world within itself," and had hinted cautiously that in this little world "we seem to be brought somewhat near to that great fact—that mystery of mysteries—the first appearance of new beings on this earth." As much as any biologist could lay claim to a piece of turf, Darwin had laid claim to the Galápagos. But his commentary on their implications had been limited and coy. So when he read Wallace's statement about the Galápagos phenomena having "not hitherto received any, even a conjectural explanation," he probably had to bite on a knuckle. It was too true.

Darwin still hadn't explained how that "mystery of mysteries" might be solved. After twenty years, his big evolution book was unfinished. Now this younger fellow, this collector and seller of Malayan beetles, this self-educated nobody who had not spent even one term in a Cambridge or Oxford college, was threatening impertinently to do the explaining himself.

"The Galapagos are a volcanic group of high antiquity," wrote Wallace, who hadn't laid eyes on them, "and have probably never been more closely connected with the continent than they are at present." Many literate Victorians had read Darwin's *Journal*, but few had pondered its Galápagos chapter as minutely as Wallace. "They must have been first peopled," he continued in the Sarawak paper, "like other newly-formed islands, by the action of winds and currents, and at a period sufficiently remote to have had the original species die out, and the modified prototypes only remain."

These "modified prototypes" were the endemic species that Darwin had reported—the tortoises, finches, mockingbirds, iguanas, and others. The finches in particular represented an instance of Wallace's "law," a whole cluster of closely allied species coincident both in space and in time. Wallace was using Darwin's facts to suggest an idea that

Darwin himself hadn't dared mention: that isolation plus time yields new species, by a natural process of evolution. Wallace couldn't say what made the process proceed. But the Sarawak paper announced plainly that he was working on it.

Around this time Wallace and Darwin became correspondents. Wallace wrote the first letter, hoping to strike up a long-distance conversation about the wild idea hinted at in his paper. Though he was a diffident young man, Wallace wasn't too shy to fish for a reaction from an eminent stranger. That first letter is lost. Darwin received it but didn't save it. All we know is that it was dated October 10, 1856, and mailed from the island of Celebes, where Wallace's island hopping had by now taken him.

Seven months later Darwin replied. "By your letter & even still more by your paper in Annals, a year or more ago, I can plainly see that we have thought much alike & to a certain extent have come to similar conclusions." In fact, Darwin said, it was unusual for two scientists to agree so closely on a point of theory.

By now Darwin had evidently reread the Sarawak paper more carefully. He no longer dismissed it as a rehash of creationist arguments. Wallace's percipient approach toward the mystery of mysteries had made him nervous. Sounding just a bit jealous, and tossing in a self-pitying exclamation point, Darwin told Wallace: "This summer will mark the 20th year (!) since I opened my first note-book, on the question how & in what way do species & varieties differ from each other.—I am now preparing my work for publication, but I find the subject so very large, that though I have written many chapters, I do not suppose I shall go to press for two years." It was a halfhearted way of claiming proprietorship of an idea without saying what that idea was. If you'll only be patient, Darwin implied, I'll presently solve this scientific puzzle for both of us. Two paragraphs later he wrote: "It is really *impossible* to explain my views in the compass of a letter on the causes & means of variation in a state of nature; but I have slowly adopted a distinct & tangible idea.—Whether true or false others must judge." Distinct and tangible, maybe it was; but that idea couldn't be judged, not by Wallace or by anyone else, until Darwin finally deigned to unveil it.

If he had no intention of sharing his thoughts, why did Darwin reply to Wallace's letter at all? Well, Charles Darwin was a polite man. He was a tireless correspondent, and having received a letter from a scientific colleague (even an obscure junior colleague), he would characteristically reply.

Beyond that, there was an ulterior reason. "I have never heard how long you intend staying in the Malay archipelago," Darwin wrote. "I wish I might profit by the publication of your Travels there before my work appears, for no doubt you will reap a large harvest of facts." And if Wallace's foreseeable travel book was still years from publication, maybe he would loan Darwin some facts in the meantime. Darwin was especially interested, he allowed, in the distribution of species on oceanic islands.

14

CHANCE HAD PLAYED a large role in bringing both Wallace and Darwin to the study of islands. Since chance is part of the whole story—an important component of both the process of evolution and the process of extinction, as you'll eventually see—let's take a peek at its contributions to these two scientific careers.

Charles Darwin hadn't left England aboard the *Beagle*, back in 1831, with biogeography on his mind or the Galápagos Islands in his sights. His seizure of that subject, his arrival in that place, reflected a fortuitous convergence of circumstance and whim on a feckless young man. He had signed on for the voyage largely to cure himself of boredom and indecision, to postpone the dreaded but seemingly inescapable prospect of becoming a country clergyman, and to escape from a parental glower—roughly the way some twenty-year-old American lad of a later era might have filled a backpack and flown Icelandic to Europe. After four years of surveying the South American coast, the *Beagle* headed westward around the world, and the Galápagos were a convenient port of call. Until then, Darwin had concerned himself as much with geology as with biology, and the Galápagos, being volcanic, may have interested him first as a good place to continue that study.

Alfred Wallace as a young man was more single-mindedly devoted to biology. He started traveling not to elude oppressive expectations or to "find himself," but to find something larger and more interesting—a solution to the mystery of origins. Despite that early clarity, he came to scientific fulfillment on a roundabout path. When he left England on his first international expedition, in 1848, he didn't set out for the Malay Archipelago. He didn't even set out in a different direction and arrive in the Malay Archipelago by accident. He went purposefully to the Amazon.

The saga of Wallace's Amazon journey is a high point in science history for anyone who relishes disaster, perseverance, high adventure, stiffness of lip, and black irony. It began in the circumstances of his adolescence. Forced to leave school at age fourteen by his family's financial troubles, young Alfred spent a short time in London, shadowing after a brother who was apprenticing as a joiner. Then for six years, throughout his middle and late teens, Alfred himself served an informal apprenticeship under another older brother, who had set himself up as a surveyor. Surveying was fine, it was agreeable, because it put him outside on the landscape (especially in Wales, which he loved) and allowed him to learn a bit of math and geology. At the time he wasn't ambitious for more. Then the surveying business slumped and his brother laid him off. Briefly he taught school—reading, writing, arithmetic, a little surveying and drawing—in the town of Leicester. Leicester was where his life veered.

In the town's public library he read Alexander von Humboldt's *Personal Narrative of Travels in South America*, which spoiled him forever against living a life bounded within Britain. He read Prescott's *History of the Conquests of Mexico and Peru* and Robertson's *History of America*. He read Malthus's essay on population, which was fated years later to go off like a detonator in his brain, as it did in Darwin's. Also at the Leicester library he made a friend, a fellow his own age, named Henry Walter Bates. Bates had a hobby that Alfred found intriguing: He collected beetles.

Beetle collecting was common in those days among boys and men with a fancy for nature. Beetles were pretty, highly diverse, and easy to preserve. To an ambitious or competitive naturalist, they were tokens for keeping score. Darwin himself had started with beetles. For Alfred Wallace, at this stage in his development, the collection of beetles was a wonderful new game. He had never imagined that there were so many different kinds—hundreds of beetle species within just the Leicester vicinity and three thousand within the British Isles. His new friend, Bates, owned a hefty reference book that cited such numbers. "I also learnt from him in what an infinite variety of places beetles may be found, while some may be collected all the year round, so I at once determined to begin collecting," Wallace would write in his autobiography, looking back across a sixty-year gap. He spent part of what he made as a schoolmaster, a few hard-earned shillings, on his own *Manual of British Coleoptera*. He and Bates became thick, a pair of demented young coleopterists scrabbling across the Leicester mead-ows. "This new pursuit gave a fresh interest to my Wednesday and Saturday afternoon walks into the country," he wrote. But it didn't re-main just a passionate hobby. Other ideas were getting inside Alfred's head, by way of other books, and causing too much excitement.

One of those books was Charles Darwin's *Journal*. Another was William Swainson's *A Treatise on the Geography and Classification of An-imals*. Swainson described patterns of species distribution and men-tioned some of the theories that had been offered to explain them; like Charles Lyell, though, he took refuge in a pious agnosticism about underlying processes. "The primary causes which have led to different regions of the earth being peopled by different races of ani-mals, and the laws by which their dispersion is regulated," wrote Swainson, "must be forever hid from human research." Alfred wasn't so sure.

Probably the strongest influence on him was a notorious book, *Ves-tiges of the Natural History of Creation*, that had been published anony-mously in 1844 and later attributed to a man named Robert Chambers. *Vestiges* was a mixed piece of work that drew a mixed pub-lic response. It proposed the general idea of biological evolution, and it supported that idea with some provocative facts and arguments; but it offered no reasonable theory of how evolution might work, and it was tainted with a good bit of credulous nonsense. For instance, Chambers described experiments in which insects had supposedly been spontaneously generated by sending an electrical current through a solution of silicate of potash. He claimed that the larvae of a creature called *Oinopota cellaris* live nowhere but in wine and beer, suggesting that the species must have arisen sometime after humans invented fermentation. He asserted that habitual lying in parents can become an inherent (essentially genetic) predisposition toward lying in children, especially notable among the poor classes, and that the Chinese language is a primitive form of communication because it consists of ideographs. Besides, Chambers noted, if the Chinese peo-ple were more advanced, they wouldn't have so much trouble pro-nouncing the letter *r*.

Vestiges was a big success among uncritical readers, turning evolu-tion into a blurred, titillating topic for parlor chatter, like a Victorian version of UFOs. But among rock-steady creationists, on the one hand, and among many careful scientists, on the other, *Vestiges* was

dismissed as a piece of sensationalistic junk. Charles Darwin considered it worse than useless, because it gave the whole subject of evolution an aura of lurid implausibility and made his own task that much
harder. To young Alfred Wallace, though, *Vestiges* was a catalyst. His
thinking at that point was much less developed than Darwin's, and
Chambers's reckless book helped him focus on a single crucial question. If species had arisen by natural transformations, one from another, as *Vestiges* argued, then what was the mechanism by which it
happened?

By now Alfred Wallace had left the schoolmaster job in Leicester
and resumed surveying. He wrote to Henry Bates:

> I have rather a more favourable opinion of the "Vestiges"
> than you appear to have. I do not consider it a hasty gener-
> alization, but rather as an ingenious hypothesis strongly
> supported by some striking facts and analogies, but which
> remains to be proved by more facts and the additional light
> which more research may throw upon the problem. It fur-
> nishes a subject for every observer of nature to attend to;
> every fact he observes will make either for or against it, and
> it thus serves both as an incitement to the collection of
> facts, and an object to which they can be applied when
> collected.

He spoke for himself. An incitement it was.

Big-eyed with youth and energy, Alfred and Henry began plotting
an audacious enterprise. They would go out to the tropics as biological collectors. They would finance the trip by collecting not just for
themselves but also for the commercial market. They would ship bird
skins and mounted insects back to a dealer in London, for sale to museums and to the affluent private collectors who, in that era, bought
and displayed natural-history specimens as others bought and displayed Pre-Raphaelite art. Selling off their own surplus specimens as
curios for the idle toffs, Alfred and Henry would use their field studies and their personal collections to solve the mystery of mysteries.
Yes, by God, they would discover the origin of species—or anyway,
they'd try.

But where exactly should they go? To the Malay Archipelago? No,
the attractions of that region don't seem to have grabbed Alfred's notice until several years later. To the Galápagos? No, that was too far

away for a pair of unsponsored amateurs, and besides, it had already been done. The decision was settled by another book. Near the end of 1847, Alfred read *A Voyage up the Amazon*, just published, by an American named W. H. Edwards. Bates read it too and they both were enthralled. They moved quickly to patch together their arrangements, which included an understanding with a well-connected London agent, Samuel Stevens, who would receive and broker their specimens. Within just four months they sailed from Liverpool on board a ship bound for Pará, at the mouth of the Amazon.

They arrived in Brazil on May 26, 1848. Wallace would stay four years. Bates would stay longer. Neither of them foresaw the length of time or the depth of troubles ahead.

15

SOON AFTER reaching Brazil, Wallace and Bates found a house on the outskirts of Pará, hired a cook, started learning Portuguese, and made some preliminary excursions into the forest.

Wallace later recalled the "fever-heat of expectation" he felt. "On my first walk into the forest I looked about, expecting to see monkeys as plentiful as at the Zoological Gardens, with humming-birds and parrots in profusion." But after several days of seeing no monkeys and hardly any birds, he "began to think that these and other productions of the South American forests are much scarcer than they are represented to be by travellers." Anyone who has ever stepped into a rainforest, head full of images from glossy nature photography, has had roughly the same disappointment, which derives from confusing diversity with abundance. Wallace discovered at once that it wasn't easy to collect insects and birds in the tropics. The biological diversity was great, true enough, but the abundance of each species was not great, and most animals didn't offer themselves readily to the eye, let alone to the net or the gun. He was surprised by how empty the forest seemed. The insects weren't so numerous as he'd expected. He saw no rhinoceros beetles. He had to be satisfied with some butterflies (including a few big metallic-blue specimens of the genus *Morpho*, reassuringly gorgeous and Amazonian), some mantids, and some giant bird-catching spiders of the genus *Mygale*. Mosquitoes and ticks were finding him more successfully than he was finding collectable arthropods. The birds likewise seemed sparse and unspectacular. Within a month, though, when he had gotten the knack of looking more carefully and more knowledgeably, Wallace began to see what was actually there. Other problems didn't evaporate so quickly.

After two months of collecting in the vicinity of Pará, he and Bates sent a first batch of specimens back to Samuel Stevens. The shipment reached London safely, and Mr. Stevens, a reliable man who would continue serving Wallace for some years, began the task of marketing. That first shipment consisted largely of insects: about 450 species of beetles and almost as many of butterflies. By October, after a trip up the Rio Tocantins, which flows northward into the mainstem Amazon not far upstream from Pará, they had packed off another batch—again mostly insects, but now also including a few bird skins and shells. This batch was accompanied by a letter, in which Wallace described the logistics of the trip:

> We hired one of the heavy iron boats with two sails for the voyage, with a crew of four Indians and a black cook. We had the usual difficulties of travellers in this country in the desertion of our crew, which delayed us six or seven days in going up; the voyage took us three weeks to Guaribas and two weeks returning. We reached a point about twenty miles below Arroya, beyond which a large canoe cannot pass in the dry season, from the rapids, falls and whirlpools which here commence and obstruct the navigation of this magnificent river more or less to its source; here we were obliged to leave our vessel and continue in an open boat, in which we were exposed for two days, amply repaid however by the beauty of the scenery, the river (here a mile wide) being studded with rocky and sandy islets of all sizes, and richly clad with vegetation; the shores high and undulating, covered with a dense but picturesque forest; the waters dark and clear as crystal; and the excitement in shooting fearful rapids, &c. acted as a necessary stimulant under the heat of an equatorial sun, and thermometer 95° in the shade.

With the Amazon venture still in its early months, Wallace could afford to find compensation for physical hassles in the scenic grandeur and the thrill of running rapids. "Our collections were chiefly made lower down the river," he added. Samuel Stevens did his best to generate interest in those collections by forwarding Wallace's letter to *The Annals and Magazine of Natural History*, which obligingly published an excerpt.

That seems to have been Wallace's first appearance in the *Annals*, which would later carry his Sarawak paper to the attention of Lyell and Darwin. In the same issue, Samuel Stevens placed his own advertisement as "Natural History Agent," offering specimens from New Zealand, India, the Cape, and, most prominently, "*Two* beautiful Consignments of INSECTS of all orders in very fine Condition, collected in the province of Pará" by Messrs. Bates and Wallace.

After he and Bates parted ways (amicably, so that they could collect different species in different parts of the Amazon), Wallace left Pará and moved upriver, spending much of the next four years on a series of arduous canoe expeditions along various tributaries. First he ascended the mainstem as far as the city of Santarém, about five hundred miles inland, where the huge Rio Tapajós drains in from the south. "I have got thus far up the river, and take the opportunity of sending you a few lines," he wrote Stevens on September 12, 1849. "To come here, though such a short distance, took me a month." Now he was waiting to go on, "but the difficulties of getting men even for a few days are very great." Although the landscape around Santarém was dry and scrubby, he did find some lush patches of forest, "and in these, Lepidoptera are rather abundant; there are several lovely *Erycinidae* new to me, and many common insects, such as *Heliconia Melpomene* and *Agraulis Dido*, abundant, which we hardly ever saw at Pará: Coleoptera I am sorry to find as scarce as ever." The butterflies kept him cheerful despite the scarcity of beetles. And maybe, he guessed, on the thousand-foot hills farther upriver near Montalegre, the beetles might also be abundant.

He was even contemplating a push up into Bolivia, or possibly toward Bogotá or Quito. He asked Stevens to advise him: Were those places still untapped, so far as the trade in natural-history novelties was concerned? Throughout all the scientific travels of his early career, Wallace was a market-driven collector. He couldn't afford not to be. "Pray write whenever you can, and give me all the information you may be able to obtain, both as to what things are wanted in any class or order and as to localities." Darwin, a son of wealth, had never faced the necessity of such grubbing. But to Wallace that necessity would eventually prove a blessing because of the particular way it forced him to collect, gathering numerous individual specimens of the showiest and most marketable species.

At Santarém he appreciated the local felicities. "The Tapajoz here is clear water with a sandy beach, and the bathing is luxurious; we bathe here in the middle of the day, when dripping with perspiration, and you can have no idea of the excessive luxury of it." Oranges were cheap in the Santarém markets, only fourpence a bushel, and good. Pineapples were also available. Wallace had come a long way from Leicester, and he was accustoming himself to this new world. "The more I see of the country, *the more I want to*," he told Stevens. The diversity of butterflies seemed endless. He was living out his dream, and he was finding it satisfactory, though also a bit scary and lonely. Remember me to all friends, he wrote.

Five hundred miles above Santarém he reached the trading capital of Barra, a town of mud streets and little red-roofed houses at the junction with the Rio Negro. From there he ascended the Rio Negro to its source in Venezuela. Crossing over a divide in those swampy Venezuelan uplands, he visited an area that Alexander von Humboldt had explored fifty years earlier. Wallace also made two ascents of the Rio Uaupés, a tributary of the Negro that was blocked by a chain of cascades, falls, and steep rapids, which required portaging. He reached places where no other European had ever waved a butterfly net. He learned to subsist on *farinha* (a crunchy flour made from manioc), fish, and coffee. Sometimes there was nothing but *farinha* and salt. For preservative fluid he carried *cachaça*, a famed moonshine brandy distilled from sugar cane, but his sometime crewmen from the local villages tended to view the use of *cachaça* for pickling snake specimens as a perverse misappropriation, and on occasion they drank up what Wallace had carefully hoarded. When *cachaça* was unavailable, the crewmen could drink home-brewed *caxiri*, a sort of beer fermented from manioc. Wallace had less flexibility; he doesn't report ever preserving a specimen in beer.

He adjusted to sleeping in a hammock. He became a gourmet of turtle. In wholesome Victorian moderation, he shared the *cachaça* at village ceremonies. Once in a while he admired the shape of a young Indian girl as she bathed in a stream, and that's all his later account ever said about romantic longing and horniness. The chigoe fleas burrowed holes in his feet, which festered and made him limp. At one point he thought he might die of dysentery. Another time he went down with malarial fever. The cure for the first was to stop eating manioc gruel, go hungry, and wait it out; for the second, quinine. His younger brother Herbert, who came over from England to join him and learn the life of a biological collector, turned out to have less appetite than Alfred for the tropical wilds and a less hardy constitution. Herbert returned downriver to Pará, unhappy with jungle life and eager to escape back to England on the next boat. Before he could sail home, yellow fever killed him. Alfred was a thousand miles away when it happened. Much later in life, he would still carry some weight of guilt about Herbert. At the time, far up the Rio Negro and unaware of what was happening at Pará, he suffered his own bouts of homesickness.

He thought about the long summer evenings of England and the long winter nights with the family gathered around a blazing hearth. He ate roasted plantains and spoke almost nothing but Portuguese. The worst fits of loneliness came and went. Once every year or so he went downriver as far as Barra and hoped for a packet of mail. Herbert is very sick, he learned from a letter one year. Herbert died from his sickness last year, he learned later. He endured the sand flies and the biting gnats and the vampire bats that fluttered quietly into a hut to nip a sleeping person on the nose or the toe. He slept with his toes covered and once woke with a painless but bloody wound on the tip of his nose. He collected and sketched 160 species of fish.

One day in the forest he saw a black jaguar—that is, a melanistic aberration from the spotted pattern of ordinary jaguars—and gaped at it without raising his gun. "This encounter pleased me much," he remembered afterward. "I was too much surprised, and occupied too much with admiration, to feel fear. I had at length had a full view, in his native wilds, of the rarest variety of the most powerful and dangerous animal inhabiting the American continent." In that melanistic jaguar, he caught a vivid glimpse of how certain individuals within one species could vary from the norm. The word "variety" itself, as he and others used it, would later take on new meaning.

He went back up the Rio Uaupés, portaging past dozens of cataracts to a point where the river traders didn't go, in search of a bird he had heard rumors about: a stunning white version of the black umbrella-bird, *Cephalopterus ornatus*. Was the white umbrella-bird an undiscovered species, or was it just a variant individual, an anomaly, a sport, like the occasional black jaguar born to a pair of normally spotted parents? Or was it, perhaps, only legendary? He never did see a white umbrella-bird. In lieu of collecting a specimen, he collected testimony. Some of the local Uaupés people had no knowledge of any such creature; others told him that it was real but rare. Wallace came away frustrated and "inclined to think it is a mere white variety, such as occurs at times with our blackbirds and starlings at home, and as are sometimes found among the curassow-birds and agoutis." But the episode wasn't fruitless. It helped expand his awareness of variation within species.

We all recognize that *Homo sapiens* encompasses variation among individuals, but it is easy to forget that *Cephalopterus ornatus*, *Panthera onca* (the jaguar), and other species routinely encompass variation too. It was easier still to forget in Wallace's day, when the typological view—the assumption that a species consists of one ideal model, represented by exact copies—predominated. The typological view is neat but fallacious. Every species comprises a spectrum of actualities. Just as some humans are taller than others, some ravens are blacker than others. Some giraffes' necks are not quite so long as the giraffic average. Some blue-footed boobies find themselves cursed with feet that aren't quite so resplendently blue as the feet of their relatives, neighbors, and sexual rivals. These small increments of intraspecific variation, these minor individual differences, became immensely important to Alfred Wallace as he continued to think about evolution.

Wallace had reason to notice such variation more clearly than most other naturalists. As a commercial collector, he collected redundantly—taking not just one specimen each of this parrot and that butterfly but sometimes a dozen or more individuals of a single species. Lovely dead creatures were his stock-in-trade, literally, and he grabbed what he could for the market. But after grabbing, he preserved, inspected, and packed his creatures with a keen eye, so he saw intraspecific variation laid out before him in a way that other field biologists (including even the best of the wealthy ones, like Darwin) generally didn't. It was a trail of clues that Wallace would follow to great profit.

There were continual problems with preserving and protecting all these specimens. During the wet season, Wallace found it nearly impossible to dry a bird skin or an insect. If a fleshy specimen was left on a floor, on a table, anywhere within reach, it would quickly draw ants. Tucked away into Wallace's drying box, it would grow mold. Exposed to the sun, it would attract egg-laying flies and then be gobbled by the maggots when they hatched. The only way he could preserve a skin, Wallace found, was to hang it above his fire every morning and evening, like smoking a ham. The smoked specimens were sealed into wooden crates.

Another problem was shipping. While he was working the lower Amazon it had been possible, if not easy, to send batches of specimens back to Stevens in England. Now it wasn't. While Wallace was upstream on the Rio Negro, his specimens from that phase of work accumulated at Barra. Those boxed collections didn't go downriver, they didn't precede him to England, because customs officials wouldn't allow it—not until Wallace himself filed declarations and paid duty. At the time Wallace was able to cover his expenses without additional revenue from Stevens, so the delay didn't seem to matter. Later it would prove to have mattered grievously.

Finally he'd had enough. Stay in the Amazon longer, he figured, and he would probably die. Stored down at Barra were six precious crates of his specimens. He also had four years' worth of journals and drawings and notes, equally precious. And now he was coming down from the upper Uaupés with an ungainly assortment of live animals, including monkeys, parrots, and other large and small birds. His concern was to get as many as possible of these animals, and himself, back to England in good health.

Descending the Uaupés with that cargo was a challenge. He had been delayed by the difficulty of finding an Indian crew, men who knew the river well enough to paddle those rapids yet were willing to be hired for a white man's foolish enterprise on a white man's foolish schedule. Almost everyone had something better to do: repair a house, attend a ceremonial dance, stay home with a young wife. Wallace was tormented by the sense, not uncommon among impatient adventurers in that age and later, that he just couldn't get good help. But he settled for the men who offered themselves, and together they did survive the river. "On April 1st we passed a host of falls, shooting most of them amidst fearful waves and roaring breakers," he wrote. During an earlier descent of the upper Uaupés he had felt "some little fear of the passage of the falls, which was not diminished by my pilot's being completely stupefied with his parting libations of caxirí," but his pilot in this case was evidently sober. Among other small mishaps, he reported that two of his captive birds were eaten by a captive monkey, and that "one of my most valuable and beautiful parrots (a single specimen) was lost in passing the falls." The losses still left him with thirty-four live animals, both furred and feathered, "which, in a small canoe, were no little trouble and annoyance." The menagerie now comprised five monkeys, two macaws, twenty parrots and parakeets of twelve different species, five small birds, a white-crested Brazilian pheasant, and a toucan. From Barra, with the customs officials placated, he continued downriver in a larger canoe. It was June of 1852. Not far out of Barra, his tame toucan went overboard and drowned. Wallace and the rest of his treasures reached Pará safely. He was back at the mouth of the Amazon, where he had started in 1848.

On July 12 he took a last look at the white houses and feathery palm trees of Pará, then boarded a brig called the *Helen*, bound for England. Commanded by a Captain John Turner, the *Helen* held a cargo of rubber, cocoa, broom fiber, and resinous balsam from the capivi tree. The balsam was packed in small kegs. As a precaution against shifting and breakage, some of the kegs were stowed in sand; others were stowed in rice chaff—a very bad idea, given the possibility of spontaneous combustion. Wallace knew nothing about the balsam-in-chaff situation when he stepped aboard. His crates of specimens, his journals and sketches, and his menagerie all went onto the *Helen*.

For three weeks they had light winds but otherwise good weather. Wallace suffered a fever. He thought at first that it might be the same disease that had killed Herbert, now finally claiming him too. He recovered but, still weak, spent his time reading and resting below deck.

By August 6 they had reached mid-Atlantic, somewhere about seven hundred miles east of Bermuda. Captain Turner walked into the cabin and told Wallace, "I'm afraid the ship's on fire; come and see what you think of it." Smoke was oozing from that part of the hold where the balsam casks sat in the rice chaff.

The crew chopped an opening in order to douse the area with water. This was another bad idea, since it gave the smoldering balsam a better draft of oxygen. Hours passed, with the crew throwing buckets of water and the situation still being misread. The ship's carpenter cut a hole in the cabin floor, no doubt further improving the draw, and the calm but incompetent Captain Turner grew pessimistic. Eventually came a moment when Turner went looking for his chronometer, sextant, compasses, charts. Wallace watched all this in some sort of numb paralysis. Turner ordered down the lifeboats. The cabin by now was full of smoke, intolerably hot, close to breaking into flame, but Wallace groped his way in and grabbed a small tin box that held a few of his shirts. Not many feet away, kegs of balsam were burbling like lava. Monkeys and parrots were screeching and flailing and dying.

Into the tin box Wallace tossed whatever drawings and other papers he could find. He managed to save one diary, some pencil sketches of palm trees and of fish, and a packet of notes for a map of the upper Rio Negro. Then he jumped into one of the lifeboats and found that, weathered and cracked, it was leaking. When he chose the *Helen*, Wallace had booked himself onto a loser.

Shortly the *Helen* went up in flames. Wallace, Captain Turner, and the crew sat nearby in their two porous lifeboats, bailing incessantly as the ship's rigging caught fire. They watched the sails burn like paper. They watched the masts topple. The main mast went first, but "the foremast stood for a long time, exciting our admiration and wonder," Wallace recalled cheerfully. A few monkeys and parrots were loose on the wreck, clambering hysterically toward nowhere. He saw several animals disappear in the flames. One parrot fell into the sea and was rescued by Wallace's boat. Coleridge would have loved it.

For ten days they lived on raw pork, biscuit, water, and a few other niceties that they'd managed to save. The cook had thoughtfully brought some corks from the galley to help plug the leaky lifeboats. They set their little sails for Bermuda. For the first day or two the wind was good, and they thought they might reach Bermuda in a week, or at least be spotted by a ship en route to the West Indies. Then the wind shifted, and they had to turn north. No sign of another ship. Wallace's face and hands were now badly blistered from sunburn, but he seems to have kept his preternatural cheeriness. "During the night I saw several meteors," he wrote later, "and in fact could not be in a better position for observing them, than lying on my back in a small boat in the middle of the Atlantic." The castaways ran short of water, presumably also of food. Wallace found his own consolations. He admired the metallic blues, greens, and golds of the dolphins that circled the boats. He never lost the instincts of a naturalist. "We also saw a flock of small birds fly by, making a chirping noise; the sailors did not know what they were." The sailors, stuck in a lifeboat with this nutcake birdwatcher, may have considered turning *him* into raw pork.

On the tenth day they were rescued—rescued from the lifeboat dilemma, at least, though not from all oceanic perils. They were picked up by a ship called the *Jordeson*, an English trader bound homeward from Cuba. The *Jordeson*'s captain welcomed them warmly. The first luxury they were given was drinking water. The second was tea.

That night Wallace didn't sleep. He thought about England and his family. His brain crackled with hopes, new fears, frustrations, and he seems to have felt more vulnerable than ever to the caprices of destiny. This was his well-deserved emotional letdown, a brief one, which seems to have run its course on board the *Jordeson*. Besides the near pass at death, the sudden rescue, and the four years' worth of homesickness, there was the matter of his collections, burned and sunk with the *Helen*. "It was now, when the danger appeared past, that I began to feel fully the greatness of my loss. With what pleasure had I looked upon every rare and curious insect I had added to my collection!" Recalling it later, he wrote a litany of laments. How many times, sick with fever, had he crawled through the forest to find a new species, exclamation point. How many weary days and weeks, how many fond hopes, how many unexplored places, exclamation point. Wallace by then owed himself a little scream therapy, at least in the form of strident punctuation. "And now everything was gone," he wrote, "and I had not one specimen to illustrate the unknown lands I had trod, or to call back the recollection of the wild scenes I had beheld!" This outburst occupies less than a page near the end of *A Narrative of Travels on the Amazon and Rio Negro*, his 350-page account of the whole misadventurous adventure. "But such regrets I knew were vain, and I tried to think as little as possible about what might have been, and to occupy myself with the state of things which actually existed." One of Wallace's great character strengths was his capacity to do that: to focus on the demands and the consolations of immediate reality, rather than dwelling on what might have been.

He had a further chance to exercise that capacity in early September when the *Jordeson* nearly sank in a heavy gale. The sails tore apart, the decks were buried in foam. The pumps were kept working all night. A huge wave broke through the cabin skylight, soaking Wallace where he slept. But the ship, and his pluckiness, both survived. Several weeks later, in the Channel, there was a last slap of foul luck in the form of another ferocious gale, severe enough to send ships to the bottom. But the old, crippled *Jordeson* sneaked through.

On October 1, 1852, Alfred Wallace stepped ashore at Deal, a town on the southeast coast of England. He carried the small tin box containing the diary and some drawings; he was alive; and he had learned a few things. He went to bed with swollen ankles.

Four days later, still laid up, he wrote to a friend back in Brazil: "Fifty times since I left Para have I vowed, if I once reached England, never to trust myself more on the ocean. But good resolutions soon fade, and I am already only doubtful whether the Andes or the Philippines are to be the scene of my next wanderings." Like good resolutions, frustration and despair soon fade, at least in a certain kind of person. Alfred Wallace was that kind of person—persistent and incorrigibly optimistic. There was still the mystery of mysteries to be solved. After four days of convalescence, he was thinking about mountains and islands.

<h1 style="text-align:center">16</h1>

ALTHOUGH HIS specimens and most of his notes were gone, Wallace's Amazon years hadn't been wasted. Some of his observations could be retrieved from memory and from the letters he had sent home to Samuel Stevens. By the sheer process of collecting redundantly, he had sharpened his eye for intraspecific variation. He had developed field techniques and survival skills suitable to the tropics. And he had gained a seminal insight: that patterns of floral and faunal distribution are significant.

Distribution, Wallace had seen, is commonly delineated by some sort of geographical barrier—a ridge of mountains, a wide river, a discontinuity of vegetation reflecting a discontinuity of geological substrate. He had noticed that two similar species of animal, closely related, often occupy opposite sides of such a boundary. These patterns are eloquent, Wallace intuited. They say something, he knew, though he didn't know what.

Based on this much, he had begun gathering biogeographical data. That is, he had started concerning himself not only with the identity of each specimen but also with where he collected it. Most other naturalists of the period were virtually oblivious to that dimension, and museum specimens generally carried no more than the vaguest geographical labeling. Bird skins and insects and reptiles in alcohol might be fit tightly within taxonomic categories—this genus, this species—but matched only loosely to place of origin. A stuffed anteater might be localized no more precisely than "South America." A pinned rhinoceros beetle might be traceable only to "southeastern Asia." It wasn't good enough, Wallace realized; valuable information was being tossed aside. Charles Darwin himself, during his Galápagos stopover, had been careless about keeping his specimens geographically sorted. Later he regretted having lumped together his finch specimens without tagging them as to *which* Galápagos island had yielded each. Which birds had come from Chatham, which from Albemarle? Darwin didn't know. Since he had admitted his mistake in the *Journal*, maybe Wallace in his turn had been duly alerted by the older man's forthright self-criticism. In any case, Wallace was more fastidious. The specimens that he lost on the *Helen* had been carefully tagged. With his Amazon methodology, Wallace had laid a foundation for biogeography as a systematic science.

He had especially noticed the boundaries of distribution among monkeys. Both the mainstem Amazon and the Rio Negro are gigantic rivers, miles wide where they converge near the town known to Wallace as Barra. A single species of bird might be able to straddle such a big river, its population remaining genetically undivided so long as individuals flew back and forth. A species of insect might encompass both sides too. A species of monkey, not likely. To a monkey, the mainstem Amazon seems almost as wide as an ocean. A paca will swim, a capybara will swim, a tapir will swim, even an ocelot will cross at least narrow stretches of water. "I have never heard of monkeys swimming over any river," Wallace wrote, "so that this kind of boundary might be expected to be more definite in their case than in that of other quadrupeds, most of which readily take to the water." And it isn't just that monkeys seldom swim; they seldom walk either. More than most other groups of terrestrial animal, monkeys require a continuous canopy of interlinked tree branches, making arboreal locomotion possible. They wouldn't likely cross the Amazon River even if it were a great brown highway of dirt.

Back in London after his shipwreck, Wallace presented his views on Amazon monkeys in a paper read to the Zoological Society. He had seen twenty-one species: howler monkeys, spider monkeys, woolly monkeys, capuchin monkeys, uakaris, marmosets, and one peculiar genus of bushy-tailed, bulky little creatures that Wallace called "sloth monkeys." The sloth monkey genus was *Pithecia*, of which he had recognized two species, *Pithecia irrorata* and another that was new to science. The new species was remarkable for having muttonchops and a beard of bright red.

"During my residence in the Amazon district I took every opportunity of determining the limits of species," Wallace told the Society, "and I soon found that the Amazon, the Rio Negro and the Madeira formed the limits beyond which certain species never passed." Those three great rivers, forming a chicken-foot pattern with their convergence near the town of Barra, divide the upper Amazon basin into four separate wedges of landscape. Wallace labeled them the Guiana district (northeast of the Amazon-Negro junction), the Ecuador district (northwest of that junction), the Peru district (south of the Amazon, west of the Madeira), and the Brazil district (southeast of the Amazon-Madeira). Already he was a maker of biogeographical schemas. Already he was viewing the Amazon basin as an archipelago: four big islands separated by water.

Some species of monkey overlapped the big rivers, especially toward the headwaters, where crossing was easier. But the four-district schema, oversimplified though it was, showed how a welter of biogeographical facts could yield meaningful pattern. The general message of his monkey data was that these animals knew boundaries.

One species of spider monkey occurred only in the Guiana district, northeast of the Amazon-Negro junction. Another species of spider monkey occurred in the Peru district, south of the Amazon and west of the Madeira. Three species of marmoset each occupied only a limited tract of country, and Wallace was able to trace the limits. His interest in distribution had begun to elucidate—here in the Amazon, long before Sarawak—that closely allied species are coincident both in space and in time.

The two species of *Pithecia* also seemed to be isolated by rivers. *Pithecia irrorata* was found on the south bank of the Amazon, west of its confluence with the Madeira. The unnamed red-bearded species lived north of the Amazon and west of the Rio Negro. Today we can guess that the red-bearded species was the one presently known as *Pithecia monachus*. According to Wallace, it didn't extend eastward into the Guiana district. Never? If not, why not? Here falls an interesting gap in his data on Amazon monkeys. If the red-bearded *P. monachus* never showed itself east of the Rio Negro, was some other species of sloth monkey resident over there? Wallace said nothing more about *Pithecia*. He was tireless and thorough, but he couldn't know everything.

The gap has been filled. East of the Rio Negro lives a third species of sloth monkey, now known as *Pithecia pithecia*. East of the Rio Negro we reenter the twentieth century.

17

THE SLOTH MONKEYS today are called sakis. There are white-faced sakis, brown buffy sakis, and a handful of others with no English names. In Portuguese slang they're *macacos velhos*, old monkeys, because some species show grizzled coloration. Up in Colombia their Spanish nickname is *monos voladores*, flying monkeys, in honor of their acrobatic agility. Names change and vary, the rivers keep flowing, the forest has its own scale of time. Barra, on the northeast bank of the Rio Negro not far from its junction with the Amazon, is now a big ragged city called Manaus.

Decades after Wallace departed, Manaus grew into a boomtown, the gilded and garish capital of the rubber trade. Great fortunes were made and spent. A gold-domed opera house, the Teatro Amazonas, was built. No roads—except the watery road of the river—connected this city to anywhere. Manaus was the hub of its own world. Then the market for Brazilian rubber collapsed, the opera house closed, and the jungle city had to find other sources of vigor. That has been a struggle, and the struggle continues.

Fifty miles north of Manaus, surrounded by rainforest, is a cluster of cattle ranches—*fazendas*, as the Brazilians know them. Across large stretches of gently rolling terrain, the forest has been leveled with chain saws, left to go crisp through a dry season, then burned. The transformation is stark; the jungle has turned to charcoal. Beneath the cinders there's precious little soil, mostly just red clay, but on some of the open land a thick mat of grass nevertheless grows, at least temporarily. The grass, alien to rainforest, has been planted by hand. In other areas, especially along the double-rut tracks that function as access roads, the sun-cooked clay has turned almost ceramic. The pecuniary logic of ranching cattle by chain saw and fire, on half-sterile ground in the heart of the Amazon, has been explained to me but remains unfathomed. It has something to do with the Manaus Free Zone Authority, a government body whose mission has been to subsidize and encourage private enterprise. Desperate measures of official benevolence were decreed, several decades ago, to compensate for the bust of the rubber trade. Investors arrived to harvest the overripe incentives, cutters and burners took down the forest, *vaqueiros* came in from elsewhere, and finally a few white cows appeared and were fed. Evidently it all served its purpose, at least to some marginal degree, because the Teatro Amazonas has lately been refurbished. During my stop in Manaus I gawk at it from the plaza in front, like any tourist, but I can't stay for an opera. I'm headed north to Fazenda Esteio.

Fazenda Esteio is one of the ranches carved out of the rainforest within the Manaus Free Zone. At its center, in the midst of a great clear-cut, a small patch of forest survives.

There are a number of such patches, both on Fazenda Esteio and on neighboring ranches, most of them neatly shaped squares. They were plotted and left standing as part of an extraordinary scientific enterprise to which I'll return later in this book. The square patches, each known by number, each with a case history, constitute a grand outdoor laboratory in which Tom Lovejoy and his colleagues study the phenomenon of ecosystem decay. It was Lovejoy's own vision that launched the enterprise. His jawboning skills, in collaboration with some farsighted Brazilians, have given those patches their rectilinear exactitude, their meaning, and (under a provision of Brazilian law) their protection against cutting. The small patch to which I'm headed is an exception among this group of exceptions. It's irregular, not square—shaped more like a boomerang than like a box—and has been spared because it buffers a tiny stream.

Within the patch, surrounded on all sides by pasture and sunlight and the perversities of subsidy economics, lives a small population of *Pithecia pithecia*. These individuals belong, more precisely, to *Pithecia pithecia chrysocephala*, the golden-faced subspecies of the white-faced saki. Just after dawn on a fine Amazon morning I get my first glimpse of them.

The name *monos voladores* is apt—they do seem to fly between branches. They don't swing, they don't clamber, they run out a limb and keep going, launched. For airborne monkeys they appear surprisingly hefty, with their thick body fur, thick tails, thick heads of hair falling forward in bangs. Despite all the thickness, they're nimble— the nimbleness evident as I see one of them, a dark bundle overhead, low-slung and horizontal in posture, make a ten-foot lunge between trees. It looks like a badger shot through the forest by catapult.

"That's the female," says Eleonore Setz.

Setz is a Brazilian ecologist who has been studying this population of sakis since 1985. Her original intent was to work in the square reserves, but the stragglers in the stream forest diverted her. By now she knows them as individuals. There are currently just six—a patriarch, a matriarch, a young female, two juveniles, a yearling. Setz can draw

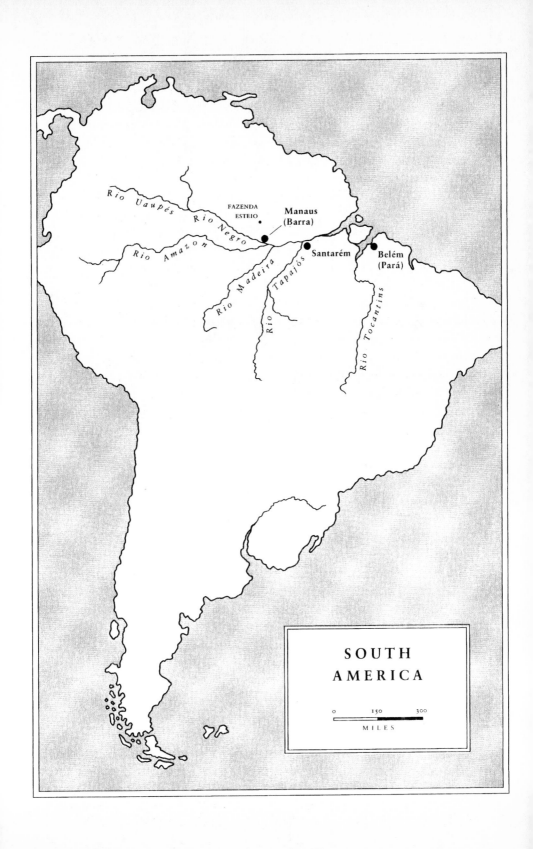

Rio Uaupés

Rio Negro

FAZENDA
ESTEIO

Manaus
(Barra)

Rio Amazon

Rio Madeira

Rio Tapajós

Santarém

Belém
(Pará)

Rio Tocantins

SOUTH
AMERICA

0   150   300

MILES

their family tree. She has chronicled births, deaths, matings, and dis-appearances. She wants to document just how they allot their time and their energy while struggling to extract a livelihood from their fragment of habitat. This morning she has allowed me to tag along. We were awake at five, out of the hammocks, wolfing down guava jelly and crackers and stout Brazilian coffee beneath the light of a gas lantern until it was time to walk. The sakis slept somewhat later. Now they are beginning to stir.

"The male is black," Setz adds, "with an orange face."

Coloring is far from uniform within *Pithecia pithecia*, another case of intraspecific variation. Alfred Wallace might have seized on it if he had known. Five hundred miles northeast, up in the forests of Suri-nam, males of this species show white facial fur, sharply contrasting with the dark body color and giving them their name, white-faced saki. Down here along the Rio Negro, the face of a male can be russet or yellow or (as with Setz's group) orange. Females are more drab than males, though they have their own range of color variation. The female above us is brownish gray, not quite so dark as the male. Like an agouti, to Setz's eyes, or a muskrat to mine. Faint streaks of white and yellow highlight the female's somber fur. Her yellowish crown and her size are distinctive enough for individual identification, at least among this little group. She's known to the humans who watch her as Wilza Carla. Her namesake is a portly blond actress on Brazil-ian TV.

"We've arrived," says Setz, meaning that the morning's first spate of monkey seeking is over. "They're going to eat."

Settling in the crown of a tree, they start cracking the seed pods of a liana tangled there in the canopy. They eat the seeds and drop the husks contemptuously toward us. I hear birdsong, the crunch of teeth, the rustle of those falling husks, and no other sounds. It's still too early for heat-driven insects. Setz is ready with her notebook and her digital watch, her binoculars, her infinite patience and curiosity—all the requisite tools of a behavioral ecologist in the field. I sit on the ground, making myself comfortable except for the crick in my neck from staring upward.

There's great tranquility to be found beneath the Amazon canopy when you quiet yourself to watch an animal at its business. If the mos-quitoes aren't biting, if the rain happens to have paused, you can dream away hours, you can space-walk the universe. But the tranquil-ity of this particular place, where Eleonore Setz does her watching, is meager and pinched. Dawn has destroyed the illusion of a vast forest. I can see daylight pouring in along the clear-cut edge twenty yards away, and beyond that, blue sky above man-made pasture.

The cutting was done in 1980, at the height of the ranching-incentives policy. Ever since, the sakis in this fragment of forest have been marooned. Their isolation isn't absolute (an occasional young male from the outer forest may have come in across that clear-cut, and several members of Wilza's group seem to have left), but it's isolation enough to define their lives and their future. Ecologically, they are hampered, with only a small area of forest in which to find food. Some of their favorite fruits probably don't grow here at all. Behaviorally, they have had to compensate, adjusting their diet to what is available. Socially, they have suffered from the absence of others of their species. They don't have neighbors, they don't have rivals, they don't have visitors from whom they might learn things. Genetically, too, they are disadvantaged—condemned to inbreeding and to the potential health problems that inbreeding entails. With all these factors set against them, their overall prospects are gloomy. The species *Pithecia pithecia* isn't immediately endangered, but this little population certainly is.

Their habitat fragment is 9.2 hectares in area. That constitutes crucial information.

A hectare is a metric unit that serves as the standard measurement of beleaguered rainforest. All over the world, tropical biologists and conservationists talk in hectares. Me, I pride myself on despising metric-system evangelism as much as the next red-blooded and obdurate American, but the hectare does happen to be handy, and the conversion is simple. One hectare is the area encompassed within a square measuring 100 meters on each side—just enough space for a football game, counting the end zones and benches. One hectare equals 2.47 acres. At 9.2 hectares, less than 25 acres, Setz's fragment is roughly the size and shape of two par-five golf holes laid side to side with a sinuous water hazard between. The water hazard is that little stream I mentioned. Her *Pithecia* inhabit a narrow world.

The previous evening, as we talked across a camp table, I asked Setz whether the fragment has a name. Well, not much of a name, she told me. In Portuguese it would be *a matinha do igarape do acampamento Colosso*. Sounds like plenty of name, I said, what does it mean? *Acampamento Colosso* was recognizable to me as the field camp where we sat, called Colosso, though it's just a few thatched roofs furnished with hammocks and a lantern. An *igarape*, I knew, is a trickle of water. Which gave me "the *matinha* around the stream at Camp Colosso." *Matinha* itself was slightly more difficult to translate.

It's the diminutive of *mata*, meaning forest, Setz said. Small forest. Tiny forest. But that doesn't capture the nuance. *Matinha*. It's familiar, it's affectionate, it's condescending, as you might speak of a child, she said, groping.

I said, What about "cute little forest"?

Yes, she said. Yes, that's the idea.

A matinha do igarape do acampamento Colosso: the cute little forest at the heart of the Amazon. Ecologically, it's an island. Like Bali, it was once rich in species—at the onset of its insularity—and since then has become impoverished. Jaguars are now absent, peccaries are absent, tapirs are absent. The patch is not only isolated but simplified. Of *Pithecia pithecia chrysocephala*, the golden-faced saki, only six individuals remain. Their chances of perpetuation are low. Before many years pass they will probably disappear, and still other species will disappear after them. The fragment will grow steadily less Amazonian and more diminutive, more diminished, more leached of its biological richness, without growing smaller. That's the dire magic we're calling ecosystem decay.

I asked Setz to repeat the Portuguese phrase. I asked for the spelling. Here, I said, write in my notebook. Foreign languages tend to spill off my brain like rain off a cheap plastic umbrella, but *matinha* was a word I didn't want to forget. Setz wrote out the phrase in her neat Portuguese print.

That was last night. Now breakfast is over for the sakis. Having had enough of the liana seeds, they travel on. Their horizontal leaps, their snatched landings, stir the branches. Their big puffy tails provide balance. They cover distance quickly and then every so often they stop, they freeze, becoming invisible among the shadows of the high understory. I sweep my binoculars this way and that, I squint, I wonder where the devil they are. I try hard not to make noise, not to spook them, not to come too near, spacing myself a half-dozen yards behind Setz, who follows them calmly. She isn't concerned about losing them. Her eyes are keen, and besides, these animals can't go far.

18

BALI, Lombok, Madagascar, the Guiana district, the *matinha* on Fazenda Esteio, Alfred Wallace, Sarawak—these elements are not assembled here randomly. There's a connectedness among them that's more than casual. They all bear on the elemental yin and yang of island biogeography: the study of evolution and of extinction.

Wallace himself traced that connectedness with the path of his life and his thought. His greatest scientific accomplishment was to discover and elucidate the transcendent significance of insularity. But he was born a century too soon to see how far that insight would lead.

Where it led in his case was to the Malay Archipelago. This was in some measure, as I've said, a matter of chance. If the brig *Helen* hadn't caught fire, if his Amazon notes and his Amazon specimens had survived, if he had successfully harvested the data from his first expedition, Wallace might have stayed home for the rest of his life, as Darwin did after the *Beagle*. If certain other things had happened differently, he might have gone back out to the Andes, as he mentioned considering in the sickbed letter. He might have matured into an eminent Victorian naturalist, nothing more, who devoted his years to observing and collecting the flora and fauna of the mainlands. He might have become vastly, conventionally authoritative. Meanwhile he might have missed the profound patterns. He might have overlooked the significance of islands.

He might have, he might have . . . But history is only what it is.

19

WALLACE NEEDED money and adventure more than he needed rest. He also needed biological data. He was still looking for a solution to that mystery of mysteries, the origin of species. Within the next eighteen months, while in England, he published two books—his *Narrative of Travels on the Amazon and Rio Negro* and a lesser volume titled *Palm Trees of the Amazon and Their Uses*, derived from notes he had salvaged from the *Helen*—both of which left his obscurity largely unbreached. Then he was ready to go back out collecting. The worst of his bad luck was over, and the good luck began with his choice of a new destination. For scientific reasons, but also because of those market pressures facing a purveyor of natural-history specimens, he chose a world of tropical islands.

Islands in general were repositories for the rare and the peculiar, he knew, and this particular set of islands especially so. "During my constant attendance at the meetings of the Zoological and Entomological Societies, and visits to the insect and bird departments of the British Museum," he wrote later, "I had obtained sufficient information to satisfy me that the very finest field for an exploring and collecting naturalist was to be found in the great Malayan Archipelago." Through a benevolent elder at the Royal Geographical Society, he was granted free passage by the Peninsular and Oriental steamship company. He sailed via Gibraltar to Alexandria, traveled overland from there to Suez, and then boarded another ship. By April of 1854 he was in Singapore.

Singapore suited him as a base for six months of collecting on mainland Malaysia. The mainland was rich in species, though not so rich in the sort of bold patterns and geographical barriers that he needed. In November he moved over to Sarawak, on the northwestern coast of Borneo. For the next seven years he would wander among islands.

In reaction against his earlier carelessness and against Darwin's Galápagos blunder, he made a resolution. "From my first arrival in the East I had determined to keep a complete set of certain groups from every island or distinct locality which I visited for my own study on my return home, as I felt sure they would afford me very valuable materials for working out the geographical distribution of animals in the archipelago, and also throw light on various other problems." Foremost among the other problems was evolution. He was smarter and more wary now, so besides tagging his specimens by locality he also shipped them off regularly in batches, rather than letting them pile up. He didn't foresee that the big problem would be solved—solved twice, in fact, once by him and once by Mr. Darwin—before he ever returned home.

Sarawak was fruitful. Wallace did well in beetles, butterflies, and birds. He shot (with the callousness of a professional hunter, which he was) several orangutans, and adopted (for the short time it survived) an orphaned orangutan infant. He saw a so-called flying frog. He spent time among the Dyak people of the interior. He also wrote and mailed off his paper on the "law" of closely allied species. He interrupted his fieldwork long enough to spend Christmas as the guest of Sir James Brooke, rajah of Sarawak, who was glad for some good conversation with a clever young Englishman. After more than a year, this part of Borneo must have felt almost like home. But home was a luxury for later. So Wallace left, looping back briefly to Singapore and then traveling east, as boat traffic allowed, toward the remote and little-known eastern end of the archipelago.

He sailed down to Bali, then across to Lombok, where he made his observation of the surprisingly drastic difference between the bird faunas of those two adjacent islands. From there, up to the trading port of Macassar on the island of Celebes.

He spent three months on Celebes, collecting as best he could despite sickness, rainy weather, the paucity of good forest immediately surrounding Macassar, and other logistical difficulties. The sickness may have been malaria again, though he blamed it on tainted water. The birds, the mammals, the insects of Celebes were distinct from what he had seen in Sarawak, and an animal called the babirusa impressed him especially. *Babyrousa babyrussa* is an endemic wild pig whose upper canine teeth (in the male) grow long and curled, pushing out through the flesh of the snout and arching back over the forehead like a pair of horns. Although it faintly reminded Wallace of an African warthog, he admitted that "the Babirusa stands completely isolated, having no resemblance to the pigs of any other part of the world." How did it eat? he wanted to know. How did it use those monstrous, inconvenient-looking tusks? What factors of growth or behavior explained this anatomical modification? Wallace was not one to accept a swoopy-toothed Celebesian pig as the whim of God. He also collected two species of hornbill peculiar to the island, and a handsome cuckoo, and three specimens of birdwing butterfly. The birdwings were magnificent creatures, seven inches across, their big black wings decorated with patches and streaks of white and satiny yellow. "I trembled with excitement as I took the first out of my net and found it to be in perfect condition," he wrote later. He matched his three specimens against the species *Ornithoptera remus* in a faunal treatise he carried, but the match was imperfect. Possibly they were a distinct Celebesian variety of *remus*, he suspected. Aside from these items and a few others, Celebes disappointed him.

The diversity of insects and birds seemed meager compared with what he'd seen on Borneo and mainland Malaysia. He found no barbets, no trogons, no broadbills, no shrikes. "Whole families and genera are altogether absent," he complained in a letter to Samuel Stevens, "and there is nothing to supply their place." Celebes, standing alone near the midpoint of the archipelago, was interesting but not very productive. And now the weather had begun to change.

The steady, strong eastern winds, which blew warm and dry onto Macassar during October and November, were giving way to variable breezes that signaled a shift of seasons, and storm clouds were sliding in from the west. It was the approach of the western monsoon—serious rain, hysterical rain, Wallace was warned. Collecting would soon be impossible. This was December 1856. The rains began. The dried-out rice paddies, which spread for miles across the flat coastal plain surrounding Macassar, filled again with water. The outskirts of town became impassable except by boat, or by tightroping along the muddy banks between paddies. Wallace could expect about five months of this weather if he hung around here in southern Celebes, he was told. So he moved again. Better to travel out ahead of the western monsoon and escape its reach, he figured, than to spend all those months hunkering indoors. He took space on a prau that was sailing east to the Aru Islands.

Wallace had heard tales about Aru. It was a tiny cluster of islands lying a thousand miles eastward, minuscule in area but famous to the seafaring native people of Celebes, the Bugis, for its precious biological commodities. Those commodities included pearl shell and dried sea cucumber as well as bird of paradise skins and other rare items, and a voyage out to Aru could be the central event of a Buginese trader's lifetime, like a Moslem's long pilgrimage to Mecca. The merchants of Macassar, both the Buginese and the immigrant Chinese, dealt in a wide variety of natural products from throughout the Malay region—rattan from Borneo, sandalwood and beeswax from Flores and Timor, sea cucumber from the Gulf of Carpentaria, cajuputi oil from Bouru, as well as staples such as rice and coffee. "More important than all these however," Wallace would write years after his trip, "is the trade to Aru, a group of islands situated on the south-west coast of New Guinea, and of which almost the whole produce comes to Macassar in native vessels." Aru, for the Buginese, was at the farthest extreme of imaginable travel.

He added: "These islands are quite out of the track of all European trade, and are inhabited only by black mop-headed savages, who yet contribute to the luxurious tastes of the most civilized races. Pearls, mother-of-pearl, and tortoise-shell, find their way to Europe, while edible birds' nests and 'tripang' or sea-slug are obtained by shiploads for the gastronomic enjoyment of the Chinese." Wallace, more so than most other Victorian travelers, was capable of empathy toward native peoples in the lands that he visited—even toward these folk he called "mop-headed savages"—but he wasn't faultless on that count, and his language reflected his times. The "mop-headed savages" were actually Papuans who had colonized Aru from the coast of New Guinea, a hundred miles away; savage or not, they added something to the aura. Throughout the eastern end of the big archipelago, from Celebes to the progressively more remote islands—Gilolo, Bouru, Timor, Seram, Banda, and Ké—the native peoples were distinctly Malayan, not Papuan. That is, they were brown-skinned, with thin features and straight hair. The presence of Papuans on Aru represented another signal, along with the gaudy bird skins and the nacreous shells, that this little set of islands was extraordinary.

December or January, at the start of the western monsoon, was the right season for riding the winds to Aru. July or August was the season for riding the winds back. At other times the route was impossible instead of just dangerous, since Buginese praus were no good at tacking upwind. "The trade to these islands has existed from very early times," Wallace added, "and it is from them that Birds of Paradise, of the two kinds known to Linnaeus, were first brought." The two kinds in question were *Cicinnurus regius*, a small scarlet species commonly called the king bird of paradise, and the greater bird of paradise, now known (with a slight correction of Linnaeus's original spelling) as *Paradisaea apoda*. Wallace was a trader himself, of course, and the commercial man as well as the naturalist in him was keen on getting a shot at those species. Among the well-heeled English collectors to whom Samuel Stevens sold Wallace's yield, a fine specimen of *P. apoda* was probably worth as much as any bird or insect on the planet.

But not every Buginese trader was daring enough to make the Aru pilgrimage. "He who has made it is looked up to as an authority," Wallace wrote, "and it remains with many the unachieved ambition of their lives. I myself had hoped rather than expected ever to reach this 'Ultima Thule' of the East; and when I found that I really could do so now, had I but courage to trust myself for a thousand miles' voyage in a Bugis prau, and for six or seven months among lawless traders and ferocious savages—I felt somewhat as I did when, a schoolboy, I was for the first time allowed to travel outside the stage-coach, to visit that scene of all that is strange and new and wonderful to young imaginations—London!" He did trust himself to the prau, which was a two-masted wooden cargo ship in roughly the shape of a Chinese junk, lashed together with rattan and finished out with bamboo and palm thatch. The sea gods were friendlier this time than they had been to him on the *Helen* and the *Jordeson*. In the second week of January 1857, he arrived safely in Aru.

Few other naturalists, few other Europeans, had ever seen the place. Throughout all his own Malay travels, he would never get farther east. Aru was, in more than one sense, his turning point.

20

THE SCIENTIFIC literature on the subject of island biogeography is formidably abundant. Most of that literature has been published as journal articles. How many? Damned if I know. It's like trying to say how many articles related to warfare have appeared in the *New York Times* since 1914. How *much*, then? Well, think of it this way. In Montana, where I live, people ballast the back ends of their pickup trucks in winter by loading up with sandbags or river rocks. An empty pickup, light over the rear wheels, gets lousy traction on snow. Instead of sandbags, those people could use island biogeography—photocopies of journal articles, bound up with twine. On my desk at this moment is a carefully culled pile, a modest selection of what's available. According to the bathroom scale it weighs eighteen pounds, counting the staples.

This pile includes "The Origin of Species Through Isolation" (1905), "Biological Peculiarities of Oceanic Islands" (1932), "Area and Number of Species" (1943), "Pygmy Elephant and Giant Tortoise" (1951), "Isolation as an Evolutionary Factor" (1959), "Galápagos Tomatoes and Tortoises" (1961), "Animal Evolution on Islands" (1966), "Small Islands and the Equilibrium Theory of Insular Biogeography" (1969), "Mammals on Mountaintops" (1971), "The Numbers of Species of Hummingbirds in the West Indies" (1973), "The Main Problems Concerning Avian Evolution on Islands" (1974), "Island Colonization by Carnivorous and Herbivorous Coleoptera" (1975), "Patch Dynamics and the Design of Nature Reserves" (1978), "The Application of Insular Biogeographic Theory to the Conservation of Large Mammals in the Northern Rocky Moun-

tains" (1979), "The Statistics and Biology of the Species-Area Relationship" (1979), "The Equilibrium Theory of Island Biogeography:
Fact or Fiction?" (1980), "Should Nature Reserves Be Large or
Small?" (1980), "The Virunga Gorillas: Decline of an 'Island' Population" (1981), "Trees as Islands" (1983), "The Evolution of Body Size
in Mammals on Islands" (1985), "Conservation Strategy: The Effects
of Fragmentation on Extinction" (1985), "Two Decades of Interac-
tion Between the MacArthur-Wilson Model and the Complexities of
Mammalian Distributions" (1986), "Extinction of an Island Forest
Avifauna by an Introduced Snake" (1987), "Extinct Pygmy Hippopotamus and Early Man in Cyprus" (1988), "An Environment-
Metapopulation Approach to Population Viability Analysis for a
Threatened Vertebrate" (1990), "A Dynamic Analysis of Northern
Spotted Owl Viability in a Fragmented Forest Landscape" (1992), "A
Population Viability Analysis for African Elephant (*Loxodonta
africana*): How Big Should Reserves Be?" (1993), and "Conservation
of Fragmented Populations" (1994). Whoa, if you're skimming, stop
here.

That's just a desktop sample. Another forty or fifty pounds of it
lurks in my file cabinets. These are all journal articles, as I said, of
which my collection is far from exhaustive; and I haven't even mentioned the books. How does a person deal with such abundance? I
don't own a pickup truck anymore and what I'm doing with this stuff,
God help me, is reading it.

Truth is, in small doses it's fascinating.

You may have noticed that most of those titles date from the 1970s
and later. There's been an explosion of island studies, island talk,
island-framed theoretical thinking, in the years since the publication
of MacArthur and Wilson's little book, *The Theory of Island Biogeography*, in 1967. A smattering of work was done earlier. Ernst Mayr gave
special attention to island birds in the 1930s and 1940s. David Lack
studied the Galápagos finches. Several scientists in the 1920s focused
on relations between the number of species found on an island and its
total area. There were a few interesting squeaks as far back as 1858.
And the ancestor of the whole lineage appeared in *The Annals and
Magazine of Natural History* for December 1857. That one was "On
the Natural History of the Aru Islands," by Alfred Wallace.

Wallace's article derived from the trip he had made eastward, by
native prau, to escape the monsoon season at Macassar. If this wasn't the
first great contribution to the modern science of island biogeography,

then its predecessors have escaped my attention. Darwin had earlier described the Galápagos, but he hadn't seen the significance of insularity until after he had sailed back to England, and the disorderliness of his Galápagos observations reflected his groping, pre-evolutionist confusion while he was there. Wallace's Aru paper, combining fastidious observation with sound analysis, was something different.

But "On the Natural History of the Aru Islands" doesn't show up nowadays in the scientific bibliographies. Working biologists are unaware of its existence, and all but a few historians have ignored it. It hasn't been recently (if ever) reprinted in unabridged form, so a determinedly curious reader is forced to track down an original volume of the journal. This is typical of the way Wallace's work has been overlooked, dismissed, and forgotten. The Aru paper lies interred, like a Gnostic gospel, in a cave of oblivion.

Wallace had found his own version, in Aru, of what Darwin saw belatedly in the Galápagos: synecdoche. What I mean by that is, here was a set of small facts that hinted toward big truths.

2 1

ON January 8, 1857, Wallace's prau landed at Dobbo, a seasonal trading settlement that functioned as the commercial capital of Aru. Dobbo stood on a low arm of beige coral sand jutting north off the end of a small island, "just wide enough to contain three rows of houses," as Wallace saw it. "Though at first sight a most strange and desolate-looking place to build a village on, it has many advantages," he recalled later. It was accessible along a clear channel through the bordering coral reefs, and it offered calm harborage in the lee of the arm—the lee being alternately one side or the other, depending on which way the monsoon winds were blowing. Those sea breezes, raking steadily over the arm, also discouraged mosquitoes and kept the town relatively free of malaria. The buildings were thatch sheds of a single basic pattern, serving as both lodgings and warehouses, with a partition to divide the living space from the merchandise.

Wallace's boat had arrived early, before most of the traders had appeared. The town was virtually empty, with a number of houses standing vacant, so he laid claim to one. He spread out his gear: a few boxes and mats, a table, a cane chair, his guns and nets, and a drying rack, the legs of which he planted in containers of water as a barrier against ants. He had a small entourage of young assistants brought along from Macassar—one boy to do his cooking, two others to help with collecting and preserving specimens. A window chopped through the palm-leaf wall gave light to his table. Although the house was gloomy and miserable, he felt "as contented as if I had obtained a well-furnished mansion, and looked forward to a month's residence in it with unmixed satisfaction." Next morning he was out collecting in the forest.

His initial take of butterflies from the Dobbo environs seemed promising. On the first day he netted thirty species, a better one-day total than he'd gotten anywhere since leaving the Amazon. Many of those were handsome and rare, known to British lepidopterists by only a few specimens that had come out of New Guinea. He caught a spectre butterfly that was identifiable as *Hestia durvillei*, a pale-winged peacock butterfly, *Drusilla catops*, and others. Several days later he spotted another huge, strong-flying, iridescent birdwing of the genus *Ornithoptera*, similar but not identical to the three he had caught in Celebes. This individual looked at first to be *Ornithoptera poseidon*, a New Guinea species, though closer inspection would suggest that it might represent a distinct Aru form. As he saw it flying majestically toward him, Wallace couldn't have foretold its role in the development of his ideas. But he shuddered with excitement and then "could hardly believe I had really succeeded in my stroke till I had taken it out of the net and was gazing, lost in admiration, at the velvet black and brilliant green of its wings, seven inches across, its golden body, and crimson breast." He had seen specimens of *Ornithoptera poseidon* in collections back home, but "it is quite another thing to capture such one's self—to feel it struggling between one's fingers, and to gaze upon its fresh and living beauty, a bright gem shining out amid the silent gloom of a dark and tangled forest. The village of Dobbo held that evening at least one contented man."

His contentment faded quickly as he realized that the Dobbo vicinity offered poor pickings for birds. Wallace had his wish list of Aru avifauna, famous species that would bring good prices in London or be useful to his own studies, but the most desirable of them turned out to be locally unavailable. Dobbo was sited on Wamma, a satellite island just off the northwest coast of the main Aru formation, and though the sea access and harborage were good, Wamma's extreme smallness was reflected in low biological diversity. The island con-

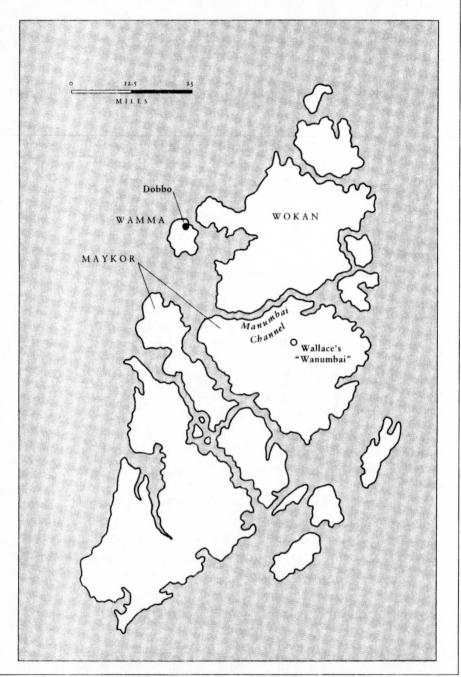

A R U   I S L A N D S

MILES
0   12.5   25

Dobbo

WAMMA

WOKAN

MAYKOR

Manumbai Channel

○ Wallace's "Wanumbai"

tained only a meager subset of the species of greater Aru. The black cockatoo, for instance, wasn't found on Wamma. Neither was the brush turkey or the cassowary. Worse, the two bird of paradise species weren't either. "It soon became apparent that in this small island there was a very limited number of birds, and I determined to go as soon as possible to the large island," he wrote.

What he called "the large island" is actually a half-dozen distinct islands that fit together like pieces of a puzzle. The islands are separated only by narrow channels, which cut through the Aru landmass, mostly along east-west lines, with an uncanny resemblance to rivers. Despite that resemblance they hold seawater, not fresh water, and the direction of flow changes with the tides. The effect is peculiar: surging channels of brine that wind between forested banks. Wallace craved to get across from Wamma and explore one of those channels, which would carry him into the richer interior of the cluster.

But pirates had been marauding near Dobbo, and none of the Wamma natives was willing to risk a crossing. This fear was grounded in experience, not paranoid fantasy, since pirates were known thereabouts to stop ships, plunder cargo, murder crew members; sometimes they even came ashore to attack settlements, burning, killing, dragging away women and children. "Not a man will stir from his village for some time," Wallace wrote, "and I must remain still a prisoner at Dobbo." He bought himself a boat. Finally, after two months of frustration in and around the town and "an enormous amount of talk and trouble," he cajoled a couple of local men to guide him.

They crossed from Wamma and ascended a stream (not one of the channels) that twisted through mangrove swamp into the center of Wokan, the northernmost island of the larger Aru group, until the swamp gave way to dry land and the stream brought them to a single native house. The house was a woebegone little shed, crowded with a dozen people and two cooking fires. In exchange for a chopping knife, Wallace was granted permission to move in with the others and stay for a week or so. He took a space at one end, rolling out his bed and hanging a specimen shelf. The accommodations were rough, the blood-sucking sand flies were pestiferous, and his first few days at the site were too rainy for fruitful collecting. He began to lose heart. Then one of his boys brought in a specimen "which repaid me for months of delay and expectation."

It was a small bird, no bigger than a thrush but extravagantly decorative. "The greater part of its plumage was of an intense cinnabar red, with a gloss as of spun glass. On the head the feathers became short and velvety, and shaded into rich orange. Beneath, from the breast downwards, was pure white, with the softness and gloss of silk, and across the breast a band of deep metallic green separated this color from the red of the throat. Above each eye was a round spot of the same metallic green; the bill was yellow, and the feet and legs were of a fine cobalt blue, strikingly contrasting with all the other parts of the body." The bird carried odd little gray-and-green fans beneath its wings and a pair of long, slender quills sticking out beyond the other feathers of the tail. About four inches down, the quills spiraled tightly inward to form a pair of tufted iridescent green buttons. "These two ornaments, the breast fans and the spiral tipped tail wires, are altogether unique," Wallace noted, "not occurring on any other species of the eight thousand different birds that are known to exist upon the earth; and, combined with the most exquisite beauty of plumage, render this one of the most perfectly lovely of the many lovely productions of nature." He recognized it as the king bird of paradise, now known to science as *Cicinnurus regius*. Here in Aru the species was called goby-goby, he learned. To the local folk, it was no more remarkable than a goldfinch would be in Kentucky.

After two weeks, notwithstanding the goby-goby and a large black cockatoo that he also counted "a great prize," Wallace decided to move. Either he doubted that this region of Wokan would yield any other spectacular specimens or else the crowded house and the sand flies were getting him down. The flies were worst at night, "penetrating to every part of the body, and producing a more lasting irritation than mosquitoes," he complained. "My feet and ankles especially suffered, and were completely covered with little red swollen specks, which tormented me horribly." By now, too, one of his Dobbo crewmen was sick with fever and wanted to go home. In sum, it seemed time to vacate. He hired a replacement for the sick man, retraced his path through the mangrove swamp, sailed southward along the west coast of the Aru cluster, and turned inland again at the mouth of the first major channel, which cut between Wokan and the large island south of it, Maykor. There at its mouth the channel was wide like an estuary, but a short distance inland it narrowed to just the width of the Thames at London. It meandered between parallel banks like any river, with only the salt taste and the tidal flux to hint that it was actually a strait. The breezes were helpful, pushing Wallace's boat along with no more than occasional work on the oars. His party spent that day and another exploring eastward. After scraping their way over some reefy shallows, they exited the channel's eastern mouth and were back at sea. They had crossed straight through the middle of Aru and out the far side.

But the east coast of the cluster was not what Wallace wanted; he wanted the interior. So they turned around again and, now finding themselves in danger of being driven offshore by a squall, rowed like hell. They barely regained the channel. Over the next two days they fought their way back westward to its middle reaches and then entered a small stream flowing in from the south, which allowed them to row up into the heart of the island of Maykor. At the point where the stream became impassable, Wallace came upon "the little village of Wanumbai, consisting of two large houses surrounded by plantations, amid the virgin forests of Aru." He liked the look of the place.

He found a decent thatch house that could be rented for a reasonable price in trade goods (ten yards of cloth, an ax, a few beads, and some tobacco), and he settled there for a stay that eventually lasted six weeks. On his first hike to reconnoiter the neighborhood, he caught sight of another big butterfly. This one was black with large panels of iridescent blue, which may have reminded him of the morphos he had seen in the Amazon. It was a swallowtail, recognizable to Wallace as *Papilio ulysses*. He also heard enough varied song to suggest that the bird collecting should be good. The local men and boys, crackerjack archers who routinely killed wild pigs and kangaroos and a wide assortment of birds, caught on fast to the fact that this eccentric, pale-skinned stranger would pay them handsomely for certain dead animals, and not just edible ones.

On his second evening in the village, Wallace was brought an interesting ground-thrush of a species previously known only from New Guinea. Soon after that he got a racquet-tailed kingfisher, bright blue and white with a coral red bill, assignable to the genus *Tanysiptera* on the basis of its two elongated tail feathers, which were lobed at the ends like little squash rackets. This kingfisher turned out to be a new species, distinct from the *Tanysiptera* that occurred farther west in the Moluccas, and also distinct from those farther east in New Guinea. From the start, then, Wallace had promising signs that the village of Wanumbai and its surrounding forest might satisfy his high expectations for Aru. The local hunters continued to bring him all sorts of nifty creatures. Wallace himself was out on the trails each day—while his health remained good, anyway—and so were his boys, who by then had gotten enough on-the-job training to make them useful as wing shots and bird skinners. At the end of a week, the more assiduous of the two boys returned exultantly with a specimen of *Paradisaea apoda*, the greater bird of paradise.

Although the bird hunting was good at Wanumbai, the living and working conditions were still difficult. The house was highly permeable to vermin. "Instead of rats and mice there are curious little marsupial animals about the same size, which run about at night and nibble anything eatable that may be left uncovered. Four or five different kinds of ants attack everything not isolated by water, and one kind even swims across that; great spiders lurk in baskets and boxes, or hide in the folds of my mosquito curtain; centipedes and millepedes are found everywhere. I have caught them under my pillow and on my head; while in every box, and under every board which has lain for some days undisturbed, little scorpions are sure to be found snugly ensconced, with their formidable tails quickly turned up ready for attack or defence. Such companions seem very alarming and dangerous, but all combined are not so bad as the irritation of mosquitoes," he wrote. The mosquitoes were especially thick out in the sugar cane fields and the vegetable gardens surrounding the village, through which he made his daily walks. And these were diurnal mosquitoes not crepuscular ones, unfortunately, so their schedule of activity coincided with his. They seemed to take particular delight in attacking his feet, which were still speckled with wounds from the sand flies of Wokan. "After a month's incessant punishment, those useful members rebelled against such treatment and broke into open insurrection, throwing out numerous inflamed ulcers, which were very painful, and stopped me from walking. So I found myself confined to the house, and with no immediate prospect of leaving it. Wounds or sores in the feet are especially difficult to heal in hot climates, and I therefore dreaded them more than any other illness."

The weather had turned clear and dry, perfect for insect collection, if he could only get out and about. He couldn't. It was maddening. "When I crawled down to the river-side to bathe, I often saw the blue-winged Papilio ulysses, or some other equally rare and beautiful insect; but there was nothing for it but patience, and to return quietly to my bird-skinning, or whatever other work I had indoors." The sheer physical discomfort was the least of the aggravation. To be "kept prisoner" by his running sores in such a tantalizing piece of country, where wondrously unfamiliar creatures "are to be met with in every forest ramble—a country reached by such a long and tedious voyage, and which might not in the present century be again visited for the same purpose—is a punishment too severe for a naturalist to pass over in silence." Some consolation came from the work of his two boys, who were still bringing in specimens of the greater bird of paradise, the king bird of paradise, and other impressive species.

By the end of six weeks at Wanumbai, he had spent more than half his time stuck in the house, nursing his festering feet. His supplies were nearly gone. His specimen boxes were full, despite the lost days of mobility. He saw no immediate hope of beating the infections, as long as he was living in these field-camp conditions. Furthermore the bird hunting had lately turned slow, probably owing to seasonal behavior as much as to the fact that Wallace and his helpers had already killed quite a few. It was time to retreat to Dobbo, he decided. He shared out his last stock of salt and tobacco among the villagers, gave a flask of palm brandy to his landlord, said goodbye, climbed into his boat, descended the stream, and caught a fair wind westward on the channel.

It wasn't entirely easy, sailing away from such a congenial village within such a rich, unusual forest. In hindsight Wallace would see that his months in the Aru Islands were the most profitable period of his whole eight-year sojourn in the Malay Archipelago, and that the weeks at Wanumbai had given him the best of Aru. "I fully intended to come back," he wrote later, "and had I known that circumstances would have prevented my doing so, should have felt some sorrow in leaving a place where I had first seen so many rare and beautiful living things, and had so fully enjoyed the pleasure which fills the heart of the naturalist when he is so fortunate as to discover a district hitherto unexplored, and where every day brings forth new and unexpected treasures." It turned out to have been doubly productive—not just in terms of the pinned butterflies and skinned birds but also in terms of his developing ideas. He never returned.

Back in Dobbo, Wallace spent another six weeks indoors before his feet healed enough for hikes in the forest. By then it was late June, the dry weather had broken, the forest had gone sumpy, and the little island on which the town stood, Wamma, seemed even poorer in species than before. The monsoon had shifted again, bringing wind and storms out of the east. "There was a damp stagnation about the paths," he wrote, "and insects were very scarce." He scrounged a few odd species of beetle and made the best of them.

The commercial season at Dobbo was meanwhile reaching its crescendo, which wouldn't last long. Loads of bananas and sugar cane were arriving in boats from the other islands, to be traded for tobacco or some other imported luxury. The piles of dried sea cucumbers were being stowed into sacks. Bundles of pearl shell, tied with rattan, were being loaded onto the big praus. Every trader who hoped to get home to Macassar was busy with his final arrangements, aware that if he missed the eastern monsoon he'd be stuck here for another year. That would be lonely. Dobbo, the temporary town, would soon disappear like a circus. Only the empty thatch sheds would remain. "The Chinamen killed their fat pig and made their parting feast," Wallace wrote, "and kindly sent me some pork, and a basin of birds'-nest stew, which had very little more taste than a dish of vermicelli." Careful observer, he was still recording impressions while everyone else bustled. And he had his own share of precious commodities to pack: nine thousand individual specimens, representing sixteen hundred species, a sizable number of which were new to science.

He was pleased with his gamble on Aru. "I had made the acquaintance of a strange and little-known race of men; I had become familiar with the traders of the far East; I had revelled in the delights of exploring a new fauna and flora, one of the most remarkable and most beautiful and least-known in the world; and I had succeeded in the main object for which I had undertaken the journey—namely, to obtain fine specimens of the magnificent Birds of Paradise, and to be enabled to observe them in their native forests. By this success I was stimulated to continue my researches in the Moluccas and New Guinea for nearly five years longer." On July 2 his prau left Dobbo, sailing west.

Within ten days he was back in Macassar, having crossed a thousand miles of ocean. Sometime soon afterward, while the material was still fresh in his mind, he wrote "On the Natural History of the Aru Islands." He sent off the manuscript aboard a mail steamer, presumably to Samuel Stevens, and it reached London quickly enough to appear in *The Annals and Magazine of Natural History* for December. No one in England seems to have paid it much notice. The Aru Islands may have been famously enticing among Buginese sea traders, but for most British naturalists they were distant and alien beyond imagining.

So was young Wallace's train of thought.

22

YEARS LATER, he would tell the story of his Aru expedition as a travel narrative, full of adventure and exotic tableaux, lonely frustrations, moments of elation, all recalled through a filter of nostalgia. That account would appear in the book, *The Malay Archipelago*, that became his most famous literary achievement.

The earlier short paper, "On the Natural History of the Aru Islands," was less personal and more conceptual. It was loosely organized into three sections. First, in a few paragraphs he dispensed with the narrative outline of his trip. Then he presented his observations, covering the unusual geology of the Aru cluster as well as the roster of bird and mammal species that he had collected or at least seen. The species roster was important as an empirical basis from which, in a third section, he teased out some patterns.

What creatures did Aru harbor? Besides the two birds of paradise, *C. regius* and *P. apoda*, Wallace had counted more than a hundred other species of bird. His tally included twelve pigeons, four falcons, nine honeyeaters, thirteen flycatchers, ten parrots or parrotlike species, a wood-swallow, three species of shrike, two pittas, five sunbirds, and eleven species from the kingfisher family. Many of those Aru birds were known also in New Guinea. The overlap was remarkable for two reasons: because New Guinea seemed rather far away (one hundred miles of ocean is a formidable gap for a parrot), and because the same species did *not* occur farther west in the archipelago. He had found the great black cockatoo, with its rose-colored cheek patches, which was native to New Guinea and to Australia but which had never spread westward as far as Celebes. He had found the sulphur-crested cockatoo, another New Guinea species. He had found a giant flightless bird known as *Casuarius casuarius*—more familiarly, the Australian cassowary. And it wasn't just the birds that suggested a link eastward. He shot, saw, or heard reports of a half-dozen species of marsupial, including several cuscuses and a kangaroo.

There were interesting absences too. Wallace mentioned six families of birds (the woodpeckers, the trogons, the hornbills, the broadbills, the puffbirds, the bee-eaters) that were abundant throughout Java, Borneo, Celebes, the whole western end of the archipelago, but missing from Aru. Among mammals, a similar pattern. With the exception of pigs (which have often been transplanted to new islands by humans) and bats (which manage sea crossings well on their own), not one major group of mammals was common to both ends of the archipelago. There were no carnivores, no rodents, no primates, and no ungulates on Aru, Wallace reported. Set aside bats and pigs, and all other Aru mammals were marsupial.

The most obvious conclusion to be drawn was that Aru belonged with New Guinea and Australia, not with the rest of the Malay Archipelago. Somewhere in the course of his boat journey eastward, Wallace had passed out of one zoogeographic realm and into another—into a realm where possums replaced monkeys, cockatoos replaced trogons, cassowaries replaced babirusas, and birds of paradise replaced who knows what. The transformation of biological communities that began in Lombok was consummated here, as reflected in the remarkable similarity of the faunas of Aru and New Guinea. "Such an identity occurs, I believe, in no other countries separated by so wide an interval of sea," Wallace wrote. Ceylon is closer to India than Aru to New Guinea, he noted, but the zoological distinctness between Ceylon and India is greater. Tasmania is no farther from mainland Australia, and again Tasmania's animal life is more distinct. Likewise with Sardinia and Italy. Bali and Lombok are much closer together, but much more distinct zoologically, than Aru and New Guinea. The only thing that could explain it, Wallace felt, was dry land. Not many millennia earlier, the Aru cluster must have been attached to New Guinea by a land bridge.

Geologically, it was plausible. West of Aru the ocean floor falls away steeply; to the east, toward New Guinea, the ocean is shallow; Aru sits on the edge of the New Guinea shelf. In addition to the ocean-floor topography, there's a more striking bit of evidence within Aru itself: those channels, transecting the landmass in three riverlike slashes. Wallace couldn't escape his feeling that "they seem portions of true rivers," though nowadays running with salt water pushed by the tides. No other set of islands on Earth, so far as he knew, was similarly transected. The only way to account for it, he argued, was "by supposing them to have been once true rivers, having their source in the mountains of New Guinea, and reduced to their present condition by the subsidence of the intervening land." He took them to be the truncated lower ends of old New Guinea drainages, left behind when geological slumping caused the upriver reaches to subside below sea level and leave Aru insularized on the shelf. If he was correct, then there had been a time when river water flowed downhill from

New Guinea to Aru while forest birds and marsupials passed back and forth.

Having gathered and sorted the facts about which species lived where, he asked a crucial question: Why? It's important to recall that the theory of special creation was still widely embraced at this time, and that any other view seemed subversively heterodox. But if God simply designed particular animals for particular types of landscape, as orthodoxy decreed, why should the Aruese fauna differ so drastically from the faunas of other Malay islands? If the landscapes and the climates were similar, why not the animals? Why *didn't* the cockatoos and the possums occur in Bali? Why *didn't* the monkeys and the woodpeckers inhabit Aru or New Guinea? Was there a scientific explanation, or was it simply God's whim? At this point Wallace's Aru paper turned to ideas and patterns much larger than Aru.

23

His real target was the special-creation theory.

Species go extinct, Wallace wrote. They live out their respective stretches of time, usually measured in millions of years, and then they vanish. That much is known from the fossil record. Over the eons it has been a steady process, steadily counterbalanced by the appearance of new species. Even pious creationists—at least the more scientific-minded pious creationists, such as Charles Lyell—acknowledged this balance between old species dying and new ones being born. But what's the origin of the new species? What's the mechanism by which they arise? Special creation, said the orthodox view. "The general explanation given is, that as the ancient species became extinct," Wallace wrote, "new ones were created in each country or district, adapted to the physical conditions of that district." Created, that is, by God's direct intervention. God put the cockatoos on those islands where we find them, which are precisely and solely the places where they belong. God in His wisdom decided that Aru was wrong for woodpeckers. God shaped each species for its appropriate context, fabricating and distributing the world's flora and fauna by divine ecological fiat. Or so says the special-creation theory.

A corollary to that theory, Wallace pointed out, is that similar species should be native to similar landscapes (presuming that God was consistent about what was appropriate to each type of landscape), and that dissimilar landscapes should contain dissimilar species. One problem with the theory is that its corollary just doesn't hold true.

For a test case, he cited New Guinea and Borneo. Those two great islands are remarkably similar in size, location, physical features, and climate. "In neither is there any marked dry season, rain falling more or less all the year round; both are near the equator, both subject to the east and west monsoons, both everywhere covered with lofty forest; both have a great extent of flat, swampy coast and a mountainous interior," and both New Guinea and Borneo, he added, are rich in tree species of the palm and pandanus families. Despite all the similarities, their faunas are utterly different.

To put the contrast between New Guinea and Borneo into relief, Wallace also compared New Guinea with Australia. Here the categories of similarity and difference are reversed. The topographical and climatic conditions differ drastically (New Guinea being steeply mountainous, mostly covered with rainforest; Australia being flat, mostly covered with desert and savanna), but the resident faunas are similar. Australia and New Guinea harbor the same groups of bird and mammal. They also lack the same groups. "If kangaroos are especially adapted to the dry plains and open woods of Australia, there must be some other reason for their introduction into the dense damp forests of New Guinea," Wallace argued. And if the dry eucalyptus woodlands of Australia are somehow inhospitable to woodpeckers, that fact fails to explain why woodpeckers should also be missing from New Guinea.

He was putting biogeography into service as a conceptual tool of great reach and leverage. Wallace knew too much about where species occurred, and where they didn't, to swallow special creation. Finding the observed data in conflict with the received theory, he was forced to discard the theory. "We can hardly help concluding, therefore, that some other law has regulated the distribution of existing species," he wrote. And he had in mind not just their distribution but also their origin.

His own candidate for that other law was the one he had proposed in his earlier paper, from Sarawak. "In a former Number of this periodical," he reminded readers of the *Annals*, "we endeavoured to show that the simple law, of every new creation being closely allied to some species already existing in the same country, would explain all these anomalies." Wallace was moving toward a great discovery, but his claim of having arrived was premature. His Sarawak paper had *not* explained those anomalies; it had only set them within an orderly generalization.

In the Aru paper he went a step further. He pursued his consideration of New Guinea and Australia, the two huge and climatically different islands that were formerly connected. New Guinea supports a dwarf species of cassowary, distinct from (though closely allied to) the Australian cassowary. New Guinea also supports tree-climbing kangaroos, distinct from (though closely allied to) Australian kangaroos. How do such circumstances arise? During the period since Australia and New Guinea became separated, Wallace wrote,

> new species have been gradually introduced into each, but in each closely allied to the pre-existing species, many of which were at first common to the two countries. This process would evidently produce the present condition of the two faunas, in which there are many allied species,— few identical.

Furthermore:

> The species which at the time of separation were found only in one country, would, by the gradual introduction of species allied to them, give rise to groups peculiar to that country.

Although the language was vague, these statements were both accurate and important. He was coming nearer to an explanation for the patterns of closely allied species coincident both in space and in time, as described in his Sarawak paper.

Two intellectual steps remained to be taken. The phrase "gradual introduction" had to be more clearly defined before it would transmogrify into "evolution." And Wallace had to answer another big question: By what mechanism does evolution occur? The second of those two was the mystery of mysteries, still unsolved.

24

TOWARD THE END of that year, 1857, Charles Darwin began to take Wallace seriously, though not yet so seriously as he should have.

Wallace had written Darwin a second letter, mailing it from Macassar when he returned there after his Aru expedition. Like Wallace's first letter, it has disappeared from the Darwin files, and the only evidence of its contents is reflected in Darwin's reply. Wallace's earlier letter had sat unanswered for months, but to this second one Darwin responded promptly.

Writing on December 22, Darwin declared himself "extremely glad to hear that you are attending to distribution in accordance with theoretical ideas. I am a firm believer, that without speculation there is no good & original observation. Few travellers have attended to such points as you are now at work on; & indeed the whole subject of distribution of animals is dreadfully behind that of Plants."

Wallace's letter, besides mentioning his work on distribution, had evidently contained a moue of disappointment that his Sarawak paper had been resoundingly ignored, because Darwin offered consolation: "You say that you have been somewhat surprised at no notice having been taken of your paper in the Annals: I cannot say that I am; for so very few naturalists care for anything beyond the mere description of species. But you must not suppose that your paper has not been attended to: two very good men, Sir C. Lyell & Mr. E. Blyth at Calcutta specially called my attention to it." This last little bit would have come as big news. Sitting lonely and frustrated beneath the tropical sun, Wallace must have felt a cool gust of encouragement as he read Darwin's comment that the celebrated Sir Charles Lyell had noticed his work.

It was a small milestone in the development of Alfred Wallace's scientific confidence. And it raised a possibility of which Wallace would later take advantage—that Mr. Darwin might be willing to serve as an intermediary between Wallace and Lyell.

"I have not yet seen your paper on distribution of animals in the Arru Islands," Darwin added. The Aru paper had only just been published. Conceivably he hadn't even heard about it, except from Wallace's own letter. "I shall read it with the *utmost* interest," Darwin said politely.

25

Darwin had been warned. Besides the hints in the two *Annals* papers, suggesting that Wallace was closing in, a more explicit warning had come from Lyell himself.

Back in April of 1856, the previous year, Lyell had spent a few days at Darwin's house in the village of Down, southeast of London. One morning during that visit Darwin had described to Lyell the whole cockamamie theory he had been developing secretly over the past eighteen years: that species evolve, one from another, through a process whereby natural variations within a population become selected and amplified by the differential survival and differential reproductive success of various individuals. Lyell was probably only the third or fourth person with whom Darwin had shared any part of this heretical notion. Although Lyell had been startled and skeptical, his scientific composure and his great intellectual fairness allowed him to record in his scientific journal an unjudging outline of what Darwin had said.

Darwin's term for the process, Lyell noted, was *natural selection*. It was applicable even to the islands that Lyell knew well. "The number of species which migrated to the Madeiras," he wrote, recording Darwin's view rather than his own, "was not great in proport. to those now there, for a few types may have been the origination of many allied species." Unlike "natural selection," the phrase "allied species" seems to have been one that Lyell chose himself—or maybe he picked it up from someone other than Darwin. This journal entry suggests that Lyell's immediate instinct, upon hearing Darwin's theory, was to connect it with the "law" paper by that young fellow named Wallace.

After three paragraphs summarizing Darwin, Lyell made the connection explicit, stating that "Mr. Wallace['s] introduction of species, most allied to those immediately preceding in Time," as revealed in the fossil record, now "seems explained by the Natural Selection Theory."

Lyell himself was dubious of the theory, but he gave his friend Darwin a piece of advice: Publish. Publish your big idea as quickly as possible, before someone else publishes it as *his* big idea. If your full treatment of natural selection isn't finished, Lyell urged, then put out at least a simplified sketch of your views, so as to establish priority.

Darwin wasn't ready. In addition to his notebooks from the *Beagle* voyage, he had two decades' worth of data on such intricately interrelated topics as barnacle taxonomy, hybridism among plants, the survival of seeds in seawater, rudimentary organs, the absence of mammals and frogs from oceanic islands, and the variation that turns up within breeds of domestic pigeon. Ever fastidious, he didn't like being hurried. He granted that Lyell had a point, but he wanted more time to shape his data into an argument. He planned to produce a huge, thorough, irresistibly persuasive manuscript, and to let that opus be his first public statement on the subject. A shorter and more slapdash effort rushed into print earlier, he worried, would only provoke resistance to the theory. Poor Darwin was a cautious man burdened with a bold idea, which made him sometimes seem paralyzed by ambivalence. He was like a hen struggling to lay a grossly oversized egg. If he rushed, it might kill him. It might kill him anyway.

Several weeks later he told Lyell by letter: "With respect to your suggestion of a sketch of my views; I hardly know what to think, but will reflect on it," adding immediately, qualification upon qualification, "but it goes against my prejudices." A good short summary was impossible, he wrote, since the theory encompassed multiple propositions each of which required abundant factual support. Maybe he could describe just the main mechanism, natural selection. Or maybe he could mention a few other features and anticipate some of the criticisms. "But I do not know what to think," he admitted. "I rather hate the idea of writing for priority, yet I certainly should be vexed if any one were to publish my doctrines before me."

He consulted another of his eminent friends, Joseph Hooker, mentioning by letter that he'd had a "good talk with Lyell about my species work, & he urges me strongly to publish something." Hooker, his closest confidant, was one of the few people to whom Darwin had already described the theory. "If I publish anything it must be a *very thin* & little volume, giving a sketch of my views & difficulties; but it is really dreadfully unphilosophical to give a resumé, without exact references, of an unpublished work." By "unphilosophical" he meant professionally shoddy. "But Lyell seemed to think I might do this, at the suggestion of friends . . ."—and so Hooker was coyly invited to second the suggestion.

Hooker declined. He disagreed with Lyell's advice. Save all your thunder and lightning for the one big bolt, is what Hooker seems to have told Darwin.

Darwin waffled. He fretted about it to others. In a letter to his cousin, he reported that Lyell had prodded him to write a sketch. "This I have begun to do, but my work will be horridly imperfect & with many mistakes, so that I groan & tremble when I think of it." By the autumn of 1856 he had changed his mind again, concluding that a sketch wouldn't do. Instead he had shoved aside Lyell's warning and begun writing the long version. He had chosen to gamble that this opus would be both the definitive treatment of natural selection and also, if his luck held, the first.

His luck didn't hold. Lyell had given him good advice.

26

FOR Wallace, the Aru expedition had been successful but physically punishing. Back in Macassar, he took a short respite from the difficult, lonely conditions of fieldwork. Presumably he was eager for word from home. Because the Dutch mail steamers called at Macassar, this would have been the place and the time for him to receive and answer half a year's worth of letters. We know he wrote to Darwin. And he faced other chores: replenishing his supplies, getting his guns repaired, crating and shipping his specimens, writing up his latest thoughts, mailing the manuscripts. In addition to the Aru paper, he sent off several others that would soon appear in English journals. He was an ambitious, productive young man, rich in field data and full of ideas.

One of those other papers dealt with the puzzling form of birdwing butterfly that he had collected in Aru—the same insect that had left him "gazing, lost in admiration," at its velvety black and iridescent green wings, its golden body, its crimson breast. Comparing that specimen with published descriptions, he couldn't make a positive identification. It wasn't quite *Ornithoptera poseidon*, the New Guinea species, and it wasn't quite not. In shape and size it seemed identical, but the color pattern was slightly different.

Possibly it was *Ornithoptera priamus*, a closely related species known from the island of Amboyna. But according to Wallace's lepidoptera book, his specimen didn't match *O. priamus* either. *O. priamus* showed four black spots on the green of each hind wing; *O. poseidon* showed two black spots. The birdwing from Aru showed three.

This ambiguity raised some questions. Was the Aru form a distinct species? All the other animals that Wallace had found on Aru were indistinguishable from New Guinea species. Why should one butterfly be different? But maybe it was. On the other hand, maybe it was just an eccentric variety of either *O. poseidon* or *O. priamus*—but if so, an eccentric variety of *which*? It seemed to stand midway between. Could it be in some sense a variety of both? That would require a new understanding of the categorical relationship between varieties and species. Under the current understanding, a variety could no more stand midway between two species than a child could stand midway between two mothers.

The three-spotted birdwing was drawing Wallace into unexplored theoretical terrain. Either it was a new species, unique to Aru—which tended further to make nonsense of the theory of special creation—or else it was a variety that lay midway between two species, thereby challenging the conventional concept of species itself.

What *was* a species? What were the defining limits of the category? Was a species immutable? Was it crisply demarcated? The three-spotted birdwing seemed to suggest it was neither.

Still another short paper that Wallace wrote during his Macassar layover was titled "Note on the Theory of Permanent and Geographical Varieties." This one was more abstract, containing no explicit mention of the birdwings, though addressing precisely the conundrum they raised—the distinction between those two categories, species and variety. Naturalists had been treating the two as fundamentally different. A species was fixed, ordained, permanent; a variety was a transient aberration from normality, often associated with the geographical isolation of some lesser population away from the main population of its species. Now Wallace had data suggesting that the difference was just one of degree.

The conundrum seemed irresolvable. But maybe he could at least clarify it. That was the stated purpose of his "Note on the Theory of Permanent and Geographical Varieties." Let's be a little more careful and logical, Wallace proposed, in how we think about species versus varieties.

Was the difference between the two categories quantitative? Were two species simply *more* different from each other than were two varieties within a given species? If so, then in some cases the threshold between a species gap and a varietal gap would be only a tiny increment, and the categories would blur into each other. If they blurred, Wallace wrote, it would be "most illogical" to believe that species but not varieties were created by divine intervention.

Or was the difference between the two categories qualitative? If so, then what special quality set species on a higher categorical plain? Permanence? No, that wouldn't do. Varieties, notwithstanding the common notion, weren't always transient aberrations from normality; they could be virtually permanent too, especially in cases (such as the sakis of the Amazon or the birdwings of the Malay Archipelago) where permanent geographical boundaries kept the populations separate. If not permanence, then what? There was no other reasonable answer. Variation within species was so common and so multifarious that it encompassed all the same types of differences that distinguished one species from another. Wallace himself was well qualified to say so. In his years as a commercial collector, gathering specimens redundantly, he had seen more intraspecific variation than most other naturalists would in five lifetimes.

If there's no qualitative factor distinguishing species from varieties, he wrote, then "that fact is one of the strongest arguments against the independent creation of species, for why should a special act of creation be required to call into existence an organism differing only in degree from another which has been produced by existing laws?"

The mail steamers were slow, but the lead time for journals, in those days, was short. Within six months Wallace's "Note on the Theory of Permanent and Geographical Varieties" had appeared in the *Zoologist*. It was just two pages long; Darwin, preoccupied with writing his huge opus, may not have noticed it. For Wallace it was another small step forward. But now he had come to a point where he needed to leap.

Late in the year he left Macassar again, this time on a mail steamer himself. He was headed back eastward, into a remote region of the archipelago that was known to Buginese and Dutch traders but not to naturalists of the English-speaking world: the northern Moluccas. After brief stops on the islands of Timor, Banda, and Amboyna, he fetched up at a place called Ternate. It was January 1858.

27

EXACTLY WHAT happened next is a small nutlike riddle that science historians have never cracked. At its core is the issue of who deserves credit for one of the greatest scientific achievements in history—the first statement of the theory that Darwin called natural selection, specifying not just that species evolve but that they evolve through a very particular process, with natural variations selected and amplified by differential survival and differential reproductive success. Did Charles Darwin cheat and lie in order to protect his proprietary claim on that theory? Did he manipulate Lyell and Hooker into helping him? Did he falsify certain documents and destroy others, so as to prevent a young upstart from grabbing what Darwin considered his own rightful glory? Did Alfred Russel Wallace, the upstart in question, allow himself to be shabbily treated by eminent scientists whose integrity he trusted too much and whose patronage he wanted too badly? Some scholars, though only a few, think so. "This is the hardest story in science, and one day it is going to blow up," claims a dogged researcher named John Langdon Brooks. Brooks's book, *Just Before the Origin*, published in 1984, is one of several that take a damning view of Darwin's behavior.

Or did the independent formulation of the theory of natural selection by two scientists working separately—one in a house outside London, one in the Malay Archipelago—represent a whopping coincidence with a cheery, edifying denouement? That's the conventional version. In that version, Darwin and Wallace were both scrupulous and goodhearted men who, having converged on the great idea, reached an amiable agreement to publish it jointly. Far from clutching jealously at the recognition to which his two decades of effort entitled him, in that version, Darwin made punctilious efforts to share it with a latecomer. So the episode was an instance of "mutual nobility," in Loren Eiseley's opinion, and "a monument to the natural generosity of both the great biologists," according to Julian Huxley. "Never was there a more beautiful example of modesty, of unselfish admiration for another's work, of loyal determination that the other should receive the full merit of his independent labours and thoughts, than was shown by Charles Darwin on that occasion," in the orotund estimation of Sir E. Ray Lankester as he addressed a commemorative celebration fifty years later. That conventional version, as polished by Lankester and the others with the chamois cloths of selective recollection, is deceptive.

It omits and distorts significant facts. It's a scientific fairy tale, almost as fanciful as the one about Isaac Newton being conked by an apple. The truth is less noble and more complicated.

Part of the complication derives from uncertainty about what really happened. Because of gaps in the historical record, we cannot know whether Darwin acted disgracefully or just unimpressively. But whatever Darwin did, notwithstanding Julian Huxley, it certainly was not a "monument" to his "natural generosity." The real issue is more negative: Was Darwin guilty of scummy behavior, or wasn't he? Hard to say. Crucial bits of physical evidence have been destroyed or lost. Testimony from the principals is partial, discreet, inconsistent. Only Darwin knew for sure, and he never told.

Wallace himself left behind a half-dozen published accounts of the episode. But even Wallace's perspective had been limited by circumstance, since he was ten thousand miles away and nearly incommunicado at the time of the controversial events. Each of Wallace's tellings differed slightly from the others. My favorite is one that appeared thirty years after the fact, as an explanatory note in a collection of his scientific essays. Since his co-discovery with Darwin had attracted a bit of historical interest, he wrote, "a few words of personal statement may be permissible."

During the period after completing his Sarawak paper, Wallace recalled, "the question of *how* changes of species could have been brought about was rarely out of my mind." But the question had stumped him, and

> no satisfactory conclusion was reached till February 1858. At that time I was suffering from a rather severe attack of intermittent fever at Ternate in the Moluccas, and one day while lying on my bed during the cold fit, wrapped in blankets, though the thermometer was at 88° F., the problem again presented itself to me, and something led me to think of the "positive checks" described by Malthus in his "Essay on Population," a work I had read several years before, and which had made a deep and permanent impression on my mind.

Malthus had catalyzed Darwin's thinking too, though Wallace didn't know that at the time. Like Darwin, he had simply recognized that the population-limiting "checks" mentioned in Malthus's essay had their parallels among nonhuman species.

> These checks—war, disease, famine and the like—must, it occurred to me, act on animals as well as on man. Then I

thought of the enormously rapid multiplication of animals, causing these checks to be much more effective in them than in the case of man; and while pondering vaguely on this fact there suddenly flashed upon me the *idea* of the survival of the fittest—that the individuals removed by these checks must be on the whole inferior to those that survived. In the two hours that elapsed before my ague fit was over I had thought out almost the whole of the theory, and the same evening I sketched the draft of my paper, and in the two succeeding evenings wrote it out in full, and sent it by the next post to Mr. Darwin.

The next post was a mail steamer that left Ternate on March 9. A letter sent via that boat should have taken about twelve weeks to reach England. But did it turn out, in the event, to be twelve weeks exactly? Was it fourteen? Was it just ten? The timing of that boat delivery has become a critical issue for those who suspect Darwin of a despicable deed.

Wallace had titled this paper "On the Tendency of Varieties to Depart Indefinitely from the Original Type." It was a lucid, forceful outline of the theory of natural selection—though it didn't contain that phrase—and therefore one of the most revolutionary scientific manuscripts that had ever been tucked into an envelope. He could have sent it directly to the *Zoologist* or to the *Annals*, and either of those journals might have published it quickly, as they did his other papers. Then again, they might well have rejected this one as too reckless, too speculative, too outrageous. Wallace had wondered about that himself. Would his idea sound crazy to the average biologist? Maybe. So instead of offering the paper directly to a journal, he had elected to foist it confidingly on Mr. Darwin. Drawing on their tenuous relationship as scientific pen pals, he asked a favor: If Mr. Darwin saw any merit in this fever-born act of iconoclasm, would he pass the manuscript along to Sir Charles Lyell, who had said those nice things about Wallace's earlier work?

The conventional version says that Darwin's mailman brought the bad news on June 18, 1858, and that Darwin promptly complied with Wallace's request. He mailed the manuscript to Lyell, explaining that Wallace "has to-day sent me the enclosed & asked me to forward it to you." The crucial tenet of this version is that Wallace's manuscript had reached Darwin only that very day, and was being passed on immediately. In a plaintive tone, Darwin added: "Please return me the M.S., which he does not say he wishes me to publish; but I shall of course, at once write & offer to send to any Journal. So all my originality, whatever it may amount to, will be smashed." And in case Lyell's memory needed refreshing about Darwin's prior claim to the theory: "Your words have come true with a vengeance that I should be forestalled. You said this when I explained to you here very briefly my views of 'Natural Selection' depending on the Struggle for existence."

The conventional version goes on to recount how Lyell and Joseph Hooker rescued Darwin from the most drastic extremes of his own sense of honor. They calmed Darwin down. They seized responsibility. No no, they told him, your originality isn't smashed, because we stand as witnesses to your priority. It was Lyell and Hooker, functioning as high-minded scientific adjudicators, who constructed a delicate compromise. They arranged that Wallace's paper and Darwin's own treatment of natural selection would be read jointly at a meeting of the Linnean Society several weeks later. That would put Darwin and Wallace onto the record simultaneously and leave the division of glory to be settled later. What could be more fair?

An alternative telling—call it the revisionist version—asserts that Darwin lied about the letter's date of arrival. Rather than having received it "to-day," as he claimed to Lyell on June 18, he'd had Wallace's manuscript for weeks. The implication is that Darwin, desperate and furtive, had spent those lost weeks trying to decide what to do. Should he destroy the manuscript and pretend he never received it? Should he return it to Wallace? Should he pass it along to a journal and see his life's work preempted? Besides the matter of honor among scientists and the matter of fairness to himself, there was the matter of what he might or might not be able to get away with. Just how long did Darwin stew over those matters?

The most sedulous of the revisionists, John Langdon Brooks, has gone to great lengths researching Dutch steamship schedules and British postal connections between Ternate and London as they existed in 1858. On that evidence, Brooks concludes that Wallace's letter was probably delivered no later than May 18, a month earlier than Darwin admitted. Brooks even posits that Darwin used Wallace's manuscript to help him complete an unfinished part of his own opus—a key section on a principle known as divergence, which explains how natural selection has produced a great abundance of biological diversity, not just a smallish number of well-adapted species.

Let's ignore the particulars of that explanation and simply focus on the issue of its authorship. The point Brooks makes about divergence is that Darwin stole the idea from Wallace and incorporated it into his own working draft as a forty-one-page insert that was added in early June.

Both of these versions, the conventional and the revisionist, are plausible, and a person could spend years sorting through the intricately ambiguous record. Several revisionists have done that. Brooks pored over documents from the Linnean Society, the British Museum, the Peninsular and Oriental Steamship Navigation Company, the Netherlands Postal Museum, and elsewhere, before offering his accusations in *Just Before the Origin*. H. Lewis McKinney's *Wallace and Natural Selection* is another scholarly volume that uses dry language and tiny footnotes to present the revisionist version. Arnold Brackman's *A Delicate Arrangement* is more melodramatic, less restrained by the limitations of evidence. The most persuasive revisionist treatment is the least sensational: a fastidious sixty-page article titled "Wallace, Darwin, and the Theory of Natural Selection," which a historian named Barbara G. Beddall published in the *Journal of the History of Biology*. These four writers make a few points in common. They all say or imply that lies were told, that connivings were committed, and that Charles Darwin was the willing beneficiary of those lies and connivings. They agree that the stink of Darwin's conduct has been retroactively perfumed by misleading accounts of what really happened.

A small instance: Darwin wrote in his autobiography, years later, that "I cared very little whether men attributed most originality to me or Wallace." It's a wishful fib that he told to himself and to posterity, contradicted by those self-pitying yelps in his June 18 note to Lyell. He damn well *did* care, more than a little. Another small instance: Seven months after the joint presentation, Darwin told Wallace by letter that "I had absolutely nothing whatever to do in leading Lyell & Hooker to what they thought a fair course of action." This lie is contradicted by a letter he wrote to Lyell on June 25, volunteering that "I should be *extremely* glad *now* to publish a sketch of my general views in about a dozen pages or so. But I cannot persuade myself that I can do so honourably." Lyell accepted that hint and helped to persuade him.

These little discrepancies only begin to suggest the case against Darwin. The decisive evidence would be Wallace's letter, along with the envelope showing its postmark of arrival in London. But that letter disappeared. It has never been published. The historians can't find it. The postmarked envelope disappeared too.

Also missing from Darwin's papers are six other letters from Wallace, as well as the ones that Lyell and Hooker wrote while they were brokering that delicate arrangement. Charles Darwin was customarily careful about saving his correspondence; aside from those lacunae, his papers are fairly complete. Barbara Beddall, the most judicious of the revisionists, has said, "Somebody cleaned up the file."

But the absence of crucial evidence doesn't prove Darwin's guilt. Notwithstanding the zeal of John Langdon Brooks, neither do schedules from the Netherlands Postal Museum. The private truth of this episode probably can't be established with certainty, because the private behavior and motivations of Darwin are beyond knowing. We're left with what's public.

On the evening of July 1, 1858, the Linnean Society held its meeting. Only thirty or forty people attended. They began with routine business. Then a letter from Lyell and Hooker was read, introducing the Darwin and Wallace contributions and explaining the unusual circumstances. Next came Darwin's contribution, given priority on the grounds of prior composition—although it was only a set of excerpts cobbled together from his unpublished drafts and correspondence. Wallace's paper followed. Five other papers, unrelated to natural selection, were then also read. It must have been a long meeting, on what may have been a hot summer night. According to Hooker's recollection, there was no follow-up discussion of the Darwin-Wallace theory. Maybe its revolutionary implications hadn't sunk in. Or maybe everyone was just desperate to get out into the evening breeze.

So the thing happened quickly and unobtrusively, like a shotgun wedding, to be falsely glorified later as a decorous nuptial event. The "mutual nobility" mentioned by Loren Eiseley was a fantasy. Alfred Wallace showed some nobility, sure enough—but that came later, after he'd been told of the arrangement, and it took form in his lifelong loyalty and deference to Darwin. Darwin himself had behaved weakly and selfishly at best; how he may have behaved at *worst*, we don't know. Amid all the uncertainty and the dark suppositions, at least this part of the conventional version is correct: It was the aftermath of a whopping coincidence.

Two men, on opposite sides of the world, had made the same great discovery at the same time. All they had shared was a language, a sense of the question to be answered, a chance to travel, a willingness to exchange letters, a passing familiarity with Malthus, and an appreciation of the significance of island biogeography.

28

CHARLES Darwin didn't attend the Linnean Society meeting on that evening in July. He stayed home in the village of Down, coping with an outbreak of serious illness among his family. Possibly he was also hiding his face.

Alfred Wallace didn't attend either. He had heard nothing about any meeting. The conventional version claims that "Wallace heartily agreed to" the joint presentation (as Gertrude Himmelfarb wrote recklessly in an otherwise careful book, *Darwin and the Darwinian Revolution*), but that's bullshit. Nobody waited for his permission. He was out of the loop. On July 1, 1858, he was collecting birds and butterflies in New Guinea, completely unaware of the fuss he had provoked in England.

Darwin shoved aside the huge opus and started fresh on a different manuscript. He had learned his lesson about the dangers of delay. Instead of presenting the whole body of theory and evidence, he would dash off a shorter book, sketching just the essentials. He referred to this new effort as an abstract. Today we would call it the quick-and-dirty treatment. He produced a draft within ten months and rushed that to the printer. The shorter book was published, in November 1859, with a long title: *On the Origin of Species by Means of Natural Selection, or the Preservation of Favoured Races in the Struggle for Life*. History knows it as *The Origin of Species*. History tends to consider it the alpha point of evolutionary biology, but history is notoriously careless.

Alfred Wallace spent 1859, and the next year, and the year after that, continuing his work in the Malay Archipelago. There was still plenty that islands could teach.

ϾϿ

SO HUGE
A BIGNES

ϾϿ

29

WITHIN a landscaped corral on a small tropical island, eight giant tortoises rest beneath a ramada, jumbled against one another like a bad freeway accident among Volkswagens. Their scutes are polished smooth by time. Their breaths come and go slowly, very, slowly. The shell of each tortoise is four feet long and arches to a high rounded dome. They seem to belong to the lost age of magisterial reptiles. Most animals this large, nowadays, are not green. I gape at them, trying to soak in a bit of meaning along with the pure visual image.

In the elemental yin and yang of island biogeography—that is, the study of evolution and of extinction—giant tortoises stand as emblems of the yin: evolution. Their bigness reflects the bigness of the idea. Their ancient calm suggests its chronological scope. Their presence on this remote island, as well as in a few other far-flung places, is a reminder of its breadth. If the mechanism of natural selection can produce such a monumental improbability as these beasts, it would seem to be capable of anything. The abiding question is *how?*

Wallace and Darwin's great insight, announced at that Linnean Society meeting in 1858, only began the modern era of asking. The mystery of mysteries had been solved, at least in rough outline; then came the task of elaboration. The colleagues and successors of Darwin and Wallace have now been at it for more than a century and a quarter, and throughout most of that effort biogeography has been their paramount tool. The patterns of species distribution have provided clues about the ways in which species originate, change, and diverge, and the question *how?* has remained inseparable from the question *where?* A giant tortoise, by its mere being, poses the first of those questions rather vividly: How did such an outlandish creature evolve? The task of addressing that one can't proceed far until the second question has been addressed too: Where is this creature found? The linkage between geographical circumstance and evolutionary development is embodied in the very word "biogeography."

The answers to both questions are mortally significant, as well as interesting, because yin gives its shape to yang. By this I mean that evolution is best understood with reference to extinction, and vice versa. In particular, the evolution of strange species on islands is a process that, once illuminated, casts light onto its dark double, which is the ultimate subject of this book: the extinction of species in a world that has been hacked into pieces.

The other factor recommending yin to our attention is that, unlike yang, it's full of grandeur and cheer. These giant tortoises, for example—they have their own sort of majesty, even as captives in a pen. That they have survived at all, into the late twentieth century, is grounds for a guarded bit of joy.

They are crowded together here, beneath their ramada, to escape from the tropical sun. It's midday. I'm sweltering too, but for them more is at issue than comfort. Giant tortoises don't sweat, don't pant, don't possess any convenient physiological mechanisms for regulating their internal temperature, and overheating is therefore a lethal threat. So they seek shade as a tactic of self-preservation, even when it means a pileup of bodies.

Each of these eight must weigh at least 250 pounds, roughly the same as a middle linebacker—but a tortoise is gentler than a linebacker, with a longer and more graceful neck. The average age is probably upward of a century. Darwin himself had only recently been buried in the floor at Westminster Abbey when they were hatchlings. They belong to the genus *Geochelone*, a famous lineage of giants. No, this is not the Galápagos.

30

"*Très jolies*," says Hamid as I stare at the tortoises. Hamid is the eager rogue who has attached himself to me as guide. He doesn't really find them so very beautiful, I suspect, he's just sucking up to a client.

We stand in the Pamplemousses Botanical Gardens on the island of Mauritius, in the Indian Ocean, two seas and two continents away from the Galápagos. The main focus of my visit to Mauritius is another of its biological prodigies—the dodo—and for me the Pamplemousses park is just an afternoon's diversion. I hadn't expected tortoises. But even Hamid can see that I'm more engaged by them than by the clove tree from Zanzibar, the betel-nut palm from Asia, the cork tree from Brazil, the Ceylonese nutmeg, the huge Amazon water lilies floating in exile on a murky pond, and all the other exotic plants he has been showing me. I have stopped walking near the little corral. I resist moving along toward the next item on Hamid's memorized roster of botanical tourist attractions—a camphor tree or some such, transplanted from China or somewhere. Screw the camphor tree, Hamid, I want to linger with these tortoises.

Très jolies? "Yes, they are," I agree. They barely move. They just loom. They hulk. A coldhearted witness might dismiss them as lumpish, but I gawk as though they were prancing panthers.

A giant tortoise seen in the flesh seems wondrously implausible, like a holdover from the Cretaceous period reappearing after eighty million years. It seems to flout all the limits and imperatives that govern our own period of time on the planet. But the limits and imperatives it does flout are only those with which we happen to be most familiar—those governing biological normality on the mainlands. On an island, the limits and imperatives of evolution and survival are somewhat different; and that's why islands are where giant tortoises can be found.

I've seen *Geochelone* before, but I'm especially intrigued by these eight. Where do they come from? To what species do they belong? What's their history? Mauritius once supported two species of *Geochelone*, both giants, but both were extinct by the nineteenth century. So the animals on display here can't be natives. Like the clove tree and the betel-nut, the tortoises of Pamplemousses must be transplanted from elsewhere.

"Seychelles," says Hamid.

The Seychelles are a cluster of islands, some granitic and some coralline, roughly a thousand miles north of Mauritius in the Indian Ocean. The granitic Seychelles once harbored their own giant tortoises, a native population that had lived there since long before humans. Hamid could be right.

But from a different source, later, I will hear a contradictory claim: that the Pamplemousses animals come from Rodrigues, a smaller island in the Mauritius vicinity. Still later, a sheaf of scientific and historical articles will tell me that the question of provenance for the tortoises of the Pamplemousses Botanical Gardens can't be easily settled. The various scattered islands of the Indian Ocean, at the time European sailors first came exploring and pillaging, contained more than a few populations of *Geochelone*. Some went extinct and others were forcibly relocated. Hybridization occurred. Tortoise herds were established, for meat production, on islands to which they weren't native. In certain cases an individual tortoise survived for a hundred years, a hundred fifty, outliving human memory of its origin. The record is richly confused.

That confused record of *Geochelone* in the Indian Ocean is valuable for what it says about evolution. It says, among other things, that the Galápagos Islands are not so extraordinary after all.

Though the marvels of Galápagos biology are an old subject by now, familiar to anyone who can find PBS with a remote control, those peculiar finches and iguanas and tortoises have become widely renowned without being well understood. People have been led to believe that the Galápagos archipelago is somehow mystically different from other places—different even from other oceanic islands. One author of a popular book about Darwin and his *Beagle* voyage, Alan Moorehead, has given strong but not atypical voice to that misleading view of the Galápagos: "The fame of the islands was founded upon one thing; they were infinitely strange, unlike any other islands in the world." No. The real significance of the Galápagos is almost exactly opposite: their fundamental resemblance to other places. They are representative. They are prototypically ordinary. What made them instructive to Charles Darwin, and what makes them instructive to us, is that they are strange in precisely the same ways that other islands tend to be strange.

Islands in general are biologically anomalous. The Galápagos, being anomalous, conform to pattern. They are unique and therefore they are normal.

The clearest evidence of this truth is that you don't have to go to the Galápagos to find a native population of giant tortoises. You do have to go to an island. Several centuries ago there was an array of alternatives, including Mauritius, Rodrigues, and the granitic Seychelles. But today you would look toward a place called Aldabra.

31

THREE DISCRETE groups of giant tortoises survived at the start of the sixteenth century. The three groups together accounted for more than a half-dozen species. They all grew as big as bears and they were all closely related, assignable to the genus *Geochelone*. In addition to those three sets of giants, the same genus also encompassed a number of smaller tortoises native to the South American mainland, Africa, Madagascar, and southern Asia. The giants, by no coincidence, were entirely confined to small islands.

One group inhabited the Galápagos. Another group lived on the Mascarene Islands of the Indian Ocean, the cluster of forested volcanic nubs that includes Mauritius, Rodrigues, and Réunion. The third group occurred in the Indian Ocean too, but remote and distinct from the Mascarene animals. This third group was scattered across seven hundred miles of oceanic wilderness, from the granitic Seychelles southwestward, beyond Farquhar, beyond Cosmoledo, to the tiny coral atoll known as Aldabra. Aldabra sat lost in its own singularity, far off the coast of Tanzania.

Aldabra was nowhere, hundreds of miles from anywhere. It was not a destination or even a stopover point on any preferred sailing route. It consisted of a low ring of limestone, a corona of coralline sand, and some scrub vegetation, all surrounding a shallow lagoon. The noon sun was blistering and the dry season was longish and grim. Not many kinds of animal could live there. With no high land, no decent harbor, hardly any food or fresh water or timber that might attract human voyagers, it was the most desolate and unapproachable of all tortoise islands. Unlike some other small oceanic outcrops, it didn't even contain guano resources worth harvesting. Scarcely anyone coveted Aldabra, scarcely anyone visited. So its tortoises survived. By the start of this century, all other giant tortoises of the Indian Ocean were virtually extinct. Some of those other populations may have left a few hybrid offspring, in translocated meat herds or as mascots held captive at botanical gardens, but they were gone from the wild. On Aldabra, meanwhile, a real population of giant tortoises persisted.

The Aldabran population belongs to the species that taxonomists now call *Geochelone gigantea*. It's probably the same species as those lost populations from the granitic Seychelles. Down in the Mascarenes, by contrast, each island seems to have harbored a distinct species, and in some cases there were two distinct species coexisting on the same island.

Mauritius had *Geochelone inepta* and *Geochelone triserrata*. The anatomical differences between these two species were subtle; the ecological differences can only be guessed at. *G. inepta* may have been more tolerant of dry coastal habitat, while *G. triserrata* may have favored the wetter zones of the island's interior. Both species seem descended from the same ancestral species, possibly having arrived on Mauritius in two separate episodes of colonization. Each colonization may have been accomplished by just a tiny number of individuals, maybe only one—a single pregnant female who found herself washed off a beach in Rodrigues, say, and floated passively across to the neighboring island. Giant tortoises are capable of oceanic crossings, riding the waves like an inflatable raft, head up, patient, enduring

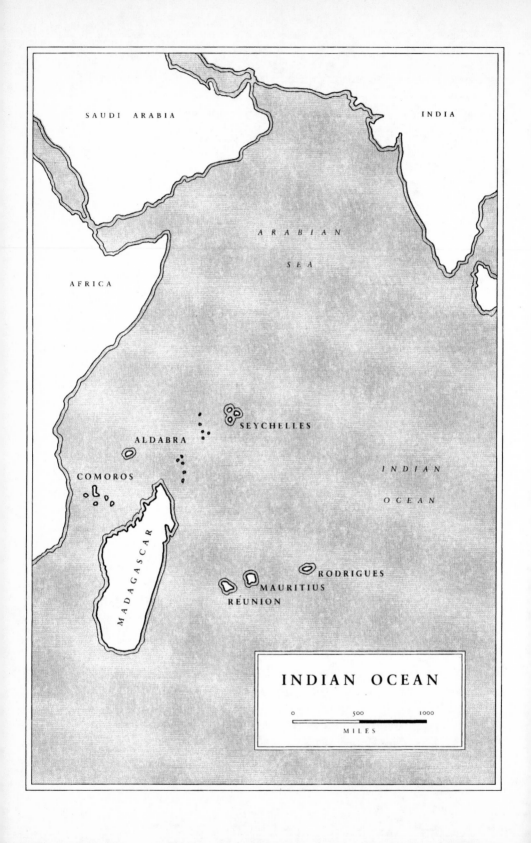

SAUDI ARABIA

INDIA

A R A B I A N

S E A

AFRICA

SEYCHELLES

ALDABRA

I N D I A N

COMOROS

O C E A N

MADAGASCAR

RODRIGUES

MAURITIUS

RÉUNION

INDIAN OCEAN

0   500   1000

MILES

days or even weeks adrift. Maybe the Rodrigues population was itself descended from colonists who, centuries earlier, had been washed off a beach in Mauritius. Such back-and-forth colonization is what makes the patterns of evolution on archipelagos so complicated.

Still, it's safe to say that the tortoises of Mauritius, Rodrigues, and Réunion were all closely related. Taxonomists now lump them within a single subgenus, *Cylindraspis*, but I prefer to think of them simply as the Mascarene cousins. Various factors contributed to their evolutionary history on the Mascarene Islands, of which one was crucial: absence of humanity. Each of those three volcanic nubs was spared the presence of *Homo sapiens* until just a few hundred years ago. By some quirk of chance (probably related to distance and to the prevailing currents of water and wind), no ocean-voyaging humans had arrived there during earlier millennia. While the islands of Polynesia and Melanesia were harboring rich Neolithic cultures, while places as remote as New Zealand and Hawaii and Easter Island were being colonized by adventurous men and women in canoes, the Mascarene Islands (like the Galápagos) remained uninhabited. That lack of two-legged predators allowed the tortoises to survive and evolve.

But it couldn't last. Portuguese ships touched at Mauritius as early as 1507. In 1598 a Dutch expedition landed, beginning an era of frequent contact, and by 1638 there was a Dutch settlement on the island. Eighty years later Mauritius was taken over by the French. The French, the Dutch, and the Portuguese all seem to have used the place as a resting point during long voyages, and as a larder. The Portuguese introduced goats, pigs, and chickens, presumably hoping those meat animals would go feral and multiply; they or the Dutch, by accident, also introduced rats; cats too were introduced, probably by the Dutch during their settlement period in a misguided attempt to control the rats. Nobody knows who introduced monkeys, and nobody can fathom why. In addition, the Dutch slaughtered native animals, most notably dodos and tortoises. Seagoing explorers in those days spread ecological mayhem without dreaming it made any difference, and they ate what they found.

A visiting Englishman in 1630 was impressed by the size of the tortoises, "so great that they will creepe with two mens burthen," but he was less impressed by their culinary appeal. He called them "odious food," and in light of English cuisine he presumably knew odious when he tasted it. The Dutch seem to have found tortoise meat tolerable as expeditionary fare. The French, of course, found ways to make delicacies of what the Dutch took for granted and the English despised. French settlers on Mauritius butchered thousands of tortoises, salting the flesh or rendering it for fat. This was before the world's navies and whaling fleets had made a discovery that doomed *Geochelone* further: Giant tortoises could be stored alive. Their reptile metabolism and their behemoth endurance allowed the animals to linger for months, without food or water, in the hold of a ship. To prevent wandering and induce stoic surrender, they could be turned upside down. Their physiological dormancy made up for the lack of meat freezers.

That phase of exploitation came later. A more immediate style of butchery prevailed during the seventeenth century and into the eighteenth on Réunion and Rodrigues as well as Mauritius. One traveler reported from Réunion: "The Land Turtles are also some of the Riches of the Island. There are vast Numbers of them: Their Flesh is very delicate; the Fat better than Butter or the best Oil, for all sorts of Sawces." Another witness, a Frenchman who spent perhaps too many weeks on Rodrigues in the course of an expedition, remembered "soupe de tortue, tortue en fricassée, tortues en daube, tortues en godiveau, oeufs de tortue, foie de tortue," and grumbled that everything he ate seemed to be only more tortoise stew. By about 1780, the tortoises of Rodrigues, Réunion, and Mauritius were reduced to the point of extreme rarity, most of them having been eaten. Within another generation they were extinct in the wild, possibly extinct altogether. No one knows the fate of the last purebred *Geochelone triserrata*, or of the last ghost of the species preserved in a translocated hybrid. In 1836 Charles Darwin himself visited Mauritius, on his homeward journey aboard the *Beagle*. With his Galápagos experience so recent, giant tortoises must have been fresh in his mind. If he had spotted any in Mauritius, presumably he would have said so. But there's no mention of a Mauritian tortoise in his *Journal*.

The Aldabra-Seychelles species, *Geochelone gigantea*, was not quite so closely related to the Mascarene cousins as the cousins were related to each other. Probably *G. gigantea* had descended from a separate branch of the genus, and taxonomists now place it within its own subgenus, *Aldabrachelys*, another piece of expendable nomenclature that you have my blessing to forget instantly. The subgenus also includes some giant tortoises that once lived on the great island to the southwest, Madagascar. Those Madagascan species have been extinct for some time, possibly several millennia, and are known today only from shell remains, which look very similar to the shell of *G. gigantea*. So it may be that this species originally colonized Aldabra along sea currents from the north coast of Madagascar. Its distinctness from the Mascarene group shows in a few curious traits of skull structure. *G. gigantea* has a more pointed snout. It has a peculiar arrangement of nasal passages, in which a flaplike ridge seems to stand in defense of the olfactory chamber, leaving a clear channel from the nostrils to the esophagus. The ridge could support a fleshy valve (though apparently no scientist has dissected an Aldabran tortoise to find out), which might seal off the olfactory chamber at will. Those structural traits, combined with the field observation that *G. gigantea* sometimes draws water in through its nostrils, suggest an interesting possibility: Maybe the species is adapted to survive droughts by quenching its thirst in deep narrow potholes—stretching its neck down, drinking through its nose. The compacted coral rock of Aldabra happens to be pitted with deep narrow potholes.

Droughts were part of its evolutionary experience, but nothing prepared *G. gigantea* to survive the arrival of humans. Aldabra was protected by its remoteness; the other sites were more accessible, so the other populations were more vulnerable. In 1609 an expedition discovered the granitic Seychelles and their most conspicuous species. A participant in that expedition, one William Revett, recorded "lande turtles of so huge a bignes which men will think incredible; of which our company had small luste to eat of, being such huge defourmed creatures and footed with five claws lyke a beare." He meant that the tortoises, not the members of his company, were "huge defourmed creatures," I think. Revett was another squeamish Englishman whose distaste for reptile meat wasn't shared by the people who came later. Although the Seychelles didn't support a permanent settlement until 1778, by the end of that century the colony's biggest export was tortoises. *G. gigantea* turned out to be just as toothsome as the Mascarene species. In fact, a large portion of the exported animals went to Mauritius, where folks maintained a traditional hunger for tortoise but had nearly exhausted their own supply. Customs records from Mauritius testify to shiploads of tortoises brought in from the Seychelles during the peak of the trade, totaling five thousand animals or more. Additional thousands were taken off the Seychelles by naval ships. Making matters worse, cats and rats had come ashore with the humans, and those animals preyed on tortoise eggs and hatchlings. *G. gigantea* soon disappeared (at least as a wild and unhybridized species) from the granitic Seychelles. Big shipments of tortoises were still passing from the Seychelles to Mauritius in the early nineteenth century, but by then the Seychellois tortoise mongers were just middlemen. The granitic Seychelles had become an import destination, and the imported tortoises must have come from Aldabra; they could *only* have come from Aldabra. In this ocean, there was no other remaining source.

A further threat arose around 1870, when someone got the notion to lease Aldabra for woodcutting. Severe habitat loss, combined with direct harvest of animals, probably would have extinguished the Aldabran population. Even without the woodcutting, that population was already badly depleted—as indicated by the fact that in 1878 a party of sailors spent three days hunting and found only one tortoise. Scientists back in England grew so concerned over what they were hearing that an illustrious group, including Charles Darwin and Joseph Hooker, signed a letter to the governor of Mauritius asking that Aldabra be protected. Some protective measures were put into force, but the tortoise population continued its decline. Rats and cats had infested the atoll by now, raising the mortality rates of tortoise hatchlings and eggs, and those rates had been naturally high under even the prior conditions. A naturalist who visited just before World War I, spending four months camped there, found the tortoises scarce and concluded that "it would be possible to live for years on Aldabra and never see a specimen." The reason, he felt, was simply that they were shy and retiring. The governor of the Seychelles was less upbeat: "No plan will effectively prevent the final extinction of these curious survivals in a wild state in their natural habitats." Not long afterward, as though to prove the governor wrong, the tortoise population began a comeback.

That trend has continued during this century. From its historic low, the tortoise population on Aldabra has increased dramatically. Why? Well, they were left mostly alone, protected from the meat harvesters, their scrub-vegetation habitat still intact. Probably they were also helped by the arid, severe conditions of the atoll, which discouraged a population explosion of cats or rats. After a passing threat in the 1960s that the British government might put an air base on Aldabra, and a public outcry against that bad idea much like the outcry that Darwin had joined earlier, the Royal Society of London assumed protectorship of the atoll. An Aldabra Research Station was built; other developments were kept out. In 1976 Aldabra was incorporated within the Republic of Seychelles, which had emerged as an independent nation, and in 1981 the Seychelles government designated Aldabra a special reserve. According to recent reports, the population of G. *gigantea* is doing well.

Possibly it's doing too well. About 150,000 tortoises now live on Aldabra, and they may be destined for a natural crash in the near future, when they have succeeded in disastrously overeating their own resources.

The story of G. *gigantea* is complicated, but at its core is one simple fact: geographical isolation. In the case of Aldabra, that isolation is extreme. And it isn't measured only in mileage. In the era of the shrinking planet, Aldabra's margins of isolation haven't shrunk. Its remoteness is uncompromised. Today you can board an airliner in Miami and step off within hours in the central Amazon; you can fly out to the Galápagos, connect with a cruise ship, and wander the individual islands on the coattails of a licensed guide; you can book an "adventure travel package," that oxymoronic commodity, to Antarctica; you can even reach Krakatau, cinderlike Krakatau, for a handful of rupiahs paid to a Javanese fishing-boat captain. Aldabra, no. Not so easily attained. Aldabra stands in a different category. It's the place you can't get to from here.

If you're a scientist—or even just a plausible scientific journalist, so I'm told—you might talk your way onto the biennial Aldabra expedition of the Smithsonian Institution, and have the privilege of spending a sun-stricken month watching serious people collect polychaete worms from the potholes. Alternatively, you might charter a somewhat reliable boat in the granitic Seychelles and venture out on the seven-hundred-mile sea journey toward Aldabra, which your compass-guided Seychellois captain might succeed in finding, with luck, though perhaps not on the first try. That charter would cost about ten thousand dollars, I'm told; and when you did reach the atoll, if you did, you might not be legally entitled to step ashore. Having heard all these things, I elect not to test my own plausibility or my own luck.

The published record is obscure but it's substantial, and this one I'll take on faith. I can content myself with having seen G. *gigantea* in a botanical garden in Mauritius. The wild tortoises of the Indian Ocean have been harried enough. Let them have their refuge, their privacy, on that one little desolate atoll. If Aldabra is the epitome of isolation, I figure, then what better way to highlight that fact than by staying the hell away from it myself?

32

GEOGRAPHICAL isolation is the flywheel of evolution. It's also a controversial subject, and has been for more than a century. The controversy has developed along with the development of biology itself, as classic Darwinian evolutionism became fused with paleontology and twentieth-century genetics, resulting in an amalgam of theory known as the modern synthesis, which has lately been further enriched by molecular biology. In recent decades, geographical isolation has been variously considered (1) extremely important to evolution, (2) surpassingly important to evolution, or (3) just flat-out essential. The parties to this controversy spend much of their zeal brandishing data from islands.

Ernst Mayr, a German-born ornithologist and taxonomist based at Harvard, stands as the preeminent authority on the subject. Mayr was born in 1904 and by the late 1920s had made his first ornithological expedition to New Guinea. His career has been long and wide. He is one of the most respected biologists of our time, author of an influential book titled *Systematics and the Origin of Species* (published in 1942), and a co-creator of the modern synthesis that brought genetics and Darwin together. He holds an unfair advantage over virtually everyone else in the debate about geographical isolation because, besides being a researcher and a theorist with his own strong opinions, he is also a historian of science. Still more unfair, he's a good writer.

This combination of skills put Mayr in a powerful position. He could decree consensus, or declare victory for one school of thought over another, with a stroke of his pen. In the historian's role he comes across as judicious, assured, and objective, even while describing his own contributions with third-person detachment: "It was not until the period of the new systematics that Rensch, Mayr, and others demonstrated the populational origin . . ." As Ernst Mayr has told the story in various papers and books, the insight that geographical isolation might be crucial to evolution dates back a long time, well before Darwin and Wallace.

There *were* evolutionary theorists in those years, though not many. One was Robert Chambers, author of *Vestiges of the Natural History of Creation*, the confused but provocative book that energized Wallace. Another was Charles Darwin's own grandfather, Erasmus, a randy and freethinking physician who wrote *Zoonomia*, a medical treatise on "the laws of life and health" that contained intimations about the transformation of species. The most famous pre-Darwinian evolutionist, and probably the most influential, was Jean Baptiste Pierre Antoine de Monet, Chevalier de Lamarck, familiar to history by the last in that list of names. Lamarck put his first groping thoughts about evolution into a university lecture delivered on May 11, 1800, and then presented a full (but misleading) theory in his *Philosophie Zoologique*, published nine years later. Each of the early theorists—Lamarck, Erasmus Darwin, Robert Chambers, and a few others—had recognized at least vaguely the reality of organic evolution. They saw evidence of it in comparative anatomy, in fossil sequences, and in biogeography. They just couldn't explain how the transformation of species was effected. The mechanism that Charles Darwin and Alfred Wallace later discovered, natural selection, hadn't occurred to these predecessors. But they weren't dolts, they weren't blind, and they managed occasionally to offer some valuable observations and ideas. One of them, Leopold von Buch, wrote in 1825:

> The individuals of a genus strike out over the continents, move to far-distant places, form varieties (on account of the differences of the localities, of the food, and the soil), which owing to their segregation cannot interbreed with other varieties and thus be returned to the original main type.

For the word "segregation," according to Ernst Mayr, who supplies the translation of this passage, we may read "geographical isolation." Von Buch continued:

> Finally, these varieties become constant and turn into separate species. Later they may again reach the range of other varieties which have changed in a like manner, and the two will now no longer cross and thus they behave as "two very different species."

The phrase about "far-distant places" takes sharper focus when you know that von Buch based his ideas on the flora and fauna of the Canary Islands.

The process that von Buch described now goes by the label *allopatric speciation*. *Allo*, different; *patric*, father—speciation in different fatherlands. The simple definition of allopatric speciation is: One

species splits into two species while the two sets of individuals are living in two different places. That stands in contrast to *sympatric speciation*: One species splits into two species even though both sets of
individuals are living in the same place. Allopatric speciation hasn't
lost any relevance in the century and a half since von Buch wrote. It's
still considered the way in which virtually all new species (or at least
new species of sexually reproducing animals) arise. Some plants present special complications. Parthenogenetic, hermaphroditic, and
asexual creatures present complications. But among sexual animals
that must pair to reproduce, and among most plants, allopatric speciation is the rule. Sympatric speciation is either a rare exception or an
illusion, depending on which argumentative biologist you consult.
And the prerequisite to allopatric speciation, as von Buch said so long
ago, is geographical isolation.

I suspect you can see where we're headed. We're headed toward
understanding the whole planet as a world of islands, and evolution
itself as a consequence of insularity.

33

DURING THE nineteenth century, the foremost proponent of the view
that I've just described—geographical isolation as the flywheel of evolution—wasn't Charles Darwin or Alfred Wallace. It was a fierce-
minded German naturalist named Moritz Wagner.

Born in 1813, Wagner came to the subject by the same route as
Darwin and Wallace. He spent a crucial part of his young manhood
traveling, collecting specimens, and charting the patterns of species
distribution. Those patterns filled his brain with wonder and wild
ideas. Also like Darwin and Wallace, he recognized that the theory of
special creation was untenable. One factor setting Wagner apart from
the two Englishmen was that he focused his attention on mainlands—
Asia, the Americas, northern Africa. He didn't match the island-
hopping journeys of Darwin or Wallace, but he found plenty of
geographical isolation on the continents, just as Wallace himself
found it among the monkeys of the Amazon basin. For three years
during his early twenties, Wagner explored Algeria, including a region along the north coast where two rivers, draining the Atlas
Mountains, demarcated the limits of distribution for some species.

Species of beetle, for instance. Certain flightless beetles of the genus *Pimelia* were restricted by these Algerian rivers. On one bank of a river lived one *Pimelia* species; on the far bank, a different *Pimelia* that was closely related. And among various other types of animal, Wagner noticed, a mountain range often formed the same sort of boundary between closely related species. Deserts likewise formed boundaries between nondesert species. Geographical isolation of one sort or another seemed everywhere to be causing—or at least marking—the division between pairs of related species. Wagner published that observation as early as 1841.

After 1859, having read *The Origin of Species*, Wagner became an ardent evolutionist and struck up a correspondence with Darwin. He expanded his own ideas into a bold, eccentric edifice that he called *die Separationstheorie*, the separation theory. On some points he was too bold and eccentric for even Darwin's taste. Geographical isolation was one of those points. Wagner wrote: "The formation of a genuine variety which Mr. Darwin considers to be an incipient species, will succeed in nature only where a few individuals transgress the limiting borders of their range and segregate themselves spatially for a long period from the other members of their species." In other words: no speciation without geographical isolation. Wagner added that "the formation of a new race will never succeed in my opinion without a long continued separation of the colonists from the other members of their species." The categorical absoluteness, the words "only" and "never," made it a stronger statement than Darwin could accept.

Darwin had changed his mind on this issue. In his own early notebooks, and in two unpublished essays on natural selection that he drafted long before *The Origin of Species*, geographical isolation figured importantly. At one point he had confided by letter to Hooker: "With respect to original creation or production of new forms, I have said that isolation appears the chief element." Having been inspired by what he saw in the Galápagos—the finches, the tortoises, and especially the different species of mockingbird, no two of which are found on a single island—Darwin might naturally have concluded that geographical isolation was instrumental. But other data, from other places, seemed to suggest otherwise.

On the South American mainland, Darwin had seen two different yet closely related species of *Rhea*, those ostrichlike birds, both species occupying the plains of Patagonia. One was smaller than the other; in addition to size, the two species were presumably distin-

guishable in more subtle ways. Whatever the full range of their differences, they seemed to have developed those differences through a
process of sympatric speciation—they were presently sharing habitat,
after all, and there was no reason to assume that they hadn't always.
Based on the evidence that Darwin saw (which was partial and probably misleading), the divergence between the two *Rhea* species might
have occurred without geographical isolation. Late in his life, Darwin
wrote: "I remember well, long ago, oscillating much; when I thought
of the Fauna and Flora of the Galapagos Islands I was all for isolation,
when I thought of S. America I doubted much." After the first edition
of *The Origin of Species* had been published, his oscillation swung him
back toward the South American fauna and the doubts, and away
from geographical isolation.

The mature Darwin could still accept the idea that geographical
isolation might sometimes contribute toward evolution. But it wasn't
necessary, he thought. Large areas of habitat on the mainlands accounted for many more species than islands did, didn't they? The evidence seemed to suggest that although isolation of *some* sort might be
a factor, maybe behavioral or ecological isolating circumstances could
effect the same results as outright geographical isolation.

Darwin and Moritz Wagner argued this point. Their debate, conducted by mail and through their publications, grew more strained as
years passed. Part of the problem was that Wagner's separation theory contained some insupportable fancies about the role of migration
in evolutionary change. The act of migration itself, according to
Wagner, had an almost magical power. He held that a species would
never evolve unless it migrated into a new locale, and that geographical isolation merely consolidated the changes that migration triggered. Darwin, on the other hand, saw nothing magical about a
species moving from one place to another. In 1875 he read a copy of
Wagner's latest paper and scribbled "Most Wretched Rubbish" across
the front. Another part of the problem between them was that they
sometimes misread each other's work, and the misreadings were exacerbated by confusion over terms and concepts. The confusion wasn't
surprising, since this whole realm of thought was so new. Crucial bits
of evolutionary theory were still inchoate, and others were missing
altogether. For example, both Darwin and Wagner were groping to
understand inherited change without benefit of the concept of the
gene. Other intricate concepts that had yet to be clearly articulated
were *phyletic evolution*, as distinct from speciation, and *reproductive iso-*

lation, as distinct from geographical isolation. I'll return to those two in a moment.

But first we need to ask: What is a species? That question, perplexing to Alfred Wallace when he pondered the birdwing butterflies of Aru, perplexing likewise to Darwin and Wagner, is still somewhat perplexing. The answers that have been offered are various, and each in its way is tendentious. Plants, microbes, and marine invertebrates don't all respect the neat species category that we commonly use to sort lions and tigers and leopards. So a botanist may not subscribe to the same definition of species as a field zoologist does, and a geneticist, a paleontologist, or a taxonomist may each use still another definition. For simplicity, in considering evolution, we probably do best with the famous definition that Ernst Mayr proposed in 1940: "Species are groups of actually or potentially interbreeding natural populations, which are reproductively isolated from other such groups." There's a lot of careful hairsplitting worded into that.

Two populations qualify as reproductively isolated if they can't interbreed, or if the first generation of interbred offspring is sterile. A horse can breed with a donkey, producing a mule; but two mules together can't produce offspring. Therefore the horse and the donkey constitute separate species. Mayr's definition begs the question of how the two populations *become* reproductively isolated. Can it occur only as a consequence of geographical isolation, or is there an alternative way?

Phyletic evolution is another fundamental process that most of us have never set straight in our minds. It's distinct from speciation, the process whereby one species splits into two. Each time a speciation event occurs, the number of species on Earth increases by one. A population of equine animals splits into two, reproductively isolated from each other, and in place of one ancestral species there exist two descendants—let's say Burchell's zebra (*Equus burchelli*) and Grevy's zebra (*Equus grevyi*). That constitutes speciation, yes; evolution at work, yes. But evolution and speciation aren't synonymous. Rather, they are nested categories. Speciation is just one aspect of the whole. Phyletic evolution is another.

Phyletic evolution is the process whereby one species changes gradually, over time, into a species that looks or acts differently. In phyletic evolution, the single species adapts to altered circumstances without splitting in two.

Speciation occurs mainly in the spatial dimension. That's why "allopatric" and "sympatric" refer to positions in space. Phyletic evolution occurs mainly in the temporal dimension. A certain species is not what it used to be, but it's still just a single species. Speciation demarcates new species, while phyletic evolution tends to increase the differences between species once they have been demarcated. The difference between Burchell's zebra and Grevy's zebra is speciation. The difference between a zebra and an elephant is speciation plus many millions of years of phyletic evolution.

This bit of schoolmarmy noodling brings us back to the debate between Darwin and Wagner. Does phyletic evolution require geographical isolation? No. Does the continuing divergence of the zebra lineage from the elephant lineage require that they be spatially segregated from each other? No. Despite their proximity on the same savanna, zebras and elephants aren't going to interbreed. Likewise a species of crow and a species of woodpecker can continue to diverge evolutionarily, even while sharing the same forest. They do it by responding differently to different aspects of the environment—that is, by adapting themselves to different ecological niches. But Moritz Wagner failed to recognize that point; he missed phyletic evolution.

With phyletic evolution set to one side, we can more clearly address the tougher question: Does speciation require geographical isolation? There is disagreement even today. Ernst Mayr argues persuasively that speciation does require geographical isolation, at least among sexual animals and most plants. Reproductive isolation is the hallmark of distinct species, but reproductive isolation can't arise spontaneously, says Mayr. It can only result from genetic mutations and other differences that build up between two populations while they are physically separated. This fact eluded Charles Darwin. Don't take my word for it, I'm no Darwin scholar, but I think we can trust Dr. Mayr. By his account, Darwin confused reproductive isolation (an ultimate result) with geographical isolation (a requisite circumstance). What are the actual causes of reproductive isolation? They include mutations at the gene level, changes in patterns of gene linkage, and changes in ecology or behavior, any of which can hinder gene-mixing throughout a population. Those factors do the direct work of establishing reproductive isolation. In barest terms, then: mutation as cause, geographical isolation as circumstance, reproductive isolation as ultimate result. That much will produce speciation. Long stretches of time, competition, and ecological perturbation will add a measure of phyletic evolution. Speciation is the sharp end of the wedge, driven between two populations, and phyletic evolution is the wedge's widening butt.

The essential point about the Darwin-Wagner dispute is that Darwin won. But he won on debating points and on reputation, not on the true merits of his case, and his victory led biology astray. Darwin persuaded his colleagues that geographical isolation wasn't necessary to evolution. Alfred Wallace was among those who agreed with him—despite Wallace's own field experience, and despite also the fact that he didn't *always* agree with Darwin on major details of evolutionary theory. Moritz Wagner had come closer to being right. But because Wagner was right for the wrong reasons, within the wrong theory, he probably discredited the role of geographical isolation more than he promoted it.

In the early decades of the twentieth century, Gregor Mendel's classic work on inheritance among pea plants was rediscovered and the science of genetics burgeoned like duckweed on a warm pond. As the geneticists in their labs became engrossed with the phenomenon of mutation (which seemed to suggest that new species might arise suddenly and sympatrically), they too overlooked the significance of geographical isolation. The concept went out of fashion. Karl Jordan, a German entomologist based in England, was one of the few scientists of that era who stressed geography. The trend was against him. With theoreticians having granted their sanction to the concept of sympatric speciation, more and more field biologists began finding (or believing they had found) evidence of its occurrence. For the length of a scientist's lifetime—from the 1880s until the 1940s—geographical isolation was dismissed. Prevailing opinion held that species arose frequently by sudden mutations or from ecological specialization, without need of geographical separation between populations. The geneticists and the sympatric-speciation school were in ascendance. This was the context in 1942, when Ernst Mayr published the first of his great books, *Systematics and the Origin of Species*.

Mayr was skeptical of sympatric speciation. Because of the prevailing bias in its favor, he took particular pains to examine certain cases in which sympatric speciation had supposedly occurred. They were illusory, he concluded. Allopatry was what lay behind the appearances. There *had* been geographical isolation, but it was subtle.

Mayr reaffirmed the view of Moritz Wagner and Karl Jordan: that geographical isolation must precede all other forms (reproductive, ecological, behavioral) of isolation. Maybe the geographical boundary is virtually undetectable, but it has to be there. It has to block the mixing of genes between two groups—at least initially, until behavioral or ecological differences arise that are themselves sufficient to block gene mixing. One implication of this view was that field scientists should carefully reexamine those putative cases of sympatric speciation. Somewhere between the geographical range of Burchell's zebra and the geographical range of Grevy's zebra—as those two were distributed at the time of their speciation—lay a physical boundary. Somewhere between the two *Rhea* species, which so puzzled Darwin, had been a divider. Mayr's summary assessment was: "Geographic isolation alone cannot lead to the formation of new species, unless it is accompanied by the development of biological isolating mechanisms which are able to function when the geographic isolation breaks down. On the other hand, biological isolating mechanisms cannot be perfected, in general, unless panmixia is prevented by at least temporary establishment of geographic barriers." *Panmixia* is the two-dollar term for unhindered mixing of genes throughout a population. Mayr's view was almost as categorical as Wagner's: no speciation ("in general") without geographical isolation.

Decades later, Mayr looked back at his own work from the historian's perspective, like a soul afloat in an out-of-body experience. "It was one of the major theses of Mayr's *Systematics and the Origin of Species* (1942) that there is a fundamental difference between the two kinds of isolating factors," he wrote, "and, as Wagner and K. Jordan had previously insisted, that geographical isolation is a prerequisite for the building up of intrinsic isolating mechanisms." Taken out of its historical context, the whole fracas over geographical isolation might seem minor. But to one of the godfathers of modern biology, it was a major battle fought and won. Mayr added: "Since 1942 the importance of geographic speciation, as worked out by the naturalists, has not been denied." As worked out by the naturalists such as Ernst Mayr, he meant, in opposition to those myopic geneticists and the other enthusiasts of sympatry. Mayr himself, in offering this historical judgment, wasn't quite a disinterested party, but most biologists today would probably agree on the point of science.

One other fact about Mayr should be mentioned. Long before he became an august historian, a philosopher of biology, and a founder of the modern synthesis, he was an island biogeographer. His *Systematics and the Origin of Species*, so compendious and exacting on the issues of how species originate and how evolution proceeds, was based in large part on bird specimens and distributional data collected by the Whitney South Sea Expedition. Operating continuously between 1921 and 1934, that expedition visited most of the bigger islands of the southern oceans. Mayr himself focused his fieldwork on the Solomon Islands and New Guinea.

This is where we've been headed: If geographical isolation is the flywheel of evolution, then the Galápagos Islands are not "infinitely strange, unlike any other islands in the world" (as Alan Moorehead carelessly claimed in his book about Darwin and the *Beagle*). And evolution as it occurs on islands—all islands—is not unrepresentative of evolution at large. On the contrary, it's paradigmatic. It just seems unrepresentative because it tends to be so antic.

34

SOME CREATURES native only to islands:

Dendrolagus ursinus, a tree-climbing kangaroo. New Guinea. *Dendrolagus goodfellowi*, another tree kangaroo, restricted to another part of New Guinea. *Deinacrida megacephala*, the big-headed flightless cricket. New Zealand. *Amblyrhynchus cristatus*, the surf-diving iguana. Galápagos. *Hapalemur aureus*, recently discovered, known as the golden bamboo lemur. Madagascar. *Sarcophilus harrisii*, the Tasmanian devil. Tasmania, of course, though at one time it also occupied the Australian mainland. *Leucopsar rothschildi*, the white starling. Bali. *Casuarius bennetti*, the dwarf cassowary. New Guinea. *Sauromalus hispidus*, the giant chuckwalla. Angel de la Guarda, in the Gulf of California. *Strigops habroptilus*, the overgrown flightless parrot, commonly known as the kakapo. New Zealand. *Hypogeomys antimena*, the giant jumping rat. Madagascar. *Hyomys goliath*, the white-eared giant rat. New Guinea. *Solenodon cubanus*, the carrot-nosed, ratlike thing that isn't really a rat but categorically unique. Cuba. *Varanus komodoensis*, the so-called Komodo dragon, actually a giant monitor lizard. Komodo, Flores, and a few other tiny islands of central Indonesia. *Ornithorhynchus anatinus*, the duck-billed platypus. Australia—which is insular, after all, despite its representing the mainland to Tasmania. *Zaglossus bruijni*, the three-toed spiny anteater. New Guinea. *Dasyurus viverrinus*, the marsupial native cat. Tasmania. *Thylacinus cynocephalus*, the marsupial wolf, presumed extinct since the 1930s. Tasmania. *Mi-*

cropsitta keiensis, the yellow-capped pygmy parrot. Native to Aru, where Wallace may have glimpsed it, as well as to the Kei Islands and New Guinea. *Apteryx haasti*, the great spotted kiwi. New Zealand. *Apteryx australis* and *Apteryx oweni*, two other kiwis, also flightless and confined to New Zealand. *Macropanesthia rhinoceros*, the giant burrowing cockroach. Australia. *Labidura herculeana*, the giant earwig. Saint Helena. *Crotalus catalinensis*, the rattleless rattlesnake. Santa Catalina, in the Gulf of California. *Halmenus robustus*, the flightless grasshopper. Galápagos. *Nestor notabilis*, the carnivorous parrot that has taken to preying on sheep. New Zealand. *Brachymeles burksi*, the legless skink. Philippines. *Nannopterum harrisi*, the flightless cormorant. Galápagos. *Rallus owstoni*, just one of the many species of flightless, island-dwelling rail. Guam. *Anas aucklandica*, the flightless duck. Auckland Island. *Dimorphinoctua cunhaensis*, the flightless moth. Tristan da Cunha. *Schoinobates volans*, also known as the greater glider, a marsupial with winglike flaps that allow it to glide through the air like a flying squirrel. Australia. *Petaurus australis*, the yellow-bellied glider, another airborne marsupial. Australia.

And of course:

Aepyornis maximus, the thousand-pound bird, now extinct. Madagascar. *Hippopotamus lemerlei*, the pygmy hippopotamus, also extinct. Madagascar. *Geochelone gigantea*, the giant tortoise. Still alive on Aldabra. *Geochelone elephantopus*, the other giant tortoise. Still alive in the Galápagos. *Astrapia mayeri*, the ribbon-tailed bird of paradise. Presently at large in New Guinea. *Hemicentetes nigriceps*, P. J. Stephenson's biting tenrec. Alive and feisty in Madagascar. *Raphus cucullatus*, the overgrown, stupid-looking flightless pigeon, famously extinct. Mauritius.

Within all the antic particularity, we can begin to trace some patterns. I've mentioned more big islands than small islands. I've mentioned more old islands than young islands. The reptiles and the birds I've mentioned are associated with both big and small islands, whereas the mammals are confined to big islands. I've mentioned only one sea bird, the cormorant. (Islands are full of sea birds, but with their large wings, their great stamina, their ability to rest on the water's surface, and their consequently vast ranges, sea birds tend not to be isolated enough for localized speciation.) I've mentioned no carnivorous mammals except from Tasmania. What's special about Tasmania? It's a near outlier of Australia, to which it was connected by land bridge in the recent past, and Australia is evolution's largest and most extravagant demonstration of alternate realities. Carnivorous mammals are notably absent from most islands no bigger than Tasmania. We'll get to the reason for that in its turn.

Other patterns? I've mentioned more tropical islands than temperate islands. I've used certain mainland words, such as "cat" and "wolf " and even "anteater," not in a literal way but as metaphors. There has been no mention of frogs or salamanders. I've alluded repeatedly to gigantism, dwarfism, flightlessness, and a few other characteristic forms of divergence from mainland ancestors. So we have glimpses of order amid chaos.

Island species tend to be different. Island communities tend to be different. But throughout the world they manifest their differences in a handful of similar ways.

35

I ADMIT it's the faintest bit confusing. Islands resemble one another in that each is unique. Huh? Islands are distinct from mainlands in that they represent simplified, exaggerated versions of . . . of exactly those evolutionary processes that occur on mainlands. Okay, we're in the realm of paradox. Differences of degree graduate into differences of kind, and conceptual borderlines shimmer.

Some scientists have offered neat lists of the phenomena that characterize island biology, of which gigantism, dwarfism, and flightlessness are only the most obvious. Those lists tend to confuse matters further, though, when they mix two categories of phenomena together: species attributes and community attributes. To penetrate the confusion and get a clear look at the various phenomena, we need first to undo that mixing of categories.

The first category—species attributes—reflects evolutionary forces impinging on species and lineages. The second category—community attributes—reflects ecological forces influencing the ways communities are assembled. The gigantism of *Aepyornis maximus*, on Madagascar, is a species attribute. The flightlessness of *Nannopterum harrisi*, in the Galápagos, is a species attribute. The fact that Madagascar contains no native felines is a community attribute. Small islands like the Galápagos tend to be strikingly empty of big predators, of mammals, and of amphibians—each of those facts is a community attribute. Every species attribute and every community attribute has a biological explanation, and at a certain point the explanations connect. But we'll do best to start by considering the two categories separately.

In discussing the signature features of island biology, the experts offer lists in this vein:

> dispersal ability
> size change
> loss of dispersal ability
> endemism
> disharmony
> relictualism
> loss of defensive adaptations
> impoverishment
> archipelago speciation
> adaptive radiation

For convenience I'll call that the Insular Menu. Each of these items is especially relevant either to island species or island communities. But the species attributes and the community attributes are jumbled together like raisins and peanuts. Better to think of them this way:

> I. Species attributes
>
> dispersal ability
> size change
> loss of dispersal ability
> endemism
> relictualism
> loss of defensive adaptations
> archipelago speciation
> adaptive radiation
>
> II. Community attributes
>
> disharmony
> impoverishment

For now let's ignore the community attributes. To understand evolution on islands, we need to start at the level of species.

Species change size. Species disperse. Species lose their dispersal ability or their defensive adaptations. Species are endemic or relictual. Groups of closely related species, not entire communities, show adaptive radiation.

Dispersal ability is prerequisite to the rest. At peril of sounding moronically obvious, I'll dare to note that evolution doesn't start until the first living creatures arrive. How they arrive, when they arrive, which kinds of creature turn up earlier and more often than which others, are matters of consequence that shape the biological history of an island.

36

PICTURE AN island that's totally empty of life. It's just a mound of bare rock, surrounded by water.

Why would this island be empty? One possible reason: because it's volcanic and newborn, a mountain of lava lately risen from a vent on the ocean floor. Steaming and sterile, it might be a recent addition to the Hawaiian chain, forty or fifty miles southeast of Mauna Loa. It might be the island of Surtsey, freshly erected near Iceland in November of 1963. It might be one of the Galápagos group, which were young and uninhabited just a few million years ago.

There's another possibility: It might be Krakatau, an old island newly sterilized.

Krakatau is a shrine to island biogeographers because its ecosystem was obliterated and founded anew within scientific memory. This gave it the significance of a vast natural experiment on the dynamics of recolonization. Of course, Krakatau's recolonization wasn't so carefully controlled or so thoroughly monitored as experimentalism ideally demands. But since evolutionary biology and island biogeography are both descriptive sciences more than experimental ones, and since even descriptive scientists covet the hard validation that experimentalism seems to provide, the Krakatau case has been extremely valuable.

The cataclysm took place in a series of blasts during late August of 1883, throwing six cubic miles of igneous rubble into the sky above the Malay Archipelago. The crescendo came in a single stupendous explosion on the morning of August 27. They heard that one in Perth. The sky went dark, every barograph in the world winced, the sun appeared eerily filtered—looking green, then later blue—and thirty-six thousand people were killed, mainly by tidal waves hitting the coasts of Sumatra and Java. One wave was a hundred feet high, moving as fast as a train. Ships were pushed onto beaches in Ceylon, and a change in sea level reached Alaska. Fire engines were called out on false alarms as far away as New Haven and Poughkeepsie, and peculiar sunsets and other atmospheric effects went on for months afterward. The dust veil in the atmosphere cooled the planet, which didn't warm back to normal for five years.

When the smoke and the terror finally cleared at the site of Krakatau itself, thirty miles off the west coast of Java, a small crescent of cauterized rock remained where the island had been. That cauterized remnant was called Rakata. It was a truncation of the original name, K-rakata-u, a gentle etymological reminder of the ungentle geological truncation. Two other small islets, which stood nearby but hadn't been part of Krakatau itself, were also scorched. Although nobody can be certain, scientific opinion holds that not a single living thing on either Rakata or the other two islets had survived the eruption—no plant, no animal, no egg, no seed, no spore. Nine months afterward, a French expedition to Rakata found nothing alive there except a single spider.

The spider of Rakata is emblematic of the fact that spiders in general are good dispersers. Devious beasts, they are wingless but still manage to fly. A thread of silk is paid out from the silk glands, it billows, it rises, it somehow attains purchase on an ascending column of air, and like a hang glider on a windy ridge, it lifts the spider away. This trick is dependent on forces that act at small scale. It wouldn't enable a full-grown tarantula to float through the skies of Arizona, thank God, but it does allow daintier spiders to go ballooning from one place to another. I've seen baby black widows, no bigger than poppy seeds, waft away on the thermals from a tensor lamp. The Rakata spider must have ridden a breeze out from Java.

The first botanical expedition, led by a Professor Treub, reached Rakata in 1886. Treub's team found mosses, blue-green algae, flowering plants, and eleven species of fern. The algae, consisting of slimy dark smears that coated the ground with a gelatinous matrix not unlike agar, had probably served as a welcoming mat for the spores of the ferns and the seeds of the flowering plants. The ferns were especially hearty and diverse. Among the flowering plants, four species belonged to the Compositae family (a group that includes dandelions, among other aggressive airborne dispersers) and two species were grasses. It's likely that the Compositae and the ferns had been delivered to Rakata by wind. There were also some species whose seeds could have washed in on the surf.

The arrival of other life forms was quick. By 1887 Rakata supported young trees as well as dense stands of grasses and an abundance of ferns. By 1889 it harbored not just spiders but butterflies, beetles, flies, and at least a single large monitor lizard of the species *Varanus salvator*, closely related to the Komodo dragon.

Varanus salvator, like the ferns and the Compositae, is notable for its ability to travel widely and colonize new habitat. It swims well, and on land it's a versatile opportunist, quick afoot, stealthy when it needs to be, capable of climbing and burrowing. Carnivorous but not fussy, it eats crabs, frogs, fish, rats, rotting meat, eggs, wild birds, and the occasional chicken from an unguarded coop. Flesh-eating animals tend to fare poorly on small islands and on new islands, where the pickings are slim, but *V. salvator* on Rakata enjoyed two advantages: It was a generalist, and it was a reptile. A generalist can eat less selectively, and a reptile can eat less often.

But even *V. salvator* depended on the presence of other animals, and those other animals depended on the presence of plants. By 1906, Rakata supported almost a hundred species of vascular plants, with a carpet of green on the summit and a grove of trees along the shore. The grove included the tamarisk-like species *Casuarina equisetifolia*, a good traveler across tropical oceans, as well as the coconut palm, *Cocos nucifera*, which turns up on virtually any balmy beach. Another beach-loving plant, the morning glory *Ipomoea pes-caprae*, had also appeared. Several years later there were fig trees and a few other species characteristic of secondary forest. The sun-loving ferns that had been so prevalent earlier were now retreating to high ground, forced out of the lowlands by grasses and shade-casting trees.

In 1934, a half century after the new beginning, Rakata and its companion islets held 271 species of plants. One botanist has given us an informed guess as to how each of those species arrived. About forty percent came on the wind. Almost thirty percent floated across the sea. Most of the others had probably been carried by animals. They all possessed good dispersal ability, but the means were various.

Ferns do their traveling as spores, one-celled reproductive capsules that serve them in place of seeds. Spores are durable genetic packets, self-contained, resistant to drying, and tiny enough to be carried on a sneeze. With their spores riding the breezes in every direction, it's no wonder that ferns get around. Coconut palms achieve widespread dispersal because the coconut, at the opposite size extreme from a fern spore, is such a seaworthy seed. Some other plant species (such as the tropical vine *Entada*, also known as the sea bean) produce seeds with an air space between the embryo and the seed coat, suited for long-distance flotation. Darwin himself, during the years of work that led to *The Origin of Species*, did experiments to gauge the dispersal ability of various plant species. He put seeds, fruits, and sections of dried stems into seawater to see which species would float for how long and whether the seeds would retain their viability afterward. "To my surprise I found that out of 87 kinds, 64 germinated after an immersion of 28 days, and a few survived an immersion of 137 days." He also learned (and reported, in *The Origin*, with his usual half-barmy attention to detail) that ripe hazelnuts sank immediately and that asparagus floated much better if the plant was first dried.

The colonization of a new island isn't simply a matter of getting there. Dispersal is just the first of two crucial steps, the second being what ecologists call *establishment*. This distinction is especially germane for creatures dependent on sexual reproduction. Having hit the beach safely, a spider or a monitor lizard still faces the problem of establishing a self-sustaining population. It needs to find food, protection, and (unless it's a pregnant female) a mate, all of which demand both adaptability and luck. If dispersal is difficult, establishment is difficulty squared. Among vertebrate animals, a reptile has the advantage of a relatively starvation-tolerant metabolism. And some reptile species (the gecko *Lepidodactylus lugubris*, for instance, widely distributed among small islands in the western Pacific) have even picked up the trick of parthenogenesis—single parenting taken to its logical ultimate. Parthenogenesis obviates the problem of finding a mate on every new island where a solitary pioneer might arrive.

Another mode of dispersal for terrestrial creatures is accidental transport on natural flotsam. The vehicle can be an old log, a newly uprooted tree, even a tangled mat of branches and vines washed out to sea from the mouth of a river or blown offshore by hurricane winds. Such flotsam can carry a colony of termites, an orchid bulb, a clutch of gecko eggs, a snake, maybe even a terrified rat. If the flotsam eventually washes ashore on some other coastline, the passengers have achieved dispersal. One biologist has called this sweepstakes dispersal, because the odds against success are so high. Over the great reaches of geological time, though, it seems to have happened often.

In rare cases the flotsam may even be massive and durable enough to support growing plants. A biogeographer named Elwood Zimmerman has collected testimony about "floating islands" of vegetation washed out to sea and adrift in the blue wilderness between Sulawesi (Wallace knew it as Celebes) and Borneo. "These mats of vegetation were lush and green, and palm trees of 20 to 30 feet high stood erect on the floating masses. A survey of these rafts probably would reveal that numerous plants and animals were riding them." Wallace himself, in *Island Life*, reported sightings of floating islands among the Moluccas. He added that, in the Philippines,

> similar rafts with trees growing on them have been seen af-
> ter hurricanes; and it is easy to understand how, if the sea
> were tolerably calm, such a raft might be carried along by a
> current, aided by the wind acting on the trees, till after a
> passage of several weeks it might arrive safely on the shores
> of some land hundreds of miles away from its starting
> point. Such small animals as squirrels and field-mice might
> have been carried away on the trees which formed part of
> such a raft, and might thus colonise a new island; though,
> as it would require a pair of the same species to be thus
> conveyed at the same time, such accidents would no doubt
> be rare.

Occasionally there is even eyewitness evidence of animal transport. Wallace went on to mention the case of a large boa constrictor that rafted its way to the island of Saint Vincent in the West Indies, almost two hundred miles off the South American coast. The snake arrived "twisted round the trunk of a cedar tree, and was so little injured by its voyage that it captured some sheep before it was killed"—an instance of successful dispersal followed by failed establishment.

And sometimes the natural flotsam might even be mineral, not vegetable—which brings us back to Krakatau. Among the various types of geological debris ejected during the explosions was pumice, a lightweight and sponge-structured volcanic glass. In its frothier form, pumice will float, and rafts of the stuff littered southern seas for as long as two years after the Krakatau eruption. Some of those pumice rafts drifted together into jams, clogging inlets on the coast of Sumatra; some washed ashore in South Africa, five thousand miles to the west; some floated eastward beyond Guam. One traveler described the concentration of floating pumice offshore from Java, with individual lumps clustered together over acres of ocean, each lump as big as a sack of coal. Another man, a ship's captain named Charles Reeves who encountered pumice on the Indian Ocean, lowered a boat for a closer look. "It was curious and interesting to note how it had been utilized by animals and low types of life as habitations and breeding places," he reported. "There were creeping things innumerable on each piece." Although Reeves confessed himself insufficiently learned to list them all by name, he did notice crabs and barnacles, and he saw small fish gathered underneath for feeding. Obviously the crabs, barnacles, and other "low types of life" had climbed aboard after these lumps of ballistic pumice achieved splashdown; seeds, eggs, and adults of various terrestrial creatures had no doubt gotten onto them too, carried short distances by wind or wing or else picked up when the rafts made passing contact with a shoreline. A modern study of the Krakatau event suggests that similar eruptions over the centuries, tossing out huge quantities of floating pumice, have been important factors in the dispersal of species.

A more obvious mode of dispersal is long-distance flying. But even this isn't so straightforward as it might seem. Many species of bird and insect are reluctant to cross even modest stretches of sea. They will fly anywhere in a forest, but they won't commit themselves offshore. In the Solomon archipelago east of New Guinea, for instance, three species and several subspecies of white-eye (the genus is *Zosterops*) remain isolated from one another on closely neighboring islands. Likewise on Salawati and Batanta, two small islands off the western tip of New Guinea—they stand less than two miles apart, but the gap seems to have been wide enough to forestall the dispersal of seventeen bird species from one to the other. The gap between mainland New Guinea and the large island of New Britain is somewhat wider, forty-five miles. That has been distance enough to keep about 180 species of New Guinea birds from colonizing New Britain. And there's the case of Bali and Lombok, where Wallace noticed that some bird species had made the short crossing while many others had not.

It's not simply that the strong-flying species travel and the weak-flying species remain sedentary. Ecological or behavioral factors are also involved. Sea birds such as albatrosses, shearwaters, frigate birds, and pelicans make long ocean journeys, of course; since they can soar effortlessly for miles and rest on the water's surface, those species aren't much dependent on terra firma. Among land birds it's a trickier matter. Some species and groups of species are more inclined than others toward reckless or accidental ocean transits. High on the list of good travelers is the pigeon family.

The typical pigeon is a slightly plump bird with a small head and strong wings, adapted to a diet of seeds and fruit, for which it might be accustomed to make seasonal migrations. Those traits seem to predispose it toward transoceanic journeys. True pigeons and pigeon descendants are disproportionately well represented on some of the most remote islands. Sâo Thomé, a small nub of land offshore from West Africa, supports five species of pigeon. Anjouan, in the Indian Ocean north of Madagascar, also claims five different pigeons. Samoa has the tooth-billed pigeon and the white-throated pigeon. Palau has the Nicobar pigeon and the Palau ground-dove. New Guinea and its surrounding islands harbor forty-five pigeon species, roughly one-sixth of the world's total. And the Mascarene Islands have known their own generous share of pigeons and pigeonlike birds, among which the dodo is only the most famous. On Mauritius, a beautiful red-white-and-blue creature called the pigeon Hollandais (*Alectroenas nitidissima*) had followed the dodo into extinction by about 1835, and the pink pigeon (*Nesoenas mayeri*) is in jeopardy of extinction at this moment. The other Mascarenes, Réunion and Rodrigues, each harbored a large, flightless species of solitaire (*Ornithaptera solitaria* on Réunion, *Pezophaps solitaria* on Rodrigues) that, like the dodo, had pigeon affinities.

What's unusual about this roster of endemic pigeons is not just the breadth of dispersal but the breadth of diversity. The pigeon ancestors traveled commonly enough to colonize many islands—but they traveled *rarely* enough that, once they had colonized, they were likely to be sufficiently isolated for evolutionary divergence. In many cases, the divergence entailed loss of their ability to disperse.

The dodo itself stands as the best emblem of this general truth—that insular evolution often involves transforming an adventurous, high-flying ancestor species into a grounded descendant, no longer capable of going anywhere but extinct. It's our reminder that insular evolution, for all its wondrousness, tends to be a one-way tunnel toward doom.

37

ON THE CAUTERIZED nub called Rakata, formerly Krakatau, sea birds must have been among the first visitors: shearwaters and petrels, pelicans and boobies, tropicbirds, frigate birds, terns and noddies. We can't know for sure, we can only assume.

We can safely assume that the sea birds didn't come alone. Flying animals, like rafts of flotsam, carry passengers. Tangled among their feathers, infesting their skins, glued to their feet with dried mud, passing through their intestinal tracts, would have been all manner of hitchhikers: spores of ferns and fungi, propagules of mosses and algae, seeds of flowering plants, mites and lice, and the small sticky eggs of snails, land crabs, centipedes, earthworms, and insects. The visiting sea birds would have helped replant an ecosystem.

Land birds and rafting reptiles, as well as larger arthropod species, would have reached the island later. The first systematic survey was done in 1908, by which time Rakata harbored thirteen species of resident land birds. By 1921, the roster had grown to at least twenty-seven species. This increase was typical for a new island: a high rate of immigration during the early years, when niches were standing vacant and everything was happening anew. By 1934, the number of land bird species had shown no further increase (one account puts it again at twenty-seven, another account puts it slightly lower), suggesting that the process of replenishment had leveled off. Again this was typical: a decreasing immigration rate later, as niches became filled. Individual birds were still arriving, but the number of species seemed to have stabilized. Rakata was a small island, so there could only be room for a modest amount of avifaunal diversity. Maybe the place was already full. Maybe it had reached some sort of equilibrium.

Immigration rate is a crucial parameter in determining the ecological destiny of all islands, as MacArthur and Wilson would explain in their pathbreaking book. And the notion of equilibrium—a dynamic balance between the factors tending to increase and those tending to decrease the species diversity within an ecosystem—would lie at the center of their work. We'll return to that notion later. For now it's enough to say that MacArthur and Wilson made much of Rakata, with its twenty-seven bird species.

A further geological development had meanwhile added itself to the case file on Krakatau. During the late months of 1930, a new island rose from the underwater stump of the old volcano. This sudden new pile of lava emerged into daylight not far from Rakata. As it steamed and solidified and began to look permanent, it was given the name Anak Krakatau—Indonesian for "Child of Krakatau." Today it stands almost six hundred feet high, a cinderous black cone garnished sparsely with green. As one of the world's youngest tropical islands, it represents another important test case for addressing some of the basic questions of biogeography. How do species disperse? Which species disperse most readily? What happens within a new ecosystem as it begins to get crowded? Does the number of resident species really settle into an equilibrium? If so, what are the factors that tend to increase and, opposingly, to decrease the species diversity?

In 1952 another eruption killed off all (or nearly all) life on the new island of Anak Krakatau, setting the process of biological colonization back to zero. Since that time, a handful of devoted biogeographers have visited intermittently, paying careful attention to the ebb and flow of species. But for most scientists, Anak Krakatau is still obscure and remote. There's no research station. There's no government outpost. It's not on the routes of ordinary tourism. There isn't so much as a dock, an old hut, or a Fotomat. The summit is still venting hot gas, the slopes are still as barren and dark as a slag pile, when Bas van Balen and I wade ashore.

38

As we walk up the beach, a five-foot-long monitor lizard sprints furtively over the sand—insofar as a five-foot-long lizard can be furtive.

We find the beach zone dominated by *Casuarina equisetifolia*, that tamarisk-like ocean traveler. We also notice a small fig tree, a single lonely sapling, presumably brought here as a seed within the gut of some fig-eating bird. The sapling's prospects of establishing a colony are bleak, since fig trees reproduce by sexual pairing and depend on a family of tiny wasps, the Agaonidae, to accomplish their pollination. In the absence of other figs to mate with, and of agaonid wasps to intermediate that mating, a fig tree faces a celibate, fruitless future. Maybe another fig-eating bird will shit out another seed and supply a potential mate before this sapling dies of old age; but there would still be the need for wasps. As I grope through the casuarina grove, sketching leaves in my notebook and contemplating the romantic desolation of a Krakatau fig tree, Bas with his keen ornithological attunement begins to identify birds.

The collared kingfisher is here, he tells me. *Halcyon chloris*. Bas carries a bird book in his head. The collared kingfisher is a notorious tramp, turning up anywhere it can find a warm coastal forest of mangroves or casuarinas, from India to Samoa. The mangrove whistler is here, Bas announces. *Pachycephala cinerea* to him, to me a glimpse of brown. And the flyeater, *Gerygone sulphurea*. Where? I say. Somewhere in those trees, explains Bas, who has heard it but not seen it. He often makes his identifications solely by ear, which in dense tropical forest is the effective way to survey birds. The forest of Anak Krakatau is tiny and not very dense, a sparse crescent fringing the beach like a Hutterite beard, so in this case Bas's encyclopedic memory of bird song is largely wasted. We pass through the forest quickly and begin climbing the mountain of black cinders.

Up here on the slope we find ferns lurking in gullies. The ferns no doubt arrived early in the colonization process, as they did on Rakata a century ago, and now that the casuarinas are hogging the beachfront, they have gone to high ground. The gullies offer an occasional trickle of water, a few hours of daily shade, on the otherwise inhospitable hillsides. Some of the ferns are nonetheless wilted and reddish brown, like filigreed iron sculptures turned rusty. Although we've come to Anak Krakatau on a day of weather—rolling Turneresque fogs, small storm clouds, warm rain—the sun, when it's out, is severe. In a shady crevice I see mosses. On the black slope itself, exposed, are some brave clumps of grass. Bas has his own focus. Savanna nightjar? he says as a question. Either he caught only a flash of the bird or he heard only an ambiguous tweet. He's just thinking aloud; he knows better than to consult me. Fastidious and professional, Bas records *Caprimulgus affinis* in his notebook with appropriate marks indicating uncertainty.

Onward we climb. The upper slope is a crust of friable lava. It crunches like peanut brittle beneath our steps. We reach a crest, which turns out to be the lesser of two camelback summits within the overall conic shape of Anak Krakatau. The other summit, higher, is the rim of the active crater. We head for it, descending into a valley between the two.

Deep in the valley we stop. There is a peculiar feel down here, an extraordinary lifelessness, different from elsewhere on the island. I've seldom been in a place so silent and sterile. No bird song, no insects, no casuarinas, no grasses, no figs. The rock is gunmetal gray, squeaking and snapping underfoot. This must be what it was like for Neil Armstrong.

Then I notice a few tiny ferns, their half-unfurled heads peeking up from beneath rocks, breaking the deathly spell. These adventurous little plants have claimed the terrain that no other creature wants. After savoring the place for a moment, Bas and I decide that *we* don't want it either.

We begin climbing the far slope. Halfway up toward the crater's rim, we halt over the skeleton of an animal.

The sun-bleached bones, clean and austere after months in this tropical autoclave, show nicely white against the dark lava. They lie in a graceful cluster, like a still life arranged for a Japanese garden. Once they belonged to a mammal.

If I haven't said much about mammals in this discussion of island colonization, it's because there is not much to say. Mammals don't travel as well as most other vertebrates. Their dispersal ability across salt water is generally low. They are burdened with urgent physiological needs and blessed with only modest endurance. Starvation and drought can kill them quickly. So can drowning. If they do manage to make a crossing, their prospects of establishment are still poor. Since they reproduce sexually, give birth to live young, and suckle those young, they don't enjoy the same adaptive advantages as many plants, insects, and reptiles. An adult mammal needs a mate; an infant mammal needs a mother. All these factors reduce their chances of colonization. Rarely a species of mammal does reach an island and establish itself, but more commonly an island remains empty of mammalian fauna despite the passage of eons. As reptiles and ferns and pigeons tend to be disproportionately present on islands, mammals tend to be disproportionately absent.

Then again, to every biogeographical pattern there are exceptions.

Standing over the skeleton, I ask Bas: What is it? The bones are delicate. Several are long and slender, like jackstraws. I wonder how on earth the little beast got here.

"A bat," he says.

39

BATS ARE obvious. Bats go everywhere. New Zealand, even, has bats.

Alone in its region of the Pacific, a thousand miles southeast of Australia, New Zealand is the planet's most isolated major landmass. One testament to that isolation is the paucity of mammals. Until humans sailed in with their pestilential entourage of domestics and camp-followers and pets, New Zealand had harbored no nonflying mammals, not even a species of rodent. But even then, before the first ships arrived, there were native bats. The short-tailed species with the long ears, *Mystacina tuberculata*, seems to have evolved from ancestors that arrived forty or fifty million years ago, probably blown across from Australia in an epochal storm. The long-tailed species with the short ears, *Chalinolobus tuberculatus*, is a comparative newcomer. Its nearest relatives live in Australia and New Caledonia (another largish island, a thousand miles to the north). The case of New Zealand lends support to a general postulate: that wherever there is an island endowed with plentiful insects and fruit, no matter how remote, a bat species will eventually find it.

Rats and mice too, despite their absence from aboriginal New Zealand, turn up on most other islands. They are susceptible to accidental rafting on natural flotsam, and they will stow away on any sort of boat that humans put in the water. The Galápagos Islands supported rice rats of the genus *Oryzomys* even before European ships brought in the infamous black rat, *Rattus rattus*. The Philippines have their cloud rats, genus *Crateromys*, plus about forty other endemic species of rodent. Sulawesi has its soft-furred rat and its Margareta's rat, among others. Tasmania has its broad-toothed rat, shared with mainland Australia, and Australia itself is full of hopping mice. Flores, like New Guinea and Madagascar, has its own endemic species of giant rat. The tiny island of Rinca, near Komodo, has an eponymous mouse. The Pacific is bestrewn with populations of *Rattus exulans*, informally known as the Polynesian rat, and the Caribbean, before humans invaded, was rich with many species of ratlike hutias.

The dispersal of rats and mice by rafting, like the dispersal of bats by wing, is straightforward enough to pass without explanation. A more curious aspect of mammalian dispersal is swimming—especially as practiced by elephants.

40

HIPPOPOTAMUSES, deer, and elephants are the naiads of terrestrial mammals. These are the groups of land animals most notable for swimming to new territory—not all the way to New Zealand, not to Hawaii or the Galápagos, but to less remote islands. On this there is a scientific consensus. People with straight faces and academic jobs have published papers, in respectable journals, about seagoing hippopotamuses, deer, and elephants. You could look it up.

You might look first at some work by a fellow named Donald Lee Johnson, published a few years ago in the *Journal of Biogeography* and elsewhere. Johnson considered the long-distance swimming ability of elephants a well-supported fact, which most biogeographers had chosen blindly to ignore. He gathered a pile of interesting data on the distribution of fossil elephants (more properly, elephantids, a group that also includes mammoths) and on the aquatic exploits of living elephants. The fossil evidence shows that elephantid species once occupied the Mediterranean islands of Malta, Sardinia, Sicily, Crete, Cyprus, Rhodes, and Delos; the Indonesian islands of Sulawesi, Timor, and Flores; the Philippine islands of Mindanao and Luzon; and a few smaller islands elsewhere in the world. The bones have been unearthed. The animals were indubitably present. But how did they get there?

The conventional assumption was that they came across land bridges during low-water ages. How else *could* they have come, given that elephants don't swim the seas?

But modern behavioral evidence suggests that the conventional assumption is faulty. Elephants do occasionally swim the seas, according to Donald Lee Johnson. He has offered both eyewitness testimony and a handful of persuasive photographs.

On the afternoon of January 17, 1958, an elephant walked into the water from Sober Island, just off the coast of Ceylon, and began swimming toward Ceylon itself. The sea depth went to thirty-three fathoms, and the distance was a third of a mile. The elephant swam steadily for seventeen minutes, occasionally raising her trunk to breathe. Two friends named Charles Rowe and Cecil Hooper watched her go, timed her, and captured her performance in a series of photos. Eventually, Mr. Rowe testified, "on finding ground beneath her feet, she paused to take a look at Cecil and me and then ambled off in the direction of Nicholson Cove."

Several years later, another elephant made the same swim in the reverse direction, accompanied by her calf. She too was watched and photographed. One of the shots shows her taking a breath, chin down, face submerged, only the graceful arc of her trunk and the dome of her head visible—looking very much like a Loch Ness monster pursued by an amorous humpback whale. There are more accounts from that part of Ceylon (or, as presently known, Sri Lanka), where it's common, evidently, for elephants to visit the small islands just offshore. In 1970 an elephant swam over from Sober Island to the naval dock. Shooed away, that one turned and swam back to the island.

Similar reports come from India, from Cambodia, from Kenya and its coastal islands. Johnson quoted a nineteenth-century account by an Englishman in what is now Bangladesh: "Full grown elephants swim perhaps better than any other land animals. A batch of seventy-nine that I dispatched from Dacca to Barrackpur, near Calcutta, in November, 1875, had the Ganges and several of its large tidal branches to cross. In the longest swim they were six hours without touching the bottom; after a rest on the sand-bank, they completed the swim in three more; not one was lost."

If it's common across short distances, it's possible across longer ones. Back in 1856, according to a newspaper story cited by Johnson, a ship was approaching the South Carolina coast when it was hit by a gale, and an elephant on deck was blown overboard. Although the ship was thirty miles offshore, sometime later the elephant came swimming into a local harbor. "Its feat of riding out the storm is, we suppose, the most remarkable instance of animal strength and endurance on record," according to a contemporary report that originated with the *Charleston Evening News*. Newspapers being what they are, credence to this anecdote can be considered optional. Most of Johnson's data is more firmly documented, though less heroic.

He examined the case of fossil remains found on three of the Channel Islands, twenty miles off the coast of southern California. These fossils represent *Mammuthus exilis*, a pygmy species derived from *Mammuthus columbi*, which was one of several mammoths that roamed North America within the last million years. The offshore presence of *Mammuthus exilis* had for a century been taken as proof, Johnson wrote, that the northern Channel Islands were once connected to mainland California by a land bridge. That hypothetical land bridge, in turn, had figured in subsequent biological studies of the islands. But Johnson himself, having looked at the geological and bathymetric data, concluded that apart from the bones of *M. exilis*, there was no cause for supposing that such a land bridge ever existed. The sea gap of the Santa Barbara Channel, in his judgment, probably never narrowed to less than four miles. Furthermore, the northern Channel Islands had yielded no fossil evidence of bears, sloths, saber-tooth cats, cougars, bobcats, coyotes, raccoons, badgers, wood rats, squirrels, chipmunks, rabbits, or toads. All of those animals had in-habited southwestern California during the Pleistocene, and if the mammoths had walked out on a land bridge, presumably they could have too. "Of the 38 extinct mammalian and herpetofaunal forms found in the Rancho La Brea asphalt deposit," Johnson wrote, "why is only one, the water-loving elephant, found on the Channel Islands?" Because, he argued, it swam there.

Johnson was right about the reluctance of other scientists to accept the notion of swimming elephants. The islands of Timor and Flores are a test case. They have yielded elephantid fossils of two different species within the genus *Stegodon*, and the presence of both ste-godonts on both islands has encouraged experts to assume that the two islands were formerly linked by a land bridge. To understand Johnson's challenge to that assumption we need to delve just a little deeper into the subject of land bridges.

Land bridges linking continental islands to their nearest mainlands have come and gone repeatedly throughout deep time. One mecha-nism producing the phenomenon, as I've mentioned, has been the major shifts in sea level that accompany episodic glaciation. Most re-cently, the Pleistocene epoch (from about two million years ago until about twelve thousand years ago) encompassed a number of such episodes—the ice ages, as we casually call them. In each instance, as a vast volume of seawater was sucked away from the oceans and piled up into glaciers and ice caps in the high latitudes, the sea level fell. Land bridges consequently surfaced at what had been the shallowest zones of sea bottom. Britain was connected by land bridge to Europe. Sicily was connected to Italy. Bali was connected to Java, which was connected to Sumatra, which was connected to the Asian mainland. How far did the sea level fall during a major episode of Pleistocene glaciation? Cautious geologists cite a figure of 130 meters—that is, more than four hundred feet.

The problem with using this mechanism to posit a Pleistocene land bridge between the islands of Timor and Flores is that even a four-hundred-foot fall in sea level wouldn't be sufficient. The Savu Sea, between those two islands, is more than ten times that deep.

The Pleistocene glaciations may be relevant nonetheless. A modest drop in sea level would have reduced the distance between Timor and Flores, at a place called Ombai Strait, to perhaps five or ten miles. That still doesn't allow the land-bridge school of experts to push a stegodont across with dry feet. But the Ombai gap wasn't too wide, possibly, for a stegodont to swim. Swimming stegodonts? It seemed unthinkable until Donald Lee Johnson spoke up. Two decades ago, shortly before Johnson began airing his unorthodox hypothesis, a pair of reputable stegodontologists declared: "Because an elephant (and presumably a stegodont) could not swim across the Savu Sea and Ombai Strait, the existence of a land connexion between Flores and Timor during the Pleistocene must be postulated." They found it easier to ignore the nine-thousand-foot depth of the Savu Sea than to accept the evidence that elephants do swim.

Why would an elephant go for an ocean swim? Johnson posits three contributing circumstances. First, famine or drought in the home territory might drive some animals to search desperately for new habitat. Second, an elephant with good long-distance vision might see an island on the horizon. Third, with the wind at its face, an elephant might smell that the island offered edible greenery. And if an elephantid could swim four miles (from Pleistocene California to the northern Channel Islands) on purpose, it might well swim ten miles (from Pleistocene Flores to Timor) in some extraordinary, accidental situation. The notion of sweepstakes dispersal, carrying birds and tortoises to islands that they would otherwise never reach, is dependent on exactly such accidents.

Johnson had no button-down proof, but his hypothesis does seem more plausible than imagining a whole series of vanished land bridges in what is now deep water. It seems still more plausible as part of the worldwide distributional pattern of elephants, hippos, and deer.

Other scientists besides Donald Lee Johnson have noted that pattern. Hippopotamus fossils occur on the Mediterranean islands of Sicily, Crete, Malta, and Cyprus, as well as on Madagascar. Fossil deer have been found on Sardinia, Corsica, Sicily, Crete, Malta, some of the smaller Aegean islands, and on the remote Japanese islands of Ishigaki, Miyako, and Okinawa. Elephantids too have left fossils on most of those islands. And their presence is set in relief by the absence of other large mammals. A land bridge should have offered equal entry to all four-footed creatures; but the distributional pattern requires an explanation admitting three groups of mammal while excluding the rest. What the pattern suggests is that hippos and deer, like elephants, must be adventurous swimmers.

The pattern is notable for another reason: It coincides with a pattern of dwarfism. Geographical isolation on continental islands had a peculiar effect on these hippos and deer and elephants. Over time, by way of phyletic evolution, it made them dainty.

41

TAKE THE Pleistocene elephant of Sicily, *Elephas falconeri*. It had descended from a mainland ancestor about the same size as a modern elephant, but on Sicily the animal was shrunk by three-quarters. Its skull was rounder than the skull of its ancestor, with proportionately more room for brain and less bone area for attachment of large muscles. *Elephas falconeri* was also more stocky. Short-legged, tiny, bright, it stood as tall as a pony. In adapting to life on its island, the Sicilian elephant had changed into a dwarf.

The same phenomenon affected other big mammals on other islands. *Mammuthus exilis*, descended from mainland mammoths that stood fourteen feet tall at the shoulders, grew only six feet tall on the Channel Islands. The elephantids of Timor and Flores were remarkably small. The elephantids of Crete, Malta, and Cyprus were dwarfs. The deer of Kasos and Karpathos, in the Aegean, were miniaturized. The Mediterranean hippos were no bigger than county-fair pigs.

There are some straightforward explanations that could begin to account for all this evolutionary shrinkage. An island-trapped elephant might have less to eat; an island-trapped deer might be relieved of the threat of predation; either circumstance might yield a trend toward dwarfism. But is the trend consistent? Do small patches of landscape uniformly produce small beasts? No. What makes the pattern more mysterious is that while insular evolution has made some species smaller, others it has made drastically larger.

Whereas mammals tend toward dwarfism on islands, reptiles tend toward gigantism. The list begins with those huge tortoises. It includes both groups of endemic Galápagos iguanids, the marine iguana *Amblyrhynchus cristatus* and the land iguanas of the genus *Conolophus*.

Geckos, skinks, night lizards, wall lizards, and other types of lizard are also inclined to be larger on islands. The night lizard *Xantusia riversiana*, found on three of the southern Channel Islands, is considerably longer and burlier than its closest mainland relative. Australia supports some remarkably hefty skink species, and on the Cape Verde Islands occurs an endemic genus, *Macroscincus*, whose name translates to "large skink." The anguid lizard *Celestus occiduus*, native to Jamaica, is unusually big for an anguid. The gecko *Phelsuma guentheri* and the skink *Leiolopisma telfairii*, formerly found on Mauritius but now extirpated there and confined to a small nearby islet known as Round Island, are both foot-long giants. And as big as it may seem by the standards of worldwide skinkdom, *Leiolopisma telfairii* itself is barely half the size of the skink *Riopa bocourti*, on New Caledonia, which grows to a lunkerish twenty-three inches.

Bigness is relative. These oversized species represent only the low end of the scale of reptilian gigantism. Even that twenty-three-inch New Caledonian skink seems puny the moment a person steps onto the island of Komodo.

42

I ARRIVE aboard an old cargo boat, chartered for twenty-five bucks out of a port town in western Flores. At the boat's modest pace, Komodo lies four hours west of Flores across what at this time of year is a gentle sea. I've managed to miss the monsoon. My captain is a small wiry man named Sultan, with an honest smile full of crooked teeth. The crew consists of two young boys, evidently his sons. Sultan wears a sarong and a strap undershirt and no shoes, and he speaks even less of my language than I speak of his, which would be problematic if not for the presence of Nyoman the Balinese Tailor, whom I've hornswoggled into another excursion. During our motor crossing from Flores, Nyoman has slept the sleep of the innocent on the boat's wooden deck, with his transistor radio plugged into his ear, while I've gaped at the sea and pestered Captain Sultan with questions in my baby-talk Indonesian. It's midafternoon as we reach Komodo.

The bay where we anchor is a lucent coralline blue. The island's volcanic peaks look like gumdrops: steep-sided, round-topped, and soft. The sign at the end of the pier says TAMAN NASIONAL KOMODO. To me and you, that's Komodo National Park. The forested valley stretching upland beyond the beach is called Loh Liang. With its thatched-roof park office, its few austere cabins, its open-air cafeteria, this is the part of Komodo in which tourists are welcome. This is where people come to gawk at the dragons.

"Komodo dragon" is a cartoonish label. What we English-speakers call by that nickname is actually a gigantized species of monitor lizard, *Varanus komodoensis*. It's native only to Komodo, to an area along the west coast of Flores, and to a few smaller islands nearby. The traditional name, in the local Mangarrai dialect, is *ora*, nicely homonymic with "aura," which these animals certainly possess. But the Mangarrai name seems to be fading as new cultural influences blow through the region, and local people are more inclined nowadays to refer to the animals as "komodos." So it's a species synonymous—in more than one language—with a place.

Other species of monitor lizard are native to Africa, Asia, and Australia, and a few of those (notably *Varanus salvator*) grow big, though not nearly so big as *Varanus komodoensis*. A full-size komodo, ten feet long and dragging a recently filled belly, might weigh 500 pounds. A lean nine-footer, on the other hand, might weigh only 120 pounds.

Unlike a crocodilian, a komodo is terrestrial. Unlike an Aldabran tortoise, it's slow-moving only in some circumstances—and shockingly quick on its feet in others. A further difference from the big tortoises is that *V. komodoensis* is carnivorous. These are significant ecological factors that make the komodo's marvelousness more marvelous still. Huge crocodilians such as the Nile croc (which spends its life in water, where mass is less weighty) and giant tortoises of the genus *Geochelone* (which move slowly from one grazing site to another, their metabolism geared low) both enjoy adaptive advantages that mitigate the problems of gigantism. The komodo doesn't share those advantages. It maintains its gigantic size the hard way: with great bursts of overland ferocity, punctuated by meat-wolfing gluttony.

The fact that *V. komodoensis* inhabits not just Komodo but also the nearby islands of Rinca and Gilimotang and the western end of Flores is a detail that doesn't figure in the popular image of the species. What figures is that this is some kind of crazybig reptile. What figures is that it's a fearsome predator with an immoderate hunger for raw flesh. On both points the popular image is accurate, if not precise.

Nyoman and I present ourselves at the office and buy our park permits, which at a thousand rupiahs (fifty cents) represent one of the great bargains of international tourism. The ranger in charge is a dour man in a blue uniform shirt, polite but godawful stiff. He seems to be psychologically stooped beneath the weight of his duties. Maybe my nattering stream of questions has put him in a grumpy mood. Or maybe it's the fact that a ferryload of additional tourists will be arriving at dawn tomorrow, making his life difficult for a few hours while they attend the regular semiweekly feeding. The feeding is a staged spectacle wherein komodos from the surrounding forest are baited in with a slaughtered goat. A small clearing suddenly fills up with slathering reptiles, doltish tourists, and a frenzy of amateur photography, and anything could go wrong. A child could wander off and be scarfed. A blue-haired American lady could lose her forearm while trying to feed one of the dragons a banana. The dour ranger's head seems aswim with such visions, each of which promises his professional doom. He holds his chin low as we stand suppliant before his desk, and looks at us with only his eyes lifted, as though he were peeking over his glasses. He doesn't wear glasses. It's easy to imagine that Sunday and Wednesday, when the feedings are held, might be the most dreaded days of his week. Today is Saturday. The feeding will be at nine, he tells us. Here are your room keys. Sign the register, please. Now get out of my sight, he implies.

Is it permissible, I ask him, that we *jalan-jalan*—that is, take a little hike—after we've dropped our bags? His eyes lift again, though not his chin. Ya ya, he says, *jalan-jalan* is okay. But only along the beach. Don't walk into the forest, he says.

The beach is safe, and the forest is not, because komodos are ambush predators. They lurk in the undergrowth, in the brush along trails, in the high grass of the savannas. They lurch out at unsuspecting prey. They are incapable of sustaining a long-distance chase, but they can be wicked at a sprint. Occasionally they jump a human. So Nyoman and I *jalan-jalan* along the beach, where the sight lines are long and there is no danger of ambush. Maybe, I'm hoping, the forested valley can be explored later.

This valley, Loh Liang, is famous in the herpetological literature. Twenty-some years ago, before the visitors' camp was built, Loh Liang was the main site of the first thorough ecological study of *V. komodoensis*, conducted by an American scientist named Walter Auffenberg. Much of what is known today about the species and its habitat comes from Auffenberg's book, *The Behavioral Ecology of the Komodo Monitor*. It's a weighty beige volume that you won't find in B. Dalton's, but it's lucid, intriguing, clearheaded on the distinction between komodo rumor and komodo fact, and encyclopedically useful to anyone concerned with the subject. Along with my mosquito net, my head lamp, my malaria pills, my binoculars, my Mercurochrome, and my rot-proof sandals, Auffenberg's book is one of the few precious items I've brought on this leg of my journeys. Late in the evening, sweltering in my little cubicle at the tourist compound while the mosquitoes try to get at me, I browse through *The Behavioral Ecology of the Komodo Monitor* again.

"Much of the notoriety attached to the Komodo monitor is related to the fact that it is the world's largest living lizard," Auffenberg wrote. "As might be expected, this circumstance has led to confusion, exaggeration, and misinformation." After more than a year in the field, taking measurements as well as watching behavior, he was qualified to rectify some of those misapprehensions.

The species had become known to science only in 1910, when a Dutch colonial official named van Hensbroek visited the island and collected the first specimens that were passed along to a biologist. Van Hensbroek heard tales about fearsome granddaddy komodos up to twenty-three feet long, but the animals that he saw and killed were no greater than ten-footers. In 1923, an adventurous German duke

shot a komodo that he claimed was twelve feet long, though when the
skin reached the Berlin Zoological Museum it measured only seven
feet ten; maybe it had dried and shrunk. The longest specimen on
record, according to Auffenberg, was a male in captivity at the St.
Louis Zoo. It went just over ten feet two inches from its snout to the
tip of its tail—or anyway it was "reputed," in Auffenberg's careful
wording, to have attained that length. One other source has reported
that it weighed 365 pounds, a figure Auffenberg himself didn't mention. The animal was stuffed after its death in 1933, and it now stands
on display at a park in California, with its tongue dangling out like a
barbecue fork.

Concerning the size of komodos, Western travelers have been
more inclined to exaggerate than the native people of Komodo and
Flores, who see their giant lizards realistically. The twenty-three-foot
claim, for instance, came from a pair of breathless Dutch pearl fishermen who prevailed on the credulity of van Hensbroek. Even biologists have found their own visual estimates (as opposed to their
methodical measurements) unreliable. Auffenberg quoted one account from researchers who, watching a komodo feed on a carcass,
took it to be fourteen feet long. Later they measured the same komodo, found it four feet shorter than they had imagined, and admitted that "in this case we had been deceived by the creature's air of
power and strength." Trout fishermen are familiar with that phenomenon. When an animal is vividly impressive—struggling and leaping
at the end of a fly line, say, or ripping apart a water buffalo with its
teeth—it assumes a certain enhanced bigness, which mysteriously dis-
sipates as soon as the contextual terms change and a tape measure is
brought into play. The twenty-three-foot komodo is the one that got
away.

Among the ecological questions addressed by Auffenberg, one is
more tricky than it seems: Why are komodos as large as they are? It's
tricky because it encompasses two antipodal questions: (1) Why aren't
they small, like other lizards? and (2) Why aren't they even larger
than they are? The second of those two, seemingly idle, is worth asking in light of certain prodigious facts that we'll come to shortly.

"That size is related to predation is obvious," Auffenberg wrote.
"In almost all organisms, optimal predator size is largely determined
by the interaction of both the abundance of different prey size classes
and the relative energy extractable from them by predators of a given
size." Roughly translated: Evolution enforces efficiency, so if larger

body size allows a predator to victimize large-bodied prey, and to do that more efficiently than it could victimize small-bodied prey as a small-bodied predator, then evolution may well produce gigantism, up to a point. Up to what point? Up to the point where the predator is *too* big and its efficiency again starts to decline. The degree of gigantism will depend on the size and abundance of the various prey animals available, and on how they behave in response to predatory threat.

Full-grown komodos, according to Auffenberg's study, prey chiefly upon deer and wild boar. They have also been known to eat snakes, ground-nesting birds, other komodos, rats, dogs, goats, water buffalo, horses, and humans. They commonly attack animals twice their own weight, and in the case of an adult buffalo it's more than ten times their weight. The ambush technique allows them to take prey of larger sizes than would be possible otherwise. The rusa deer (*Cervus timorensis*), a species found in dry habitats across central Indonesia, is currently their main food item. Rusa deer travel in small herds, which makes them apt victims for the lurk-and-lurch tactics of *V. komodoensis*. One deer skitters up a game trail, another deer follows, and a komodo in hiding comes alert; it might strike the third or the fourth deer, or any unfortunate straggler that passes by too slowly or too unwarily. In connection with herd behavior among the rusa, Auffenberg wrote that "the evolution of body size in ambush predators is strongly influenced by this pattern of prey distribution." The herding pattern might allow a big ambush predator to feed itself efficiently with deer, whereas solitary deer scattered evenly throughout the habitat might not sufficiently reward the predator's patience. That sounds logical, yes? Whether it accounts for the evolution of gigantism among *V. komodoensis* is another matter.

Having offered the herding-pattern notion, Auffenberg undercut it by noting that rusa deer were probably not present on the island until recent millennia. Neither were water buffalo or, most likely, wild boar. The fossil evidence suggests that deer, buffalo, and boar were introduced onto Komodo by humans just a few thousand years ago. That was too late to account for komodo evolution.

Then what does account for their having evolved as flesh-eating giants? If *V. komodoensis* lived on the island for thousands of years before their current prey were imported, what did they eat? Auffenberg's answer, behind its dry wording, is wonderfully lurid: "Thus the evolution of large size in the Komodo monitor—due perhaps completely to the energetic benefits derived from feeding on large prey—was dependent on the available prey found in this area at that time, i.e., two species of miniature elephants."

43

BY NINE the next morning, the ferry has anchored and Kamp Komodo (as I can't help but think of it) is dizzy with tourists, a sudden crowd of Americans, Brits, Australians, Canadians, Dutch, and Japanese, dozens of giddy people manifesting a wide range of differences in age, economic status, cultural boorishness, and depth of tan. There may even be a few Indonesian travelers among them—I mean a few besides Nyoman, who appears so augustly official in the souvenir hat that I've bought him, with a TAMAN NASIONAL KOMODO patch on the brow, that white folks are asking him for directions. We're all here in expectation of seeing giant lizards gobble a sacrificial goat.

The crowd is called to assembly and the dour ranger addresses us, in English, from the steps of his little office. He supplies a few fundamental facts for those who got off the boat without Walter Auffenberg in their luggage: the komodo's scientific name, its diet, its status as the world's biggest lizard. The total population of *Varanus komodoensis*, he tells us, is not great. In the wild are living no more than five thousand, perhaps. In captivity, a few. Only to Indonesia, he explains proudly, is the species being native. A recent field census, conducted by researchers of his department, says that 2,571 komodos live here on Komodo itself. On the island of Rinca, another 795. The small island of Gilimotang is containing 100. And in western Flores has been established a wildlife refuge, Wae Wuul, in which are found 129. The ranger is a conscientious man and his numbers are precise. If a few young komodos have hatched within the past three or four years, if a few old ones have died, he hasn't heard about it.

The species is under protection by Indonesian law, he states. Everything of the park is under protection. Do not pick up the coral. "We are keeping this island for children to enjoy," he adds feelingly, and I'm glad to hear it. The cost of the goat, he says, is such that an equal share for each of you, seventy people, will be an amount of 750 rupiahs. Another shameful bargain, I think. You will please pay the man beside the trail, says the dour ranger. To the feeding area is the distance of two kilometers. You are wearing comfortable shoes? Stay together. Good. Now we go.

I've begun to like this guy for his earnestness, but still I wish he would lighten up. Feeding dead goats to Komodo dragons for the edification of sun-addled tourists could be construed, after all, as a sick sort of fun. There are worse jobs in the world than his.

The feeding area consists of a chainlink corral overlooking a bare dirt gully. At first glimpse I misconstrue the purpose of the corral: I assume the komodos will be baited into it and that we tourists will gawk at them from the perimeter. But the rangers herd *us* into it. That's nicely symbolic of who belongs to the landscape, who doesn't, but of course the practical rationale is safety. As we gather inside the fence, the komodos begin emerging from their secret retreats. They have been drawn by the goat smell, I suppose, or by the general human hubbub, or maybe by their conditioned sense of the calendar: *Hey, it's Sunday! Let's go down to the corral for some goat.*

One animal comes shuffling out of the shrubbery, then another. Then, by God, a whole gang of them. The largest are nine feet long and must weigh about 200 pounds. They walk high on their legs with a steady, surging stride. They surround the corral. Expectant but calm, they pose obligingly for everyone's camera. They nose up to the fence like puppies. This calmness, combined with their pleasant, old-boot faces, is misleading; they appear amiable and harmless. Their tongues, which are bile yellow, flap out languidly and then withdraw, tasting the air for aromas. Then the goat carcass suddenly flies overhead, heaved out of the corral by a pair of rangers. It lands with a meaty wallop in the gully. Instantly these lummoxy reptiles become fast and scary. They pile onto the carcass like NFL linemen attacking a fumble.

They snarf and chomp. They gorge. They thrash, they scuffle, they tug and twist. They stir up one helluva ruckus. Within a few seconds they have composed themselves into a neat radial pattern with the goat now invisible at its axis; elbow to elbow, jaws locked on the meat, tails swinging, they resemble a monstrous nine-pointed starfish. Their round-snouted faces, which looked as gentle and dim as a basset hound's until just a moment ago, have gone smeary with blood. When the goat rips in half, they split into two mobs over the severed halves and the tussling continues. They have each seized a mouthful but the mouthfuls are still held together, barely, by bone and sinew. They wrestle. They lunge for new jaw-grips and clamp down, straining greedily against the tensile limits of mangled goat.

It all happens fast. The lucky ones snatch away big bites of meat, swallow hastily, and dive back for more. They climb over each other for second helpings. Their teeth are terrible little knives, thin and serrated, perfectly suited for slicing out great whonks of meat, yet despite the wild scramble they manage somehow to avoid mutilating one another. They compete madly but they don't fight. They ignore the five dozen Nikons and Minoltas that crackle above them like Chinese firecrackers. They polish off the goat—flesh and offal, skull and backbone, hide and hooves—as thoroughly as if it were a hamburger. Only about ten minutes pass, maybe fifteen, until two of the more tenacious animals are scuffling over a last slimy bone. The others splay out onto their bellies, relaxing on the cool dirt of the gully in patches of shade. They aren't sated, I'm sure, but they seem content for the moment with their little snack. The calm has returned as abruptly as it left.

This staged feeding, however artificial, has been no disappointment to anyone's expectation of garish drama. At the same time, it has revealed certain significant aspects of the anatomy, physiology, and behavior of *V. komodoensis*. It has revealed that they are voracious, strong, and fast-moving when they care to be; that their cutting-edge teeth and their flexible jaw structures let them carve off and swallow huge, greedy gobbets; that they share a sharklike sociability when convened over food; that they waste nothing—not bones, not fur— and that their digestive tracts are capable of coping even with hooves. Altogether, the spectacle of the gobbled goat has been an educational experience of the sort that Caligula might have appreciated.

While the komodos rest and digest, the tourists lose interest and drift away, reconfirming still again the truth that *Homo sapiens* in general has the attention span of a cricket. The corral empties. The mob scene is over, the Fujichrome has been shot, and now I've got the overlook to myself. After another few minutes the komodos too wander away, all except one placid individual who continues basking. Finally a pair of rangers approach to pry my hands off the chainlink fence. They inform me in polite Indonesian that the show has ended. They want to finish their morning's work by escorting me back down the trail. They carry forked poles, like shepherds' crooks, a nice low-tech precaution against any lingering komodo that might lurch out of the brush. A komodo comes at you and you hold it off with the pole, is the idea. But no komodo comes at us during our little walk. These animals are just savvy enough to choose their chances advisedly.

By the time we reach Kamp Komodo, an exodus is in progress. The tourists have had enough of this island and its overgrown lizards; they're streaming back to their ferry. The boat departs. The dour ranger is presumably happy to see it leave, and so am I.

The camp is blessedly deserted. Besides myself and Nyoman, the rangers, and the camp staff, almost no one has stayed. I settle myself at a table in the open-air cafeteria with a glass of dreggy Indonesian coffee, prepared to while away another hour on the scientific literature. Tucked into my Auffenberg is the photocopy of a one-page essay, published in *Nature* a few years ago, that I want to review. It was written by an ecologist named Jared Diamond. The title is "Did Komodo Dragons Evolve to Eat Pygmy Elephants?"

44

THEY DID INDEED evolve on a diet of elephants, Diamond argued. This idea, he acknowledged, had been suggested by Walter Auffenberg.

Diamond's supporting data were paleontological and circumstantial: (1) the fossil remains of the two elephantid species from Pleistocene strata on the island of Flores; (2) the fact that komodos presently occur on Flores as well as Komodo itself; and (3) the absence of fossil remains from any alternative candidates for the role of prey. If the ancestral komodos didn't eat elephants during that earlier epoch, there was no evidence of what they *did* eat. The smaller of the fossil elephantids, *Stegodon sompoensis*, stood only about five feet high and weighed no more than a buffalo. The juveniles would have been even smaller and more vulnerable. Diamond reasoned that ancestral komodos, during their evolutionary progress toward gigantism, may have fed on these pygmy elephantids, possibly victimizing the juveniles in particular. They *must* have, in fact, since there was no other food supply. Deer, boar, buffalo, and humans had not yet arrived.

When exactly did *V. komodoensis* achieve its gigantism? Nobody knows. Had it grown large enough to attack a small elephant during the era when those elephants still lived on Flores? Nobody knows. The fossil record from the two islands—being tantalizingly incomplete, as fossil records generally are—doesn't say much about the stages of evolutionary change in *V. komodoensis*, and Jared Diamond didn't say much about it either. Evidently the komodo ancestors haven't shown themselves in that record, as the elephants have. But Diamond offhandedly mentioned another fact that holds me transfixed: "A clue to the evolution of oras on the Flores island group is the former existence in Australia of an even larger monitor lizard, *Megalania prisca*, which was up to 6 m long and 2,000 kg in weight." That is, a komodoesque dragon roughly ten times as hefty as the present one.

It lived during the Pleistocene. It wasn't a dinosaur, it wasn't a crocodilian; it was a lizard. A predator, presumably terrestrial and quick on its feet, like *V. komodoensis*. What did it eat? Giant kangaroos, giant wombats, probably anything it wanted. In the Kamp Komodo cafeteria I scan that sentence of Diamond's paper again, and pause to consider the conversion from metric. Six meters long, two thousand kilograms? It doesn't take much mental arithmetic to comprehend that *Megalania prisca* was one badass beast.

45

NYOMAN INTERRUPTS me with good news. He has cadged us an invitation to hike over the mountains with a backcountry ranger and explore a remote valley. On that side of the pass there will be no contrived spectacles, no chainlink fences, no tourists, and plenty of komodos. The ranger is a friendly young Floresian named David Hau, and the valley is called Loh Sabita. Within an hour, having stuffed our packs with provisions from the cafeteria—malt biscuits, sardines, and bottled water—we start walking. For reasons of safety we want to reach Loh Sabita before dusk.

A mile up the main trail—not far from the feeding corral—we branch away and strike north on a fainter track. The brush is thick on both sides, good cover for an ambush. It occurs to me that if I were a komodo, this is where I would be, waiting for witless Americans. David Hau doesn't carry a forked pole. But he seems to know what he's doing. When we move from the zone of dense brush into forest, where the understory is sparse and the sight lines are longer, I feel slightly more comfortable. We cross a dry streambed, in the dust of which we see tracks—the sinuous mark of a dragging tail with clawed footprints along each side. The stance is narrow, the stride short.

"Komodo?"

"Ya," David says.

"*Kecil?*"

"Ya, ya."

In the Indonesian language, *kecil* means small. In Komodo specifically, it means too small to bite off your leg. We pause to watch cockatoos. We sample the serikaya fruit, cobbled green globes filled with sweet pulp like vanilla yogurt. We pass a tall earthen hump that looks like a giant anthill. It's a nest built by a mound-builder, a species of big-footed ground bird that erects such a great heap of dirt and compost to incubate its eggs. The nest is defunct, dug open by some egg-eating predator.

"Komodo?"

"Ya," David says.

"*Kecil?*"

No answer. Maybe he didn't hear me.

We climb out of the forest onto a grassy slope and follow the trail up toward a mountain saddle between the two valleys. It's not a long climb, but the sun is blastingly hot. My shirt is soaked; Nyoman, a city boy, looks ill. At the crest we pause for breath and water, savoring the view back across the Loh Liang drainage and Kamp Komodo down near the beach. A white wooden cross stands here on the pass, propped with a cairn. A plaque says:

IN MEMORY OF
BARON RUDOLF VON REDING, BIBEREGG,
BORN IN SWITZERLAND THE 8 AUGUST 1895
AND DISAPPEARED ON THIS ISLAND
THE 18 JULY 1974
"HE LOVED NATURE THROUGHOUT HIS LIFE."

I've read about the baron. He stopped to rest somewhere hereabouts while his companions hiked on. Two hours later, when the others came back, there was nothing left but a Hasselblad with a broken strap. His disappearance was put down to hungry komodos.

We top over the saddle and descend through savanna into the valley called Loh Sabita, where the deer are not tame, the water is not bottled, goat carcasses don't fall from the sky, and the komodos still live by their skill as hunters.

The ranger post at Loh Sabita consists of two thatched-roof cabins on stilts, an outdoor kitchen, a rough wooden table, and a kerosene lamp. Nearby is a spring. In front of the cabins, a broad estuarine

mudflat stretches off toward a fringe of mangroves. We arrive just before dusk, with the distant tree line beginning to go dark and the
cockatoos looking grayish pink as they move toward their nocturnal
roosts. David's three colleagues—Ismail, Johannes, and an avuncular
fellow named Dominikus—greet us warmly and insist on sharing
their dinner. Dried fish and rice have never tasted better.

Then Dominikus serves tea, a mosquito coil is lit, and they all roll
clove cigarettes and sit back to talk. I follow as best I can with my
forty-word Indonesian vocabulary. It emerges that a friend of Ismail's,
back in his home village, was attacked just two weeks ago by a komodo. Yes, the man survived. He's in a hospital on Flores, Ismail says.

Such attacks are uncommon, but they happen. Auffenberg mentioned a handful, including the case of a fourteen-year-old boy who
encountered an especially bad-tempered komodo in the forest. Auffenberg got his account from the boy's father. Trying to run away, the
boy had tangled himself in a vine. "The vine stopped the youngster
for just a moment, and the ora bit him very severely in the buttocks,
tearing away much flesh. Bleeding was profuse and the young man
apparently bled to death in less than one-half hour." If the immediate
attack isn't fatal, the aftereffects may be. Some victims succumb to
massive infections of pathogenic bacteria, which komodos carry in
their mouths like poisonous halitosis.

David tells of another incident. This one occurred seven years ago
at a village called Pasarpanjang on the island of Rinca. As a family finished eating their midday meal, a six-year-old child sprang up from
the table and ran down the outside steps. A komodo, having skulked
into the village, was lurking beneath those steps. It stopped the child
somehow, maybe with a swat of its tail, and then it pounced. It had
the child half-swallowed by the time help arrived. The whole village
turned out. They pried open the animal's jaws, got the child free,
killed the komodo. But the child, David says, was already dead.

The after-dinner conversation eventually dies out, and I retire to
my sleeping mat with a belly full of rice and a head full of menacing
komodos. After a halcyon night's rest, the next day I hear still another
story in the same vein. Some fishing people have beached nearby to
take on fresh water from the rangers' spring, and in thanks for the wa-
tering privilege, the women are fixing us lunch. One of those women
has a komodo-attack story—but she's too shy to tell it to an American
journalist. Her name is Saugi. She wears an orange sarong and a self-
conscious grin. While Saugi cleans and fries fish, hiding herself back

by the fire, gentle Dominikus teases the narrative out of her, to be translated in fragments by Nyoman. It all happened to Saugi's mother.

Her mother was cutting thatch, Nyoman reports. "Suddenly, the komodo come from the hill." For a fateful instant Saugi's mother thought that the komodo was attacking a puppy, which had been playing nearby. "But dog is too quick," Nyoman says. "Komodo is get there, and maybe angry or disappoint." Switching targets abruptly, it struck the woman. It clamped its jaws onto her arm and held. She struggled. "Is like a dance," Nyoman tells me, adding his own Bali-flavored simile to the raw tale. In a far corner of the kitchen, Saugi works vehemently on a fish. She mumbles. "The mouth of the komodo is already bite, and stop, and just stay there," Nyoman says. Saugi's mother pushed a sarong over the animal's eyes, which must have seemed horribly near and intimate, as she did her best to wrench free. The jaws wouldn't unlock. She tried to pull herself up into a tree. "And she is, swoosh, like this"—Nyoman sweeps back his arm, as though pulling a rabbit from a hat, and glares at me wide-eyed and questioning. *Why? Why must we relive this horrible story? Why must we torment this woman?* is what his sensitive Balinese heart wants to know. "And the meat of mother is already in mouth of komodo." Half the flesh of her arm had been stripped away.

But Saugi's mother was luckier than some others. She survived. She spent a month in the hospital and managed to keep her arm; she even recovered the use of it, more or less. As Nyoman recites this coda, Saugi goes silent. The cold-minded stranger with the notebook, namely me, gapes demandingly at his translator. More. What else does she say?

"Now is still can see . . . ," says Nyoman, hesitating. "Still can see . . . how you say? Spots? If you cut, and can see later?"

Scars, I say. Saugi finishes her task, washes her hands, wraps an end of her sarong onto her head like a turban, and stalks away.

46

LATER IN THE DAY David Hau and I hike out into the habitat. We follow a set of komodo tracks along a dry streambed through the forest. The streambed leads up toward the head of the valley. We come to a wall of lava, the vertical face of a volcanic bluff rising out over the treetops. At the bluff's base David shows me several small caves where komodos have denned. He shows me a pile of komodo dung, its grayish white color indicating a high concentration of digested bone. Other scats nearby suggest that this is a heavily occupied piece of terrain. We circle around the bluff and emerge from the forest into sunlit savanna, then climb to a flat spot above. Here we find bits of femur or humerus, dry as tinder and light as balsa, half-crushed by mastication.

"Deer. Komodo is here one time," David says, lifting a bone fragment. "He can eat all."

From this quiet moment, events lurch forward rapidly.

As I inspect the bones, David glasses the opposite slope of the valley with my binoculars. "Ah, komodo!" he chirps. "Komodo!" With his guidance, yes, I can barely spot it: a dark elongated form, half a mile away, on a light patch of dirt. It doesn't budge. It's basking. Or else I've trained the binoculars on a komodo-shaped log. All right, I console myself, that's better than nothing: a backcountry komodo seen at long range. I can't expect the same artificial immediacy as produced by a sacrificial goat at the corral, can I? Not here, not in Loh Sabita, where *V. komodoensis* lives under natural conditions, no. We climb off the rock and ascend toward another large bluff.

We notice a commotion just ahead in the brush. Then a very large komodo breaks into view, spooked by our trespass, and scrambles straight up the vertical face of the bluff, like an alligator scaling a four-story building. Lumps of rock crumble and fall. My jaw drops like the lid of a dumpster.

We watch the komodo go. Exactly how big is it? I couldn't say. If not for the likelihood that I've been deluded by its aura of power and strength, I would tell you twelve feet, maybe fourteen, roughly four hundred pounds. But I won't tell you that, because nine feet and two hundred pounds is about the maximum that Walter Auffenberg will allow me. Nine feet and two hundred, then—but for God's sake the beast is *galloping up a cliff!*

I raise my binoculars in time to see the komodo take the summit. It pauses there, a giant reptilian silhouette against the bright sky. Then it tops over, disappearing beyond view. At that moment I hear David scream as another big komodo charges out of its hiding place just behind us.

Yaaaggh. We whirl, caught flat.

But this animal makes a split-second decision against cutting us off at the knees. Maybe it isn't hungry. Maybe we smell unappetizing. It has gotten exactly the opportunity for which all komodos supposedly yearn—ambush advantage against walking meat—but for some reason it has chosen not to attack. Instead of snatching a mouthful of my buttocks or biting away David's left calf, it has peeled a sharp turn and set off downhill, moving as discreetly as a rhinoceros. We give chase at precisely the right pace to assure that we won't catch up. We see it plunge into a gully, and from there we let it go.

When our giddiness has subsided and our breathing is normal again, we begin working our way off the grassy hillside. At the head of another dry streambed, near the forest edge, David stops. He picks up a bone. This one is no fragment of balsa. It's heavy and fresh, tacky with dried blood and saliva.

"Too late," he says. "Late one day. Eating in the night." It's a recent kill, David means, and the komodo has only just finished eating.

Then again, maybe the komodo *hasn't* finished eating. A strong, sweet, bad smell floats in the afternoon air.

Twenty steps on, we find the severed head and neck of a rusa deer, a three-point stag, attended by a million hysterical flies. It has been torn away from itself at the withers, the upper rib bones left dangling from a stub of spine. The stag's eyes are sticky and black. Its body is gone. It looks like it was hit by a train. No, correct that. It looks like it was hit by a Pleistocene monster named *Megalania prisca*.

I recall what David has just said about the komodo: *He can eat all.* Elephants, goat hooves, buffalo bones, humans, and certainly a morsel like this. Chances are good, then, that the komodo will be back. And if not the same one that made the original kill, then others that catch the scent. Leaving the fly-blown head to its various claimants, we clear out.

47

UNDER PECULIAR conditions, evolution decrees peculiar specifications for survival. Islands provide peculiar conditions. Size change among insular species is one signal of this fact. There are other signals too, but size change is conspicuous in a way that the others aren't. Size change is almost linear, proceeding largely in two directions: toward gigantism, toward dwarfism. That might sound simple. In fact, the subject is complicated, and it becomes more complicated the longer scientists study it.

The observational data can be sorted into two sets. These critters get bigger; those critters get smaller. Elephants, we know, are among those critters. Hippopotamuses too are inclined toward smallness—there's no case on record of a giant insular hippo, but there are more than a few species of dwarfs. Rabbits and deer and pigs and foxes and raccoons also tend toward dwarfism on islands. Snakes tend toward dwarfism, with some interesting exceptions. Rodents, on the other hand, like iguanas and tortoises and geckos and skinks and monitor lizards, tend toward gigantism. Birds are often larger on islands than elsewhere, though in some cases they're smaller. Notwithstanding all the giant flightless birds that have trodden the forests and plains of islands throughout the world (the moas of New Zealand, the emus of Australia, the cassowaries of Australia and New Guinea, the Madagascan *Aepyornis*, and of course the dodo), insular ducks tend to be smaller than mainland ducks. Evolution on islands has never, it seems, produced a giant flightless duck. Why not? A question to lie awake over.

Insects are noticeably inclined toward gigantism. Australia has its giant cockroach. New Guinea has walking sticks as big as a Havana cigar. That giant Saint Helena earwig, which I so love to invoke, was about three inches long. But insects too, like birds, are in some instances miniaturized. The sphinx moth of the Galápagos is smaller than its mainland relatives.

Midgets and giants, behemoths and runts—it's all a confusing welter of upscaling and downscaling. Emblematic of the welter is that Pleistocene elephantid on the northern Channel Islands, oxymoronic in itself: *Mammuthus exilis*, the dwarf mammoth. Can anyone make sense of these data?

J. Bristol Foster is one biologist who tried, back in 1964, and his work is still treated as a minor landmark. Foster limited himself to explaining the data for mammals. He published a short journal paper with one modest chart, in which he summarized his own survey of size differences between insular mammals and their mainland ancestors in roughly a hundred cases. Foster's survey suggested that rodents on islands are more often gigantized than dwarfed; that carnivores show the opposite trend, more often dwarfed than gigantized; and that artiodactyls (even-toed ungulates, including hippos and deer and pigs) are usually dwarfed. The generalized pattern was that big mammals tend to get smaller, under conditions of insular evolution, while small mammals tend to get bigger. His generalization was sound enough and intriguing enough that other biologists labeled it the "island rule." But since island biology is full of patterns and codifiable trends, any one of which might be considered an island rule, we'll do better to think of this one as "Foster's rule."

More significant than the rule itself was the explanation of what factors enforced it. For size change occurring in two directions—toward bigness and toward smallness—Foster proposed two distinct causes. Small mammals such as rodents might evolve toward larger size, he thought, if they found themselves in circumstances of reduced predation and reduced interspecific competition. With fewer predators trying to eat it, and fewer competitors drawing on the available food and shelter, an insular species of rodent can afford to be larger. The larger rodent might store fat better, store water better, survive better through hungry times, stay warm more efficiently, give birth to larger babies, live longer, and throw its weight around more decisively in competitive interactions with others of its own species. Consequently, the largest individuals born in each generation might succeed best at raising offspring. Gigantism would be the evolutionary result.

For dwarfism, different causes might apply. While rodents become automatically less prolific when they are crowded, Foster wrote, artiodactyls generally don't. The result is that deer, hippos, and other even-toed ungulates "are especially susceptible of reaching the limit of their food supply resulting in malnutrition and stunting of the young." In the absence of intrinsic population control, overpopulation among such a species could result in stunting. Starvation-caused stunting isn't the same as dwarfism, but over many generations it might lead to dwarfism, if smaller individuals achieved better reproductive success than big individuals.

Foster's hypothetical explanation for Foster's rule, then, involved these two discrete mechanisms. Put them together and you have the two-way pattern of change that appeared in Foster's chart. It's simple, it's logical, and it fits the silhouette of the empirical data.

Foster made one other interesting point. "Insular adaptations are likely attained more rapidly and more precisely due to the smallness of the gene pool and the restriction of gene flow to other populations." No doubt he was right about that. Whatever we might say about evolution on the mainlands, we would not call it rapid or, in most cases, precise. Rapidity and precision are part of what give insular evolution its special loony charm.

Foster's own fieldwork focused on size differences between two species of deer mice on the Queen Charlotte Islands, forty miles off the coast of British Columbia. But in a larger sense he was writing about the shape of the world.

48

IN THE PAST thirty years, J. Bristol Foster's little paper has been cited frequently by biologists for whom size change on islands is an important matter. Generally they have mentioned "Foster (1964)" in order to critique it, to refine it, or to use it as a point of departure. Like most rules in ecology and evolutionary biology, Foster's rule was too simple to keep the experts happy for long. It represents a sort of prerevolutionary innocence, standing on the distant side of a major upheaval—the one that broke out in 1967, when MacArthur and Wilson published *The Theory of Island Biogeography*. Things were different after that.

A fresh generation of ecologists and biogeographers came along in the footsteps of MacArthur and Wilson. The newcomers were fiercely intelligent young grad students and assistant professors, dissatisfied with the natural-history tradition in biology and its descriptive approach. These younger scientists were mathematically sophisticated, and they aspired to replace description with quantification. They were eager to push the theoretical side of biology to new levels of statistical rigor and formulaic intricacy. They preferred to answer such questions as Foster had addressed—Why do rodents get bigger on an island, whereas elephants get smaller?—with differential equations, five-parameter conceptual models, and, eventually, Apple computers. Like the physicists of Einstein's generation as they groped toward relativity, and then toward quantum mechanics, and then toward a unified field theory, the new-wave theoretical ecologists believed that arcane mathematical analyses of quantified observations could lead them to deeper truths about nature—in this case, deeper truths about the character of species and communities. Robert MacArthur himself was their first and greatest prophet.

Dying young in 1972, after having produced a small but provocative body of work, MacArthur became the James Dean of theoretical ecology. Nobody will ever know exactly how great he might have been, and so the imaginations of his survivors have waxed generous. In the course of his short scientific career he did leave behind some very concrete, very far-reaching accomplishments. Besides working to inform ecology with mathematics, he had worked to merge mathematical ecology with biogeography, and his last book, written hurriedly as he died, is the imperfect but noble culmination of that effort. It's his scientific cenotaph, titled *Geographical Ecology*, with a subtitle that speaks even more plainly: *Patterns in the Distribution of Species*. Like other great conceptual naturalists such as Wallace and Darwin, MacArthur knew that patterns are what make factual data *signify*. He said so in the book's introduction: "To do science is to search for repeated patterns, not simply to accumulate facts, and to do the science of geographical ecology is to search for patterns of plant and animal life that can be put on a map."

Many leading practitioners of this new approach are MacArthur's former students, or his appreciative former colleagues and collaborators, or collaborators with his former collaborators. A direct or secondhand association with MacArthur is now a form of credential. It's like being able to say you once had a cup of coffee and a political argument in Zurich with an irascible Russian expatriate named V. I. Lenin. But Robert MacArthur, unlike Lenin, has been remembered lovingly as well as reverentially. And because of his charm as a teacher and a co-worker, his own personal genius for applying mathematics to ecology has perhaps been overstated—amplified by affection and nostalgia. Discount the overstatement, and he might or might not be left with a portion of genius. His *methodological* influence on his field of science, though, needs no discounting. He changed the way ecologists think. He changed the way they ask and answer questions. He established a new M.O. For twenty years now, MacArthurites have been doing much of the most interesting work in ecology, evolutionary biology, and biogeography. Among this postrevolutionary generation is one Ted J. Case, who may be as brilliant and demented and insightful as any.

Case studies giant lizards on the islands of the Gulf of California. But don't be misled. The real subject, again, is the shape of the world.

In the late 1970s, during the early phase of his career, Case published a lucid but intricate paper, in the journal *Ecology*, on size change among island animals. It was his own improvement on Foster's explanation of Foster's rule. Case called the paper "A General Explanation for Insular Body Size Trends in Terrestrial Vertebrates," though the generality promised by that title was elaborately qualified with particulars. Running to eighteen pages, it was both long and full. It included a survey of the empirical data, a discussion of various conceptual and mathematical models describing the factors that determine body size, a discussion of how those models might discriminate between mainland and island situations, a section on testing the models' predictions, a section on apparent exceptions to what Case himself had proposed, a section subtitled "A Further Corollary," and, back beyond its three-page list of cited sources, a mathematical appendix for the delectation of the zealous. It was a fine example of what theoretical ecology has come to look like in the postrevolutionary age. Whereas the little paper by J. Bristol Foster had said commonsense things like this:

	Smaller	Same	Larger
Marsupials	0	1	3
Insectivores	4	4	1
Lagomorphs	6	1	1
Rodents	6	3	60
Carnivores	13	1	1
Artiodactyls	9	2	0

the paper by Ted J. Case said things like this:

Model 1:
1A) $dR/dt = rR(1 - R/K) - WRC/(M + R)$
1B) $dC/dt = -dC + WRC/(M + R)$

and this:

Model 3:
3A) $dR/dt = F - WRC/(M + R)$
3B) $dC/dt = sC(1 - C/JR)$

in which R represents the food resources on an island, C represents the species consuming those resources, and dR/dt represents the fact that life is too short for you or for me to be expected to comprehend calculus.

Fortunately for us, Ted J. Case also thinks and writes clearly in English. Yes, he declared in the paper, it is possible to discern a pattern among the size changes that affect island species, and it is possible to discern biological causes behind the pattern. Gigantism of small animals and dwarfism of large animals each occurs for distinct reasons, as Foster suspected, and not because there is anything magically right about being medium-sized. Many kinds of animal are likely to grow larger on islands, yes, except under exceptional circumstances, which instead make them grow smaller. But to the exceptional circumstances there are other exceptions, which might again make them grow larger or, on the other hand, smaller. Case's rule, unlike Foster's, wasn't simple. The preeminent factor from which all else proceeds, he argued, is the net amount of energy that can be gained by an individual of a particular species in a given amount of time.

Measuring that crucial parameter involves asking some questions. How much food is available? How many predators and competitors are interfering while the animal tries to get a meal? How efficiently can the animal gather and consume nutriment? With predators and competitors absent, on an island, the net energy harvest for each animal might be high, and so, yes, the species might evolve toward gigantism. Unless. Unless, said Case, the absence of predators and competitors leads to overpopulation by the particular species, in which case the net energy harvest might be lower than on the mainland, and the particular species might become dwarfed. Unless. Unless, said Case, the particular species happens to be territorial, each individual seizing an area for its own use and defending that food supply against others of its species, in which case the harvest might still be exceptionally good and the species might get bigger. Unless. Unless of course there are inherent physical constraints that make bigness disadvantageous. (A bird that's too large can't fly; a gecko that's too large can't climb vertical surfaces with its suction-cup toes; a burrowing rodent that's too large can't dig a reliable burrow, since a bigger-diameter burrow is more prone to collapse.) Or unless the species has customarily suffered, on the mainland, from the sort of predator that focuses its attacks on larger individuals. In which case, having escaped from that constraint against largeness, the insular population might be freed to evolve toward gigantism. Unless for some other reason it wasn't, it didn't, or it couldn't. And there are further exceptions to the exceptions. For instance, the rattlesnakes of Angel de la Guarda.

Angel de la Guarda is an island in the Gulf of California, about a third of the way south along the east coast of the Baja peninsula. It supports two species of rattlesnake that occur also in mainland Mexico. As found on the mainland, *Crotalus ruber* is about twice as big as *Crotalus mitchelli*. On the island, their relationship is reversed: *Crotalus mitchelli* has gotten larger, *Crotalus ruber* has gotten smaller, to the point where the "small" species is twice as big as the "big" species. Now why should that happen? Case hypothesized that *C. mitchelli*, small on the mainland, had colonized Angel de la Guarda before *C. ruber* did. The chronological order may have been a mere accident, but accidents have consequences. By arriving early, *C. mitchelli* could have gotten such a head start toward gigantism that *C. ruber*, when it followed, was obliged to take the opposite path, evolving toward dwarfism. In other words, with the big-rattlesnake niche already occupied by *C. mitchelli*, *C. ruber* had to settle for the small-rattlesnake niche.

Case's paper, "A General Explanation for Insular Body Size Trends in Terrestrial Vertebrates," is a formidable piece of scientific synthesis. Anyone reading it might take the author to be a serious, methodical, highly cerebral fellow. That impression is largely true but, like other generalities, it has exceptions.

49

JUST A DOZEN miles beyond the Mexican border, towing Ted Case's boat on an iffy trailer, we pull off the highway for margaritas and lobster. Tijuana has barely disappeared in the rearview mirror. But it's a hot afternoon in May, Dr. Case has only just made his escape from his university office and his university persona at UC–San Diego, and the start of a new fieldwork season needs to be toasted festively. The margaritas at this place are great, he says. The lobsters are small but cheap, you'll want at least two, he advises. His boat is a new one, a seaworthy twenty-three-foot panga of a design developed for Mexican commercial fishermen, and he has affectionately christened it *MacArthur*. The previous boat disappeared last year in an embarrassing mishap, leaving Ted and a grad student named Ray Radtkey stranded for days on Angel de la Guarda. Things happen when a person does fieldwork in remote places, and Ted is philosophical about the costs and the risks; but he has vowed this year to be better equipped. After multiple margaritas and multiple lobsters we're on our way again, but the trailer breaks a leaf spring in Ensenada.

So we lose a night and most of another day, roaming the streets of Ensenada in search of a blacksmith who does leaf springs. With twenty seasons of field experience on the Baja peninsula and the islands of the Gulf of California, Case has good Spanish and just the right demeanor of easygoing persuasion. We consult Tito, foreman at the local boatworks, and his man Togo, Welder of Trailers. Tito and Togo are sorry, so profoundly sorry, but it is impossible to be of assistance for they are quite busy. Down another street we locate Temo the Spring Man, who is occupied pounding a U-bolt into shape on an anvil, having softened the length of straight bolt to a glowing orange with heat from a simple charcoal forge. Mesmerized by this process, I watch Temo work. Who would have thought you could do it so perfectly with a hammer? The iron age isn't dead. My own Spanish being nearly nonexistent, I'm a passive and resigned spectator to the trailer-repair quest, determined to amuse myself as best I can. Temo cannot help us, alas no, not until later this afternoon, he also is burdened with commitments, but perhaps his brother. His brother is to be found at a different smithy, not far away, down this street and then another, an alley, a sign, et cetera. *No es difícil*, supposedly. Ted's memory captures Temo's directions, somehow, and we follow them. "Broken Trailer in Mexico," by Franz Kafka, I'm thinking. El Hermano de Temo is miraculously obliging, however, and to my very great surprise a trailer spring is crafted from scraps of old metal within two hours. The new field season for island biogeographers will not, after all, be spent in downtown Ensenada. We roll.

By dawn the next morning, our expedition has reached its base-camp, a motel in the village of Bahia de los Angeles, on the east coast of Baja. Angel de la Guarda looms offshore, twenty miles out, beyond a handful of smaller islands.

Like the smaller islands, Angel de la Guarda is a lump of baked rock supporting a meager coverage of Sonoran desert vegetation, a few populations of reptiles, even fewer of birds, and no permanent human presence. It became isolated as a result of sea-floor spreading, which broke it loose from the east coast of Baja about a million years ago. Sliding away from the peninsula on its tectonic solo journey, it must have taken some plants and animals along with it. Other species have presumably arrived by cross-water dispersal in the millennia since. The fact that Angel de la Guarda is such a severely inhospitable place—offering no fresh water, fierce temperatures, scarcely any shade, and virtually nothing you would call soil—has protected it against human settlement. Though it's only three hundred miles by direct line from San Diego, it may be one of the more primitive and pristine islands on Earth. A million years of isolation, with only a little intermittent human tampering in the past century, has made it an excellent place to study island processes, especially for anyone who happens to like reptiles and hot sun. Ted Case loves both.

The main scientific focus of Case's annual expeditions to Angel de la Guarda is a long-term demographic and ecological study of the giant chuckwalla, *Sauromalus hispidus*, a homely reptile that weighs about two pounds and roughly resembles an iguana. As I learn more about his background and his disposition, though, I begin to suspect that Case returns here year after year in part for the sheer joy of sweating profusely and getting a good tan.

"This is Angel," he says, placing a tortilla chip on an imaginary map. By now we're seated at an outdoor table in the best restaurant in Bahia de los Angeles, a humble establishment of beans and rice and cold beer, and Case is briefing us on the geography of the Gulf of California. He's using freshly fried chips as tokens, and Angel de la Guarda is represented by a large one with an elegant curl. "Here's Esteban," he says. Another island, another chip. "And these are the Lorenzos"—two other islands on which he has also done fieldwork—"and Partida." We stare down at five insular bits of crispy tortilla, awash in a sea of tabletop. The bottles of Pacifico beer could be a fleet of fishing boats. "The problem is," says Case, "all these islands create crazy patterns of tide and wind moving between them. It can get very squirrelly."

"Are you done with this one?" asks a large, hungry grad student who is along on this expedition to help with the fieldwork. Yes, Case has finished with the Angel de la Guarda chip, so the student dips it in salsa and eats it.

Squirrelly tides and winds are an occupational hazard for island biogeographers in the Gulf of California, as are some other forms of bad weather and stormy luck. Although the air temperature tends to be warm and the sky tends to be blue and the Gulf itself is narrow and long, travel here by small boat can still be perilous. Case has been caught in high choppy seas that made the sixty-mile crossing to San Esteban into a seven-hour ordeal out of Joseph Conrad. He has broken suddenly through fog and nearly rammed the cliff at Partida. He has been afloat for three days at a stretch, unable to land safely on any of those rocky islands, trapped on the boat, using a bucket for his latrine, running short of gas, putting life jackets on the carboys of drinking water in anticipation of shipwreck, and then finally limping back to Bahia without having captured a single chuckwalla—which for him represents the penultimate indignity. He's a stubborn researcher, a field-tested professional, and herpetological data constitute the currency of his profession. After that three-day nightmare on the verge of shipwreck, he spent a day or two resting in Bahia, took a shower, drank a Pacifico, declared "I'm not going home till I get the data," and sailed back out, much like Alfred Russel Wallace after the *Helen*. Of course the ultimate indignity, worse even than going home without chuckwalla data, is death.

But Case has dodged the perils so far. He didn't die from his first stingray wound, which was merely unpleasant, and he didn't quite die from his second, though that time his leg swelled up like a sonofabitch and the doctor later warned him about acquired sensitivity to stingray toxin, suggesting emphatically that Case should not find out what would happen in the case of a third. Nor did he die last year, when he and Ray lost the old boat and spent four days marooned on Angel de la Guarda.

They hadn't done anything really foolhardy in order to get themselves marooned. Again it was just the malignly unpredictable Gulf, with its winds and its squalls and its pinball-machine surges of water. He and Ray were at sea off the northeastern end of the island, not far from an infamous spot named Punta Diablo, when a storm came up. They took refuge in a small cove nearby. Case knew the cove because it was the landing point for Cañon de las Palmas, a site of his long-term chuckwalla work. They beached the boat and began dragging supplies up to the shelter of a cave. But the boat blew offshore while their backs were turned, and it drifted away so quickly that even Case, a strong swimmer, couldn't catch it. They were left with almost no food. What they did have were seven days' worth of water, a supply of gin, and two folding aluminum lawn chairs. They sat down to contemplate their chances of survival.

They lived for four days on water, gin, and foraging. They gathered mussels from the cove. They caught and ate a few rattlesnakes. They tried chuckwallas too, but chuckwalla meat turned out to be more disagreeable than starvation. For shelter, they had the cave. For enter-

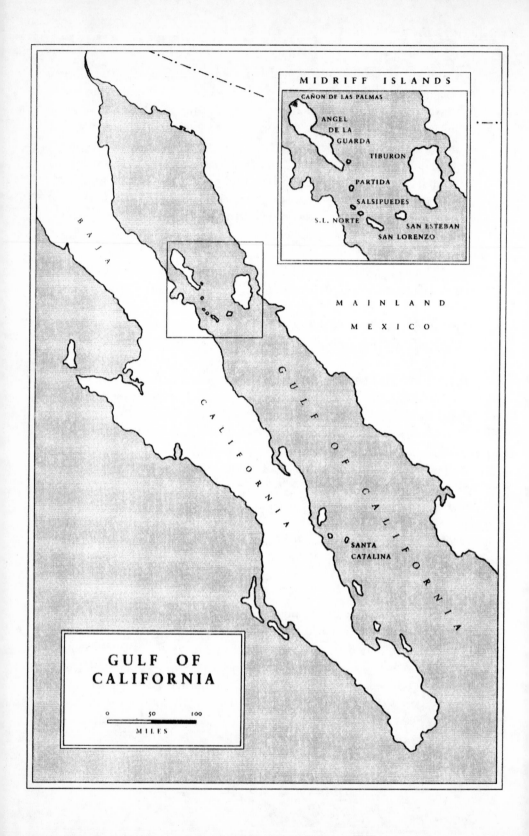

MIDRIFF ISLANDS

CAÑON DE LAS PALMAS

ANGEL
DE LA
GUARDA

TIBURON

PARTIDA

SALSIPUEDES

S.L. NORTE

SAN ESTEBAN
SAN LORENZO

BAJA

MAINLAND

MEXICO

CALIFORNIA

GULF OF CALIFORNIA

SANTA
CATALINA

GULF OF
CALIFORNIA

0   50   100

MILES

tainment, they had two books: Doris Lessing's *Briefing for a Descent into Hell* and something titled *Endurance*, about a disastrous South Pole expedition. Case found the South Pole book rather cheering, on the grounds that he and Ray weren't nearly *that* desperate. Finally they flagged down some fishermen who, unfortunately, were not headed back toward the port of Bahia de los Angeles. Ted and Ray accepted this rescue anyway and rode with the fishermen to another island, slightly less remote, where they were dropped off and had to wait to be rescued again. The second set of rescuers was headed back to Bahia. Case ordered a new boat and, with this one, a large anchor. A grant from the National Geographic Society allowed him to make the purchase.

The new boat should ride better through choppy seas, Case believes. If the design is good enough for shrimp fishing, he figures, it's good enough for island biogeography. At six in the morning, on a day of clear sky and unknowable prospects, Ray and Ted and I are aboard, the gear is stowed, the *MacArthur* is ready. Leaving the other expedition members behind for this foray, we cruise out through the glassy bay toward Angel de la Guarda.

In good weather it isn't far. We reach the southern tip of Angel while the morning sea air is still chilly, and from there we head north along the island's east coast. Suddenly the air becomes warm. Though we're a half-mile offshore, we can feel the heat of Angel de la Guarda as the island's east wall radiates back the stored solar energy of yesterday's sunshine, like a brick oven after the fire has gone out. We pass Punta Diablo with the weather still fair. Along this north-south axis, the island is fifty miles long. Mostly it's bare sterile stone—naked mountains of reddish gray, scrub canyons, cliffs, places unknown even to Ted Case. Three hours out of Bahia, we land at the same cove where Ted and Ray were marooned.

We set up our camp near the cave. Inland, at the mouth of Cañon de las Palmas, is a flat sandy plain covered sparsely with giant cardon cacti, chollas, smoke trees, agaves, elephant trees, palo verdes, and a small stand of the blue fan palms that give the canyon its name. The plain is alluvial, consisting of small-grain sediment washed out of the canyon by runoff from the island's very infrequent rains. Case calls the plain a *bajada*. It slopes gently seaward. There aren't many places like it on Angel de la Guarda; elsewhere the terrain is too steep to support even the most drought-tolerant vegetation. With an ironic laugh Case says, "This is the garden spot." He has surveyed the chuckwalla population here every summer for twelve years.

For the past three years he has seen the place suffer a drought. Pre-
cipitation, always scarce, has been virtually zero. The desert-adapted
animals and plants, though accustomed to extremes of aridity, have
been sorely stressed. Trees are dying. Reptiles are dying. One of the
palms near the mouth of the canyon recently tried to produce fruit,
but failed. Our mission here is to get a sense of the drought's effect on
chuckwallas. During earlier seasons, Case established a survey grid on
the *bajada*, marking the grid points with strips of blue plastic tape tied
to branches. The grid is roughly rectangular, covering seventy-five
acres. As in past years, Ted and Ray go to work on the *bajada* beneath
a searing sun, walking from point to point, counting and measuring
chuckwallas, making notes. As in past years, they find dozens of specimens of the giant species, *Sauromalus hispidus*. What's different this
year is that most of the specimens are dead.

Censusing dead chuckwallas turns out to be much easier than censusing live ones, Case comments. Some types of animal tend to hole
up when they are dying; they hide themselves and croak privately, one
result being that you'll never find a carcass. And the dying animals
that don't hole up—in a mainland situation, at least—are likely to be
killed and devoured by mammalian predators. It's different with the
chuckwallas of Angel de la Guarda, where there are no mammalian
predators. These animals expire in the open, unconcerned about privacy, unconcerned about deathbed predation, leaving themselves to
be posthumously scavenged by ravens. The ravens roost on the tops
of giant cardons. They carry dead chuckwallas back to their roosts,
get a few bites of belly meat, then drop the remains. At the bases of
many cardons on this *bajada*, consequently, are the picked-over carcasses of drought-killed chuckwallas. To a visiting herpetologist it's
grim but convenient.

Some of the dead chuckwallas bear numbers on their backs,
painted there during earlier field seasons by Case. "I just got a recapture on a dead one I first banded twelve years ago," he announces at
one point. Given the condition of these specimens, "recapture" is a
word used loosely. Case can read the history of a recaptured animal
out of his notebook: growth rate, changes in physical condition, minimum longevity. Now with these recaptures he can add year of death.
He can also cite a probable cause of death (or at least an important
contributing circumstance): drought.

He seems to feel affection for *S. hispidus* as a species, though no un-
due sentiment toward individuals. His past interactions with each of

them have been brief—catch a chuckwalla, weigh it, measure it, determine its sex, paint a number on its back, release it. The present travail, this murderous drought, is bad for the chuckwalla population but morbidly fascinating for the herpetologist. It's a natural environmental perturbation that brings a natural set of consequences. Case's job, his opportunity, is to record those consequences. If he worries about anything, he might worry that *S. hispidus* could suffer not just decline but extinction if the drought lingers on. Even that would be an eloquent biological event, well worth having documented. Death in the desert is dry.

My own job on Angel de la Guarda, a fool's errand for the new guy, is to survey the variety and abundance of ants. Case has designed an experiment and asked me to run it while he and Ray chase dead chuckwallas. Fine, I'll be glad to, I've said, not foreseeing the tedium of trying to collect ants on an island so sere as this one.

Following Case's instructions, at various points on the grid I bury small plastic vials—pitfall traps, if you're an ant—up to their rims in the desert gravel. Into the bottom of each vial I squirt a lethal little reservoir of ethanol. The ethanol, besides killing ants that fall into it, will preserve them. Near each vial I plant a marker flag. And then at four compass points around the vial I place tiny piles of what Case in his wisdom has prescribed as an irresistible smorgasbord of ant bait: seeds in one pile, shredded coconut in another, canned catfood in another, and in the fourth pile those little candy sprinkles made for decorating the tops of Christmas cookies. What are they called? Sprackles, shakums, edible sequins, glossy sugar deedeebobs, I don't know. Instead of sprinkling them onto a cookie, I sprinkle them onto Angel de la Guarda. The ants, as it develops, are not amused.

The midday temperature on this *bajada* is 110. Humidity, zero. For days I dutifully walk the grid carrying my little bag of vials, my ethanol, my deedeebob shaker, my seeds and coconut, and my open can of catfood, the last of which stinks about as nicely as you'd imagine. Anything for science. I capture no more than a half-dozen ants. As an insular myrmecologist I am a failure, though Ted Case knows it isn't my fault. Angel de la Guarda is simply impoverished of ants.

In the cool of the late evenings we sit in our folding aluminum lawn chairs, drinking wine and allowing the *jejenes* to feast on us. Jejenes are the local version of tiny biting flies. They sneak up invisibly and then deliver a jolt like a poppy seed shot from a cyclotron, but Case doesn't seem to mind. He sips his wine, swats at jejenes, smiles contentedly beneath his LOS ANGELES ROD AND REEL CLUB cap, and talks. Passingly, he mentions the schism between population biology and molecular biology—which concerns him because he is the incoming chairman of the Biology Department at UC–San Diego, and therefore he will have to function as peacemaker. He makes a few observations about the ecology of the giant chuckwalla. He says something about cycles of hardship and decline among small populations. But mostly the evenings are relaxed, after our wilting though not very strenuous days of fieldwork, and the talk isn't serious. During these hours I see a Ted J. Case who seems rather different from the author of "A General Explanation for Insular Body Size Trends in Terrestrial Vertebrates." This one grew up on the beaches of southern California, a bright kid who misspent his youth on a surfboard and would never regret it, turning himself finally into an eminent scientist within whom still beats the heart of an incorrigible nineteen-year-old.

As a teenager Ted traveled to surf the hot beaches, got very tan, and did predictably poorly in high school—"because I wasn't there half the time," he explains. "I was surfing." Wide grin of satisfaction at that memory. It wasn't a sport, says Dr. Case, it was a subculture. The music was loud, the sun was golden, the women were girls, and things were bitchin'. A zealous historian of that subculture could find photos of young Ted in old copies of *Surfer* magazine. I ask whether he still surfs. I know that swimming is part of his daily regimen, a ten-speed bicycle clutters his office at UCSD, and his midsection doesn't seem to carry more than a hamburger's weight of middle-aged fat. "I don't have the body for it anymore," he says. "Last time I got up on a board was three years ago. It made me feel old. I'm in good shape, I can handle it aerobically. But I don't have the sheer agility. Quickness and balance. It's very frustrating to be out there on a wave, and have your head know exactly what you want to do, and your body just can't do it." After pipelining his way through high school with lousy grades but some success in acting and modeling, he went off to the University of Redlands on a drama scholarship. But at Redlands he was chastened. "*Everyone* at that place knew more than I did about *everything.* It was demoralizing. So I needed something I could excel at. I got interested in biology, and I worked like a demon." By the time he graduated, Case had a National Science Foundation grant and a Woodrow Wilson Fellowship that would take him to grad school wherever he wanted. He chose Stony Brook, on Long Island. A mistake, he says now. The California surfer in him hated the place. After a few months he left, heading back to the West Coast for an alternate graduate program and, eventually, an alternate life. Long Island was decidedly not the kind of island that interested him.

Here on Angel de la Guarda he's in his element. With each passing day he wears less clothing and turns darker brown. If island biogeography involved only Greenland and Spitsbergen and the Falklands, I suspect that Ted Case would be in another line of work.

Each morning we walk the grid. Then lunch, a siesta in the shade of the cave, and in late afternoon we walk it again. Reeking of sun-poached catfood, I follow the little flags of my ant survey from one point to the next, a separate path that leaves me still within earshot of Ted and Ray. The air of the *bajada* is so aridly pure that I can hear either of them clear his throat at two hundred yards. Every hour or so I find an excuse to converge with them because, compared with my tedious task, their tedious task seems electrified with drama and import.

They visit cardon after cardon. They follow the ravens. They also peek under drought-withered bushes and slabs of rock. The number of dead chuckwallas rises steadily, each carcass receiving its statistical obituary in Case's notebook. The number of live chuckwallas captured and released remains drastically low. For every live animal spotted, Ted and Ray find roughly ten carcasses, and the severity of the population crash may be even greater than that ratio suggests, because of another trend showing in the data. Ray tells me, "We're having a demographic catastrophe—all the live ones we're finding are males." His point is that the female mortality rate is especially high, and females of a polygamous species are often more precious, demographically, than males. If the number of females drops too low, the number of males will be irrelevant. And if the females perish totally during this drought, the surviving males will constitute a colony of walking dead, foredoomed to extinction.

The chuckwalla mortality is alarming, but another alarm rings somewhat more stridently. We straggle back into camp one evening to find that Case's anchor chain has failed. His nifty new boat is gone from the spot where we parked it.

It has broken free of its anchor and drifted off—though only to the reach of a long backup line with which Case had tethered it. Some distance offshore, it bobs insolently on the swells. The anchor is lost, five fathoms down, but at least we're not marooned. Ray is relieved. "We'd a been eating rattlesnakes again," he moans. Worse still, this time there's no gin. Case swims out to retrieve the *MacArthur*. Then he free-dives the cove. Finally he surfaces, gasping, with the anchor in his arms and a burst vein in one nostril. We help him out of the water. He plugs his hemorrhaging nose with a wad of paper towel and puts on a red Hawaiian shirt, which won't show the dribbles of blood. And I begin wondering whether Ted J. Case is fated with bad luck, or simply working the forward edge of a hazardous profession. Nothing is dull about this expedition, nothing is uneventful, with the possible exception of my ant survey.

As for ants: All I've established is that (a) hardly any seem to be present on Angel de la Guarda, and (b) among those that are, most prefer catfood to shredded coconut.

There are no further serious mishaps or major revelations. The ravens raid our food supply one afternoon, gorging themselves on two days' worth of bread and lunch meat while we're away, but Case is indulgent toward these birds, since they offer company on a lonely island and help with the gathering of chuckwalla carcasses. The three days on Angel de la Guarda pass quickly. Only the heat makes them feel long and slow. We encounter the gigantized insular form of the small rattlesnake, *Crotalus mitchelli*, and the dwarfed insular form of the big rattlesnake, *Crotalus ruber*, but our luck holds and (despite the anchor chain, despite the ravens) we aren't forced to eat them. Nobody steps on a scorpion. Nobody steps on a stingray. Case's nosebleed remains oddly persistent, coming and going every few hours, so that Ray and I get accustomed to seeing him with a plug of scarlet-splotched paper crammed up one nostril, but the blood loss isn't life-threatening. The boat makes no further attempts to abandon us. In due time, after three mornings and three afternoons of crisscrossing the *bajada*, we're ready to leave.

On the last evening we open our last beers, which we have fished from the blessedly coolish dregs of our last ice, and I raise my can in a toast: to Angel de la Guarda and our successful visit.

Quickly I amend that: "At least, I guess it was successful."

"Yes. It was successful," says Case. His beer salutes mine. "We got the data." The nosebleed has finally been stanched. He wears his blood red Hawaiian shirt, a pair of lemon yellow sweat pants, rubber thongs, a three-day stubble of dark beard, a deep cocoa tan, and his jaunty LOS ANGELES ROD AND REEL CLUB cap. He looks like a disreputable Malaysian saloonkeeper on the morning after New Year's Eve.

"That's the name of the game: getting home with the data," he adds. In one sense, then, the game hasn't changed its name since the era of Darwin and Wallace. Island biogeography can still be a parlous enterprise. "If you take too many risks, you don't get home," says Case. "If you take too few, you don't get the data."

50

THE DATA: Case and Radtkey, with help from the ravens and to a lesser extent from me, have found 108 dead chuckwallas. They have captured, measured, and released just 12 live ones. Among the live chuckwallas, they have seen only a single female. So the ratio this year is heavily weighted toward death and barrenness—heavily enough to surprise even Case, who later says, "The mortality is fantastic out there. Biblical. It's like World War Three." Numbers like these raise the subject of extinction. That's one of the reasons Case finds them profoundly worth tallying.

Sauromalus hispidus, like any other species chosen for study by a thoughtful ecologist, is a synecdoche, a concrete reality that signifies more than itself. The details of its life history are unique, and the statistical chance that it might at any given time go extinct is related to those details, as well as to external events such as a regional drought. But in even its most particularized details can be seen aspects of general patterns. So a few further facts about Case's giant chuckwalla will be relevant to our larger concerns—to the yang of extinction, I mean, as well as to the yin of evolution.

S. hispidus is endemic to Angel de la Guarda and a handful of smaller islands nearby. Besides being three times as large as its closest mainland relative, it differs in several other ways. It tends to live longer but to reproduce less frequently and less abundantly than mainland chuckwallas. The gigantism of the species, made possible by the dearth of competitors and predators, may in turn be a contributing factor to the chuckwalla's endurance and longevity, because gigantism offers advantages for surviving through prolonged episodes of famine and drought. Big-bodied reptiles do better at storing water and food, at hunkering down, at waiting out the bad times. Still, some climatic episodes are too stressful for even the big-bodied and durable. Then chuckwallas die—more than a few. A severe population decline, caused by an unusually long drought, is a setback from which a species with low fecundity can't quickly recover. And since reproductive efforts may be limited to the easy years, when water and food are plentiful, a bad drought is liable not only to kill many individuals but also to keep the survivors from breeding. If the drought continues still longer, or if another form of misfortune (such as a viral disease, an invasion of exotic predators, or a sudden vogue among Baja fishermen for collecting chuckwallas for the international pet trade) compounds the travails of *S. hispidus*, then an episode of "biblical" mortality could turn into a case of extinction.

Next year, or the year after, or by the time this book has reached your hands, the island of Angel de la Guarda might contain no *S. hispidus* except sun-dried and raven-picked carcasses, discarded beneath the cardon cacti. That may sound dire. On islands, such direness is routine.

If this homely chuckwalla disappears from this woebegone place, Ted Case will be the first to know. He could get just as tan on the beach at La Jolla, but he's interested in the shape of the world.

5 1

THE SHAPE of the world is complicated by more than bigness and smallness. Gigantism in particular is sometimes combined with another item from the Insular Menu: loss of dispersal ability. That combination is less characteristic of reptiles, such as *S. hispidus* or the giant tortoises, than of insects and birds.

Loss of dispersal ability is often synonymous with loss of the ability to fly. Reptiles rely on phlegmatic endurance rather than wing power in getting themselves transported across stretches of ocean, and since phlegmatic endurance increases with body size, gigantism can serve to enhance their dispersal ability. This is exemplified by those pioneer tortoises riding the waves toward Aldabra; a tortoise the size of a box turtle would be less likely to make it. But among insects and birds, dependent on flight for dispersal, the correlation is reversed. Larger wings may enhance dispersal—an albatross crosses oceans more easily than a sparrow—but a larger body is a hindrance. The principles of aerodynamics and scale dictate limitations of body size upon any species that would wander the world by air. It's no accident that most insects are small and that most birds are delicate and hollow-boned.

Evolution can challenge those limits, but the challenges tend to oc-

cur in evolutionary arenas where drastic experimentation is not severely and promptly punished. That is, in places where competition
and predation are less intense. That is, on islands.

Once a species of insect or bird has reached a new island and established a population, evolution toward gigantism does offer certain advantages: fat storage, thermal stability, and defense against predators,
if there are any. But gigantism is also a way of becoming flightless,
and flightlessness is a way of becoming marooned. These evolutionary trends have a circular momentum. A gigantic and insular species
of insect or bird may be gigantic because it's stuck on an island, and it
may be stuck on an island because it's gigantic. Whatever the first
cause, loss of dispersal ability is the result.

The giant cockroach of Australia is flightless. The giant crickets of
New Zealand are flightless. The Galápagos cormorant, larger than all
other cormorants, is also the only cormorant that can't fly. The
kakapo of New Zealand, larger than all other parrots, is flightless. On
the island of Tristan da Cunha in the cold South Atlantic lives that
species of flightless moth, *Dimorphinoctua cunhaensis*, bigger than its
mainland relatives. It looks like a size-nine hellgrammite wearing a
pair of size-three wings. The great auk, now extinct, was an overgrown member of the same sea bird family that includes puffins and
murrelets. Native to islands of the North Atlantic, it might have escaped total eradication by humans if not for the fact that, unlike
puffins and murrelets, the great auk was flightless. Cuba had its giant
flightless owl, *Ornimegalonyx oteroi*, also now extinct. Madagascar had
Aepyornis maximus, largest of what were familiarly known as the elephant birds. Flap as it might, an elephant bird couldn't fly, except in
Arabian legends. New Zealand had the moa *Dinornis maximus*, a giant
on the same scale as *Aepyornis maximus*, though skinnier—more like a
giraffe than like an elephant. And besides *Dinornis maximus*, there
were at least twelve other species of New Zealand moas.

This record from New Zealand is especially remarkable—thirteen
or more species of giant flightless birds in a country no bigger than
Colorado. Many of those species survived until the Maori people, arriving from Polynesia, first colonized New Zealand around A.D. 1000.
Despite persecution by the Maoris and later by European settlers, a
few species lived on into the eighteenth century. The remnants of
these vanished birds in the form of subfossils (ancient bones, not yet
petrified) and fossils suggest wide variation on the basic model: enormously tall moas of the genus *Dinornis*, smaller but thick-legged moas

of the genus *Pachyornis*, medium-sized moas of the genus *Euryapteryx*, pygmy moas of the genus *Anomalopteryx*. Why so many? One compendium on the subject of extinct birds explains it by citing two factors: "First, not all described species lived at the same time—forms evolved and died out over millions of years; secondly, numbers and variety were influenced by particular circumstances that applied in New Zealand but in no other land of comparable size and equable climate. Birds adapted to niches normally occupied by mammals simply because there was no competition from warm-blooded, four-legged beasts." New Zealand, remember, was empty of terrestrial mammals during most of its history. So the moa lineage found itself in hospitable conditions for an outbreak of adaptive radiation, parallel to what the tenrecs achieved on Madagascar.

Madagascar itself showed a similar avian radiation. In the millennia before human invasion, it harbored not just *Aepyornis maximus* but a handful of lesser elephant birds.

The moas of New Zealand and the elephant birds of Madagascar might seem obvious exemplars of the pattern of insular gigantism and flightlessness. But in fact they aren't obvious; their phylogeny is really pretty intricate. Both groups belong to an extraordinary category, the ratites. Even amid a gallery of anomalies, these creatures stand out as weird.

52

THE RATITES all share several traits: large size, flightlessness, strong legs for walking or running, and the absence of a keel on the sternum. The sternum is the breastbone, and the keel is where wing muscles normally attach. On a bird with no wing muscles to speak of, there is no keel. So for a bird known only from a handful of subfossils, the sternum can be an eloquent datum, revealing at a glance whether the species was capable of flight.

The living ratites, with their keelless sternums, are all big gawky wonders with useless little wings. Like the extinct ratites, they are mostly restricted to islands. The ostrich is an exception, a ratite that inhabits the African mainland. The rheas, native to South America, are not quite so exceptional, since South America itself was an island until the rise of the Panama land bridge. Other living ratites are the cassowaries of Australia and New Guinea, the emu of Australia, and the kiwis of New Zealand.

One of the intractable riddles about ratites is whether their flight-lessness came as a consequence of gigantism or their gigantism evolved as a compensation for flightlessness. The smallish size of the kiwis provides a false clue toward solving that riddle, because their smallness developed as a late evolutionary modification of the earlier body plan. In other words, a kiwi is a dwarfed version of a giant flight-less bird. Another lingering uncertainty is whether the ratites are a genuine kinship group, reflecting common evolutionary origin, or just an unrelated cluster of species that incidentally share bigness and flightlessness. That question implies still another. Did early ratites ar-rive in New Zealand, Madagascar, and Australia by wing power or did they walk?

If they flew, then the moas and the kiwis and the elephant birds and the emus and the cassowaries all lost their dispersal abilities indepen-dently. Under that scenario, the losses may have happened within just the past five or ten million years. If they walked, on the other hand, they must have done it in the more distant past, before the great southern continent now known as Gondwanaland had split into pieces. The walking-ratite scenario has certain merits over the flying-ratite scenario. For one thing, it permits the parsimonious assump-tion that the ratites inherited flightlessness as a common ancestral condition rather than acquiring it independently by a whole series of coincidences. But the walking-ratite scenario was befuddling for bio-geographers of earlier generations, before continental drift theory had been developed and accepted. How could a giant bird walk from Africa to New Zealand?

Only a few naturalists of the last century came to grips with any of the ratite questions. One was our man Wallace. He guessed pre-sciently, back in 1876, in his book *The Geographical Distribution of An-imals*, that the giant birds of New Zealand, Madagascar, Africa, and South America did derive from a common ancestor. He took them to be "a very ancient type of bird, developed at a time when the more specialized carnivorous mammalia had not come into existence, and preserved only in those areas which were long free from the incur-sions of such dangerous enemies." But he had no explanation for how such flightless giants could have colonized their remote islands. Con-tinental drift was the crucial missing piece.

The idea of giant flightless birds walking from Africa to New

Zealand, during the Mesozoic era, is nowadays a reasonable hypothesis. How could they have gotten there? By cutting across Antarctica, of course. It was just a matter of migrating from one province of Gondwanaland to another. The big continent began breaking apart about 130 million years ago, at which time the earliest birds had already evolved from their dinosaur ancestors. About 90 million years ago, when South America separated from Africa, the early ratites were probably dispersed throughout Gondwanaland, so South America could have taken some with it. Madagascar may have ripped away from eastern Africa at about the same time, and New Zealand from western Antarctica, each of those new islands carrying its own share of ratites. Australia detached from Antarctica somewhat later, drifting northward toward its convergence with New Guinea. This is all speculative, but it represents the informed speculation of credible geologists.

The ratites are both instructive and misleading. They exemplify much about island ecology, but not so much about island evolution as a person might suppose. They tell less about the process of innovation than about the process of survival. The distinction I'm alluding to now is between endemism and relictualism.

An endemic species is one that has arisen evolutionarily in the same place where it's presently found—say, on an island. A relictual species is one that has survived in a given place while disappearing elsewhere. This dichotomy is not applicable only to species; larger taxonomic groups, such as a genus or a family, can be endemic or relictual as well. The ratites, though very old as a group and previously more widespread than now, have continued to generate new species. So the ratite lineage combines aspects of both endemism and relictualism—endemism at the level of species, relictualism at the level of the group. *Aepyornis maximus* was endemic to Madagascar, having evolved from its immediate ancestor on that island. But the ratite lineage in general, with its heritage of gigantism and flightlessness, is relictual on Australia, New Guinea, and New Zealand (and was until recently on Madagascar), having held out in those places while long ago disappearing from the mainlands.

The lineage has also persisted, barely, in Africa. The single species of living ostrich, *Struthio camelus*, has survived there against seemingly bad odds—on a continent populated by lions and jackals and hyenas and other predators, as well as by huge herds of herbivorous mammalian competitors. How has this giant flightless bird avoided extinction? With some difficulty. Even the ostrich family is relictual in its present distribution, having once been represented also in Europe and Asia. *Struthio camelus* survived in Africa by adapting to a fleet-footed way of life on the great grasslands. Its leg bones are drastically different from the leg bones of *Aepyornis maximus*, those of the ostrich being suited for much greater speed. Its swiftness, its kicking ability, and several other characteristics have made the ostrich capable of coping with life on a continent.

Close ancestors of *Aepyornis maximus* probably once lived on the African mainland too. But as the mammalian predators became formidable, as the mammalian herbivores became abundant, as the climate changed and the wet forests gave way to drier and more open grasslands, the ancestral elephant birds couldn't cope. Evolution, which sometimes (but not always) supplies new solutions to new problems, didn't move quickly enough to save them. So in continental Africa the elephant birds went extinct. On Madagascar, a haven from such pressures, they became relictual.

There and in their other havens, the ratites continued to evolve. New species arose, endemic to each place. If they had enemies, those enemies were endurable. So the lineage of huge flightless birds survived, at least until humans arrived, on most of those scattered fragments of Gondwanaland that we now know as the planet's great southern landmasses—Australia, New Zealand, New Guinea, South America, Africa, Madagascar. The only big Gondwanan fragment where ratites didn't last so long is Antarctica, and the reasons for that are presumably climatic.

If the ostrich is an exceptional ratite, the ratites themselves stand as exceptions to virtually every rule except one: the rule that says islands shall show us fresh and extravagant definitions of what's biologically possible. My advice is to savor them for their oddity, then forget them. With the ratites set to one side, flightlessness as a characteristic of insular evolution can be more clearly and broadly understood. Most of the flightless birds and insects that exist on the world's islands did *not* get to those places by walking across Antarctica.

The giant owl of Cuba didn't arrive there by leg power from Gondwanaland. The flightless cormorant didn't reach the Galápagos by wading. The Guam rail wasn't flightless when it colonized Guam. The flightless beetles of the Madeiras are descended from species that made their dispersal to those islands by wing.

We can trust the Madeiran beetles. They are as reliably diagnostic on this point as any group. Why would an island-dwelling species of

Coleoptera lose its wings? Charles Darwin himself found that question interesting, and he incautiously thought he could answer it.

53

THE FLIGHTLESS beetles of the Madeiras play a curious role in *The Origin of Species*. They show Darwin at both his best and his worst. The power of Darwin's intellectual method derived from marshaling many small obscure facts toward a great singular argument, and that's how he used the Madeiran beetles. They were certainly small and obscure, and they seemed to have large implications. Whether they proved what Darwin claimed they proved is another question.

Rising in the eastern Atlantic, four hundred miles west of Morocco, the Madeiras are a pair of tiny islands, Porto Santo and the namesake Madeira, of which the larger and ecologically more rich is the latter. Madeira is a steep hump of volcanic rock that rises to six thousand feet, with a high central crest standing exposed to tropical breezes and storms, and a series of ridges dropping transversely from the crest, like ribs from the backbone of an emaciated horse. This partitioning topography probably helps to account for the fact that such a small island harbors so much species diversity, at least among certain insect groups. Lying as it does in the path of the trade winds, Madeira was one of the last stops for European sailing ships headed westward across the Atlantic. Relatively accessible for an oceanic island, it attracted early attention from scientists. I've already mentioned the visit by Charles Lyell in 1854, and the fact that Lyell noticed a high degree of endemism among Madeiran beetles. Another colleague of Darwin's, a quiet man named T. Vernon Wollaston, did a more thorough study of Madeiran entomology and in 1856 published some of his findings in a book that was dedicated to Darwin.

Wollaston had made the remarkable discovery that 200 beetle species, of the 550 native to Madeira, were flightless. The wings of those 200 species were undersized to the point of uselessness. In the more extreme cases, Wollaston found, they were no more than dinky little nubs. The ratio of flightlessness was higher still at the level of genus. Among the twenty-nine beetle genera endemic to Madeira, twenty-three genera contained only flightless species. Wollaston also noticed that certain large groups of beetles, common on the mainlands but especially dependent upon flight, were absent entirely from Madeira.

His friend Darwin seized on these facts and invoked them in connection with two separate arguments: for evolution by natural selection and for the evolutionary effects of use and disuse. In one of those arguments (use and disuse) Darwin proved to be broadly wrong. In the other (natural selection), it seems he was wrong only narrowly.

Chapter 5 of *The Origin*, devoted to "Laws of Variation," contains Darwin's mistaken assertion: "I think there can be little doubt that use in our domestic animals strengthens and enlarges certain parts, and disuse diminishes them; and that such modifications are inherited." His study of dogs, captive-bred pigeons, and racehorses had convinced him that characteristics acquired in one generation—by laborious training, for example, or by muscle-building exercise—could be passed on biologically to the next generation. And if that were true for domestic animals, why not for beetles on a wind-swept island? Darwin's logic entailed three reasonable premises: that the Madeiran beetles tend toward concealing themselves on the ground, so as to avoid being blown out to sea; that this habit of concealment leads to the atrophy of their wings; and that a female with wings atrophied by disuse would produce progeny with wings stunted by birthright. The shrunken-winged beetles of Madeira, Darwin felt, helped to prove that acquired characteristics can be inherited.

Today we know otherwise. The notion that one individual can acquire some trait during the course of its life, and then pass that trait to its offspring as an inherent biological attribute, contravenes the holiest tenets of modern genetics. If a woman's arm is severed in a car accident, will she give birth to one-armed babies? No. Will the children of a body-builder have bulging biceps? No. Yet if acquired characteristics were heritable, those phenomena would occur. Some contemporary biologists, who take Charles Darwin as their true prophet, refer to this notion as "the Lamarckian heresy"—blaming it on that earlier and somewhat confused evolutionist Jean Baptiste Lamarck. But you can find the same notion in chapter 5 of *The Origin of Species*. If it's heresy, it's also Darwin's heresy against Darwinism.

Still, Darwin didn't claim that disuse and its heritable effects were the sole factor accounting for Wollaston's data. Even more important, he thought, was natural selection.

In a typical mainland situation, natural selection will favor robustness, and one standard of robustness is the proficiency of flight. Assuming that flight is useful toward finding food, mates, and protection from predators, individual beetles that are good flyers will leave more offspring than poor flyers, and the flying ability of a species will gradually improve. But on an island like Madeira, natural selection might instead *punish* the big-winged, strong-flying individuals. How so? Because those strong flyers face a higher probability of being swept out to sea by winds. If they are swept out to sea, their genes disappear from the island's gene pool. Consequently the genes for strong flight will be winnowed away from the insular population. Only the weaker flyers, and the occasional mutants that are totally flightless, will be left on the island to breed. Embracing this hypothesis, along with the more dubious one about acquired characteristics, Darwin concluded that "the wingless condition of so many Madeira beetles is mainly due to the action of natural selection,"—on which point he was right—"but combined probably with disuse"—on which point he was wrong.

Darwin set the notion of disuse to one side and focused more confidently on his main subject, natural selection. He described how it might lead to flightlessness among an insular population of insects: "For during thousands of successive generations each individual beetle which flew least, either from its wings having been ever so little less perfectly developed or from indolent habit, will have had the best chance of surviving from not being blown out to sea; and, on the other hand, those beetles which most readily took to flight will oftenest have been blown to sea and thus have been destroyed." With or without corroborative evidence, coming from Darwin it sounds good enough to be true. Certainly it seems more plausible than the explanation by disuse. But the plausibility of a hypothesis and the high authority of the man offering it don't guarantee its correctness.

The Origin of Species is a book of encyclopedic richness and inexhaustible tendentiousness, a great potpourri of argument and fact in which a reader can find almost anything a reader might want: Lamarckism, animal husbandry, geology, ethology, experimental botany, the kitchen sink, island biogeography. But Darwin didn't answer every question, and every answer he gave wasn't infallible. Evolutionary biologists of the twentieth century have rejected his general idea about the heritable effects of disuse. And one modern researcher, Philip Darlington, has challenged his view of flightlessness among Madeiran beetles.

54

DARLINGTON WAS a museum curator and a taxonomist who specialized in the Carabidae, a family of Coleoptera commonly known as the
ground beetles. His paper on flightlessness was published in 1943, despite what must have been difficult and distracting circumstances—
Darlington served as an army entomologist during World War II,
moving from Australia to New Guinea to the Philippines to Japan.
He may have written this gentle monograph as an escape from
wartime reality. I picture him in a tent, surrounded by jungle, scribbling over a sheaf of smeary pages. "Lieutenant Darlington, what the
ding-dong you doin'?" says his noncom assistant. "We got mosquitoes and lice to kill, sir. We got three hundred GIs to dip." Maybe his
beetle manuscript smelled of DDT and half-rotted boots. He mailed
it to the journal *Ecological Monographs*, in which it eventually appeared
under the title "Carabidae of Mountains and Islands: Data on the
Evolution of Isolated Faunas, and on Atrophy of Wings." Based on
his survey of ground beetles from all over the world, Darlington
begged to disagree with Mr. Darwin.

The Carabidae family encompasses about twenty thousand known
species of mainly unspecialized, mainly predaceous beetles, abundant
on all continents (except Antarctica) and on most islands. Many
species are dark, shiny, and flat. Many live on the ground, hiding
among stones and plant debris. When disturbed, even a normal-
winged carabid is more likely to run than to fly; flight does have its
purposes for the typical carabid, but those purposes don't seem to include routine locomotion and escape. Although the entire carabid lineage has evolved from normal-winged ancestors, species with
nonfunctional wings are now common. One-fifth of the carabid
species in eastern North America, one-seventh of the species in South
America, and almost half of the Australian species are shrunken-
winged. The Carabidae lend themselves well to a study of flightlessness because it's such a recurrent part of their family tradition.

Darlington was careful to focus his analysis on observable wing
anatomy, not on presumed flightlessness, since flightlessness is a negative fact (like innocence, ignorance, virginity, or strict vegetarianism)
that can be hard to prove. Still, in the cases where wings were drastically reduced, flightlessness could safely be deduced. Darlington used
symbols (a plus sign for normal wings, a minus sign for shrunken) to

designate wing morphology, but for our purposes it's as easy to talk about full-winged and shrunken-winged species. His full-winged species presumably flew—regularly, sometimes, or at least on occasions of desperate necessity. His shrunken-winged species were undoubtedly flightless. On the island of Madeira, Darlington figured, nearly two-thirds of all the carabid beetles were shrunken-winged.

On mainlands, he had noticed, there was a similar pattern in the distinction between carabid species of mountaintops and carabids of lowlands. Wing reduction was more common on the mountaintops than in the lowlands. Since mountaintops often hold islandlike patches of habitat, isolated ecologically and swept by winds, it might seem logical to apply Darwin's explanation—that mountaintop beetle populations tend to be flightless because the good flyers have been blown away. But that wasn't correct, according to Darlington. "Darwin's explanation was too simple. The problem is a complex one which cannot be solved by generalities but only by careful analysis of exact data." So he had gathered some exact coleopterous data from the White Mountains of New Hampshire, the Appalachians of North Carolina, the Sierra Nevada de Santa Marta of Colombia, the Blue Mountains of Jamaica, and from Hawaii, Saint Helena, the Seychelles, and Bermuda, among other places. Shrunken-winged, full-winged, shrunken-winged, full-winged, percentages, percentages, percentages. To make sense of the data, Darlington argued, you've got to know a little about carabid ecology.

Some species live almost exclusively on the ground. He called them geophiles. Some species live along stream banks, slow rivers, and ponds. He called them hydrophiles. And there is an arboreal group, living in trees. These three categories helped him refine his perception of the pattern of flightlessness.

On mountains of the mainlands, most carabid species are geophiles, living low to the ground, and most of those mountain-dwelling geophiles are shrunken-winged. The two traits correlate with each other—geophilic, wings shrunken—but they don't correlate with exposure to wind. It turns out that the most mountainous areas are not necessarily the most wind-raked. Many of Darlington's mountain-dwelling carabids lived in dense, stable forests on moist mountain slopes, with good protection from wind. But if the winnowing wind wasn't serving as the mechanism of natural selection, causing flightlessness to predominate among the survivors, then what was?

At least part of the answer seemed to be that geophilic species,

within stable forests, didn't *need* wings. They lived in dense local concentrations, which eased the problem of finding mates, and since their
food also was localized they had no reason to go wandering widely.
Furthermore, they had few enemies. So why should they fly? To them
flight had become a superfluous enterprise. And because the struggle
for existence, as adjudicated by natural selection, mandates strict
economy of resources, the energy resources that had previously been
allocated to wing development were gradually reallocated toward
other ends. Shrunken-winged carabids may have been able to afford
stronger legs for running and burrowing, say, or more powerful jaws
for seizing prey. Given the lives they led, wings were useless extrava-
gances.

This is different from Darwin's notion of disuse. Disuse alone can't
cause an organ to disappear from the genetic heritage—but natural
selection, judging harshly the misallocation of resources, can.

On islands as on mainland mountaintops, the pattern was more
complicated than Darwin had guessed. Darlington teased out its puzzling components. First, he found that carabid species on islands belonged mainly to the geophile group, the dirt-lovers. Stream-loving
species and tree-loving species, by contrast, were mostly absent. Second, he found that mountainous islands (such as the volcanic
Madeira) tended to support shrunken-winged species, whereas lower
islands (such as coral atolls and cays) tended to support full-winged
species. Third, he found that warm tropical islands supported more
full-winged species than colder islands. Colder islands supported
more shrunken-winged species.

Darlington offered his own explanations for the carabid pattern,
involving such factors as wing-muscle physiology, population density,
and the instability of habitat. Sparing you those details, I'll just note
that Darlington nodded toward the fundamental yin and yang of island biogeography: evolution and extinction playing complementary
roles in determining biogeographical patterns. And he touched on an
important empirical truth I've mentioned earlier: the correlation between limited habitat area and limited species diversity. Darlington's
beetle work was another intellectual advertisement for something
called the *species-area relationship*, which I'll discuss later in the context
of the yang.

Have you forgotten the ratites? Good. Now I invite you to forget
also the geophilic carabids. In all honesty, the notion of a dirt-loving
ground beetle seems blindingly redundant even to me. But remember

the name Philip Darlington, and the fact that in 1943, while quibbling with Charles Darwin's interpretation of flightlessness among insular coleopterans, Darlington wrote: "Limitation of area often limits both number and kind of species of animals in isolated faunas." In other words, small places support fewer species than big places. He knew he was talking about more than the beetles of Madeira.

55

LOSS OF defensive adaptations, another item from the Insular Menu, shares some logical overlap with loss of dispersal ability, insofar as dispersal ability is equivalent to mobility and mobility is one means of defense. A flightless beetle, besides being hampered from dispersing across stretches of ocean, is hampered also in escaping danger. A flightless bird, likewise. But flightlessness isn't the only form of secondarily acquired defenselessness; others are loss of protective coloration, loss of warning mechanisms, prolonged infancy, and loss of wariness. The last of those is what makes some island species seem remarkably trusting in the presence of humans.

Loss of protective coloration shows up gaudily in the lizard *Uta clarionensis*, carrying bright blue markings that make it perilously conspicuous against the dark lava of its native habitat, on Clarion Island off the west coast of Mexico. Loss of a warning mechanism has occurred in *Crotalus catalinensis*, the rattleless rattlesnake of Santa Catalina. Prolonged infancy is seen in New Zealand's kiwis, whose eggs require about seventy-five days of incubation before hatching. Loss of wariness is sometimes manifest as ingenuous nesting behavior: In the Galápagos, the blue-footed booby puts its eggs onto a bare patch of ground, unprotected, unconcealed, not even cushioned by a cradle of vegetation. Another form of ingenuous nesting involves building a nest in plain view on a tree limb, where it can easily be raided by a climbing predator. The Mariana crow practices that sort of reckless behavior on the island of Guam. A more cautious bird might at least conceal the nest, or place it beyond reach at the end of a thin branch, or suspend it in an elaborate woven pouch, as the tropical oropendolas do. But oropendolas are mainland species, surrounded by predators and obliged to be more cautious. Boobies can be boobies. As for the generalized reduction of wariness—it turns up,

misleadingly called "tameness," among a wide range of insular animals throughout the world.

The Falkland Islands once supported an endemic species of fox,
but the Falkland fox was "incurably unsuspicious of man," according
to one authority, and in lieu of being cured it was extinguished.
Aldabra at the start of this century harbored not just giant tortoises
but also an endemic species of ibis that was "tame, stupid, and extremely inquisitive," and it may or may not still exist. The Galápagos
hawk also gives the impression of being tame.

The Galápagos boobies seem tame. The Galápagos tortoises seem
tame. The herons and warblers of the Galápagos seem tame. The famous Galápagos finches seem tame. In fact, virtually all of the Galápagos fauna seems tame. Anyone who visits the archipelago is bound
to be struck by this phenomenon—the tranquil indifference with
which the native animals tolerate human intrusion—and Charles
Darwin was no exception. He ended the Galápagos section of his
Journal by noting the "extreme tameness" of the birds that he had
seen.

Darwin had found this trait among all the terrestrial avifauna—the
finches, the flycatchers, the one species of dove, the mockingbirds,
and others. "There is not one which will not approach sufficiently
near to be killed with a switch, and sometimes, as I have myself tried,
with a cap or hat." It was drastically unlike the tropical forests of
mainland South America, in which a bird collector couldn't expect to
bring down many specimens by flailing his hat. The Galápagos offered shamefully easy pickings: "A gun is here almost superfluous,"
Darwin wrote, "for with the muzzle of one I pushed a hawk off the
branch of a tree." On another occasion, a mockingbird landed on the
lip of a water pitcher he was holding and proceeded to help itself to a
drink. "I often tried, and very nearly succeeded, in catching these
birds by their legs," he reported. As much as their incautious behavior
surprised him, Darwin believed that the birds might once have been
less cautious still. He cited an earlier traveler, identified only as Cowley, who had described the Galápagos back in 1684. "Turtle-doves
were so tame," Cowley wrote, "that they would often alight upon our
hats and arms, so as that we could take them alive: they not fearing
man, until such time as some of our company did fire at them,
whereby they were rendered more shy." And those that weren't rendered more shy were rendered dead.

"Tameness" is the wrong word for this phenomenon, implying that the animals have been conditioned by experience to feel friendly to-ward humans. The reality is different: Evolution, not individual experience, has leached them of wariness. Those "tame" Galápagos animals are simply unwary, not merely toward humans but toward any potential enemy. The Galápagos hawk that was so vulnerable to Darwin might have been equally vulnerable to a mongoose. The Galápagos mockingbird that trusted Darwin might have been equally trusting toward a tree snake. Why? Because that species of hawk and that species of mockingbird evolved in a fool's paradise. Throughout a million or so years of evolutionary history, their ancestors inhabited an archipelago that was blessedly empty of mongooses, tree snakes, and balding young English naturalists.

We can do better than "tameness." While we're at it, we can do better than "loss of defensive adaptations." My own preferred term for this whole set of traits (including the loss of protective coloration, the loss of warning mechanisms, the prolonged infancy, the ingenuous nesting behavior, and the loss of inherent wariness) is *ecological naïveté*. These animals aren't imbecilic. Evolution has merely prepared them for life in a little world that is simpler and more innocent than the big world.

56

DURING HIS Galápagos stopover, Darwin's attention was captured at one point by the marine iguana, *Amblyrhynchus cristatus*, a singular beast even by island standards. This is the only lizard on Earth that dives in the ocean to feed. It grows large for an iguana—four feet long, more than twenty pounds in weight—with a muscular tail that serves well for swimming. Darwin found it abundant throughout the archipelago, always near the water along rocky coastlines, never so much as ten yards inland. "It is a hideous-looking creature, of a dirty black colour, stupid and sluggish in its movements," he wrote ungenerously. But he admitted that it is graceful when it swims.

He cut open the stomachs of several marine iguanas and found nothing but "minced sea-weed" in thin wisps of bright green and dull red. "I do not recollect having observed this sea-weed in any quantity on the tidal rocks; and I have reason to believe it grows at the bottom of the sea, at some little distance from the coast. If such is the case,

the object of these animals occasionally going out to sea is explained."
Ever the astute naturalist, he was right. Marine iguanas live on a diet
of algae, which they browse from exposed rocks when the tide is low
or dive to when the tide is high. Forty feet, fifty feet down, they can
be found feeding on underwater meadows. Ordinarily a dive might
last just a few minutes, but sometimes it can stretch beyond a half
hour. Bigger individuals are capable of swimming through modest
surf, and occasionally a marine iguana has been seen a half-mile offshore. Nevertheless, Darwin noticed, "there is in this respect one
strange anomaly; namely, that when frightened it will not enter the
water." He didn't get that on hearsay. He cornered an animal,
grabbed it, and threw it into a large tide pool left by the ebbing sea.

Evolutionary biology, as I've said, is not generally an experimental
science. But there are historic exceptions.

This interaction between Darwin and a hapless marine iguana is
reported in the *Journal*. Having been rudely tossed, the iguana swam
straight back to where Darwin stood, like an apache dancer returning
to her partner. Darwin threw it again. Again it returned. "It swam
near the bottom, with a very graceful and rapid movement, and occasionally aided itself over the uneven ground with its feet. As soon as it
arrived near the margin, but still being under water, it either tried to
conceal itself in the tufts of sea-weed, or it entered some crevice. As
soon as it thought the danger was past, it crawled out on the dry
rocks, and shuffled away as quickly as it could." But the hapless iguana
wasn't dealing with some idle yahoo, some sadistic schoolboy with a
short span of attention; it was dealing with Charles Darwin. "I several
times caught this same lizard, by driving it down to a point, and
though possessed of such perfect powers of diving and swimming,
nothing would induce it to enter the water; and as often as I threw it
in, it returned in the manner above described." Despite his occasional
moments of character lapse (especially twenty-three years later, in the
matter of Wallace), Darwin seems generally to have been an admirable, unpretentious man as well as a great scientist, and I take this
iguana encounter as exemplary. He was inexhaustibly curious. He was
observant. He was gentle but firm. He was smart. And he wasn't
afraid to look like a lunatic.

What did he learn? "Perhaps this singular piece of apparent stupidity," he concluded, referring to the iguana's, not his own, "may be accounted for by the circumstance, that this reptile has no enemy
whatever on shore, whereas at sea it must often fall a prey to the nu-

merous sharks. Hence, probably urged by a fixed and hereditary instinct that the shore is its place of safety, whatever the emergency may be, it there takes refuge."

Evolution had provided that single defensive instinct: When threatened, when attacked, when discombobulated, *head for shore*. But evolution hadn't provided the wit to adjust to altered circumstance.

57

THE Charles Darwin Research Station is located today near the south coast of Santa Cruz island, in the Galápagos, just up a dirt road from the settlement at Academy Bay. Academy Bay is a natural harbor formed by ancient episodes of surging and cooling lava. A small international community of biologists uses the Darwin Station as its base. In one building is a library, filled mostly with technical sources pertaining to the Galápagos but including, not illogically, a good collection of published reports on the giant tortoises of Aldabra, that sibling ecosystem on the far side of the planet. Within walking distance of the station is a modest hotel. The shoreline outside the hotel—a stretch of surf-polished lava boulders in hues of chocolate and blood, giving way just above the tide line to a scrubby chartreuse carpet of salt-tolerant vegetation—is rife with iguanas.

They bask on the rocks and the flotsam. They stare out to sea. Their faces are dark and severe, puggish, concealing a basic gentleness. Occasionally they bully each other for position. I sit among them, in a camo cap, meditating on young Charles Darwin and his throwing arm and ecological naïveté among *Amblyrhynchus cristatus*. Because I've been at it for several days, my sunburned ears are as flaky and red as the head of a buzzard.

Darwin was unfair to these creatures, I think. They aren't really so "hideous-looking," and "dirty black" doesn't do justice to their rich coloring. They congregate here on a small patch of coast, for sunning and feeding and flirting, in a variety of sizes and shades: juveniles, dusky adult females that match the lava, big dominant males mottled with orange, olive, turquoise, and crimson. A full-grown male carries a crest of thornlike spines down its head and back; a female's crest is daintier. Generally they are more blunt-nosed than other iguanas, which may represent an anatomical adaptation to grazing thin layers of algae off rocks. Their teeth are designed for clipping and grinding. Their bodies are encrusted with salt. Salty paste leaks from their nostrils, dribbling like snot from a child; they excrete it to unload the salt unavoidably swallowed while feeding. Some of them lie belly-flat, limp as raw sirloin, soaking warmth from the scorching sun and the sun-heated lava. You certainly might call them homely, but hideous is a bit strong.

In mating season the males are territorial. One studly iguana stakes out his territory and threatens any intruding male with iguanid variants of the usual masculine signals. His crest goes erect, his body inflates with air. This is how big I am, he exaggerates. He also puffs out his throat and gapes his mouth, showing the redness inside. He bobs his head up and down. Either the intruder takes these awesome displays of virility as sufficient discouragement or, crazed by his own hormones, he does not. If not, the resident male will march toward the intruder and deliver a butt with the top of his head. Thunk, take that. It's the ultimate insult and challenge. Each male's head is outfitted with a cluster of conical scales, like rivets, adding pointedness to the blow. Maybe the intruder lowers his own head and butts back. Now the two are engaged in manly combat. They butt, they shove, they become exhausted and pause for rest. They butt again. They blast salt out their nostrils, mutually disdainful. This nonsense can go on for as much as five hours, but who are we to judge? It's no crazier than a Tennessee roadhouse on a Saturday night. At least they don't hit each other with pool cues.

Usually the intruding iguana concedes. He crawls off in search of an easier score. Nobody gets hurt.

The victorious male wins the privilege of mating with any and all females who choose to assemble on that patch of territory. Some scientists have called those assemblages harems, implying a forcible control over the females—a control that, in truth, the territory-holding male may not have. It might be as logical to talk about gigolo stations, which the females visit at their discretion.

My presence here coincides with mating season, so the iguanas among whom I sit are a little touchy about their turf and their status. One large male has established his territorial headquarters atop a huge timber, a landmark piece of flotsam, washed ashore from the wreckage of some old wooden ship. Three females surround him on the timber, which is a prime spot for basking, and others loll nearby on the lava. Juveniles come and go, oblivious to the sexual tension. When I first

EASTERN PACIFIC

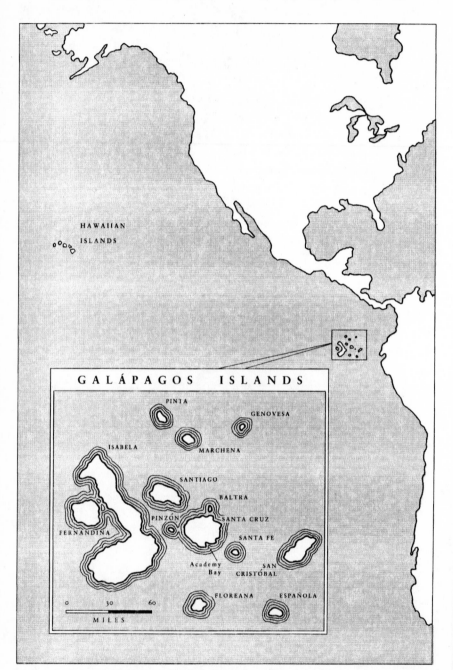

HAWAIIAN
ISLANDS

GALÁPAGOS ISLANDS

PINTA

GENOVESA

ISABELA

MARCHENA

SANTIAGO

BALTRA

PINZÓN

SANTA CRUZ

SANTA FE

FERNANDINA

Academy
Bay

SAN
CRISTÓBAL

FLOREANA

ESPAÑOLA

0   30   60

MILES

approached, moving with soft steps but slightly too close, the male gave me his head-bobbing display. He seemed to be saying: Back off, asshole, these babes are mine. I backed off. He wasn't asking much— eight feet between me and the timber seemed to suit him as adequate privacy. I settled at that distance and now he ignores me, concerning himself with more serious rivals. But during the hour that I watch today, no one mates. No one butts heads. No one goes for a swim in the ocean. So I leave, seizing a chance to visit Karl Angermeyer's more eccentric iguana colony on the opposite side of Academy Bay.

I make the crossing by boat with a guide who knows Karl. It's a re- mote and improbable household, a private refuge, not the sort of place where a stranger should wander in uninvited. The house itself is a sturdy little fortress of lava rock, perched near the edge of a cliff, overlooking the bay. Its roof is covered with live iguanas.

Karl Angermeyer is a barrel-chested man of about seventy, dressed today only in shorts, his torso and legs as dark as scorched butter. He wears a goatee. He welcomes us graciously. He resembles the sort of elderly European roué to be seen on the Spanish Riviera with a young wife. Karl is a painter, I learn. He paints the Galápagos. He has lived here for fifty years and he venerates these islands, he is a curator of their secret beauties, he has memorized whole lagoons, he knows the light. Come, he says, have a look at the studio. His veranda, like his rooftop, I can't help but notice, is populated by iguanas. A drowsy mongrel dog with some Airedale blood also patrols the area, growling at human intruders but evidently conditioned to ignore the iguanas, like a big mean hound that has been raised as a sibling of cats. Several more iguanas hang vertically on the outside walls of the house. Hang- ing vertically on a wall is no small feat for a reptile weighing ten or twelve pounds, but *Amblyrhynchus cristatus* has strong limbs and sharp claws, and lava rock is porous. Come, Karl repeats hospitably.

The studio is a cool room cluttered with canvases, cigar boxes full of paint tubes, palettes, cans and jars, turpentine, rags, matches, gim- cracks and debris of all sorts—everything a painter might need except brushes. Brushes? No, Karl doesn't use them. Doesn't bother. He prefers to paint quickly, and brushes slow him down. He paints with his fingers. Swoops and swirls and dapples, daubs, a touch here, a push there, impetuous thumbprints of blended color. For a fine line, he uses a fingernail. His hands are huge and serviceable, like the hands of a backhoe operator or Andrés Segovia.

I will show you, he says. Having grabbed some half-empty tubes of paint, he creates, in less than two minutes, a miniature landscape on the lid of a cardboard box. Unmistakably it's a Galápagos scene, featuring a lava-rock seacoast at sunset with an arboreal prickly-pear cactus. Then he tosses it aside. He wipes his hands. The studio is wallpapered with similar views of tree cacti, mangrove lagoons, erupting volcanoes, marine iguanas silhouetted against crepuscular pastel light, all of them finger-painted. His oeuvre is coherent, showing especially strong affinity for rosy sunsets and backlit forests of prickly pear. It falls somewhere in the large uncertain zone between Monet and black velvet. The pads of the giant Galápagos prickly pear have a shape that's conveniently rendered by the pad of a fingertip. Some of Karl's paintings are beautiful, some are even delicate. Some are fresh and sincere. The others are impressive mainly on their own terms—as products of twenty or thirty minutes' work, done without brushes. As I browse among them, there comes a thumping noise on the roof.

Mating season, Karl explains. It's the iguanas. The males are being fierce with one another. The season ends February 12, he says with the confidence of a longtime observer, after which they will be mild again. They will get along amiably. Meanwhile they are tearing the roof off the house.

Karl is a forthright fellow, happy to discuss his life's peculiar trajectory with a lover of reptiles and art. It all happened this way, he volunteers. He arrived in the Galápagos with his brothers in 1937. They worked as fishermen, also farmers; later they ran a charter boat business for island tours. In those years, a Galápagos settler had to be versatile. Karl built his house here on the coast decades ago, when the area was heavily populated with iguanas, and the iguanas decided that they wouldn't leave. Fine with him. Room enough for everyone. Now he dotes on them, feeding them as if they were a flock of sacred chickens, though admittedly in mating season they try his patience.

Karl's marine iguanas will eat sea algae if necessary, but they have come to prefer bread and rice. At first there was just one old male shameless enough to beg for food. The others caught on, and presently more than a hundred loiter around his house and grounds, alert for handouts. Three big males divide the rooftop into territories, he tells me, and two other males rule the veranda. Now he summons them for a snack. "Annay-annay-annay!" he hollers, and a stampede of iguanas ensues. They come down from the walls, they scutter greedily in from far corners of the veranda. He tosses them pieces of bread, throws bread up to the animals on the roof. Later in the day they will get rice. If it was once just a matter of letting them clean up his table scraps, it has latterly gone far beyond that, with a hundred hungry bellies waiting to be filled. His maid has given him notice, Karl says wearily. She is an elderly woman, roughly his own age, and she finds herself too old to be cooking rice for a herd of randy reptiles.

"Annay-annay-annay!" Bread for the darlings, yes. He gives what he has, sharing it out until he is empty-handed. All gone now. Sorry, all gone.

Karl lifts one big iguana by its tail, then grabs it deftly behind the neck, inviting me to admire its ear holes. The iguana has snagged Karl's hand with a claw, and he seems not to notice until I inform him that he is bleeding. Oh yes, that's nothing. "All these scratches," he says proudly, affectionately, showing me thin white scars down his left arm, "are from the iguanas."

I ask for a translation of *annay-annay-annay*.

There is no translation, no meaning, he says. It's just a nonsense sound that the iguanas have learned to associate with food. He has taught them to expect a certain largesse, and they have taught him to provide it.

The dog is resting peaceably, surrounded by resting iguanas. To-morrow I'll return to buy one of Karl's paintings. For now, I climb back aboard the boat by way of Karl's private dock, which juts from the lava cliff just below his veranda. He waves. Even from some distance out in Academy Bay, a person can see iguanas basking complacently on his roof. It occurs to me that these particular animals are different from most individuals of *Amblyrhynchus cristatus* in the archipelago, different from most endemic species on most islands. Karl's iguanas aren't just ecologically naïve. By God, they are tame.

58

ARCHIPELAGO speciation, still another item from the Insular Menu, is a process that drives island faunas and floras not just toward greater eccentricity but also toward greater diversity. It involves an initial colonization of an archipelago by a species of animal or plant, followed by further short-distance colonizations that establish discrete populations on the individual islands. The discrete populations diverge genetically, over time, until each island contains an endemic species, closely related to all the others. This process occurs in archipelagos where the islands are spaced just enough to make island-to-island colonizations possible but rare. If the islands are *too* near one another, there is continual back-and-forth traffic and no genetic divergence.

The Galápagos marine iguana represents an incipient instance of archipelago speciation—but only incipient, since the separate populations on their respective islands haven't yet diverged to the point of full speciation. The species *Amblyrhynchus cristatus* encompasses seven populations on seven different islands, genetically somewhat distinct but not unbridgeably so. Put the populations together again and presumably they could still interbreed.

A recent taxonomic revision recognizes those populations as seven subspecies within the species. The subspecies native to Santa Cruz island—including the animals of Karl Angermeyer's roof and of the coastline outside my hotel—is *A. cristatus hassi*. It is larger than the subspecies on the peripheral islands of the archipelago, and more similar to other large-bodied subspecies on the nearby islands of Fernandina and Isabela. *A. cristatus venustissimus*, which inhabits the island of Española at the southeastern fringe of the archipelago, is notable for its bright colors. *A. cristatus nanus*, on Genovesa up at the northeastern fringe, is small and dark. The divergence among these iguana populations doesn't constitute an extreme case—certainly not so extreme as the thirteen distinct species of Galápagos finches, or the sixteen species of weevils on the Tristan da Cunha archipelago, or the speciation among rodents in the Philippines. It's just a reminder that archipelago speciation occurs commonly, and sometimes subtly, on island groups all over the planet.

Deep-water volcanic archipelagos, such as the Mascarenes or the Hawaiian chain or the Galápagos, tend to offer the right physical spacing for archipelago speciation. They stand separate though clustered, providing a balance between isolation and colonization that allows populations to segregate and diverge. A typical result is the seven subspecies of marine iguana—or the five species of endemic Galápagos gecko within the genus *Phyllodactylus*, or the seven species of Galápagos lava lizard within the genus *Tropidurus*, or the six species of Galápagos prickly pear within the genus *Opuntia*, or the fourteen subspecies of Galápagos tortoise. All of these patterns of localized diversity are more easily appreciated in light of modern evolutionary theory. Charles Darwin himself saw evidence of archipelago speciation among the tortoises, but he knew not what to think.

59

THE MOST remarkable aspect of the natural history of the Galápagos, Darwin wrote in his *Journal* with the wisdom of hindsight, was that "the different islands to a considerable extent are inhabited by a different set of beings." What he meant by that murky phrasing was that the different islands supported distinct species and subspecies within certain lineages. "My attention was first called to this fact by the Vice-Governor, Mr. Lawson, declaring that the tortoises differed from the different islands, and that he could with certainty tell from which island any one was brought." Although the vice-governor's comment may have been the most consequential thing anyone said to Charles Darwin in the course of his life, at the time it seemed only mildly intriguing. Mr. Lawson had inadvertently alerted him to archipelago speciation, but Darwin was slow on the uptake.

Shell shape is the clearest indication of differences among the Galápagos tortoises. The shell of a tortoise from Española is saddle-shaped, with an arched front that lets the animal crane its head upward toward high vegetation. The shell of a Santa Cruz tortoise is dome-shaped, with a low rim in front. These different shapes evidently reflect adaptations to ecological differences on the two islands. On Española, which is arid, the tortoises browse on the drooping branches of arboreal prickly pears, whereas the wet highlands of Santa Cruz are lush with grass and other ground-level fodder, allowing the tortoises to keep their heads low. The big island, Isabela, contains five volcanic peaks, each of which supports a distinct population of tortoises with a distinct shape of shell. The populations of southern Isabela, which is grassy, have shells that conform generally to the dome-shaped type, while the northern Isabela populations tend toward the saddle-shaped type.

There are other clues besides shell shape. According to Darwin, the tortoises of Santiago were not only rounder but blacker, and had "a better taste when cooked." Tortoise taxonomists of today are obliged to forgo the evidence of the palate.

Darwin would later regret that he had let the vice-governor's comment about shell differences slide past him. "I did not for some time pay sufficient attention to this statement, and I had already partially mingled together the collections from two of the islands." He hadn't guessed that islands so similar in physical conditions, and so close together, could be inhabited by distinct sets of creatures.

His methodological mistake, as I've mentioned earlier, didn't apply just to tortoises; he had also jumbled some of his bird specimens together without regard to their islands of origin. "Unfortunately most of the specimens of the finch tribe were mingled together," he admitted in a later edition of the *Journal*. Bear in mind that even the first published edition of this *Journal* was written after he had arrived back in England; it was a literary composition for which his *Beagle* diary and field notebooks supplied raw material. During the actual field-work, his perspective had been different. Collecting with zeal but without time or distance to reflect, he had lumped his specimens together carelessly and ignored the suggestive patterns of archipelago speciation. Only later, back in England, with help from taxonomic specialists, did he take notice enough to bemoan the mistake and imagine the implications of data he hadn't quite gathered. Such retroactive insights were exactly what distinguished the *Journal* (especially in its revised editions) from the diary.

There exists a popular misapprehension that the Galápagos finches—famed now as "Darwin's finches," thirteen species of drab little birds with differently shaped beaks and correspondingly distinct ecological niches—inspired Darwin's great insight into evolution. But here's a bit of spoilsport historical reality: It wasn't the finches that inspired Darwin, it was the mockingbirds.

He called them "mocking-thrushes," and in the *Journal* he acknowledged their role: "My attention was first thoroughly aroused, by comparing together the numerous specimens, shot by myself and several other parties on board, of the mocking-thrushes, when, to my astonishment, I discovered that all those from Charles Island belonged to one species (Mimus trifasciatus); all from Albemarle Island to M. parvulus; and all from James and Chatham Islands (between which two other islands are situated, as connecting links) belonged to M. melanotis." The scientific names are different today (*Mimus* has become *Nesomimus*), as are the names of the islands themselves (James, for instance, has become Santiago), but the meaning remains constant. Four species of Galápagos mockingbird have diverged from a common ancestor. Within one of those species, seven subspecies are now native to seven separate islands. It's archipelago speciation in fact and in progress. Darwin saw the pattern and knew that something was up.

The pattern is relatively simple. No two mockingbird species inhabit any one island. This contrasts notably with the distributional pattern of the finches, which is a tangled skein. Among the finches there have been multiple colonizations this way and that, with as many as ten finch species now coexisting on some of the islands. Why is the mockingbird pattern simple? Possibly the mockingbirds are less adventurous than the finches, less inclined to travel between islands. Possibly they arrived in the archipelago more recently and they just haven't had time. Whatever the reason, the divergence of the mockingbirds hasn't progressed past an elementary stage. The different species and subspecies remain isolated on separate islands, and they haven't forced one another into strange new roles by the pressure of competition. The contrast between the mockingbirds and the finches, therefore, makes a good illustration of the tricky distinction between archipelago speciation and another item on the Insular Menu, adaptive radiation.

60

You know something about adaptive radiation because you know something about the tenrecs of Madagascar, yes? But there's more. Adaptive radiation can be construed broadly or narrowly. If construed broadly, the concept applies to virtually every aspect of evolution. Condors are different from chickadees because of adaptive radiation. Coyotes are different from wombats because of adaptive radiation. Wombats are different from condors . . . et cetera. Adaptive radiation is what has happened on Earth over the past four billion years. But this view is so broad that it converts a useful concept into a truism, as airy and dizzying as vodka. I recommend the narrow view, which is focused and pungent, like gin.

Here's a textbook definition:

> Adaptive radiation is the diversification of species originating from a common ancestor to fill a wide variety of ecological niches. It occurs when a single species gives rise, through repeated episodes of speciation, to numerous kinds of descendents that remain sympatric within a small geographic area. The coexisting species tend to diverge in their use of ecological resources in order to reduce interspecific competition.

Niches, sympatric, competition—these are the key words. It's a good careful statement, so compact that it begs to be annotated.

You have already heard about sympatry. Closely related species are sympatric when they inhabit the same geographical area. Niches too you have heard mentioned, but maybe it's time to match that important term with its precise scientific meaning. An ecological niche is the set of resources, physical conditions, and behavioral possibilities within which a population of organisms lives. Every species must have its own niche. Some dictionaries describe the niche as a place, but that's wrong; it's more like a role. As for interspecific competition, that's almost self-explanatory: the struggle between one species and another to exploit a limited supply of resources.

Competition and sympatry, along with an emphasis on the adaptive differences that separate closely related species or subspecies, are what distinguish adaptive radiation from archipelago speciation. Before applying that standard to the mockingbirds, let's glance again at the tortoises.

The fourteen subspecies of Galápagos tortoise arose in fourteen different insularized bits of habitat—by allopatric divergence, if you like jargon. Nine of those subspecies emerged solitarily on nine separate islands, and five others on the five volcanoes of the island of Isabela. Three of the subspecies have been driven extinct during recent centuries; the surviving eleven include all five on Isabela plus those native to Santa Cruz, San Cristóbal, Santiago, Española, Pinzón, and Pinta. There is inevitably some variety in shell shape between individuals within each subspecies, but in general a subspecies manifests its own collectively characteristic shape of shell.

Most of those eleven surviving subspecies are clearly assignable to one of the two types I've mentioned—those with saddle-shaped shells or those with dome-shaped shells. Do the shell shapes have adaptive significance? That question requires two answers, qualified and separate. An extreme saddle-shaped shell (allowing the animal to stretch its head upward) might in some circumstances provide an adaptive advantage over a dome-shaped shell (prohibiting high reach but affording better protection), and in other circumstances the advantage might be reversed. The saddle-type tortoises of Española are probably better adapted to that arid island than a dome-type tortoise would be; the dome-type animals of Santa Cruz are probably more successful amid the green pastures of that island than a neck-stretching, less-protected, saddle-type subspecies would be. For the two extremes of shell shape, then, the answer is yes, the differences are adaptive.

But do the nuances of shell difference among eleven subspecies all represent significant adaptations? Not necessarily.

This is an important truth, usually missing from simplified explanations of evolutionary processes: that not every genetic difference between closely related species or subspecies is adaptive. Such a difference might also be accidental. It might result from random differences in gene frequency between a pioneer population and the base population from which it became separated. For the subspecies of Galápagos tortoises, that's probably the case.

Such random genetic differences could result from two causes. One is known as the founder principle, described by Ernst Mayr back in 1942. When a new population is founded in an isolated place, the founders usually constitute a numerically tiny group—a handful of lonely pioneers, or just a pair, or maybe no more than one pregnant female. Descending from such a small number of founders, the new population will carry only a minuscule and to some extent random sample of the gene pool of the base population. The sample will most likely be unrepresentative, encompassing less genetic diversity than the larger pool. This effect shows itself whenever a small sample is taken from a large aggregation of diversity; whether the aggregation consists of genes, colored gumballs, M&M's, the cards of a deck, or any other collection of varied items, a small sample will usually contain less diversity than the whole. For a pedestrian example, imagine a drawerful of socks.

If you own ten pairs of black socks, nine pairs of brown socks, and one pair of flamingo-pink socks, your total supply is five percent pink. Now picture yourself stumbling to the sock drawer in the early morning darkness and hastily packing a small suitcase. You haven't had coffee. Your plane leaves in an hour. Grabbing blindly, you pack four pairs of socks. Will the color proportions of your total sock supply be precisely reflected in what you have grabbed? No. Will your suitcase contain any flamingo-pink socks? Probably not (and it's just as well, unless you're headed for Las Vegas). The founder principle works roughly the same way. The sample of genes carried by a small group of founders will not be precisely representative of the total gene pool. Sometimes, by chance, it will be quite unrepresentative.

The other cause of random genetic differences between populations is called genetic drift. This comes into play after the founders have begun to reproduce. Already isolated geographically from its base population, the pioneer population now starts drifting away genetically. Over the course of generations, its gene pool becomes more and more different from the gene pool of the base population—different both as to the array of alleles (that is, the variant forms of a given gene) and as to the commonness of each allele. The point of the drift metaphor is that these changes too occur to some extent by chance, as well as by natural selection. To help make this clear, I'll offer a more vivid definition of alleles. Turn your mind back again to haberdashery.

Think of socks as a gene. Think of shoes as a gene. Think of gloves as a gene. A gene is a rubric that can be embodied in various different ways. Imagine a selection of variants that might all fit within just one of those rubrics: black socks and brown socks and argyles and sweat socks and plaid knee socks and nylon stockings and flamingo-pink socks. Those are alleles of the socks gene. Of course socks come in pairs. So do the genetic alleles within an organism descended from two parents. Why? Because sexually reproducing creatures generally carry two separate alleles (one on each of two paired chromosomes) for every gene, a single allele inherited from each parent. If the two alleles are identical, the individual is said to be *homozygous* for that gene. A *heterozygous* individual carries two different alleles for the gene, like somebody wearing unmatched socks. These are familiar terms, no doubt, but not quite part of everyday parlance. *Meiosis*, another old word that we all learned in high school biology and all hoped to be allowed to forget, denotes the cell-division process of the gonads, in which a half share of the available alleles is randomly dealt into each egg and each sperm. Parenthood involves passing that randomized half share of available alleles to the next generation.

Here's the catch. Some alleles are common within a population, some are rare. If the population is large, with thousands or millions of parents producing thousands or millions of offspring, the rare alleles as well as the common ones will usually be passed along. Chance operating at high numbers tends to produce stable results, and the proportions of rarity and commonness will hold steady. If the population is small, though, the rare alleles will most likely disappear in the course of meiosis and fertilization, because chance operating at low numbers produces aberrations against statistical probability, and just a modest bit of aberration is enough to lose a rare allele completely. So the process of random sample-taking will tend to miss those rare alleles, and often they won't get passed along. (Random sample-taking could conceivably also make a rare allele more common, but the odds are against that.) As it loses its rare alleles by the wayside, a small pioneer population will become increasingly unlike the base population from which it derived. Its heterozygosity will decline. A greater proportion of the remaining individuals will be homozygous for a greater proportion of their genes. This is genetic drift. There is no precise analogy from the realm of socks, but only because suitcases are incapable of meiosis.

The Galápagos mockingbirds, like the tortoises, show signs of both adaptation and drift. The four mockingbird species are quite similar to one another: bold, predatory, about the size and shape of a robin, with a dark patch behind each eye, a creamy breast, dark wings, a long tail, and a slender, curved beak. Both the beak and the legs are longer than those of the South American mockingbirds to which they are most closely related. Their predatory behavior is also more pronounced than on the mainland. They kill and eat insects, centipedes, lava lizards, even young finches.

Among the four mockingbird species, one stands distinct: *Nesomimus macdonaldi*, on the island of Española. Its beak is exceptionally large, its behavior is exceptionally aggressive, and it eats eggs. Several of the other three Galápagos mockingbird species also occasionally eat eggs, but none so regularly as *N. macdonaldi*. It preys on albatross eggs, booby eggs, gull eggs, iguana eggs, virtually any eggs it can find. A possibility raised by some biologists is that the enlarged beak of this species is an adaptation for cracking eggs. Let's grant that possibility—that the beak modification of *N. macdonaldi* is more adaptive than random. The other three species manifest no such useful specializations. The differences that distinguish *N. trifasciatus*, *N. melanotis*, and *N. parvulus* seem to be more random than adaptive. In the absence of further research, those differences are best explained by the founder principle and genetic drift.

Together the Galápagos mockingbirds represent a case of archipelago speciation but not of adaptive radiation. The former is a prerequisite to the latter. That is, archipelago speciation produces differences that serve as points of departure for adaptive radiation. By way of archipelago speciation, geographically isolated populations become also reproductively isolated. Adaptive radiation begins only when the geographical isolation is breached—for instance, when a second set of pioneers reaches an island occupied by an earlier set. At

that point, if the two populations of long-lost relatives have accumulated a sufficient degree of genetic distinctness from each other, reproductive isolation will prevent them from interbreeding. Instead of
interbreeding, they will compete.

Now we're back to the matters of competition and sympatry. We're
back to the Galápagos finches.

61

THE DISTRIBUTIONAL pattern of the Galápagos finches is a puzzle
that would have pleased Metternich. It's a mess of overlapping sets
and subsets and associations and exclusions. You couldn't depict it by
painting colors on a map because too many zones would end up pur-
plish black.

The phylogenetic pattern is even worse. From a single ancestral
stock, these restless and protean finches have crossed between islands
and speciated and crossed back and competed and diverged further
and crossed again and competed. Diverged. Crossed. They have radi-
ated like fiends. Today they constitute thirteen different species (with
a closely related fourteenth species on Cocos, a solitary island about
halfway between the Galápagos and Central America), all classified
within the subfamily Geospizinae. Over the millennia since they first
colonized the archipelago, some additional species may have evolved
and gone extinct. The precise details are lost in the sea fog of Galápagos prehistory, but the surviving evidence has been sufficient to trans-
fix the attention of more biologists than just Charles Darwin. Back in
1947, an ornithologist named David Lack published an influential little book titled *Darwin's Finches*. More recently there is *Ecology and
Evolution of Darwin's Finches*, a masterful work by Peter R. Grant, who
has studied them in the field for decades. Journal articles on the subject would make a pile as high as an ottoman. Several of the islands
support ten species each, but not in all cases precisely the same ten.
Several of the species occur on a dozen islands, but not in all cases
precisely the same dozen. Consider the sheer mathematical fact that
every one of those thirteen finch species either is or is not resident on
each of seventeen major islands, and you get a sense of the potential
complexity. And to make matters worse, there are subspecies. If I
were so foolish as to expect you to contemplate these particulars, I

would pack them all into an elaborately annoying chart, with the islands along a horizontal axis, the species along a vertical axis, and alphabetical symbols in gridwork boxes to indicate the presence of different subspecies. You'd have the whole pattern at your fingertips, but your eyes would drop shut like an electric garage door.

A few salient points can be stated simply. First, the pattern of distribution of finches differs from the patterns for tortoises and mockingbirds in one crucial way: Closely related species are living sympatrically. Four species of ground finch, from the genus *Geospiza*, coexist on the island of Santa Cruz. Those four species also turn up on Marchena. Three species of tree finch, from the genus *Camarhynchus*, share the island of Floreana. Sympatry.

Second, the finches occupy a wide range of ecological niches that, on a mainland, would be filled by other bird groups. Finding those niches vacant when they originally colonized the Galápagos, the finches have seized opportunities. They have taken to the ground, to the trees, to the mangroves, and one species has developed the habits and skills of a woodpecker. They have adopted new diets. David Lack began his book by noting: "In an English garden, finches of several kinds and sizes eat seeds and fruits, tits of various sizes examine the twigs and branches, warblers of different kinds take insects off the leaves, thrushes search the ground and woodpeckers climb the trunks." His point was that, in the Galápagos, one finch species or another does it all. The common names reflect the ecological role-playing: ground finches, tree finches, cactus ground finches, woodpecker finch, warbler finch, vegetarian finch, mangrove finch. They have diverged into vacant niches.

The third salient point is that sympatry entails competition, and competition impels divergence. This is adaptive radiation. Closely related species on a single island (such as the ground finches on Santa Cruz or the tree finches on Floreana) can't coexist for long unless they find ways to be different from each other. If they are too similar, one species will die away or be forced to move on, while the other becomes exclusive proprietor of the niche. To remain sympatric, they are obliged to partition further the available resources, adapting themselves to narrower, more acutely defined niches. Where closely related species converge on the same zone of habitat, they might eat different sorts of food. Or if they persist in eating the same sort of food, they might collect it in different places. Another dimension along which they can distinguish themselves is size—size of their bodies, size of their beaks, size of the food particles they can handle.

So there appear clusters of similar species distinguished by size. Small, medium, and large ground finches. Small, medium, and large tree finches. Their beaks range from tiny and delicate, like a warbler's, to big and heavy, like a parrot's.

A fourth salient point: The larger Galápagos islands (Santa Cruz, Isabela, Fernandina) support more different species than the small islands (Santa Fe, Española, Genovesa). This is the species-area relationship as measured in Geospizinae. Large areas harbor more diversity of finches than small areas. It's not a surprising fact, but it's more complicated than it seems. And in the course of this book it will never stop being important.

62

THE Galápagos finches and their supposed importance to Charles Darwin have been woven into one of the more piquant legends in the history of science. The legend says that Darwin observed these birds in the Galápagos, saw the signs of adaptive radiation among them, and was thereby led toward his theory of evolution. Textbooks of biology, bird books, Darwin biographies, broad-stroke science chronicles, and even some serious studies of evolutionary topics have tended to embrace this legend. David Lack's landmark volume even played a part.

"Just over a hundred years ago, in 1835," Lack wrote in his preface, "Charles Darwin collected some dull-looking finches in the Galapagos Islands. They proved to be a new group of birds and, together with the giant tortoises and other Galapagos animals, they started a train of thought which culminated in the *Origin of Species*, and shook the world." The very title of Lack's book, *Darwin's Finches*, suggested an intimate link between Charles Darwin as evolutionist and the Galápagos finches as evidence of evolution.

Truth is, the link wasn't so intimate. Like most legends, this one is convenient, narratively satisfying, and largely imaginary. For textbooks, it was convenient because a classic case of adaptive radiation could be paired with a historic episode of discovery. For the rest of us, it was satisfying because Darwin in the Galápagos, watching birds, tossing iguanas, represents such an appealing image of the dotty young genius. But convenient or not, satisfying or not, it was erroneous and needed correction. Then in 1982 a science historian named Frank J. Sulloway published a quiet, careful article in the *Journal of the History of Biology* titled "Darwin and His Finches: The Evolution of a Legend," which sorted the real from the unreal.

In the years since, a few other historians (including Stephen Jay Gould and the Darwin biographers Peter J. Bowler and Ronald W. Clark) have given due credit to Sulloway's revisionist work. But most people, if they have any notion at all about Darwin and the finches, are still misled.

As elaborated in detail, the legend goes like this. The alert young Charles Darwin, during his stopover in the Galápagos, noticed some interesting differences among a closely related group of bird species. He could see they were finches. He observed gradations in the size and the shape of their beaks. He collected specimens. At first he made the mistake of not labeling his specimens according to the individual island where each was collected, but soon, having recognized the significance of that information (with help from the vice-governor's comment about tortoises), he did begin to tag them by locality. He saw that this single group of finches included a dozen or more separate species, all anatomically and ecologically distinct. He wondered what it could mean. Back in England, he turned his finch specimens over to an ornithologist named John Gould, who performed the taxonomic descriptions. When Gould's chore was finished, Darwin studied the finches again, as well as other collections and notes, until there came a revelation: that the theory of special creation was bogus. Suddenly he saw the world's fauna and flora with new clarity. Rather than having been created by a micro-managing deity, Darwin realized, species had their origin in a natural and continuous process of organic change. They evolved. Somehow. Darwin didn't yet know how. In 1837 he made an entry in his diary, recording the fact that he had opened his first notebook on the transmutation of species. Since the previous March, he wrote, he had been greatly struck by the character of South American fossils and by the native species of the Galápagos. Without bothering about verbs, he reminded himself: "These facts origin (especially latter) of all my views." The most crucial of the originating facts pertained to the finches. Yes?

No. It makes a good story, but no.

According to Frank Sulloway's assiduous tracking, bits of the legend have turned up in books as varied as *Introduction to Evolution, Evolution as a Process, Darwin and the Beagle, Darwin's Century, Introduction* to *Physiological Psychology*, *Grzimek's Animal Life Encyclopedia*, and several textbooks simply titled *Biology*. To some extent, the legend was founded in the writings of Darwin himself. The second edition of his *Journal*, published in 1845, contained a description of the finches and a pen-and-ink illustration. The illustration showed four bird heads, each with a differently shaped beak. These birds, Darwin wrote, were "a most singular group of finches, related to each other in the structure of their beaks, short tails, form of body, and plumage." The remarkable thing about them, he noted, was the perfect gradation in beak size among the different species. He added that some of the beaks also differed in shape. "Seeing this gradation and diversity of structure in one small, intimately related group of birds, one might really fancy that from an original paucity of birds in this archipelago, one species had been taken and modified for different ends." It's a genuinely resonant passage when considered in its correct historical context—an 1845 hint toward an 1859 announcement that startled the world. It shows that fourteen years before publishing *The Origin of Species*, Darwin knew where he was going.

But careless recapitulation of the finch story has stretched the passage to prove something more: that ten years still earlier, back in 1835, while he was observing and collecting in the Galápagos, Darwin had recognized the finches as a case of adaptive radiation. He had recognized no such thing. In revising his *Journal* for the 1845 edition, he coyly suggested what "one might really fancy," but it wasn't what he himself had first fancied. His treatment of the finches in that edition was an afterthought.

The first edition of the same book, published in 1839, includes only a brief mention of the Galápagos finches. No pen-and-ink drawing accompanies the passage. There is no talk about species being "modified for different ends." Yet this edition was written just after Darwin got home from the voyage, at almost the same time he hit upon his theory. More faithfully than the edition of 1845, the 1839 edition seems to reflect what the Galápagos finches contributed to Darwin's earliest evolutionary thinking. It suggests that they were an insignificant factor.

The legend of the finches, as scrutinized by Frank Sulloway, encompasses two principal themes. The first of them, Sulloway writes, "involves the claim that the different forms of the finches, along with the tortoises and the mockingbirds, first convinced Darwin that species must be mutable while he was still in the Galapagos Archipelago." An issue of timing. The legend's second theme, according to Sulloway, "holds that Darwin's observations on the finches inspired all his later theories by providing him with a decisive example of evolution in action. In particular, the finches are said to have elucidated the crucial roles of geographic isolation and adaptive radiation as mechanisms of evolutionary change." Reversing the order of the two themes, and expressing them as questions: What was it that inspired Charles Darwin, and when did it inspire him?

Sulloway's work runs to fifty-three pages of intricate argument and citation. It draws upon Darwin's unpublished diary and his specimen catalogues from the voyage, his ornithological field notes, his original bird specimens (finches and some others, still held in the British Museum), his tagging procedures, his later notebooks (compiled back in England, while he was developing his theory), his early drafts of the theory, the first and second editions of his *Journal*, his more technical volume called *The Zoology of the Voyage of H.M.S. Beagle*, his letters, his autobiography, the minutes of a Zoological Society meeting held in 1837 (at which John Gould presented the first results of his study of Darwin's specimens), and other forms of historical clue, including a few that lay neglected for a century and a half. For a sample of what Sulloway found, we can look at those two principal themes.

First, timing. Is it true that Darwin came alert to the finches' significance while he was still in the Galápagos and therefore started tagging his specimens by island? David Lack, in his finch book, insisted that Darwin did. Sulloway says no, and provides evidence that Lack and others have overlooked. The actual finch specimens brought home by Darwin do bear locality tags, Sulloway reports, but the tags aren't in Darwin's handwriting. They were added later by curators. This much isn't decisive; we might reasonably assume that the curators, following common procedure, had simply replaced Darwin's tags with standardized tags of their own, and that in making the switch they had copied Darwin's locality notations. Against that assumption, Sulloway sets the evidence of a single lost specimen, rediscovered by him among another group of birds and still holding its original tag. The rediscovered specimen (an individual of *Dolichonyx oryzivorus*, the American bobolink, which visits the Galápagos as a migrant) reveals that Darwin's labeling system was *not* adjusted, near the end of the Galápagos visit, to include notations of locality. This bobolink was one of the last birds that Darwin collected and tagged, yet its original tag says nothing about its locality.

Sulloway makes a strong case that Darwin was back in England, sorting his data, already thinking about evolution, before he was seized by the urge to consider his finch specimens on an island-by-island basis. Wasn't it too late? How could he reconstruct such lost information? Luckily for Darwin, some finches had also been collected by three other men on the *Beagle*, all of whom had been more careful about noting locality. Captain FitzRoy himself, the ship's commander, had made a sizable collection, which went (as Darwin's eventually did) to the British Museum. A shipmate named Harry Fuller had collected modestly, and so had Darwin's own manservant, Syms Covington. Soon after Darwin began work on his theory, he decided it would be useful to know which finches had come from which islands. Since his own collection and his own catalogues were inadequate, he assembled the data by inference and analogy, borrowing from the collections of FitzRoy, Fuller, and Covington. The notes made from those other collections, in Darwin's handwriting, are preserved in the Cambridge University Library. Sulloway's paper includes facsimile pages.

At the time of his Galápagos fieldwork, then, Darwin must not have cared much about the patterns of distribution displayed by the finches, since he never bothered to record their localities. He had returned to England still ignorant of whatever it was that the Geospizinae could teach.

The second theme of Frank Sulloway's paper is causality. Did the finch specimens give Darwin his inspiration? Sulloway says no. He returns to that comment in Darwin's 1837 diary: "Had been greatly struck from about Month of previous March on character of S. American fossils—& species on Galapagos Archipelago. These facts origin (especially latter) of all my views." But which facts was Darwin referring to? The crucial facts, according to Sulloway, came chiefly from John Gould.

In March of that year Darwin had met with Gould in London and heard the ornithologist's expert analysis of the Galápagos collections. Gould's conclusions were surprising. Three of the four mockingbirds were new and distinct species. In all, twenty-five of the twenty-six Galápagos land birds were considered by Gould to be new and distinct, unknown elsewhere on the planet. And besides the land birds, Gould felt he had looked at some other new species, including a gull, a heron, and a turnstone. Darwin was stunned, says Sulloway, not only by John Gould's report on the mockingbirds, but also by the fact (which Darwin had apparently not noticed himself) that these unique Galápagos species were closely related to species of the American mainland.

It was the same spooky phenomenon, too striking for coincidence, that Alfred Wallace would subsequently call a law: *Every species has come into existence coincident both in space and time with a pre-existing closely allied species.* It argued against special creation. It persuaded Darwin, as it would later persuade Wallace.

The finch specimens, according to Sulloway, played no role in this phase of Darwin's epiphany. Unlike the mockingbirds, they showed no dramatic affinity with species on the nearby American mainland. They held scarcely any importance in his thoughts or his notebooks at the time that he wrote, "These facts origin (especially latter) of all my views."

Frank Sulloway's paper isn't an attack on Charles Darwin; it's a critique of blurry thinking by those who have reinvented the record for narrative convenience. Sulloway's message is simply that scientific heroes (like other sorts) attract legends, and that legends are something distinct from science and history:

> As it turns out, Darwin made absolutely no effort while in the Galapagos to separate his finches by island; and what locality information he later published, he reconstructed after his return to England, using other shipmates' carefully labeled collections.

Sulloway adds:

> As for Darwin's supposed insight into evolution by adaptive radiation while he was still in the Galapagos, the more the various species of finch exhibited this remarkable phenomenon, the more Darwin mistook them at the time for the forms they were mimicking. Even after his return to England, when John Gould had clarified the affinities of this unusual avian group, Darwin was slow to understand how the Galapagos finches had evolved. In particular, he possessed only a limited and largely erroneous conception of both the feeding habits and the geographical distribution of these birds—information that was vital to a proper explanation of their evolution. Lastly, far from being crucial

to his evolutionary argument, as the legend would have us believe, the finches were not even mentioned by Darwin in the *Origin of Species*.

The Geospizinae, as first encountered by Darwin, were a confusing and unremarkable bunch. While he was collecting them, he hadn't even been sure they were finches. Some seemed to be grosbeaks, some wrens. Some looked more like meadowlarks or blackbirds. He had gazed at this evidence of adaptive radiation without comprehending what he saw. Although the different species occupied different niches and those niches were defined largely by diet, Darwin hadn't noticed. To his eye, the finches all appeared to be eating the same sort of food. The conclusion that later emerged from studies of their beak morphology, their distribution, and their habits—that the Galápagos finches had been forced to diverge from each other by interspecific competition—escaped Darwin during his fieldwork in the islands. He had paid more attention to the coloring of their plumage than to the sizes or shapes of their beaks. But their plumage was generally drab, and the distinctions between one shade of brown and another weren't illuminating.

He brought home thirty-one puzzling little carcasses, shook his head, and dropped them in John Gould's lap. They became "Darwin's finches" about a hundred years later.

63

THE Galápagos finches aren't the be-all and end-all of adaptive radiation. They are merely an emblematic case, made famous in textbooks and popular histories because of their association with Darwin. By the standards of science, as distinct from the standards of notoriety, they are far less dramatic than other examples from other places.

The Hawaiian Islands, in particular, have given rise to some spectacular excesses of speciation and radiation—among flowering plants, land snails, insects, and that group of birds known as the honeycreepers. What makes Hawaii so conducive to this phenomenon? A combination of factors. It's the most isolated archipelago on the planet. It's relatively old for a set of oceanic islands. It encompasses more total terrestrial area than most other remote archipelagos, and its major islands (Hawaii, Maui, Oahu, and Kauai) are individually larger than most other oceanic islands. Also, the Hawaiian archipelago is warm, wet, and topographically steep—three conditions that tend to enhance biological richness. Finally, its mountainsides are cut by stream canyons, old lava flows, and ridges, which divide it into many small strips and patches of isolated habitat. Charles Darwin never saw the Hawaiian Islands. If he had, the Galápagos might have paled by comparison.

The most vivid of Hawaii's evolutionary showpieces are the honeycreepers, an endemic subfamily of birds known to scientists as the Drepanidinae. In the Hawaiian native language they bear names such as *mamo, ou, iiwi, poo uli,* and *akohekohe.* Like the Geospizinae, the Drepanidinae have descended from a finchlike ancestor; unlike the Geospizinae, they have taken divergence to extremes. Evolution has given them bright colors and striking shapes. They include thirty or so species—the correct present number being somewhat uncertain, both because of taxonomic disagreements and because of losses by extinction. There are black-bodied birds with long beaks, red-and-black birds with hooked beaks, greenish birds with delicate scimitars, parrot-beaked birds with olive wings and yellow eyebrows, birds that resemble goldfinches except for their blue beaks and black masks, on and on to the thirtieth permutation. Some species are (or were, until their extinction) specialized to drink nectar, some to eat insects, some to rip open seedpods with their beaks. And some are less specialized, feeding eclectically on insects, fruit, seeds, even the eggs of sea birds. The Hawaiian honeycreepers as a group have been termed "the finest example known of the process called adaptive radiation," but that statement comes from a team of bird guys who undoubtedly harbor a bias. An entomologist with a Hawaii fixation would presumably say otherwise, because an entomologist would be unable to ignore *Drosophila.*

Drosophila is an astonishingly diverse genus of fruit flies. The name figures prominently in the history of genetics, owing to its association with laboratory studies of inheritance. A few select species, most notably *Drosophila melanogaster,* have been invaluable to such research because they breed quickly, they thrive in captivity, and their salivary glands contain oversized chromosomes. Those giant chromosomes, which are easily reached by dissection and easily decoded, prevent geneticists from going crazy and blind.

Even we simpler folk, who never count chromosomes through a

microscope, tend to think of drosophilas only as lab creatures. They live in Tupperware habitats, they eat puréed banana, they give fast answers to big genetic questions, and they seem to possess an abstract ontogenetic essence halfway between a white rat and an abacus. But some drosophilas are actually wild animals. A safe guess is that the genus includes fifteen hundred species throughout the world, about five hundred of which are endemic to Hawaii.

They have proliferated there by adaptive radiation. Their pioneer ancestors reached the archipelago on a wing and a chance, finding fruit, empty niches, pockets of isolation, and eventually a generous span of time. The various isolated populations diverged genetically—by way of the founder principle, genetic drift, and localized adaptation—to the point of speciation. Then they dispersed again. Newly evolved species went from Kauai to Oahu, from Oahu to Maui, from Maui to everywhere, reencountering their own close (but now reproductively incompatible) relatives. They competed. They suffered starvation and crowding. Some lines went extinct. Others survived, diverging still further under the pressure of competition into five hundred distinct modes of being drosophilous. "Of all the groups of organisms, plants or animals, that can be studied on islands, the Hawaiian Drosophilidae are supreme," writes a population biologist named Mark Williamson. Although his enthusiasm is understandable, Williamson too might be just a bit biased. But having invested the effort to master *Drosophila* diversity, which is ungodly complicated, he is entitled to indulge in superlatives.

Compared with Hawaii, the big lakes of eastern Africa seem unpromising sites for adaptive radiation. But biologists now recognize that, over the past few hundred millennia, something dramatic was happening there too. Extraordinary radiations were occurring among a family of fishes known as the Cichlidae.

These great African lakes punctuate the Rift Valley, a wide geological gash down the east side of the continent. The largest is Lake Victoria, round as a pond but almost as big as Ireland, lying just west of the highlands of Kenya. Farther south in the Rift Valley is Lake Tanganyika, a long narrow body of water forming the boundary between Tanzania and Zaire. Tanganyika is notable for its extreme depth. Still farther south is Lake Malawi, also wondrous in its own way. None of the three lakes connects by water channel to another. All three were formed when tectonic stresses caused that part of the continent to uplift and buckle and crack, trapping old African rivers within wide new basins. In capturing those rivers, not surprisingly, the basins also got fish. Included were a few representatives of the cichlid family, which turned out to possess great evolutionary plasticity.

The cichlids thrived, filling Victoria and Malawi and Tanganyika not just with an abundance of fish but with an abundance of species diversity. Critical to this explosion of diversity was the physiographic heterogeneity of each lake—its scalloped shoreline, its irregular patterns of depth and slope, its periods of drought and falling water level, all of which left some pockets of habitat either permanently or episodically isolated. A cichlid trapped in one pocket of deep water couldn't breed with a cichlid trapped in another pocket of deep water if, for instance, the two pockets were separated by inhospitable shallows. Within those pockets, the cichlids diverged and speciated.

As centuries passed, many of the habitat pockets were intermittently reconnected. Reconnection allowed the fish populations to mix—spatially, at least, if not genetically. Distinct populations, reproductively incompatible though closely related, came into competition and were therefore presented with two options: to radiate or to die. Many didn't die.

Cichlid diversity in these African lakes became not just rich but garish. Many species developed specialized anatomical tools and specialized behaviors. In particular, their mouth structures transmogrified along with their dietary habits. They partitioned the lakes into a startling multiplicity of small niches. They shifted shape and took on narrower roles: *Pseudotropheus tropheops*, rock scraper; *Hemitilapia oxyrhynchus*, plant scraper; *Lethrinops brevis*, sand-digging insect eater; *Genyochromis mento*, scale eater; *Docimodus johnstoni*, nibbler of other fishes' fins; *Haplochromis similis*, leaf chopper; *Haplochromis euchilus*, rock-probing insect eater; *Haplochromis pardalis*, fish eater; and my personal favorite, *Haplochromis compressiceps*, famed in the scientific literature for biting out eyeballs. Each of them is a cichlid, similar in its basics, unique in its particulars.

These examples are from Lake Malawi. Within its still waters lurk more than two hundred cichlid species, all but four of them endemic to that lake. Molecular geneticists suspect that they may all be descended from a single ancestral species. Two hundred different cichlids leave ecological room for not many other kinds of fish. The cichlid lineage arrived early, it seems, and seized most of the piscine niches that were available. Lake Tanganyika contains at least 126 species, every one of which is endemic. Lake Victoria once supported about two hundred cichlid species—until a cataclysm of human-caused extinctions thinned them out. If you were an evolutionary ichthyologist, you would be entitled to speak of these facts in the same sort of vaunting superlatives that a fly expert reserves for Hawaiian *Drosophila*.

The cichlids of Lake Malawi, the Galápagos finches, the honey-creepers and the fruit flies of Hawaii—they are all textbook examples of adaptive radiation. Still another example comes from Madagascar, where a single remnant of rainforest harbors three sympatric and closely related species of bamboo-eating lemur. This one you won't find in the textbooks.

Three species. Compared with the sheer numerical scope of the other cases, it sounds meager. But the bamboo-eating lemurs have their own kind of drama.

64

IN 1986, a forty-one-year-old primatologist named Patricia C. Wright led a small expedition to southeastern Madagascar. It was a long-shot scientific adventure. She hoped to locate and study *Hapalemur simus*, commonly known as the greater bamboo lemur, a beast that almost no one alive had ever seen.

Hapalemur simus was one of the two known species of bamboo-eating lemur, native only to the rainforests of Madagascar. Throughout most of this century it was presumed extinct. But in the mid-1960s a living specimen turned up for sale in a village market, where it was ransomed by a French entomologist, from whom it later escaped. In 1972 a pair of *H. simus* were captured in a small patch of forest surrounded by a coffee plantation. The pair were moved up to the Parc Botanique et Zoologique de Tsimbazaza, in Antananarivo, and they survived there as zoo captives for several years. They even produced two offspring. Then the pair died, so did the offspring, and *H. simus* lapsed back into obscurity, possibly gone forever.

Both sites where the species had appeared—the coffee plantation at Kianjavato, the village market near Vondrozo—were in southeastern Madagascar, near the great escarpment that slopes steeply down from the country's central plateau. The terrain of the escarpment is so mountainous, so rain-washed and muddy, so inconvenient for woodcutting, rice growing, and cattle pasturing, that in some places the ancient rainforest has survived. Elsewhere in Madagascar the landscape has been ravaged by high-intensity human use. The wet forests of the eastern coastal plain have mostly been cut; the savannas of the plateau have been repeatedly burned; the soil is poor, so are the people, and lemur habitat is preciously scarce. But the southeastern slope still contains some sizable tracts of forest. Those tracts, Pat Wright suspected, would be the place to find *H. simus* if it was ever to be found again anywhere.

She was looking for stands of bamboo. Little was known about the ecology or behavior of the greater bamboo lemur except that, true to its name, it ate bamboo. Earlier observers had reported that it fed on young shoots; if shoots weren't available, it gnawed on mature stalks. Several types of bamboo were native to Madagascar, including a giant species identified as *Cephalostachyum viguieri*, which grew in dense groves of thick stalks that could easily support a lemur's weight. Because of its particular ecological limits and the generalized destruction of vegetation, though, giant bamboo was now rare.

Pat Wright herself was an obscure assistant professor whose career as a field scientist had begun late. After finishing college in 1966, she signed on as a caseworker for one of Lyndon Johnson's social programs in a ghetto neighborhood of New York City. The only hints toward her later destiny, at that time, were that she had grown up as a woods-loving tomboy in western New York state and that, during her first year in the urban jungle, she had adopted a South American monkey. The monkey was a woebegone little capuchin, bought from a pet store in the East Village by a friend of hers and then cast upon Wright's fostering mercy when the original master-and-pet match didn't work out. Her friend lived with his mother, Wright explains, "and he took that monkey back home, and his mother said that she was going to get cancer within three days if he didn't get that monkey out of the house." Wright and her husband accepted it on the rebound. Later there would be others. After you've lived with a monkey, she claims, you find yourself lonely when you don't have one. She spent the next decade in Brooklyn—as a social worker for some years, then as a mother and housewife, usually nurturing a monkey or two along with her young daughter—before she went back to school, having qualified for a doctoral program in physical anthropology.

The husband, whom she had supported through years of his own graduate work, promptly walked out on her. "It's very different," she

says now, forgivingly, "being married to somebody that sort of de-
votes her whole life to you and"—when circumstances change—"being married to somebody that is a total maniac and obsessed by
questions of science." Maniac or not, Pat Wright herself was obsessed
by questions involving the ecology and behavior of tropical primates.

Fortunately, her little daughter was a hardy child, physically and
emotionally, who adjusted well to life in field camps. Mixing motherhood with science, Wright did her dissertation research on nocturnal
monkeys in the Peruvian Amazon. Then she studied tarsiers in
Malaysia and the Philippines. Her first glimpse of Madagascar came
in 1984, after a conference had brought her as far as Nairobi. She
caught an excursion flight onward to Antananarivo for a quick bit of
scientific tourism, and as her plane flew in over the island, laid out below in its fire-scarred and erosion-gouged nakedness, bleeding its
silty red rivers into the Indian Ocean, she started to cry. "I looked at
those burned hills and I decided I was never going to come back
again. It was just too heartbreaking." But a few days among the
lemurs changed her mind. These creatures were so intriguing that she
couldn't resist.

She returned in 1985 to study *Hapalemur griseus*, a small species
known as the gentle bamboo lemur. Her study site was a tiny patch of
forest on the eastern escarpment, a place she had visited and been
smitten by a year earlier. It was a special wildlife reserve known as
Analamazaotra, fated to leave an indelible mark on her professional
and emotional life. There she watched *H. griseus* through an entire
field season, finding it to be a very specialized creature that lives
chiefly on a diet of bamboo. Although earlier researchers had mentioned the bamboo connection, no one else had gathered such thorough data. Wright also grew curious about the monogamous mating
behavior of *H. griseus* and the roles played by males and females in
rearing young. Male lemurs, she noticed, seemed to be rather more
crucial to that process than, say, male humans. It was a significant observation with wide resonance: gender roles as reinvented in the isolation of Madagascar, among an alternate world of primates. But when
she later described her work to Russ Mittermeier, an eminent primate
conservationist then with the World Wildlife Fund, he suggested that
she focus on a different species. *H. griseus* was still relatively common
in the eastern rainforest, Mittermeier pointed out. Why didn't
Wright study something that was more immediately endangered, so
that he could justify supporting her with a grant from WWF?

"I thought about it," Wright remembers, "and I said, 'Well, you think I should go and look at *Hapalemur simus*, the greater bamboo lemur?' " Mittermeier knew, as she did, that this species hadn't been seen since the early 1970s. "And he said, 'Yeah.' And I said, 'It's probably extinct, you know. Nobody is going to give me money to study an extinct animal.' And he said, 'I'll give you money until you find it.' "

So she went back to search for survivors of *H. simus*. She shifted her effort to a region of rainforest toward the southern end of the escarpment, not far from that coffee plantation where the last known pair had been captured a dozen years earlier. She was taking a flier— setting out to gather ecological data on a species that might no longer exist. But she had taken fliers before.

This time she was accompanied by a small team: two American students, a field-tempered expedition coordinator who had worked with her elsewhere, and a teenage Malagasy naturalist named Joseph Rabeson, known to all by the nickname Bedo. This youngster, Bedo, had grown up near the Analamazaotra reserve, farther north, and become friendly with Wright during her work there the previous year. In his own local forest, he was a prodigy of woods lore and zoological knowledge. But the southeastern region was new to them all.

Their field vehicle was an old Toyota Land Cruiser. "We went up and down these roads that were just so godawful. Worst roads I'd ever seen," Wright recalls fondly. "We'd see bamboo clumps and we'd stop. We'd ask the people if there were bamboo lemurs. We'd go off and explore. We did a lot of off-road walking, for days, that kind of stuff, which was really great fun. None of us knew anything about anything." At one point they drove east from the city of Fianarantsoa on a mucky two-lane road that led toward the coast, following switchback by switchback along the Namorona River where it cascades down the escarpment. The more they distanced themselves from Fianarantsoa, the more wet and lush were the hillsides. Some of that lushness seemed to be undisturbed rainforest. Just outside the small town of Ranomafana they sighted clusters of giant bamboo on the slopes above the river. "This looked like the kind of place," Wright says, "where the greater bamboo lemur might still live."

They hiked up a mountain and pitched tents. It started to rain. For five days the rain continued—and this was supposedly the dry season. Wet and mud-covered, Wright and her colleagues explored anyway, climbing the slick slopes and traversing the foggy ridges, taking unfunny pratfalls, struggling to keep binoculars dry, moving from one grove of bamboo to another, searching for *H. simus*. They had been lucky to meet a pair of forest-savvy local villagers, Emile and Loret, who became their guides. Although not quite such prodigies as the boy Bedo, these two Malagasy men carried intimate traditional knowledge of the sort by which village people had once connected themselves, physically and emotionally, with the forest. Throughout most of Madagascar that knowledge has been lost, as the forest itself has been lost. But Ranomafana was still on the cusp between primordial landscape and agricultural settlement. With help from Emile and Loret, and from the extraordinarily keen forest vision of Bedo, Wright and the others saw plenty of wildlife: red-bellied lemurs, brown lemurs, and a spectacular species of leaping black lemur known as *Propithecus diadema*, the diademed sifaka.

They even got a brief view of *H. griseus*, the gentle bamboo lemur, same species that Wright had studied up north. They watched it eat leaves of bamboo. They noticed that this particular bamboo (later identified as *Cephalostachyum perrieri*) was different from the giant bamboo species. All these small facts would eventually emerge as important.

The sighting of gentle bamboo lemurs, here in the southeastern forest, raised a troubling question. Was the niche for a bamboo-eating lemur preemptively filled? If *H. griseus* was present, did that mean the rare species, *H. simus*, would be absent? Wright and her team kept looking. They found evidence that some stalks of giant bamboo had been chewed on by what seemed a larger animal than *H. griseus*. But they didn't set eyes on a flesh-and-blood representative of *H. simus*.

"I had just about given up hope," Wright related in an article published soon afterward, "when early one cold, misty morning, I caught a glimpse of a lemur clinging to a trunk, making loud tonking calls, and twirling its long, red tail in a circle, like a windmill. It was a golden red color with a short muzzle and short, furred ears that reminded me of the gray gentle lemur, but it was twice the size." She was ecstatic. She thought it might be *H. simus*.

Still, she had expected the greater bamboo lemur to be charcoal gray. Earlier descriptions of the species, based on just a few individual animals, had said nothing about this russet color that to Wright's eyes appeared golden red. But the discrepancy didn't seem too problematic, since color variation within lemur species is common. The species *Propithecus diadema*, for instance, encompasses one subspecies with a white facial ruff and a brownish gray body, another subspecies that can be pure white, and at least three other subspecies that can be black. The color difference between a gray *H. simus* and this new creature at Ranomafana, Wright figured, might be of the same sort: a subspecific distinction within the species. Maybe there was a gray variety of *H. simus* elsewhere and a reddish gold variety here in the southeast, separated by some kind of geographical boundary.

But then the reddish gold animal disappeared. Although her team kept searching, they didn't see it again. Finally they struck camp and climbed back into the Land Cruiser. They drove down the road a short distance, to that fragment of forest surrounded by coffee plantation at Kianjavato. There, only thirty miles away, they achieved the expedition's original goal: They found a dozen surviving individuals of *H. simus*, the greater bamboo lemur. Hoorah, the species was still extant—and Russ Mittermeier's grant had paid off.

But this discovery, as cheering as it was, also kiboshed the hypothesis about different-colored varieties of *H. simus* in different regions, because the dozen animals at Kianjavato were charcoal gray, just as Wright had originally expected. It was perplexing. Kianjavato and Ranomafana were too close together and too similar ecologically, Wright guessed, to harbor two distinct subspecies of a single species. So now another question had to be answered: If the reddish gold creature they had seen at Ranomafana wasn't *H. simus* and wasn't *H. griseus*, what was it? The earlier fieldworkers hadn't said anything about a third species of bamboo-eating lemur.

For a while Wright's group continued exploring. They drove south to the Vondrozo region, where twenty years earlier that poor lonesome *H. simus* had been bought in the village market. Here the landscape was grim. The forests were mainly gone, cut and burned to clear space for growing rice. Bare hills filled the foreground. Only the highest peaks and ridges still held their primary vegetation. If the riversides had ever been graced with groves of giant bamboo, those were gone too. And so was the greater bamboo lemur, as far as Pat Wright could see. Before giving up, she talked with some of the local elders. Yes, they remembered the large gray lemur. Yes, it had vanished with the forest. In addition to that one, however, they spoke of a different animal—a red bamboo lemur. But that animal too had vanished about ten years earlier.

She and her team returned to Ranomafana with renewed interest in the reddish gold lemur. By now another researcher had arrived, a

German named Bernhard Meier. Meier was following a tip about an unidentified species of lemur in the southeastern forest. During the time Wright's team was elsewhere, he too had discovered the reddish gold creature. After his first sighting, he had begun combing the forest for more individuals, more little groups, and charting their visits to various groves of bamboo. Now Wright's team and Meier were both on the trail. They weren't competing—at least not overtly—but they worked separately. Sometimes they met and shared thoughts. Both found these reddish gold lemurs so drastically shy that tracking them was exceptionally difficult.

Pat Wright and her team focused on one group. They spent two months trying to habituate that group to the presence of human observers. But the animals stayed wary, refusing to accept human proximity in even its most discreet form. Meier worked in his own style, moving from one group to another and covering more geographical area. Meanwhile, the matter of identification remained puzzling. "Bernhard and I discussed this a lot," says Wright. "Whether or not it was two species or one species." They agreed eventually that the reddish gold type must be a new species of bamboo-eating lemur, so far unrecorded by science. Its vocalizations were different from those of *H. simus*. Its color pattern was unique. And beyond these bits of behavioral and morphological evidence, there was an ecological item.

As Wright remembers it: "I think the day I was really, totally convinced was the day we found them sympatric here. In the same place. You know, you can't dispute that." The circumstance of sympatry— the fact that charcoal gray individuals of *H. simus* and reddish gold individuals of the mysterious new form were both resident in the Ranomafana forest—strongly suggested that they were distinct species. Wright and her team began calling the new form by a name of its own: the golden bamboo lemur. Later, when it was formally described for the taxonomic record by Meier, Wright, and others, it would become *Hapalemur aureus*.

Before closing her camp at Ranomafana that season, Wright needed one further piece of evidence: a photograph. She knew that a photo would be crucial in supporting the claim that this was a new species. But because of the rainy conditions, the bad light, the wariness of the animals, and other logistical hassles, Wright and her team hadn't captured an image on film. She had nearly resigned herself to that frustration. But young Bedo, the whiz kid of the forest, hadn't.

"On the last day we were there," Wright recalls, "suddenly Bedo came bursting into camp. He said, 'I've found them. They're low. And they're right in the direct sunlight.' " At first she didn't believe him; in addition to being an uncommonly good naturalist, Bedo was a mischievous adolescent with a penchant for practical jokes. Still, Wright and the others followed him back out into the forest. "And there they were. Just sitting there. So we snuck up on them and took a bunch of pictures." One of those shots became the first photograph of *Hapalemur aureus* ever published.

65

THINK OF a heavily trimmed Christmas tree. Near the top, along a single thin limb, placed so close together that their sides touch, imagine three shiny glass ornaments: a small gray one, a large gray one, and a reddish gold one. Let these represent *H. griseus*, *H. simus*, and *H. aureus*, existing sympatrically. A classic case of adaptive radiation, modest in scale, vitreous in delicacy. The tree is Madagascar, festooned with lemurs from base to crown.

Lemurs belong to the suborder Strepsirhini, sharing that category with bushbabies, pottos, and lorises—that is, with the small-bodied, big-eyed, long-nosed, insectivorous, nocturnal sort of primate. They stand distinct from the suborder Haplorhini, which includes chimps and gorillas and us. Some biologists refer to the Strepsirhini as primitive representatives of the primate line. But the word "primitive," you'll remember from the tenrecs, is considered misleading and invidious by other biologists—no less so as applied to strepsirhines. It tends to obscure the fact that lemurs have come a great distance down the path of being successfully lemuroid, adapting quite well to the circumstances they have faced throughout dozens of millions of years. Today they are native only to the island of Madagascar and to a cluster of small islands, the Comoros, off its northwestern coast.

When the Madagascar landmass first split from the African mainland, it contained no primates of any kind, for the very good reason that primates did not yet exist. The timing of the split is uncertain, but it had probably taken place by a hundred million years ago—that is, in the middle Cretaceous, still the high season of dinosaurs. The earliest fossils of lemurlike ancestors come from Eocene strata in Europe and North America, dating back about sixty million years. Some of those lemurlike ancestors, evidently, also inhabited Africa. And they colonized Madagascar. How? The water gap between the continent and the island was still narrow, and a few lemurs must have rafted across. But the gap continued to widen during subsequent epochs as tectonic forces shunted Madagascar out to sea. By the time evolution had produced the Haplorhini—that lineage encompassing humans and apes and monkeys—the channel was too wide for crossing by primates. So, by an accident of timing and opportunity, the lemurs kept Madagascar to themselves. Their distant haplorhine relatives, coming late, had no entree.

Still later, the lemurlike lineage died out on the African mainland and also disappeared from Europe and North America. We can only guess why. Possibly those lemurlike species suffered fatally from competition with monkeys. Possibly they were incapable of defending themselves against the canine and feline predators that had arisen on the mainlands. Their strepsirhine relatives (the ancestors of bushbabies and pottos, among others) did manage to survive in Africa, but at the price of restricting themselves to a narrow range of traits—remaining nocturnal, insectivorous, and solitary, with no intricate social behavior. On Madagascar, meanwhile, the lemurs were free to blossom.

They had escaped to a separate world, a land of gentle forests and empty niches. There was no competition from monkeys and no predation by jackals, hyenas, or leopards. So the lemurs thrived and diversified. In this forgiving place, they could afford to be diurnal and herbivorous; they could live together in small family groups or larger tribes; they could develop elaborate forms of social interaction. Independently, they hit upon some of the same ecological, morphological, and behavioral adaptations that monkeys and apes were inventing in Africa. They also missed some important adaptations—the enlarged brain, for one. And they explored a few possibilities that monkeys and apes mostly missed, such as social dominance by females. They gave a radically different story line to the general theme of primate evolution.

The roster of Madagascan lemurs now includes about thirty living species, and at least a dozen extinct species are known from subfossil evidence. Even that sizable list gives only a hint of lemur diversity throughout time, since the subfossils are mainly quite recent, representing species that survived until just one or two thousand years ago. The more distantly ancestral species, the ones that lived ten or twenty million years back, representing the early lemur lineage in forms we can only guess at, haven't yet been unearthed. Among the subfossil species, several stand out. *Megaladapis edwardsi* was a giant form, at least three times as big as any lemur alive today. *Palaeopropithecus ingens* was smaller but still huge by the standard of living lemurs. This species had the long arms and short legs of an orangutan; it may have moved in a belly-up, four-point dangle beneath branches, like a sloth. Whatever killed off *Megaladapis edwardsi* and *Palaeopropithecus ingens* (one suspect: *Homo sapiens*, having just colonized Madagascar about that time) seems to have doomed large-bodied species preferentially. The giant lemurs are all gone.

The surviving lemurs show a striking geographical pattern. They are spread out through a narrow ring of forests that encircles the island, with overlap among some species, others strictly separated by geographical barriers. Central Madagascar, which was probably once forested too, protrudes above the ring of woodlands like the bald pate of a medieval monk. It's a high, grassy plateau that gets modest rainfall and cold winter temperatures, and in its present condition it's inhospitable to arboreal lemurs. The segmented ring includes several forest types: rainforest along the eastern escarpment and in the north, dry forest with baobab and other thirst-tolerant species in the west, cactuslike thorn forest in the south. (The types are actually more numerous and the distribution of plant species is more intricate, but never mind. We're concerned with the gross pattern, not the welter of details.) At certain places on the ring, the transition between forest types is gradual; at other places, coinciding with ecological barriers, it's abrupt. In the far southeast, for instance, a radial chain of hills divides the eastern domain from the southern domain. Wide rivers, draining seaward off the plateau, add another sort of barrier. The segmented-ring pattern tends to promote evolutionary divergence among forest-adapted species. "In fact, the ring of wet and dry forest of Madagascar," write three lemur experts, "acts almost like an archipelago of islands, in which evolution runs more quickly than [on] either fully divided islands or continuous mainland. This is one clue to Madagascar's riches of lemurs as well as all its other burgeoning forms of life." Archipelago speciation has been the first phase of a two-phase process. Adaptive radiation—in the strict sense, involving competition—has been the second.

In 1970 an anthropologist named R. D. Martin delivered a lecture on adaptive radiation among lemurs. It's hard to imagine how Martin inflicted such an abundance of information on a live audience, but the

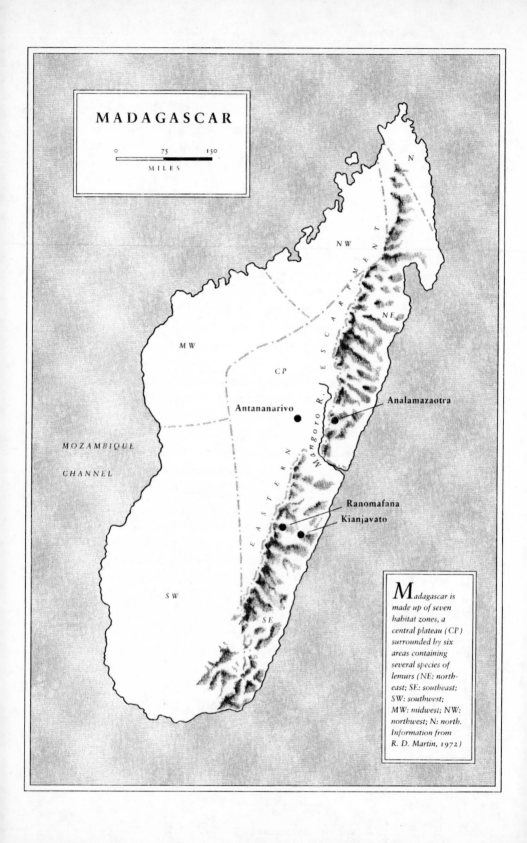

MADAGASCAR

0   75   150
MILES

N

NW

NE

MW

CP

Antananarivo ●

Analamazaotra ●

MOZAMBIQUE

CHANNEL

EASTERN ESCARPMENT

Mangoro

Ranomafana ●
Kianjavato ●

SW

SE

*M*adagascar is
*made up of seven
habitat zones, a
central plateau (CP)
surrounded by six
areas containing
several species of
lemurs (NE: north-
east; SE: southeast;
SW: southwest;
MW: midwest; NW:
northwest; N: north.
Information from
R. D. Martin, 1972)*

sixty-page published version holds a couple of points that are pertinent to what I've been describing. Martin schematized Madagascar
into seven habitat zones—six peripheral segments of forest, with the
central plateau as their hub. The plateau itself was a zone of uncertainty, having possibly once contained a rich lemur fauna that disappeared with the forests. The other six zones harbored combinations
of species. Between his northeastern zone and his southeastern zone
Martin drew a boundary line just south of the Mangoro River, which
flows eastward off the escarpment about halfway between Analamazaotra and Ranomafana. He also delineated a southwestern zone, a mid-
western zone, a northwestern zone, and a zone covering the island's
northern point. He presented a chart showing which species of lemur
occurred where. He used plus marks and minus marks to indicate
whether or not each species was present in each zone.

The fat-tailed dwarf lemur, for instance: present in the far north,
present in the three western zones, absent from both zones of the
east. The woolly lemur, also known as *Avahi laniger*: six plus marks,
reflecting its broad distribution. The diademed sifaka, *Propithecus
diadema*: present in both eastern zones, not elsewhere. In place of
Propithecus diadema, according to Martin's chart, the western and
northern zones harbored its close relative, *Propithecus verreauxi*. The
ring-tailed lemur—that species vaguely resembling an anorectic rac-
coon, familiar to all of us from zoo specimens and cute photographs—was confined to the southwest. The indri, largest and most
astonishing of all living lemurs, with black-and-white fur like a giant
panda and the head of an Afghan hound, showed itself only in the
northeastern zone, including Analamazaotra. The aye-aye, a bizarre
creature with long fingers and nocturnal habits, was also confined to
the northeast. The sportive lemur, like the woolly lemur, occurred in
all six forest zones.

This was a simplified version. Martin himself acknowledged that
the four populations of *verreauxi* might actually fall into separate
species. Other experts have added qualifications of the same sort, and
finer taxonomic distinctions are still being made, changed, and argued. The whole picture has gotten more complicated since Martin
delivered his lecture.

Even Martin's early chart wasn't entirely unambivalent. He used
question marks in place of his plus and minus signs to represent indeterminate status. Maybe the indri might survive in some unsurveyed
scrap of woodland on the central plateau. Maybe the sportive lemur

might too. For *Hapalemur simus*, the greater bamboo lemur, Martin offered seven question marks, one for each zone: "? ? ? ? ? ? ?" It suggested a note of despair as much as epistemological caution. At that time, the species was presumed extinct everywhere. And of course *Hapalemur aureus*, the golden bamboo lemur, was not even listed. Its existence was a revelation for later.

In summary, Martin judged that "the situation in Madagascar is quite similar to the situation in the Galapagos Islands, except that the barriers to migration are provided by rivers and climatic factors (rather than by expanses of sea), and that the closed system in Madagascar has probably operated for a far greater period of time than that in the Galapagos Islands." The natural boundaries between forest zones, Martin guessed, had allowed genetic divergence among lemur populations to progress to the point of no return. The segmented-ring pattern of forests had become a mandala of speciation. Later, the original boundaries had been breached, at least by a few adventurous individuals. New forms had invaded neighboring zones, where they were obliged to share habitat with their close kin. Those situations of sympatry, according to Martin, provided "the basis for subsequent specialization of new species."

He was talking elliptically to an audience of professionals. The "basis" he had in mind was competition. Closely related species, when they found themselves suddenly sharing a habitat zone and competing for its resources, would have faced the familiar imperative: to radiate or to die.

"With the lemurs," Martin noted, "it may well be the case that ecological competition has provided one of the main factors leading to the extensive radiation of this isolated group of Primates." He didn't dwell. The lecture was long enough already. For more on that subject, he said, a person could always turn to David Lack's book on Darwin's finches.

66

IN SOME CASES, adaptive radiation is subtle. External anatomy gives almost no clue. Behavioral differences seem minor and inconsequential. In the forest at Ranomafana, three closely related species within the genus *Hapalemur* were living sympatrically, having somehow partitioned the role of bamboo-eating lemur into three distinct niches. Pat Wright wondered how.

For several years after the discovery of *H. aureus*, Wright paid special attention in her field studies to the question of diet. She worked with a small group of collaborators, American and Malagasy, among whom was a plant chemist from Illinois. Together and individually, these people gathered many hours of precise behavioral observations, took weights and measurements from a few captured animals, collected samples of bamboo, and performed chemical analyses on the samples. The chemical analyses entailed dicing the plant material while it was fresh, adding it to a hot alcoholic medium, boiling, cooling, decanting, and then testing the solution with a strip of color-responsive paper. The paper would turn dark blue in the presence of cyanide, which occurs as a natural lacing in some bamboo. How much cyanide could a lemur tolerate? The answer, for one species, turned out to be: lots.

Combining behavioral ecology and biochemistry, Wright and her colleagues reached two conclusions, the first of which wasn't especially dramatic. Bamboo-eating, they found, is not a single mode of behavior but several. Each of the three *Hapalemur* species went about it differently, either by eating different parts of the bamboo plant or by eating different species. *H. griseus* ate the leaves of one bamboo species, *Cephalostachyum perrieri*. *H. simus* ate the pith from mature stalks of the giant bamboo, *C. viguieri*. The golden bamboo lemur, *H. aureus*, also ate giant bamboo but not the pith from mature stalks. Pat Wright's observations revealed that *H. aureus* confined itself to the new shoots, the leaf bases, and the pith from narrow stems.

The difference in eating habits between *H. simus* and *H. aureus* was especially conspicuous. *H. simus* seemed to spend more effort, chewing apart those thick, fibrous stalks in order to get at the pith; *H. aureus* seemed to exercise more finesse, eating those tender shoots, leaf bases, and small stems. Why shouldn't *H. simus* save itself trouble and eat the more tender parts too? The division of food resources was puzzling—at least until the chemical analyses were done.

The chemical analyses showed that the shoots, the leaf bases, and the narrow-stem pith—all eaten by *H. aureus*—were full of cyanide.

The stalks eaten by *H. simus*, on the other hand, showed no trace of cyanide. Neither did the leaves of *C. perrieri*, favored by *H. griseus*. *H. simus* and *H. griseus* were avoiding the stuff, as most animals sensibly would. *H. aureus*, newly discovered and still quite mysterious, was gobbling it recklessly.

Cyanide is ordinarily quite toxic to mammals. The lethal dose for a dog is about 4 milligrams per kilogram of body weight; for a cat, 2 milligrams. For a rat, slightly more tolerant, 10 milligrams per kilogram of body weight is lethal. As determined for a variety of mammal species, the average lethal dose is 4.3 milligrams per kilogram of body weight. Since a golden bamboo lemur weighs about a kilogram and a half, 6.5 milligrams of cyanide should be fatal—assuming that *H. aureus* is as vulnerable as other species. But it isn't. Wright and her coworkers estimated that a golden bamboo lemur each day swallows twelve times as much cyanide as should be expected to kill it.

"We do not know how golden bamboo lemurs avoid cyanide poisoning," they reported, "but there are several possibilities." One possibility involved detoxification of cyanide with help from certain amino acids. But those amino acids are almost entirely unavailable to a bamboo-eating lemur, so the detox hypothesis seemed weak. Another possibility was that maybe *H. aureus* maintains a severely acidic stomach—beyond the pale of Rolaids, beyond Mylanta, an excruciating brew of acidity—which might inhibit the release of cyanide into its system. Does the golden bamboo lemur have such a stomach? Nobody knows, not even Pat Wright, because the digestive system of the species has never been studied.

These three species of sympatric *Hapalemur*, with their niche-partitioning differences in diet and physiology, could stand as a prototypic example of adaptive radiation, more instructive than even Darwin's finches. But there are still unanswered questions. What ecological conditions originally shaped them? Unknown. What biogeographical incidents brought them together at Ranomafana? Unknown. What is their phylogenetic history? Unknown. Creating an evolutionary scenario for the bamboo lemurs would be guesswork, and scientists are chary of guesswork. They do it, but they do it in private. Guesswork in public makes them look intellectually whimsical, whereas a scientist (especially one like Pat Wright, who goes off to work in a jungle, emerges months later covered with leech bites and mud, and announces a new species of primate) needs to be taken seriously by her professional peers.

I'm not burdened with that problem, so I can afford to guess. Let's suppose that the gentle bamboo lemur, *H. griseus*, widespread and successful throughout Madagascar, was the first of its genus to occupy the forest at Ranomafana. It flourished there on a diet consisting mainly of *C. perrieri*. Maybe it didn't specialize solely on that species of bamboo;

maybe it also ate the leaves and the fruits of other plants. Anyway, at Ranomafana it held its niche solidly. Let's postulate that the greater bamboo lemur, *H. simus*, having evolved somewhere else on the island, arrived next. Though bigger than *H. griseus*, in the Ranomafana region it had the disadvantage of being a newcomer. It found itself unable to displace *H. griseus* from the tasty patches of *C. perrieri*. Still, as a heftier animal with stronger jaws and teeth, it could resort to another ecological option: feeding on the giant bamboo, *C. viguieri*. Under this arrangement, let's suppose, *H. griseus* and *H. simus* existed sympatrically, sharing the bamboo resource by browsing on two distinct species.

Now another speculative leap. Remember that plants adapt to predation, just as animals do. Imagine that *C. viguieri* adapted to the herbivorous attacks of *H. simus* by developing the capacity to concentrate cyanide in its most vulnerable parts. This adaptation helped to discourage the lemurs—but not completely. By ignoring the vulnerable parts, by instead chomping its way through the mature stalks and eating their pith, *H. simus* could still support itself on *C. viguieri*.

Then, let's say, *H. aureus* arrived—the golden bamboo lemur. Maybe it had evolved from *H. griseus* (as Pat Wright suspects) in some pocket of forest where the only available bamboo was riddled with cyanide. Maybe this new species had diverged from *H. griseus* by adapting itself, like Mithridates, to a poisonous diet. Expanding its range later to Ranomafana, *H. aureus* faced a different sort of challenge: not just poison but competition. It was obliged to find its own niche. It survived by eating what no other bamboo lemur could stomach.

67

PAT Wright's field camp at Ranomafana consists of a tight wooden building that functions as the storeroom and office, a canopied porch that serves as the mess hall, a latrine, and a scattering of tents in the forest nearby. The immediate research site is spread across mountain slopes above the Namorona River, crisscrossed with a network of irregular trails. In the wet season, the trails flow slick with mud. The fashion in field dress, for both biologists and visitors, runs to tall rubber boots and rain parkas and rain pants, with fetid polypropylene underneath. The camp is a cheerful place to anyone with an interest in lemurs and no fetish for being clean and dry.

I pitch my own tent on a flat spot along one of the trails, with a borrowed ground cloth and trampled brush underneath to keep liquidy earth sauce from seeping up through the floor. After two weeks of living there, feeding on rice and beans with Wright and her students and colleagues, slogging the trails, studying the biologists while they study the animals, washing my socks in the river, drinking cheap Malagasy rum at sunset in good company on the porch, scribbling notes late at night by the light of my headlamp, I feel almost adapted.

One morning I go out with a laconic but very competent Malagasy guide, Pierre, to look for the famous *H. aureus*. We set off toward the river, then turn uphill, climb to a ridge, and traverse it. Dropping over the ridge onto a lesser trail, we descend to a stream bottom and follow that through a thick grove of bamboo. The bamboo shafts are bowed, as though under the weight of rain. They form an arched canopy beneath which we bushwhack. Pierre, with the focus of any good tracker, stoops to pick up a severed stem. "*Ils mangent ici*," he says. "*Peut-être hier.*" The stem tells him that lemurs paused here for a bite of lunch, maybe yesterday.

After another hour of walking, down slope and up, through open understory and more thickets of bamboo, Pierre speaks again. "*Ah. Ici.*"

Where? I don't see anything. Where?

And then I spot them, just thirty feet above in the canopy of drooping bamboo, backlit against the pale gray sky: a pair of fuzzballs no bigger than house cats.

"*Aureus?*"

"*Oui*," says Pierre with proud certainty. "*Aureus.*" Golden bamboo lemurs, yes.

The morning is too drizzly and chilly for them to bestir themselves for feeding. Or else they have fed earlier and just don't happen to be hungry. Anyway, they show no interest at present in the succulent, cyanide-laced leaf bases and stems around them. They seem to be huddling together for warmth. They barely move. In objective terms it's not a dramatic sight—two reddish brown shapes of fur—but I've come a long way, I've heard much, and this *is* one of the world's rarest primates. For a long while I sit watching.

When the rain begins drumming more steadily, I raise the hood of my parka. The ground is soggy against my ass. I hunker. The golden bamboo lemurs hunker. I gape at them and, every so often, they glance pityingly down at me. An hour creeps by. The rain doesn't stop and the lemurs don't perform any memorable behavioral hijinks. A few leeches come inchworming up my legs, thirsty for blood. I flick them away without malice. I savor another day of romantic adventure in the rainforest.

Late in the afternoon, back at camp, an intestinal bug reaches ferment in my belly and I go to my tent without supper. Achy and nauseated, I zip myself into the sleeping bag. I sleep fitfully for many hours, alternating between chills and shirt-soaking sweats. Around midnight I get up, stumble out into a patch of wild ginger behind the tent, and force myself to puke. This provides some relief. I go back to bed.

All considered, I've been lucky enough to stay fairly healthy during my weeks here at Ranomafana. But I'm a short-term visitor, not immune to all local challenges—and I haven't even sampled the bamboo.

68

ADAPTIVE radiation, your good memory will confirm, was the last of the species attributes on our Insular Menu:

dispersal ability
size change
loss of dispersal ability
endemism
relictualism
loss of defensive adaptations
archipelago speciation
adaptive radiation

Comprehend those eight, and you have comprehended a fair bit of evolutionary biology, applicable on continents as well as on islands.

If this book were devoted solely to evolutionary biology, we would both, you and I, need to comprehend a lot more. It's a huge subject, encompassing more cogs and ratchets than a clock tower. One among the many ratchets is Ernst Mayr's theory of genetic revolutions. That one in particular merits attention here because it pertains to large things that happen in small places.

Mayr's theory of genetic revolutions was an advance from his founder principle, described in *Systematics and the Origin of Species*. He had argued, in *Systematics*, that evolution proceeds more rapidly among small populations than among large ones. As an example, he mentioned the paradise-kingfisher group of species—*Tanysiptera galatea* and others—on mainland New Guinea and its satellite islands. *Tanysiptera galatea*, the common paradise-kingfisher, is an arresting bird with the unmistakably huge head and the trowel-like beak characteristic of kingfishers generally. Beneath the big head is a small blue-and-white body and a long pair of white tail feathers, slender with terminal tufts, like iced-tea spoons. During his own early fieldwork in that part of the world, Mayr had studied the distribution of *T. galatea* and its allied species. *T. galatea* itself occupied New Guinea proper, having diversified there into only three subspecies, despite the great heterogeneity of available habitats. But among the smaller offshore islands, the genus *Tanysiptera* had diversified into six different forms, most of which were considered distinct species. Evolution seemed to be moving slowly on the New Guinea mainland and fast on the satellites. Mayr's inference was that new species might arise more quickly in situations where a small founder population becomes isolated.

It seemed to hold true among the bumblebees of the genus *Bombus*, and among the gall wasps of the genus *Cynips*; it seemed to hold true among various species in the West Indies, the Solomon Islands, and Hawaii. "The potentiality for rapid divergent evolution in small populations," Mayr concluded, "explains also why we have on islands so many dwarf or giant races, or races with peculiar color characters (albinism, melanism), or with peculiar structures (long bills in birds), or other peculiar characters (loss of special male plumage in birds)." He could easily have cited the Aldabran tortoise, the chuckwalla of Angel de la Guarda, the hippo of Madagascar, the beetles of Madeira, the elephantids of Timor, the iguanas of the Galápagos, the finches of Darwin and Lack, the earwig of Saint Helena, and the dodo. He was alluding to them all.

Why should such oddities appear within small populations so rapidly? Genetic drift plays a role, but genetic drift isn't notably quick. Natural selection plays a role, but natural selection is slower still. Mayr had suggested in *Systematics* that the main cause of speedy insular evolution is the founder principle. A small pioneer population tends to show abrupt evolutionary divergence from the ancestral population because its gene pool is just a tiny and unrepresentative sample from the larger pool.

Still, the founder principle didn't satisfy Mayr. The rate of evolutionary divergence seemed too fast. For some years this discrepancy remained "a great puzzle" to him. Then, in 1954, he offered a new explanation that was bolder and more speculative: the notion he called "genetic revolutions." Mayr would eventually look back on it as "perhaps the most original theory I have ever proposed."

He presented it in a paper titled "Change of Genetic Environment and Evolution." At its core was the concept of "genetic environment"—Mayr's oxymoronic term for the matrix of interdependent effects that link the alleles of a given gene pool with one another. A sudden change in that genetic environment, he proposed, could initiate a cascade of further changes in the way each allele interacts with others. One possible trigger for such a cascade of changes was the founder principle. When a small population of some animal or plant species became isolated, bearing only a small sample of the ancestral alleles, the new genetic environment might be drastically different from the genetic environment of the larger population. If so, a genetic revolution would ensue, as the small population adapted itself to its new circumstances.

To make sense of this, we've got to remember that alleles don't have absolute values. Adaptation itself is a relativistic concept, and the adaptive (or maladaptive) value of any allele is relative to context. Though a given allele might be helpful (or harmful, or possibly neutral) within one environment, within a different environment its value might be different. An allele that contributes toward gigantism in a species of bird, for instance, might be helpful on an island but harmful on a mainland.

One aspect of environment is external—the physical and biological conditions of habitat. Another aspect of environment, in Mayr's view, is genetic—the matrix of interdependent alleles within an individual and the broader matrix of alleles within a population. Alleles shouldn't be thought of as so many beans in a beanbag, he argued. They function as parts of an integrated system; when the system is altered, a particular allele's value may be altered too. An allele that has been beneficial in one genetic environment might be lethal in another. An allele that has been inconsequential in one genetic environment might be assertive in another. An allele that was only slightly advantageous when it was rare in a population might turn out to be highly advantageous when the whole population carries it. Assuming that the alleles of one gene interact with the alleles of many other genes in the same creature and with the greater gene pool of the breeding population, the cascade of changes consequent from a change of genetic environment could be

dramatic. "Indeed, it may have the character of a veritable 'genetic revolution,'" Mayr wrote.

In the 1954 paper, he returned to the example from his earlier
book: the paradise-kingfishers in New Guinea. This time he focused on
Numfor Island, a small offshore satellite less than fifty miles from
New Guinea's northwestern coast. Numfor harbors a distinct species
within the paradise-kingfisher group, *Tanysiptera carolinae*. Mayr postulated that, sometime in the past, a few pairs of *Tanysiptera galatea* happened to reach this little satellite from the mainland. They bred and
multiplied, establishing a small population. With time, in the new circumstances, the character of the population changed. They became *T.
carolinae*, distinct from *T. galatea* in having a blue breast instead of a white
breast, as well as legs and feet of bright yellow. Possibly they had also acquired some behavioral differences. What effected these changes?

The climate of Numfor is very much like the climate of the near
coast of New Guinea, according to Mayr. The flora and fauna of
Numfor are somewhat different but not drastically so. The chief
predator upon kingfishers of the coastal mainland, a species of
goshawk, is also present on Numfor. "The physical and biotic environments in these places are thus rather similar," he wrote. "A third
environment, however, the genetic environment, is strikingly different." The results of that difference in genetic environment may have
been profound: alleles performing in new ways, offering new syner-
gies with other alleles, yielding new advantages and new disadvantages in a kingfisher's struggle for reproductive success.

It would have been an unstable situation. Some common alleles, no
longer so valuable, would have quickly become rare. Some rare alleles,
suddenly useful, would have become common. Natural selection
would have enforced utility and economy. The pace of evolutionary
change would have surged. Eventually the system would have settled
into a new equilibrium. When it did, there was *Tanysiptera carolinae*,
newborn from old materials, a paradise-kingfisher with a blue breast
and yellow feet and other invisible eccentricities that ornithologists
haven't yet plumbed.

Such a genetic revolution might bring changes in anatomical structure, in size, in physiology, in habits. It might turn a giant into a
dwarf, a carnivore into a vegetarian, a flying animal into a fat,
grounded waddler, or a fat, grounded waddler into a sleek swimmer.
Oceanic archipelagos, Mayr declared, are the best places on Earth to
see the results of such upheaval.

But not the only places. He added, "I visualize all major evolutionary novelties as occurring in a similar manner."

69

IT'S NO WONDER that islands, being so conducive to genetic restructuring, have served as the open-air laboratories in which Ernst Mayr and other evolutionary biologists have learned many lessons. But the final two items on our Insular Menu are more strictly biogeographical than evolutionary. That's why I give them their own category:

II. Community attributes

disharmony
impoverishment

These two involve broader patterns, not traits characteristic of species. They pertain to the most basic questions of biogeography: Which creatures occur where, which creatures are absent where, and why?

Disharmony and impoverishment both relate to the fact that species diversity isn't uniformly distributed across the planet's surface. Impoverishment refers to a paucity of species in certain places; disharmony refers to unequal representation of species, or groups of species, in different places. Logically the two concepts overlap, impoverishment being just one type of disharmony—a type that's especially characteristic of islands.

An island generally contains less than its share of species diversity, relative to a nearby continent. That is, the island harbors fewer species than does an equal area of mainland habitat. This is impoverishment. Disharmony is trickier to define. It encompasses any discrepancies of species representation, either positive or negative, between an island and its neighboring continent. Although islands contain less than their share of some plant and animal groups, they contain more than their share of other groups. Madagascar, for example, has all the world's lemurs; Africa has none. So, at least among that group, Madagascar far exceeds the species diversity of its neighboring continent. The Galápagos are inordinately rich in finch species. They far exceed the finch diversity (per unit of area, at least) of their neighboring continent, South America. This is disharmony.

Differences of dispersal ability are one source of disharmony. Some groups of creatures are better transoceanic travelers than others, and those creatures tend to be well represented on islands. The poor travelers are poorly represented. An island will usually contain fewer mammal species, fewer amphibians, and fewer freshwater fish than the nearest continent. It will often contain fewer reptiles. Most likely, too, it will contain fewer birds, though not by such a big margin. But because of prodigious speciation, an island may well contain *more* different species within a certain group of creatures (more fruit flies or more mockingbirds, say) than the nearest continent. This is the scale at which disharmony is distinct from impoverishment.

Disharmony also reflects an island's remoteness and its age. Sri Lanka, connected to India during the last ice age, contains an assemblage of species showing no great disharmony from the species assemblage of India. Madagascar, being old and remote, shows radical disharmony from eastern Africa, and not just among lemurs and tenrecs.

Besides dispersal differences, remoteness, and the passing of time, at least one other factor contributes to disharmony: size of the island. Small islands gain fewer species by colonization than large islands do. And small islands lose more species by extinction. Mark that one well: *Small islands lose more species by extinction.* We'll come back to it, soon and often.

Patterns of disharmony on islands are often expressed as negatives. Something or other *isn't*. Something *doesn't*. For instance: Hawaii has no native species of oak tree. Nor does Hawaii have any of its own maples, willows, or elms. Though the Hawaiian archipelago harbors six thousand endemic species of insect, not one of those species is an ant.

Madagascar has no native cats and no native antelopes, despite its proximity to the African savannas.

Tristan da Cunha has no birds of prey.

New Guinea has no woodpeckers and no vultures.

New Zealand, whose nearest large neighbor is Australia, the great cradle of eucalyptus diversity, contains no native eucalyptus trees itself. Nor does New Zealand have native marsupials.

Snakes do not generally live on very small islands.

Frogs, as I've mentioned, are meagerly represented on islands. The Galápagos archipelago has plenty of reptiles but no amphibians.

Mammals are meagerly represented. And of the mammal species that do occur naturally on islands, few are large-bodied. A major portion are bats or rats.

Tasmania, though almost as rich as the Australian mainland in pigeons and quails, has only half as many species of owl.

Aldabra supports little besides tortoises, beach flies, opisthobranch mollusks, and polychaete worms.

Mauritius, until humans arrived, was devoid of terrestrial mammals. Not even rats had managed to reach it. Then the boats came. Disharmony took on a new meaning.

70

THE SUBJECT of impoverishment was touched on by Darwin in The Origin of Species. "The species of all kinds which inhabit oceanic islands are few in number compared with those on equal continental areas," he wrote.

Darwin wasn't the first student of islands to notice it, and he wasn't the last to consider it important. But the causes of this simple truth are complicated, and the implications are far-reaching and controversial. In the early years of our own century it was refined beyond Darwin's bare statement by other biologists, who noticed a corollary: Not only are islands impoverished relative to the mainlands, but small islands are more severely impoverished than large ones. That bit of insight became famed as the *species-area relationship*. During your lifetime and mine, the science of ecology has responded to it as oceans and coyotes respond to the moon.

The species-area relationship is empirical, but intellectual wars have been fought to interpret it. Mention the phrase to a biogeographer or a population biologist or a theoretical ecologist or an ornithologist who has worked in New Guinea or Indonesia or the Antilles, and your meaning is clear. No one asks: *Which* relationship? A vast corpus of relevant data and ideas is evoked as the talk turns toward fourth-degree consequences of impoverishment. The word "impoverishment" itself is less often spoken. It's too basic. The sun rises in the east, the pope is Catholic, islands are impoverished.

But *why* are islands impoverished? There are two basic reasons. One: Many species fail to arrive. Two: Some species, having arrived, having colonized, fail to endure. The first is attributable to isolation, distance, blue water, and other barriers to dispersal and colonization. The second involves extinction.

It's a point of so huge a bigness that I've left it for last. Dispersal ability, size change, loss of dispersal ability, endemism, relictualism, archipelago speciation, disharmony, and the rest—they are all characteristic of insular evolution and insular ecosystems, but nothing on that list is more characteristic than extinction. A high jeopardy of extinction comes with the territory. Islands are where species go to die.

&

RARITY

UNTO DEATH

&

71

THE UNGAINLY and gigantized species of pigeonlike bird that eventually received the name *Raphus cucullatus*, more familiar to us as the dodo, had lived on Mauritius a long time. No one knows how long. Probably the lineage had been there many thousands of years. It was an evolutionary success; it had adapted itself well to the local conditions. It existed nowhere else.

Most likely it ate fallen fruit. It may also have fed, at least occasionally, on seeds and bulbs. It grew big and husky, storing fat to a degree that true pigeons could not, and it ran through the forest on stout legs. Probably it wasn't a fast runner, not what we'd call graceful, but competent within its context. As its body had evolved toward greater size, its wings hadn't, and at some point it dispensed with flying. This was a well-measured compromise that yielded more advantage than disadvantage. A bird so hefty, with such a big head and such a wide gape, could swallow sizable fruits whole—pounding down big meals quickly and thereby stockpiling nutriment efficiently during the seasons of bounty. Bulked up, it could survive through the seasons of scarcity. The smaller species of fruit-eating birds, mincing and tentative by comparison, less able to gain and lose weight, would have been hard-pressed to compete. On the negative side, flightlessness meant surrendering one means of escape from enemies. But that was an easy sacrifice, since the dodo had evolved in an ecosystem impoverished of predators. Mauritius in its primordial state contained no terrestrial mammals—no rodents, no carnivores, no humans. There were some large reptiles, but none that could intimidate a thirty-pound bird with a beak like a claw hammer. Flightlessness dictated that the dodo put its nests on the ground. That too was no problem, since ground nesting in this ecosystem was safe and economical. The big bird may have gathered grass into an egg-holding basket, as claimed by one historical source. The same source suggests that a dodo clutch consisted of a single white egg, approximately the size of a pear. The dodo seems to have carried pebbles in its crop to help grind its food. Its sharply hooked beak may have been used for ripping morsels away from large fruits held on the ground with its claws. A recent historian has deduced that, unafraid of water, it drank and bathed freely but was unable to swim.

This much is all we really know about the ecology, behavior, and biogeography of *Raphus cucullatus*. Everything else is rumor.

The song of the dodo, if it had one, has been lost to human memory. A visitor on the island in 1638 recalled later that it made a squeak like a gosling. Another historical report, from a Dutch mariner shipwrecked on Mauritius in 1662, said: "When we held one by the leg he let out a cry, others came running forward to help the prisoner, and were themselves caught." But a scream of distress is not to be confused with the natural call of a species. More recently Errol Fuller declared in his book *Extinct Birds*: "The call was never clearly described but one school of thought inclines to the opinion that the word 'dodo'—probably coined by Portuguese sailors—is simply a rendering of it." From there Fuller digressed into etymological speculation about its name, without saying more about the bird's voice. Other modern sources on *Raphus cucullatus* generally don't go into the subject of vocalization at all. The song of the dodo, if it had one, is forever unknowable because no human from whom we have testimony ever took the trouble to sit in the Mauritian forest and listen.

As for the dodo's anatomy, we have three or four cobbled-together skeletons preserved in museums, a scattered assortment of other bones, one severed and dried foot, a few written records of the bird's appearance, and an array of paintings and engravings dating back to the year 1600. The three-dimensional reconstructions displayed in some museums, suggesting taxidermy, are only dummies made of plaster and wire and glued-on plumage from expendable poultry. The paintings and engravings form a chain of derivation and distortion, later artists having loosely copied earlier ones. What the images commonly show is a homely animal with an oversized, knob-ended beak and a bare face, behind which the feathered part of its head looks like the hood of a sweatshirt. The nakedness of the face, the gawkiness of the feathery hood, the knobbed beak, and the gross body have contributed toward making the dodo seem unreal. But it wasn't unreal, of course. Beyond the reconstructions and the half-fantasy portraits lay a real bird whose face, we can reasonably assume, was featherless. If the paintings can be trusted at all, the dodo's body plumage was gray and the ineffectual little wing feathers were either yellowish white or, alternatively, black. A few of the engravings and paintings are persuasive, evoking a flesh-and-blood animal, but others are cartoons. None of the images, it seems, was done by an artist who had ever seen *Raphus cucullatus* in the wild.

So the story of the dodo is obscured by a fog of uncertainties. We can be sure that it lived. That it was confined to Mauritius. That its shape was consistent with the remnant skeletons. That it thrived for a long time and then met disaster. That the disaster was caused, directly or indirectly, by *Homo sapiens*. Isn't this a fair bit of knowledge about an animal that's been gone for centuries? Well, no. For an instructive comparison, think of the hadrosaurian dinosaur *Maiasaura peeblesorum*, which died out about seventy million years ago. Paleontologists possess better data on that species than ornithologists possess on the dodo. There are fossilized eggs and embryos of *Maiasaura peeblesorum*. But there is no verified egg and no embryo of a dodo in any of the world's museums. We know that it was a bird. Warm-blooded. Earthbound. Conspicuously meaty and, as matters developed, vulnerable. We know that, beginning around 1600, it faced a struggle for survival in the presence of humans and pigs and monkeys. We know very little else. The main certainty about *Raphus cucullatus* is that by about 1690, if not earlier, it was extinct.

72

THE EXTINCTION of the dodo is representative of modernity in several different ways, not least of which is that the event occurred on a small island. The small size and the remoteness of Mauritius make it seem a marginal place, but that impression is misleading. Small islands are central to the matter of species extinction, just as species extinction is central to the question of how *Homo sapiens* affects its own world. Throughout the past four hundred years, as humanity achieved ubiquity and dominance on Earth, the extinction of species has been largely an island phenomenon. Recognizing and understanding that phenomenon is the best route toward understanding our present crisis of extinctions on the mainlands.

The pattern of island extinctions is most clear among birds—partly because the dossier is more complete than for other groups of animals, birds having been so carefully watched and identified and so feelingly missed. The evidence for mammals and reptiles is more complicated and not so well documented, though it tends to support the same point. The bird data are stark. Several scientists have made independent tallies, which differ slightly but all show that species extinction has been especially associated with islands.

A set of figures offered by Jared Diamond (the ecologist who speculated about ancestral Komodo dragons feeding on pygmy elephants) is probably as reliable as any. These figures come from a paper titled "Historic Extinctions: A Rosetta Stone for Understanding Prehistoric Extinctions," in which Diamond discussed recent species losses within a broader paleontological context. Let's ignore that broader context for now, and consider just the recent particulars. Since the year 1600, according to Diamond's count, 171 species and subspecies of bird are known to have gone extinct. The earliest and most famous was the dodo. Of the total, 155 species and subspecies lived and died on islands. That's more than ninety percent.

Hawaii lost twenty-four species and subspecies. The islands around Baja California, according to Diamond, lost eight. The Mascarene Islands, including Réunion and Rodrigues as well as Mauritius, together lost thirteen besides the dodo. The West Indies lost fifteen. Lord Howe Island, a lonesome little bump halfway between Australia and New Zealand, lost more species and subspecies of bird than the combined total lost in Africa, Asia, and Europe.

The numbers seem still more extreme when you consider that only twenty percent of the world's bird species are confined to islands. Nine-tenths of the historical bird extinctions, then, have occurred in habitats holding one-fifth of the species. This suggests that an island bird faces about fifty times as great a likelihood of extinction as a mainland bird. Furthermore, another tally indicates that three-quarters of the insular extinctions occurred on small islands. Large islands were much safer. So a bird on a small island, such as Mauritius, seems to face a still greater likelihood of extinction than a bird on a big island. Why?

The answer to that one is intricate.

73

PORTUGUESE sailors first reached Mauritius in 1507, during the era when Portuguese sailors were going everywhere. The expedition, under command of Alfonso Albuquerque, probably touched also at Réunion and Rodrigues. This counts as the European discovery of the Mascarene Islands, though seafaring Arab traders from the lands around the perimeter of the Indian Ocean already knew of them. There were no human inhabitants of the Mascarenes at the time Albuquerque's group visited. In 1512 came another Portuguese navigator, Pedro Mascarenhas, who left behind his name and apparently little else. Throughout the rest of the sixteenth century, the Portuguese manifested no very keen interest in Mauritius, although by one account they did introduce some domestic animals. Still, they came and went more lightly than later European visitors. And if the Portuguese ever caught sight of a dodo, they don't seem to have mentioned it to posterity.

With the Dutch, who arrived at the end of the century, it was different. An expedition commanded by Jacob Cornelius van Neck reached Mauritius in 1598. Beginning that year, Dutch sailors making transits across the Indian Ocean treated the island as a pasturing and breeding ground for livestock and as a source of wild native meat. I've already noted that they ate up a share of the native tortoises. They also ate dodos.

The earliest record of dodophagy comes from a narrative of the van Neck voyage, published in 1601. Since the dodo at that time was still an unheard-of beast, the van Neck account introduced it visually before offering a culinary critique:

> Grey Parrots are also common there, and other birds, besides a large kind, bigger than our swans, with large heads, half of which is covered with skin like a hood. These birds want wings, in place of which are three or four blackish feathers. The tail consists of a few slender, curved feathers, of a grey color. We called them Walckvögel, for this reason, that the longer they were boiled, the tougher and more uneatable they became.

It was probably the first European report of the dodo's existence, but the pattern of contumely was already set, *walckvögel* being Dutch for "disgusting bird."

In various spellings, that epithet recurs throughout later accounts as a conventionalized insult: *Walgh-voghel, walg-vogel, waldtvögel, wallighvogel, walyvogel*. A typical complaint was that "even long boiling would scarcely make them tender, but they remained tough and hard, with the exception of the breast and belly, which were very good." In the same spirit, a titled English visitor observed snottily that the dodo "is reputed more for wonder than for food, greasie stomaches may seeke after them, but to the delicate they are offensive and of no nourishment." No nourishment to his rarefied sensibility, anyway; but most early travelers to Mauritius couldn't afford such choosiness. Tough or tender, the dodo was never quite disgusting enough for its own good.

Several years after van Neck's expedition another ship reached Mauritius. The journal of its captain reported that a foraging party returned from shore with some impressively fat dodos, and "the whole crew made an ample meal from three or four of them, and a portion remained over." Ten days later, the foragers brought back another two dozen dodos, enough to suggest either that dodo hunting had quickly progressed to a fine art or that the birds weren't hard to kill. They were so big and heavy that a pair of them more than fed the crew, "and all that remained over was salted." Still later, five men went ashore "provided with sticks, nets, muskets, and other necessaries for hunting. They climbed up mountain and hill, roamed through forest and valley, and during the three days that they were out they captured another half hundred of birds, including a matter of 20 Dodos, all which they brought on board and salted." Easy pickings, like the tortoises.

This journal mentioned three different names for the big flightless bird: *dronten* and *dod-aarsen* as well as *wallich-vogel*. The names are significant insofar as they reflect fresh early impressions, later to be replaced by a received caricature. But the etymology of those names is controversial. One scholar declared that *dronten* isn't traceable to the Dutch language; another scholar claimed that it's an obsolete form of a Dutch verb meaning "to be swollen." The second scholar asserted that *dod-aarsen*, too, alludes to the bird's bulbous shape: *dod* from the Dutch for "round heavy lump" and *aarsen* as cognate to the venerable English "arse." Round-ass. Fat-ass. The swollen, lump-assed bird. But our first scholar argued that *dod-aarsen* derives rather from the Dutch word *dodoor*, meaning "sluggard." Centuries were passing as the debate went back and forth, and whole dictionaries were becoming obsolete. To confuse matters further, *dod-aarsen* (like *walckvögel*) has a number of variants, appearing in some of the early sources as *dodaersen, dottaerssen, dodderse, dodars*, even *totersten*. What can I tell you? The Gutenberg revolution was young and nobody gave a shit about spelling. *Dodoor* became *dodars*, according to our first scholar, and *dodars* in turn became *dodo*. The sluggardly bird. The slow, lazy bird. There is a third opinion, offered by the snotty Englishman of the delicate stomach, to the effect that *dodo* comes from a Portuguese word, *doudo*, meaning "foolish" or "simple." The Englishman was Sir Thomas Herbert, whose travel narrative was published in 1634. Herbert might merit some credence on this point, since he was reputedly the first person to use "dodo" in print. And there is a fourth opinion, mentioned by Fuller in his *Extinct Birds* and traceable to earlier sources: that "dodo" was an onomatopoeic approximation of the bird's own call, a two-note pigeony sound like "doo-doo."

The various conjectures all converge on a common notion: the foolish, swollen, sluggardly bird with the big butt. It may not be ornithologically accurate, it may not be fair, but it's recognizable as the dodo of legend.

What the European witnesses saw as stupidity and laziness we can take to have been only a certain inherent ingenuousness, reflecting the bird's long-term adaptation to a benign environment. In the terminology I proposed earlier, the dodo was ecologically naïve. Further testimony comes from a Dutch sailor who saw dodos in 1631: "They are very serene or majestic, they showed themselves to us with an extremely dark face with open beak, very dapper and bold in their walk, would hardly move out of our way." It's a nice sympathetic observation, "serene" and "bold" having greater biological validity ("dapper" is another matter, admittedly) than those invidious adjectives associated with the name. The dodo was no more foolish or simple or sluggardly than the marine iguanas, the giant tortoises, the mockingbird that perched on Charles Darwin's water pitcher, or the other tame-seeming animals of the Galápagos and other islands.

Even on Mauritius, the dodo wasn't unique in its ecological naïveté. Some other native birds were also disastrously trusting—including some of the small, agile species with unimpaired powers of flight. Van Neck's chronicler reports that scouts sent ashore found no human inhabitants but "only Turtle-doves and other birds in great abundance, which we took and killed, for as there was no one to scare them, they had no fear of us, but kept their places and allowed us to kill them." Another narrative, from a 1607 voyage, says that the men of that party "lived on Tortoises, Dodos, Pigeons, Turtle-doves, grey Parrots and other game, which they caught by hand in the woods." In 1611 the chronicler of still another voyage was impressed by both the quantity of wild game and the ease with which it could be killed. This witness focused especially on the dodo: "In colour they are grey; men call them *Totersten* or *Walckvögel*; they occur there in great plenty, insomuch that the Dutch daily caught and ate many of them. For not only these, but in general all the birds there are so tame that they killed the Turtle-doves as well as the other wild Pigeons and Parrots with sticks, and caught them by hand. They also captured the *Toter-sten* or *Walckvögel* with their hands, but were obliged to take good care that these birds did not bite them on the arms or legs with their beaks, which are very strong, thick and hooked; for they are wont to bite desperately hard." It's good to know that dodos weren't utterly docile and defenseless, despite their caricatured reputation. But the biting didn't save their lives.

Summarizing the historical records, Errol Fuller wrote: "Dodos were hunted mercilessly during the short time in which Europeans came into contact with them and for a period of upwards of half a century they proved a very useful source of fresh meat for travellers in the Indian Ocean." Unlike the giant tortoises, they weren't stockpiled alive on shipboard, passing weeks and months in a state of stoical dormancy. Instead, when the supply was excessive, some dodos were salted or smoked; others were eaten fresh, sating whole boatloads of men and supplying leftovers. We can imagine the shipboard menu—boiled dodo, roast dodo, pickled dodo, kippered dodo, dodo hash. The birds may have seemed disgusting because of the sheer surfeit of dodo meat.

Although the heavy harvest went on for decades, it couldn't continue indefinitely. It was insupportable for a species of bird that grew huge and laid one egg per clutch. So the direct slaughter by humans no doubt hurt the dodo population, maybe badly. Still, that's only part of the story.

Another part of the story is more subtle and indirect. Sometime during the early years of their acquaintance with Mauritius, the Portuguese had introduced pigs. And not many decades later, by a route that remains obscure, the island became infested with monkeys.

74

PIGS AND monkeys presented a bigger threat to *Raphus cucullatus* than the goats, chickens, cattle, deer, cats, and dogs that also came ashore with the Portuguese and, later, the Dutch. Goats competed for food, cats caused trouble, but pigs and monkeys were capable of deadly efficient predation upon the flightless dodo—or at least upon its juve-

niles and its nests. Having arrived in small numbers, those new enemies took hold in the forest and proliferated nightmarishly.

They were omnivorous, so their population levels weren't limited
by a single food supply. Reports from the late 1600s tell of pigs
putting an end to tortoise and sea turtle reproduction by eating the
eggs. In 1709, feral pigs were so numerous, causing "great destruction," that a posse of eighty men killed more than a thousand in one
day. Probably that slaughter had no lasting effect on the pig population, and by then it was too late for the dodo anyway.

The case of the monkeys is more puzzling. The particular monkeys that haunt the forest remnants and the roadsides of Mauritius
belong to the species *Macaca fascicularis*, commonly called the crab-
eating macaque, a notoriously adaptable primate whose original
native range stretched from upper Burma down through Indochina
and Malaysia to some of the less remote islands of the Philippines
and Indonesia. *Macaca fascicularis* is capable of spreading across a
tropical continent like a tribe of Visigoths but not of invading an
oceanic island without help; unlike the monkeys of Oz, it doesn't fly.
How it first reached Mauritius, and why, nobody knows. At one time
the species was thought to have been introduced by the Portuguese
during the sixteenth century. The only evidence for such a supposi-
tion seems to be that, first, the presence of monkeys was mentioned
in at least one of the earliest Dutch accounts and that, second, Portuguese sailors were allegedly fond of monkey meat. But the monkey-
meat allegation could have been just a slander invented by the Dutch.
Besides, the Portuguese favored Réunion over Mauritius as a way station in the Mascarenes, yet Réunion didn't end up cursed with a permanent infestation of monkeys. And there seems to be no record that
the Portuguese ever transplanted crab-eating macaques onto any
other island. Why would they? It's a wild species, skinny and smart,
not a tractable, fat-bearing domestic beast that could be conveniently
harvested on demand. Lack of evidence against the Portuguese tends
to point the blame back toward the Dutch. Furthermore, the strain of
M. fascicularis that persists on Mauritius has recently been traced to
Javanese ancestry, and Java was a Dutch point of contact as early as
1596.

What motivated the Dutch is anyone's guess. Maybe they themselves had a secret monkey-eating vice. Or possibly some Dutch ship
stopped at Mauritius with a pair of mascots, which escaped. Or the female alone escaped, but she was pregnant. Or maybe a few macaques

were intentionally discarded onto Mauritius (like those pet alligators of legend, discarded into the sewers of New York) because they had become a nuisance on shipboard. In any case, the conditions of this new island were highly agreeable to *M. fascicularis*. The population exploded. The same traveler who reported the pig eradication effort in 1709 also mentioned that he "had the pleasure to see more than 4,000 monkeys in a near-by garden."

Four thousand monkeys is a passel. Although the spectacle was a pleasure for him, it suggests a plague of frantic and clever omnivores that may have made life (or at least procreation) impossible for a species of ground-nesting birds. So my own notion, as unencumbered by evidence as the other hypotheses I've described, is that whoever did introduce the crab-eating macaque, for whatever inscrutably stupid reason, committed a very consequential act. Those monkeys have been much underestimated, I suspect, as a contributing factor in the extinction of the dodo.

75

RIVIÈRE DES Anguilles, which translates promisingly as River of Eels, is the name of a resort town on the south coast of Mauritius. It's a two-hour drive from my hotel at Rivière Noire, in the west. I've been invited along on a jaunt down to Rivière des Anguilles by a fellow named Carl Jones. The bones of the dodo at the Mauritius Institute, up in the city of Port Louis, are what brought me to this island; but falling in with Jones has added great dividends of serendipity, and now there's a chance for more. I'd be delighted to see a river of eels. I'm only a sidekick on this excursion, though, and I'll happily settle for what Jones has offered, which is a throng of crocodiles and monkeys.

Carl Jones is a tall, sarcastic Welshman with a sheepdog haircut, a weakness for bad jokes, and a manic devotion to native Mauritian wildlife, especially the birds. His reputation carries far beyond southwestern Mauritius, but I've only just met him, and I hardly know what to expect. I know that he has lived on the island a dozen years. I know that he runs a bird-rescue project in the vicinity of Rivière Noire, near the mouth of a system of steep valleys and igneous cliffs called (at least by English speakers in this polyglot culture) the Black River Gorges. I know that the Black River Gorges contain small, precious remnants of forest, enclaves of wild landscape amid the sugar cane and the urban sprawl and the beach-tourism development that carpet modern Mauritius. I know that one focus of Jones's efforts has been a severely endangered species of falcon, the Mauritius kestrel, which stood just a sneeze from extinction when Jones first arrived. And I know that Jones, though trained as a scientist, retains the admirable passion and zeal of an adolescent pigeon fancier.

After four days' acquaintance, I'm still learning about the overlap between his professionalism and his fancies, and still groping to locate the boundary between his cheerfully snide wit and the more serious zone of his personality. Rivière des Anguilles on Sunday, he told me, drag along for the ride if you like. What's down there? I asked. Owen's crocodiles, he said. Jones doesn't favor long explanations when short ones will do. So we've come to visit his friend Owen Griffiths, a snail taxonomist by training, who runs a commercial breeding farm for monkeys and crocodiles because snail taxonomy, however exciting, doesn't pay. Like writing poetry, it's a vocation, not a job.

Serendipity demands a certain abandon to circumstance, and I am curious about Mauritius at large—its monkeys included. Besides, in lieu of this outing I'd be maundering at the hotel, where someone plays "Raindrops Keep Falling on My Head" on a Hammond organ during dinner.

Owen is an Australian, cordial, grave, and slightly proper in manner, despite being chums with Jones. He and his wife have raised monkeys since 1985. At the moment they have almost two thousand head. He gives us the tour.

Owen's stock are crab-eating macaques, of course. He keeps them segregated into groups, carefully sorted, in large and small pens. Each of the larger pens holds scores of macaques in riotous togetherness. They come charging up to one end and lock their fingers through the wire, gaping curiously as we approach. The smaller pens contain his prized breeders. Owen explains a bit of the economics, and the sedulous attention to hygiene and medical record-keeping, involved in marketing monkeys. "This is *the* biomedical research monkey," he says. In the United States, the retail value of one crab-eating macaque destined for biomedical research is fifteen hundred dollars. That's for an animal caught in the wild. The captive-bred animals are even more valuable because each of them comes with precise records of its age and medical history. "The labs don't want junk monkeys," Owen says. "The demand now is for *quality* monkeys." Owen's wife, a beautiful

Mauritian creole named Mary Ann, is a microbiologist. She oversees the technical management of the farm. No commercial macaques are more healthy than those that Owen and Mary Ann offer, it would seem. They ship by the hundreds. Business is good. Revenue from the monkeys allows Owen to dabble in the more speculative, more slowly maturing business of crocodile hides.

All this scientific monkey husbandry is phase two of an enterprise that began in the Mauritian forest. Though captive-bred and health-certified macaques are what the market now wants, Owen's operation was founded on a supply of animals trapped in the wild. A snail taxon-omist with the vision of an entrepreneur, he saw a boundless, wasted resource—the island's population of *M. fascicularis*—and invented a way of connecting supply and demand. It was like making money from autumn leaves. Crab-eating macaques are a nuisance today in Mauritius, as they probably have been for three centuries. They don't count as protected wildlife. As you drive the network of two-lane highways you see them along the roadside, loitering, watching the traffic pass, fearless and opportunistic. They seem to be more com-mon than porcupines in Wisconsin or armadillos in east Texas. One of the sites where Owen did his early trapping was in the Black River Gorges, the same area of steep, forested valleys where Jones labors to rescue the kestrel from extinction. During his first year of trapping there, Owen took fifteen hundred macaques. It seemed an abundance that couldn't last. But the following year he trapped another fifteen hundred at the very same site, and as many the third year, "and it hasn't even made a dent in the population," he says. The total number of *M. fascicularis* at large on Mauritius today, and the resilience of the population, are unimaginable.

Also unimaginable is the impact that this alien species must have on the native wildlife. On Jones's endangered kestrel, for one in-stance, which at the nadir of its decline in the mid-1970s was reduced to a desperately tiny population—about a half-dozen individuals—and seemed irretrievably doomed. The kestrel nests in small cavelike holes that pock the lava cliffs of Mauritian mountains, as in the Black River area. Any predator agile enough to climb the cliffs, and inclined to eat eggs, could exterminate the species by persistently raiding its nests; the adult kestrels, even if they weren't preyed on directly, would die of old age or other causes without leaving offspring. That may have been the problem, or part of the problem, when Jones inter-vened. The decline of the kestrel population was obvious, but the cause was a mystery. A dozen years after he started work, Jones is still trying to solve it.

By the way, he asks Owen, do you have any eggs?

Eggs?

Eggs, Jones repeats. Hen's eggs, kitchen variety, raw. He had meant to bring some himself, for a certain little scientific experiment, but he forgot the bloody things. Might Owen have a few that can be spared?

Owen does. Accustomed to Jones's peremptory fancies, he supplies a couple of raw eggs from the kitchen. We walk back to one of the monkey pens. While the macaques are distracted, Jones delicately slips the eggs onto the dirt floor at one end of their pen. He steps away. Averts his gaze, cool and discreet. The question addressed by his makeshift experiment is: Will a captive-bred crab-eating macaque, conditioned to other food, eat the egg of a bird on instinct? By the time he turns around, six animals are squabbling over slimy fragments of shell.

76

THE LAST credible eyewitness account of living dodos dates from 1662. A Dutchman named Volquard Iversen was marooned on Mauritius that year, after his ship was wrecked in a storm. Though a lifetime had passed since the first Dutch arrival, a tenuous colony had failed several years earlier, and the island was uninhabited when Iversen and his shipmates straggled ashore. We can safely assume that, hungry and desperate, they scoured the vicinity for food. Despite the incentive for thoroughness, Iversen evidently found no dodos at all on the Mauritian mainland. But he did see some dodos on a small islet just off the coast, near enough that he and the others could walk to it at low tide. "Amongst other birds," Iversen reported, "were those which men in the Indies call *doddaerssen*; they were larger than geese but not able to fly. Instead of wings they had small flaps; but they could run very fast." Iversen's account, though short on detail, makes good ecological sense. The small water gap must have been just enough to protect the islet from pigs and monkeys.

At least some of those islet dodos were captured by Iversen's party. It was his testimony that I quoted earlier: "When we held one by the leg he let out a cry, others came running forward to help the prisoner, and were themselves caught." Of course the Dutchmen weren't wrestling with these birds for entertainment. A knife, a campfire, a stout green stick to serve as a roasting spit—the details are imaginary, but without doubt Volquard Iversen helped to butcher and eat a few dodos, which may or may not have been the last of the species. Probably the *walckvögel*, to this party of shipwreck survivors, didn't seem quite so *walck*. I'm reminded here of some lines of old doggerel about cannibalism by a group of castaways, among which the last victim to be sacrificed—by the poem's narrator—was the cook himself:

> And I et that cook in a week or less
> And as I eating be
> The last of his chops, why, I almost drops
> When a vessel in sight I see.

Iversen and his mates were lucky. After just five days, they were rescued by a passing ship. No living dodo was ever seen again.

77

THE VIVIDNESS of the Iversen episode is somewhat misleading. The crux of the matter of extinction—the extinction of Raphus cucullatus or any species—is not who or what kills the last individual. That final death reflects only a proximate cause. The ultimate cause, or causes, may be quite different. By the time the death of its last individual becomes imminent, a species has already lost too many battles in the war for survival. It has been swept into a vortex of compounded woes. Its evolutionary adaptability is largely gone. Ecologically, it has become moribund. Sheer chance, among other factors, is working against it. The toilet of its destiny has been flushed.

The causes of extinction are usually multiple and linked to one another in complicated synergy. But the precondition to extinction can be put in a single word: rarity. Obviously, common species become rare as a prelude to vanishing totally; and species that are naturally rare carry an extra, inherent jeopardy of extinction. This dictum—that rarity leads to extinction—might sound like a truism, but the truism has content. The scientific study of the extinction phenomenon addresses three basic questions: (1) why some species are inherently rare, (2) why others become rare, and (3) what particular sorts of trouble afflict rarities.

To be rare is to have a lower threshold of collective catastrophe. Any misfortune, even one that would seem small by an absolute standard, is liable to be a total misfortune. A modest-sized disaster can push a rare species immodestly close to oblivion. At the end, the last individual's death might turn out to be accidental, independent of the factors that shoved the species into the foyer of extinction.

Imagine, for instance, that the last dodo didn't perish on Iversen's islet. Imagine a single survivor, a lonely fugitive at large on mainland Mauritius at the end of the seventeenth century. Imagine this fugitive as a female. She would have been bulky and flightless and befuddled—but resourceful enough to have escaped and endured when the other birds didn't. Or else she was lucky.

Maybe she had spent all her years in the Bambous Mountains along the southeastern coast, where the various forms of human-brought menace were slow to penetrate. Or she might have lurked in a creek drainage of the Black River Gorges. Time and trouble had finally caught up with her. Imagine that her last hatchling had been snarfed by a feral pig. That her last fertile egg had been eaten by a monkey. That her mate was dead, clubbed by a hungry Dutch sailor, and that she had no hope of finding another. During the past half-dozen years, longer than a bird could remember, she had not even set eyes on a member of her own species.

Raphus cucullatus had become rare unto death. But this one flesh-and-blood individual still lived. Imagine that she was thirty years old, or thirty-five, an ancient age for most sorts of bird but not impossible for a member of such a large-bodied species. She no longer ran, she waddled. Lately she was going blind. Her digestive system was balky. In the dark of an early morning in 1667, say, during a rainstorm, she took cover beneath a cold stone ledge at the base of one of the Black River cliffs. She drew her head down against her body, fluffed her feathers for warmth, squinted in patient misery. She waited. She didn't know it, nor did anyone else, but she was the only dodo on Earth. When the storm passed, she never opened her eyes. This is extinction.

78

CARL Jones sits on a jetty at Rivière Noire, the small village that serves as headquarters for his bird-rescue efforts in the Black River Gorges. It's late afternoon. The shallow water just off the beach is changing color, azure to gunmetal, under the slanting light. The Indian Ocean stretches westward before us, six hundred miles to Madagascar and another few hundred beyond that to eastern Africa. The jetty is quiet. Immersed in conversation, Jones and I have walked out here to escape the demands and distractions that tug on him routinely back at his aviary compound.

Today there has been an extra distraction, not quite so routine. His ladyfriend, a bright and serious plant ecologist named Wendy Strahm, is momentarily miffed at him. She has reason to be. Within the tall adult body of Carl Jones there lurks a strain of feckless schoolboy, and that schoolboy had appropriated the shipping box from Wendy's computer to serve temporarily as a pigeon cage. Wendy's computer had meanwhile gone on the fritz and needed to be shipped away for repairs, which sent her hunting for the box. Of course the pigeons had shat in it. So there was hell to pay—a little hell, nothing dire, nothing irreconcilable, but Jones's hell-paying account was already overdrawn. Jones merely ducked his head guiltily when Wendy stormed past us through the compound, and then with a nervous smirk he suggested we walk to the beach.

Now we're on the coast, and the coast is clear.

Jones is describing his chosen homeland as it existed before humans arrived. "If you look at the early illustrations of this island, and you read the Dutch accounts, there's this overwhelming feeling that Mauritius was a paradise. There were dugongs in the lagoons, there were multicolored shells and fishes everywhere, there were turtles, there were giant land tortoises. Herds of land tortoises. There were vast colonies of seabirds, frigates and boobies, fairy terns and sooty terns and noddies. There were many species of parrot. We still argue today about how many species of parrot, and what species, were here. There was a very large species of parrot, a ground parrot, that probably couldn't fly. The largest parrot ever known. Called *Lophopsittacus mauritianus*. Herons and egrets, flamingoes, cormorants. All types of birds. Many of these, regrettably, have become extinct."

As have snakes, geckos, and skinks, he says. "There was a large skink, *Didiosaurus mauritianus*, the largest skink ever known. That was found here." Most of the endemic reptiles, like the endemic birds, are gone. Jones recites this litany of death and extinction in an energized rush. The thought of it all makes him angry, sad, frustrated. And now suddenly he has focused on the dodo.

"We still argue about when it actually became extinct, but it probably disappeared about the 1660s. It's become the sort of legendary bird of extinction. And a very important bird. There were extinctions before and there's been lots of extinctions since, but it was an important extinction because that was the first time, the first time in the whole of man's history, that he actually realized *he* had caused the disappearance of a species." The word "he" in this usage stands for *Homo sapiens* and, by extension, "we." The twirr of my tape recorder has incited Jones toward heartfelt declamation. In a tone that carries out over the water: "And it was at that moment—or in that era—when he realized the dodo was gone, that he realized the world was an exhaustible place. That he couldn't just go on pillaging and raping. So it signified a very profound moment."

Jones pauses. "I don't know who actually came up with that thought. Who actually realized it. But if you think about it, that was a very, very important time in the dawning of human consciousness."

We watch the sun drop toward Madagascar. Back at my hotel, at this hour, they'll be setting out the buffet and warming up the Hammond organ.

Carl Jones is a raving optimist, and I like him for that. His sails full of faith, he tacks upwind. The Mauritius kestrel, at one time the world's rarest bird, can be saved. His beloved and tolerant Wendy will forgive. Another shipping box can always be found. *Homo sapiens* comes to moments of realization.

79

MAURITIUS WAS a way station for another notable voyage, made by one Abel Tasman in 1642. Tasman had sailed westward from Batavia (now Jakarta) on the island of Java and then, after stopping in Mauritius for food and water, zigged back to the southeast on a long exploratory journey. Although historical sources tell us Tasman's route and that his mission was to map "the remaining unknown part of the terrestrial globe," history doesn't mention whether he salted away any dodos.

Batavia at that time was headquarters of the Dutch East India Company, and Tasman had been sent out by Anthony van Diemen, governor-general of the company, to reconnoiter the southern ocean. Fragments of testimony from earlier Dutch mariners had suggested that a large body of land lay out there somewhere—either an undiscovered continent or at least a huge island. What they were rumoring about, we know now, was Australia. To the Dutch East India Company, ambitious to expand its commercial territory, those rumors constituted operational intelligence. From the company's perspective this big southeastern landmass, wherever it was, might be worth finding and mapping. It might contain something more valuable than giant tortoises and fat flightless birds. It might contain precious commodities like cloves and nutmeg, which were yielding immense profits in other parts of the East.

The plan for Tasman's expedition was that, after his turnaround in Mauritius, he would strike southward as far as latitude fifty-four, then sail eastward along that line until he hit the mysterious lump of uncharted terrain. The plan was flawed, since latitude fifty-four slices midway between Australia and Antarctica. The execution too was flawed, and Tasman's southward leg didn't take him down that far. Maybe he preferred to stay warm. Maybe he caught an irresistible ride on the Roaring Forties westerly winds. Anyway, he missed the Australian mainland, getting not even a glimpse, and made landfall on a smaller island below it.

This smaller island was too far south for a tropical climate. It lacked the warm beaches and the inland deserts that he would have found in western Australia. It was cool, wet, and heavily forested with unique temperate-zone flora. Captain Tasman, a dutiful subordinate, named it Van Diemen's Land for the man who had sent him. His self-effacement would be corrected on later maps, which label the smaller island Tasmania.

Abel Tasman didn't give the island much attention. It looked unpromising, a wilderness without sign of advanced human culture or valuable produce. "No natives showed themselves," by one account, "although notched trees and traces of cooking-fires were seen." Another historical source tells that Tasman reported "footprints not ill-resembling the claws of a tiger." This was loose talk, overlooking the fact that claws generally don't show in the footprints of a cat and the fact that (although

Tasman couldn't have known it) the nearest actual tiger was in Bali, three thousand miles away. Still, the footprints he mentioned were presumably real and most likely did include claw marks, which would show in the tracks of a dog, a wolf, or—to the point, in this case—a wolfish marsupial. The tiger analogy gives Tasman's early report a certain foreshadowing quality. In one cursory visit to this unknown island, he had detected its major carnivorous beast, which would later be misleadingly labeled "the Tasmanian tiger."

The Dutch East India Company hadn't sent him to document zoological oddities, so he didn't linger. He sailed on across the southern ocean, groping eastward as far as New Zealand, then up to Tonga and Fiji. Later he succeeded in exploring the northern coast of Australia—not that his exploration did the company any good. He found no cloves or nutmeg there either. And Tasman himself never got back to that wild southern island with the peculiar footprints. His relationship to it was fleeting and insignificant, aside from the fact that it eventually took on his name.

80

TASMANIA HAS been inhabited by humans for at least twenty thousand years. The earliest aboriginal settlers seem to have come by foot, not by boat, migrating down from southeastern Australia during one of those ice-age episodes of lowered sea level that left a land bridge connecting Tasmania with the mainland. They lived on kangaroo, possum, shellfish, and wild plants, and they decorated their bodies with ocher.

A party of British invaders made landfall in 1803. The group included free settlers, military men, two dozen convicts, and a twenty-three-year-old lieutenant in command. Among the unloaded cargo, destined for a big role in the conquest of the island, were thirty-two sheep. The British had few things in common with the Aborigines, beyond what was forced by the landscape. They ate kangaroo when they had to, depended on shipped-in provisions when they could, raised a little wheat, and decorated their bodies with muttonchops, waistcoats, and brass buttons. Some wore leg irons. Van Diemen's Land, as they called the place, would soon become infamous as one of the harshest outposts of Britain's generally harsh Australian penal

WESTERN PACIFIC

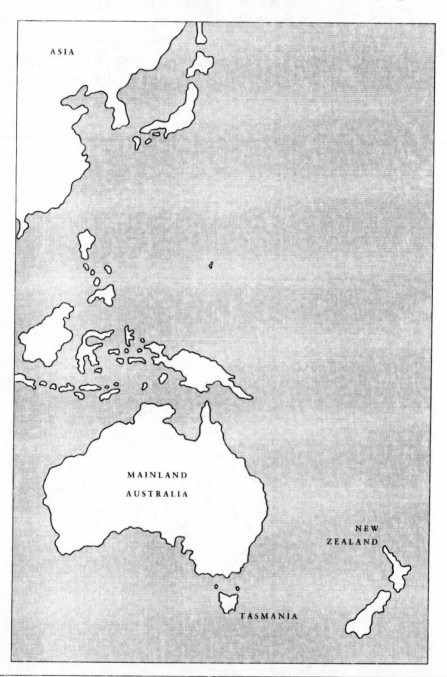

ASIA

MAINLAND
AUSTRALIA

NEW
ZEALAND

TASMANIA

complex. Devil's Island and Alcatraz would never match it. The later name change, to Tasmania, would be an act of political cosmetology intended to cover that infamy.

Within a few decades of the British arrival, sheep ranching began to dominate the landscape, the economy, and the political culture. Tasmania offered special promise in those years to pioneer-spirited sheep ranchers, with or without experience, because verdant pastureland was available and the government made a practice of furnishing cheap labor in the form of assigned convicts. It was an otherworldly colony, even by the standards normal to wilderness colonies. Outside the tiny fortified settlements, the woods were haunted by a bandit subculture of runaway convicts, as well as by a scattered population of angry, fearful Aborigines who correctly perceived that they were being robbed of homeland and mortally threatened. Also, there were strange beasts. As early as 1805, a clergyman named Knopwood, at the settlement then called Hobart Town, wrote in his diary:

> Am engaged all the morn. upon business examining the 5 prisoners that went into the bush. They informed me that on 2 May when they were in the wood they see a large tyger that the dog they had with them went nearly up to it and when the tyger see the men which were about 100 yards away from it, it went away. I make no doubt but here are many wild animals which we have not yet seen.

About the same time, a colonial official named Paterson reported that "an animal of truly singular and nouvel description was killed by dogs." To Paterson's eye, it seemed "a species perfectly distinct from any of the animal creation hitherto known, and certainly the only powerful and terrific of the carnivorous and voracious tribe yet discovered" on either mainland Australia or Tasmania. He was right: It was a predatory marsupial. Paterson's and Knopwood's reports together provide several revelations, not least of which is that dogs had already been introduced onto the island and were killing native wildlife, including even the "powerful and terrific" tigerish creature.

Several years later, another colonial official stole time from his duties to make the first scientific description of this unknown animal, based on two specimens caught in a trap baited with kangaroo. He named it *Didelphis cynocephalus*, which translates as "the dog-headed opossum."

It was a seemingly chimerical beast with a long snout, strong jaws, meat-ripping teeth, and a pouch for carrying its young. The scientific name was later revised, as marsupial taxonomy became more precise and scientists realized that whatever this thing was, it definitely wasn't a possum. It was a hunter and killer of other big mammals. It grew to the size of a Labrador retriever, though more lanky. It resembled a canine in many points of anatomy (owing to convergent evolution, not relatedness), but it could be distinguished at a glance by the sixteen or so dark, vertical stripes wrapping down off its back. Its tail was thick at the base and stiff in movement, almost as though the caudal vertebrae were fused. Its pouch opened rearward, unlike a kangaroo's. Its new and permanent scientific name was *Thylacinus cynocephalus*.

More informally, it came to be known by names that, like *cynocephalus*, used and compounded metaphor: Tasmanian wolf, marsupial wolf, zebra opossum, dog-headed opossum, opossum hyena, striped wolf. (The Tasmanian devil is another animal, smaller and stockier, not nearly so wolflike.) The most famous of the English names applied to *T. cynocephalus* was also the least accurate: Tasmanian tiger. The Tasmanian Aborigines had known the creature for ages, different tribal groups giving it names of their own: *ka-nunnah, lagunta, corinna*. For our purposes the preferable label is the one that biologists and historians use, sacrificing luridness for precision: the thylacine.

It was not a Tasmanian variant of something else, this animal. It was only itself: a sleek and prognathous marsupial flesh-eater.

Probably its total population on Tasmania was never large, not even before the British arrival. Predators of all sorts are limited to low population densities by the energy economics of the food chain; generally they are less than a tenth as abundant as the animals on which they prey. The thylacine may have been far less abundant than that. It was also elusive and nocturnal. During the early decades of the British presence, it avoided encounters with humanity, keeping its distance in the bush. By 1820, according to one tally, only four thylacine sightings had been reported since the British arrived.

Then something changed. The thylacine learned to kill sheep.

In 1824, a colonist complained that "there is an animal of the panther kind, which commits dreadful havoc among the flocks." Having discovered the taste of mutton, the thylacine altered its behavior, if only marginally. It still avoided the settled and cleared regions, but when sheep ranchers moved their animals into the hinterlands, the thylacine, thanks very much, would kill some. Many ranchers in turn altered their own behavior. They continued pasturing sheep in the wild country but threw fits over their losses and began to nurture an almost hysterical hatred for the thylacine. Their complaints became frequent and focused. "Tigers" were killing too goddamn many sheep. Meanwhile, domestic dogs brought ashore from British ships had gone feral in Tasmania, and those rampaging wild dogs were themselves killing quite a few sheep. Sheep were also being stolen, presumably by bushrangers, the runaway convicts who lived off the land. But the thylacine, an alien-seeming creature (though in ecological terms it was native and the sheep were alien), made a more satisfactory object of loathing and dread. "Where sheep run in large flocks near the mountains, these animals destroy a great many lambs." Thylacines took blame for killings they did commit and for many they didn't. Bounty hunting of the species began in 1830.

The first bounty program was a private initiative, decreed by the Van Diemen's Land Company, a sheep- and cattle-raising enterprise that had acquired huge tracts of land in northwestern Tasmania. This bounty was designed also to rid the company holdings of feral dogs and Tasmanian devils, but thylacines—"hyenas," in the company's terminology—were its primary target. It promised "5 shillings for every male Hyena, 7 shillings for every female Hyena (with or without young), half the above prices for male and female Devils and Wild Dogs." These were sizable amounts for a laboring person then, about equal to a day's wage. What made the bounty especially threatening to the whole thylacine population was a corollary provision, setting payment on a progressive scale: "When 20 Hyenas have been destroyed the reward for the next 20 will be increased to 6 shillings and 8 shillings respectively, and afterwards an additional 1 shilling per head will be made after every 7 killed, until the reward makes 10 shillings for every male and 12 shillings for every female." The company managers were smart. A progressive bounty would compensate for the fact that as thylacines became rarer, hunters would have to work harder for each kill. A progressive bounty was the right means for maintaining relentless pressure on a vanishing species until total eradication was achieved.

In addition to offering a public bounty, the Van Diemen's Land Company went so far as to create the position of "tiger man" on one of its holdings, a property called Woolnorth at the extreme northwestern cape of the island. The tiger man's job was to destroy predators, chiefly thylacines. He received food and lodging and a reward for each carcass. No records have survived to show just how many thylacines were killed during the 1830s and 1840s by the tiger men and the public at large, but a plausible guess would be in the hundreds. By midcentury, some observers who looked on the thylacine with scientific or historical interest (rather than agricultural loathing) had begun to report its decline. One writer called it "extremely rare," predicting that "in a very few years, this animal, so highly interesting to the zoologist, will become extinct." Another wrote: "When the comparatively small island of Tasmania becomes more densely populated, and its primitive forests are intersected with roads from the eastern to the western coast, the numbers of this singular animal will speedily diminish, extermination will have its full sway, and it will then, like the Wolf in England and Scotland be recorded as an animal of the past." That was an uncommonly sensible view in 1863, but it didn't require magical prescience.

Despite the reported decline, landholders in eastern Tasmania argued that thylacine depredations had grown intolerable and that the government should bloody well do something. They caterwauled in the House of Assembly at Hobart, which was now an imposing colonial capital. One man made the claim that thirty to forty thousand sheep were killed annually along the east coast by thylacines; later he revised it to fifty thousand. Another man complained of twenty percent losses among flocks that he pastured in the mountains. Many of the Assembly members, evidently, were sheep ranchers themselves. Some of them, so they said, were already paying private bounties; but they wanted the government to bear the expense. A motion to that effect passed. The government bounty program began in 1888 and continued until 1912, when it was canceled for lack of thylacines.

The surviving records from this program show a curious pattern. In the first year, eighty-one dead thylacines were redeemed for bounty; the next year, more than a hundred. Then the annual number held roughly steady for a decade. It increased again at the end of the century, peaked at 153 in the year 1900, and then fell sharply. The last year that saw more than a hundred thylacine killings rewarded was 1905. In 1908, there were just seventeen. In 1909, only two bounties were paid. In 1910, none. None ever again. Sometime between 1905 and 1909, judging from these records, the thylacine population had crashed.

Why did the numbers drop so suddenly? Up at Woolnorth, parallel records from the Van Diemen's Land Company show the same abrupt fall. Why?

One authority on the thylacine, a zoologist and historian named Eric Guiler, has commented: "This final decline, which was very rapid and occurred all over the state at about the same time, is not typical for a species which has been hunted to extinction." The hunting undeniably played a role. But some combination of other lethal factors, Guiler has suggested, must have converged on the thylacine at just that time.

What factors? Well, there was habitat loss. By the mid–nineteenth century, much of the island's most hospitable terrain—on the east coast, along the Derwent River upstream from Hobart, in the midlands, and up in the northwest—had been granted or sold to individual settlers or to the Van Diemen's Land Company. Valleys had been cleared, fences had been built, huge numbers of sheep and cattle were being put out to graze, all of which transformed large portions of Tasmanian landscape in ways that the thylacine couldn't accommodate. Additionally, there was competition from wild dogs. The dog packs were preying not only on sheep but probably too on the native wallabies and pademelons (small herbivores resembling kangaroos) that constituted the thylacine's natural diet. And there was the hunting. The government had paid bounties on more than two thousand thylacines by 1909, and tannery records suggest that further thousands had been killed to supply a thriving trade in exported hides. Besides these three horsemen of the apocalypse—habitat loss, alien competition, and hunting—Eric Guiler suspected one other: plague.

Around 1910, a distemper-like disease reportedly struck among several smaller predators, the eastern quoll (*Dasyurus viverrinus*, also known as the native cat) and the spotted-tailed quoll (*Dasyurus maculatus*), two species within the dasyurid family of marsupials. Possibly it hit the thylacine too. The disease devastated both quolls, causing such mortality that those species seemed to disappear. Later their populations rebounded. Yet the quolls had some advantages—their ecological needs were more modest than the thylacine's, their reproductive rate was higher, they were less troubled by exotic competition and by hunting. If the same disease or a similar one did hit the thylacine at a time when it was already beleaguered, that epidemic may have pushed the species so close to extinction that a population rebound wasn't possible.

This is Guiler's hypothesis. "All of these factors combined to depress the numbers below a satisfactory breeding threshold," he wrote, "and so prevented their recovery from what ought to have been a normal cyclical change." The phrase "satisfactory breeding threshold" conceals a wealth of complexity and dire implications, which will come whistling back to us later like an ebony boomerang. For now, the most remarkable aspect of Eric Guiler's view is that he stopped short of bewailing the thylacine's extinction. Was he indifferent to that event? No, he just didn't think it ever happened. Not quite. Not yet.

His book on the subject, published in 1985, is *Thylacine: The Tragedy of the Tasmanian Tiger.* The real tragedy, Guiler argued, is not that the species went extinct but that everyone too hastily assumed it did. Guiler believed that the species has survived, though barely, in the Tasmanian bush.

Conventional wisdom says that he was wrong—that the thylacine went extinct decades ago. Most biologists who have considered the question believe that it did. The year often cited is 1936, when the last captive animal died. As with the dodo, though, the timing and circumstances of total extinction in the wild are unknowable. Did a few lonely stragglers survive beyond 1936? Probably. For how long? Hard to say. All that we have to go on is the evidence of bounties paid, hides shipped, sightings made, photographs taken, and individual creatures that lived and died in captivity. Beyond this, there is the negative presumption that grows stronger each year as a missing species fails to reappear.

In 1930 a farmer named Wilf Batty shot a thylacine and posed for a smiling photo, gun in hand, with the carcass balanced upright against a fencepost. Aside from its odd stiffness and companionable calm, the victim in Batty's photo looks almost alive. That was the last killing of a wild thylacine ever reported. The London Zoo had a survivor, purchased from Tasmania for the hefty price of £150 during the post-bounty era when thylacines were zoological curios, worth more alive than dead. But the London animal died in 1931. The Hobart Domain Zoo still harbored two thylacines, which had been bought along with their mother back in 1924. The two littermates grew up—and then old—in captivity. One died in 1935. The other hung on until September 7, 1936. Its death was noted, a week later, in the minutes of a committee of the Hobart City Council, and the committee muttered parsimoniously that efforts would be made "to obtain another tiger up to value of £30 each." They were dreaming. Not at that price nor any price. In all the years since, there has been not a photograph, not a capture, not a piece of incontrovertible physical evidence that any thylacines have survived.

In July 1936, two months before the last known individual died, the government of Tasmania declared *Thylacinus cynocephalus* a protected species.

81

POLITICALLY, Tasmania is now part of Australia. Together with Queensland and Victoria and New South Wales and the other colonial territories, it joined the Australian federation in 1901, as the smallest and southernmost state. Geologically, the connection dates back much farther.

Tasmania is a continental island—it was once connected to mainland Australia by a land bridge—as distinct from an oceanic island, permanently isolated amid the sea's great depths. The distinction, as I explained earlier, is consequential. An oceanic island begins its insular existence by rising up from beneath the water's surface—either slowly, as a coral atoll, or abruptly, the way Anak Krakatau did, as a steaming, sterile mound of lava. Either way, it acquires its terrestrial ecosystem from scratch, through the slow processes of dispersal and establishment. A land-bridge island, on the other hand, begins its insular existence with a carryover ecosystem.

Thirteen or fourteen thousand years ago, during the last glacial episode of the Pleistocene, when the sea level stood several hundred feet lower than it presently does, Tasmania was the high-elevation knob of a knob-ended peninsula. Connected ecologically by way of that peninsular bridge, it shared a continuous floral and faunal community with the southeastern region of the mainland. Then the climate warmed and the great glaciers and ice caps released their water. The sea level rose, the bridge was submerged, the peninsula was severed, converting Tasmania into an island. Those plants and animals that had already migrated down the peninsula and colonized the knob found themselves now confined there. Among the animals were humans and thylacines.

The extension of its range from mainland Australia into Tasmania is what enabled *T. cynocephalus* to survive into modern times. On the mainland it went extinct much earlier. On Tasmania it was relictual— stranded but saved (at least temporarily) from the forces of doom that had swept across the larger landmass, just as the dodos were saved (temporarily) on Iversen's islet. Whatever had killed off the thylacine up north hadn't followed it, evidently, across the land bridge.

82

AT A SACRED Aborigine site called Ubirr, amid the red sandstone bluffs of Kakadu National Park in Australia's Northern Territory, a guide says to me, "There's your thylacine."

Ubirr is famous for its rock art. Sure enough, fifty feet up on a sandstone wall is an unmistakable pictograph: a slim, doglike creature with dark vertical stripes wrapping down off its back. The image is not so artistically spectacular as the x-ray-style turtles and fish for which Ubirr is renowned—I might have walked this trail without noticing it—but, once brought to my attention, it's striking enough. How old is it? The guide can only guess. Maybe five thousand years, maybe ten, maybe more. He tells me that some of the x-ray figures at Ubirr have been repainted faithfully during the passing millennia, even into recent decades, by local Aborigines devoted to keeping their culture alive. But this thylacine, executed in a more primitive style, looks to be ancient and unretouched. It's a simple two-dimensional outline in red ocher. The hind legs have been leached away by weather. While the other tourists in my party shuffle on, I stay behind, scribbling a sketch into my notebook.

The stripes. Stroke, stroke, get the stripes. Identifying marks on the real animal, they are identifying marks also on the rock-wall portrait. I want a record for my memory: Yes, definitely, there were stripes on the image at Ubirr. A feral dog carries no such markings. A dingo (*Canis dingo*, the species of wild canid that has been resident in Australia since prehistoric times) also lacks vertical stripes. Clearly the artist who created this image knew the thylacine.

Other stripe-backed figures painted on rock walls at other sites, both in the Northern Territory and in Western Australia, illustrate the same point. Aborigines on the mainland, during earlier millennia, had seen thylacines. And the artistic record is confirmed by the sub-fossil record. On the Nullarbor Plain, along the arid southern coast, a cave has yielded thylacine bones, radiocarbon-dated to about one thousand B.C. Sometime between then and when Europeans landed, *T. cynocephalus* disappeared from mainland Australia.

The causes of its disappearance there are almost impossible to determine. But we can speculate. It may have been hunted by Aborigines, who thought enough of the thylacine as a food item or a totemic beast to honor it with portraits like the one at Ubirr. No doubt it suffered competition from feral dogs, which came into Australia from Asia (presumably as camp-followers to a wave of human migration) about ten thousand years ago. In addition to the trouble from dogs, the thylacine was probably also threatened by dingoes.

Canis dingo is a placental mammalian carnivore of doglike shape and size, whereas the thylacine was a *marsupial* mammalian carnivore of doglike shape and size, and maybe they were too similar for comfort. The dingo had a long history on mainland Australia and had adapted itself well to the landscape. With its placental mammal advantages, it may have been an overpoweringly effective competitor that gradually usurped the thylacine's niche. One bit of negative evidence tends to support this notion: The dingo never reached Tasmania. Either it arrived on the mainland too late or it spread southward too slowly. Before *C. dingo* could follow *T. cynocephalus* into that southernmost refuge, the last great ice age had ended. The glaciers had melted back. The Tasmanian population of thylacines benefited from a circumstance of climate and timing. The water had risen and the bridge was out.

83

THE GAP OF SEA separating Tasmania from mainland Australia is known as the Bass Strait. Along the eastern and western edges of what was the land bridge are a few patches of high ground. Topping out above the present sea level, those patches are now islands themselves. At least four are large enough to appear on a half-serious map. Flinders, the largest, lies on the east side of the strait, and Cape Barren and Clarke are tucked just south of it. King Island lies on the west side, with less elevation than Flinders but a moister climate. Scattered nearby are several dozen other islands, some hardly bigger than a parking lot. Each of the larger four and most of the tiny islands contain at least some habitat suitable for at least some species of marsupial. Since the whole area was once a continuous landscape—the Bassmanian Peninsula, I call it—many species now found on Tasmania must have formerly occupied these small patches too. When the global temperatures warmed and the sea level rose, the patches became insular refuges. Each island could harbor only small populations, so there was a likelihood of accidental extinctions. Over the ensuing ten or twelve millennia, some islands kept many of their refugee species, some kept only a few. Consequently, the biological history of the Bassmanian hilltops has constituted a natural experiment on the extinction of insular populations.

Details of this natural experiment have been gathered by a biogeographer named J. H. Hope, about whom I know nothing except her quite interesting work, done originally for a doctoral dissertation at an Australian university. Hope focused on ten species of herbivorous marsupial, some big and lummoxy, some small and ratlike: the forester kangaroo, the red-necked wallaby, the pademelon, the southern potoroo, the Tasmanian bettong, the wombat, the brushtail possum, the ringtail possum, the brown bandicoot, and the eastern barred bandicoot. A bandicoot, to save you wondering, is the marsupial approximation of a giant shrew. A bettong is much like a potoroo, only fatter. Hope listed these species on a chart, cross-indexed against twenty-five Bass Strait islands, with the islands ranked in order of decreasing size. She was aware that island size might be a telling variable.

Using historical sources and a smattering of subfossil evidence, she established which species maintained populations on which islands throughout most of the past twelve thousand years. She also turned up a few clues about the presence or absence of marsupial carnivores. Her chart shows an eloquent pattern: The number of species surviving on each island correlates directly with island size.

Flinders, the largest island, retained seven of the ten herbivore species. It lacked only the forester kangaroo, the bettong, and the eastern barred bandicoot. King Island and Cape Barren Island each retained six species. They both lacked the same three that were missing from Flinders, plus one other, a different species in each case. Clarke Island, much smaller in area than either King or Cape Barren, retained only four species. And so on, from large islands to small, from many species to none, in a correlated decline down the columns of Hope's chart. Badger Island, just a tenth the size of Clarke, retained two species. Kangaroo Island, a tenth the size of Badger, retained one species. Little Green Island, smallest of the twenty-five, retained none of the ten herbivore species. Not even the adaptable, undemanding pademelon had managed to survive on Little Green;

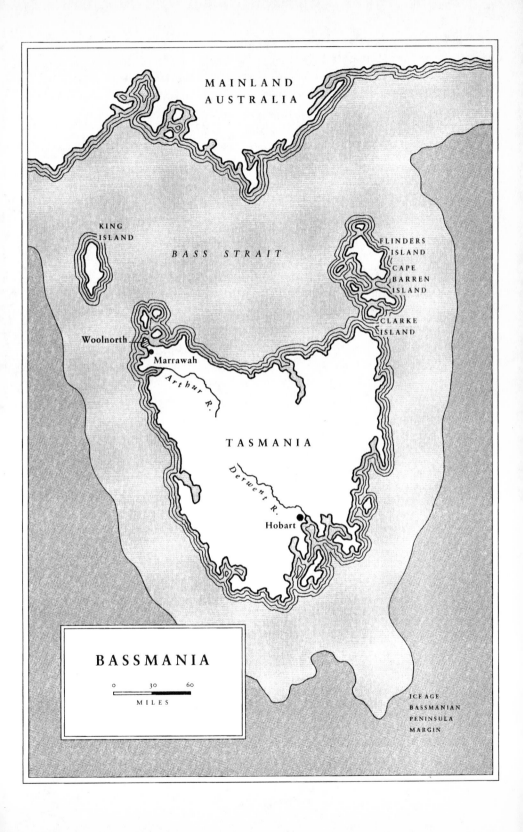

MAINLAND
AUSTRALIA

KING
ISLAND

BASS STRAIT

FLINDERS
ISLAND

CAPE
BARREN
ISLAND

CLARKE
ISLAND

Woolnorth

Marrawah

Arthur R.

TASMANIA

Derwent R.

Hobart

ICE AGE
BASSMANIAN
PENINSULA
MARGIN

BASSMANIA

0   30   60

MILES

evidently the place was too small. Overall the trend was clear: The smaller the island, the greater the incidence of local extinction.

At this point, we need a precise definition of the term *local extinction*, as distinct from total extinction of a species or subspecies. Local extinction refers to the eradication of any geographically discrete population of individuals while others of the same species or subspecies survive elsewhere. When the last forester kangaroo on Cape Barren Island died, it was a local extinction, since the forester survived on Tasmania. The same term applies to the bettong's disappearance from King Island, and to the eastern barred bandicoot's disappearance from Flinders.

J. H. Hope didn't call much attention to the theoretical implications of her Bass Strait data. But those implications were later illuminated by Jared Diamond, citing Hope in a paper he delivered at a symposium on extinctions in 1983. Describing her work as "a particularly clean study of differential extinction," Diamond summarized the findings, then offered his own conclusion: "Species varied greatly in their resistance to extinction following habitat fragmentation."

Tasmania, Flinders Island, King Island, and the other islands represent dry-land fragments of the old Bassmanian Peninsula. Why did the faunal histories of these fragments possess larger resonance? Because habitat fragmentation, by the time Diamond read his paper, had been recognized as a severe problem throughout the world's mainlands. If certain species varied greatly in their resistance to local extinction under such circumstances, not just in Bassmania but perhaps elsewhere too, then Jared Diamond wanted to know why.

He extracted three generalizations from Hope's data.

First: "Carnivores were more susceptible than herbivores." The thylacine had failed to survive on Flinders or King or any other Bass Strait island. The Tasmanian devil and the eastern quoll had likewise disappeared everywhere. Among the four large carnivores, only the spotted-tailed quoll had kept a foothold (maybe) on any Bass Strait islands. By contrast, most of the marsupial herbivores within the same size range were resident on quite a few islands. Plant eaters had survived where meat eaters hadn't.

Second: "Large carnivores were more susceptible than small carnivores." While the thylacine, the devil, and the eastern quoll had been lost altogether, two minuscule species (the white-footed dunnart and the swamp antechinus, a couple of mousy carnivorous marsupials) did maintain populations on some of the tiny islands. And the spotted-tailed quoll, as I've just noted, had possibly survived on islands where the thylacine and the devil hadn't. Large body size had been a fatal disadvantage, at least among marsupial carnivores.

Third: "Habitat specialists were more susceptible than habitat generalists." Not every island contained a full representation of all Bassmanian habitats. On the smaller islands, habitat diversity was meager. Broadly adapted herbivores—the type that could eat anything and live anywhere—tended to tolerate such meagerness, making opportunistic adjustments. But not every species was broadly adapted and opportunistic. Two small species of Tasmanian rodent (the long-tailed mouse and the broad-toothed rat) survived on none of the Bass Strait islands. Why not? Evidently because those species are too dependent upon food and shelter found only within certain kinds of wet forest and high-altitude scrub. The duck-billed platypus is another specialist, dependent on freshwater lakes and streams, which are plentiful on Tasmania but scarce on the Bass Strait islands. Such specialists had suffered severely from the Bassmanian fragmentation.

Beyond these three generalizations, Diamond offered another, more comprehensive still. The main pattern discernible within the data, he concluded, was that "risk of extinction decreased with inferred population size." Within any bounded area, a carnivorous species will usually be less numerous than an herbivorous species; a large-bodied carnivore will usually be less numerous than a small-bodied carnivore; and a habitat specialist will usually be less numerous than a generalist. The specialists and the carnivores, in particular the large-bodied carnivores, will consequently face higher risk of extinction. Big populations don't go extinct. Small populations do. It's not surprising, but it is significant.

Otherwise put: Rarity is perilous.

84

WHY *is* rarity perilous? Why *do* small populations go extinct? The answer is simple in outline and complicated in its scientific details. For now, let's stick with the outline version. Small populations go extinct because (1) all populations fluctuate in size from time to time, under the influence of two kinds of factors, which ecologists refer to as *deterministic* and *stochastic*; and (2) small populations, unlike big ones, stand a good chance of fluctuating to zero, since zero is not far away.

What are the deterministic factors? Those involving straightforward cause-and-effect relations that to some extent can be predicted and controlled. Essentially that means human activities: hunting, trapping, destroying habitat, introducing exotic animals that compete with or prey on native species, applying pesticides, swiping the eggs of birds or sea turtles, blockading the migrational routes of salmon or eels, and any other willful deed that either directly or indirectly exerts a negative influence. Because they are predictable, rational (well, arguably), and subject to control, we can eliminate such factors if we really care to, and in most cases we can rectify the damage that has been done.

What are the stochastic factors? Those that operate in a realm beyond human prediction and control, either because they are truly random or because they are linked to geophysical or biological causes so obscurely complex that they *seem* random. For practical purposes, we can consider them accidental. Weather patterns, for instance. A hard winter is an accidental factor. So is a drought. So is an ice storm in spring. So is a forest fire set by lightning. Each might cause a downward fluctuation in the population size of some species. If a hard winter were followed immediately by a drought, then by a fire, then by another hard winter, the downward fluctuation might be large. Other accidental factors? A hurricane or a typhoon might strike, battering habitat, interrupting courtship, wrecking nests. A population boom among some predator species might lead to a population bust among its prey. An epidemic disease, too, could reduce the size of a population. An infestation of parasites. A statistical aberration in the gender ratio of newborns—that is, too many males and not enough females. On the longer scale of time, an ice age would be an accidental factor. The population size of any animal or plant species fluctuates naturally, routinely, in response to this sort of accident.

Often the fluctuations are small. Sometimes they are big. And occasionally they are numerically small but big in significance.

Think of two species that live on the same tiny island. One is a mouse. Total population, ten thousand. The other is an owl. Total population, eighty. The owl is a fierce and proficient mouse eater. The mouse is timorous, fragile, easily victimized. But the mouse population as a collective entity enjoys the security of numbers.

Say that a three-year drought hits the island of owls and mice, followed by a lightning-set fire, accidental events that are hurtful to both species. The mouse population drops to five thousand, the owl population to forty. At the height of the next breeding season a typhoon strikes, raking the treetops and killing an entire generation of unfledged owls. Then a year passes peacefully, during which the owl and the mouse populations both remain steady, with attrition from old age and individual mishaps roughly offset by new births. Next, the mouse suffers an epidemic disease, cutting its population to a thousand, fewer than at any other time within decades. This extreme slump even affects the owl, which begins starving for lack of prey.

Weakened by hunger, the owl suffers its own epidemic, from a murderous virus. Only fourteen birds survive. Just six of those fourteen owls are female, and three of the six are too old to breed. Then a young female owl chokes to death on a mouse. That leaves two fertile females. One of them loses her next clutch of eggs to a snake. The other nests successfully and manages to fledge four young, all four of which happen to be male. The owl population is now depressed to a point of acute vulnerability. Two breeding females, a few older females, a dozen males. Collectively they possess insufficient genetic diversity for adjusting to further troubles, and there is a high chance of inbreeding between mothers and sons. The inbreeding, when it occurs, tends to yield some genetic defects. Meanwhile the mouse population is also depressed far below its original number.

Ten years pass, with the owl population becoming progressively less healthy because of inbreeding. A few further females are hatched, precious additions to the gender balance, though some of them turn out to be congenitally infertile. During the same stretch of time the mouse population rebounds vigorously. Good weather, plenty of food, no epidemics, genetically it's fine—and so the mouse quickly returns to its former abundance.

Then another wildfire scorches the island, killing four adult owls and, oh, six thousand mice. The four dead owls were all breeding-age females, crucial to the beleaguered population. The six thousand mice were demographically less crucial. Among the owls there now remains only one female who is young and fertile. She develops ovarian cancer, a problem to which she's susceptible because of the history of inbreeding among her ancestors. She dies without issue. Very bad news for the owl species. Let's give the mouse another plague of woe, just to be fair: a respiratory infection, contagious and lethal, causes eight hundred fatalities. None of this is implausible. These things happen. The owl population—reduced to a dozen mopey males, several dowagers, no fertile females—is doomed to extinction. When the males and the dowagers die off, one by one, leaving no offspring, that's that. The mouse population fluctuates upward in response to the extinction of owls, a rude signal that life is easier in the absence of predation. Twelve thousand mice. Fifteen thousand. Twenty thousand. But while its numbers are so high it will probably overexploit its own resources and eventually decline again as a consequence of famine. Then rise again. Then decline again. Then . . .

The mouse population is a yo-yo on a long string. Despite all the accidental disasters, despite all the ups and downs, the mouse doesn't go extinct because the mouse is not rare. The owl goes extinct. Why? Because life is a gauntlet of uncertainties and the owl's population size, in the best of times, was too small to buffer it against the worst of times.

85

ON Tasmania, by 1936, the thylacine population was nothing if not small. The travails it had suffered for more than a century—hunting, competition from feral dogs, habitat loss, disease—had reduced it to a vanishing point. But did the species go extinct? Or has it survived in the Tasmanian outback, elusive and desperately rare? Despite the strong likelihood that it's gone, there continue to be sightings.

In judicious language, *alleged* sightings. More than three hundred have been reported since 1936. A scientific article has been published under the title "Recent Alleged Sightings of the Thylacine (Marsupialia, Thylacinidae) in Tasmania," complete with tables and graphs. Many of those allegations were no doubt flaky, and if the witnesses in question hadn't glimpsed *T. cynocephalus* in the bush they might have spotted Elvis at the grocery. But many others were plausible, made by people of sober reputation who saw their animals under reasonably good conditions and volunteered details that fit the known facts of thylacine reality. These observers were accurate as to body shape and size; they put the stripes where the stripes belonged; in some cases, they even mentioned the peculiar stiffness of the tail. One report, in 1982, was especially persuasive. It came from a ranger of the National Parks and Wildlife Service, a man known among his colleagues in Hobart as knowledgeable and meticulous. His account:

I had gone to sleep in the back of my vehicle which was parked at a road junction in a remote forested area in the northwest of the State. It was raining heavily. At 2:00 a.m. I awoke and, out of habit, scanned the surroundings with a spotlight. As I swept the light-beam around, it came to rest on a large thylacine, standing side on some six to seven metres distant. My camera bag was out of immediate reach so I decided to examine the animal carefully before risking movement. It was an adult male in excellent condition with 12 black stripes on a sandy coat. Eye reflection was pale yellow. It moved only once, opening its jaw and showing its teeth. After several minutes of observation I attempted to reach my camera bag but in doing so I disturbed the animal and it moved away into the undergrowth. Leaving the vehicle and moving to where the animal disappeared, I noted a strong scent. Despite an intensive search no further trace of the animal could be found.

This ranger wasn't out for attention. He didn't sell his story to the *National Enquirer*. He didn't even want his name published.

In the wake of their own man's report, wildlife officials put a research officer on the case. Substantial amounts of money and man-hours were spent. Scat specimens were collected for chemical analysis. Sand pits were monitored for thylacine footprints. It wasn't a whimsical quest: If any thylacines had survived, the Tasmanian branch of the National Parks and Wildlife Service bore a statutory responsibility to protect them. Every other week for two years, the research officer returned to that same remote area, hoping for a repeat of the sighting. No luck.

In Hobart, I discuss the episode with a hardheaded young biologist. This fellow has spent months in the Tasmanian bush himself, gathering data for a doctoral project on another marsupial carnivore, the Tasmanian devil. Devils are fairly common and he has seen plenty. He knows the difference between a reliable observation and one that's merely intriguing. I ask him about the ranger's sighting, or alleged sighting, of a living thylacine. Is it, by any stretch, believable?

"*I* believe it," he says.

Where did the sighting happen? A remote forested area in the northwest, yes, I know—but just *where*? Up in that Arthur River country, he tells me.

86

AT THE END of the two-lane blacktop in northwestern Tasmania is a hamlet called Marrawah, populated by a few dozen humans and a few hundred Guernseys. It's just south of the old Van Diemen's Land Company property of Woolnorth. Cows have replaced sheep as the favored stock animals in these parts, at least among small operators who don't happen to hold a sprawling ancestral land grant. Municipal downtown Marrawah consists of a few houses, a store, a Baptist church, a town hall, and a pub. The pub calls itself a tavern—a semantic distinction indicative of something or other, I suspect, but I'm not sure what. THE MARRAWAH TAVERN, PROP. BUCK AND PETER BENSON, says a sign. The word "pub" still carries its British tone, and though much of Tasmania wears an old-country cultural veneer, with Dorset teas served in village hotels, Marrawah is decidedly not British. To the contrary—it has a rough, hearty, edge-of-the-universe atmosphere more suggestive of a timber town in northwestern Montana. The Guernseys are gentle introverts, but there may not be many others. Arriving on a Saturday night, I stop at the tavern for a beer. The devil biologist back in Hobart has advised me: Don't miss the Marrawah pub.

A television is playing, but the jukebox is louder. There are video games. Naughty bar-joke signs (SHIRT AND SHOES REQUIRED—BRA AND PANTIES OPTIONAL) are plastered here and there, for the undying amusement of whoever plastered them. High on a wall behind the bar hang antique pioneer rifles, a set of ox horns, a bleached cow skull, and a broken surfboard. The beers on tap are Boags lager and Foster's. Boags is the regional brand hereabouts, which is what a person drinks if a person cares to blend in.

There's a modest Saturday night crowd spread through the room. The decor includes a dartboard and several cricket team photos, so maybe I'm wrong about Buck and Peter Benson's attitude toward cultural heritage. The jukebox is whomping out "Wild Thing" by, if I'm not mistaken, the Troggs. A rangy young thug, presently in command of the pool table though drunk as snot, plays air guitar on his cue and pantomimes the vocal. Someone hollers, Take yer bloody shot, Mike. From comments tossed, from eyes rolled, I'm given to understand that Mike is an infamously tiresome character, well known if not well loved locally. He does not take his shot. He's onstage, he's transported: Waald thang, and ah wanna know fuh shuh. He's busy. Hot licks on the cue. *Take* yer bleeding *shot*, another voice suggests. But this flak from the gallery only delights Mike. His ham instinct rises like an erection. He gives a moue. He's in no bloody hurry to take his bleeding shot, mate. He prances. Feet like a Percheron, in work boots. He grabs his crotch expressively. Waald thang, you make muh heart sang, you make evvvthang, guh . . . roovy. More hollers from the crowd, acknowledged with another crotch grab. Sipping my Boags on a corner stool, I wonder idly whether I'm about to see a dandy Tasmanian fight. If so, which good citizen will it be who puts out Mike's lights with the fat end of a cue stick?

About this time, on a far wall, I notice a diamond-shaped plaque in the style of a highway sign, facetious but seemingly official, alerting motorists: TASMANIAN TIGER, NEXT ? KM. The animal pictured is unmistakable: doggish snout, stripes wrapping down off the back, stiff tail.

Taking it for a good omen, I drive south out of Marrawah the following day, into the Arthur River country.

87

IT'S A REGION of rolling hills and coastal scrub vegetation. The air is salty. The soil is poor, a sandy beige powder that gets plenty of rain but doesn't hold it. The dominant plant is the tea tree, a scrawny species capable of growing on limy dunes. This is no wonderland of biological diversity, this bush country around the Arthur River, and no spectacle of grand scenery, but the landscape is scruffy and wild, only sparsely crosshatched with dirt roads, little coveted by people, inhospitable to agriculture, yet garnished with enough native greenery to offer excellent habitat for marsupial herbivores, the wallabies and wombats and possums. Those traits make it excellent habitat also for thylacines.

Excellent potential habitat, anyway. In their day, they must have thrived here.

The permanent human population is minuscule. A few other folk, visitors from small cities such as Burnie and Smithton up on the north coast, use the Arthur River area as a poor man's beach getaway. They come to spend weekends and vacations in the ramshackle cabins and the motels offering "holiday units," to relax, to drink beer in quiet privacy, to fish, to drive recreational vehicles along the beach, to sit in lawn chairs and wiggle their naked toes, to do God knows what else. But the motels and the cabins aren't many, and fewer still as the coast road stretches southward. The hamlet that borrows the name Arthur River, a huddle of weathered houses at the northern end of a scary old wooden bridge across the river itself, makes a village like Marrawah look crowded. I drive into the hamlet, blink once, cross the bridge, scan for more buildings but don't see any, turn, and come back. Yes, this is it: downtown metropolitan Arthur River. Foolish of me to have hoped for a café.

In the dead of midafternoon no one is stirring. Eventually I find a motel. I wander up to the office, which doubles as a grocery and seems to be the only spot within miles where a person can buy food.

Behind the counter is a small, red-headed man, a woebegone fellow whose life has gone sour and who takes some satisfaction, it develops, from sharing the details with strangers. While he comes alert and commences to talk at me, I browse his cooler and his shelves. The inventory has been chosen for long shelf life and low turnover—a few dozen dusty cans and boxes—but who am I to complain, having been too eager or too stupid to get provisions before I drove down here? The red-headed man was a truck driver, he tells me. Long hauling. Burnie to Hobart and around. Izzat right, mmm, I say. My slowness to pick out a bag of beans or a jar of Vegemite passes for sociable loitering. But now, the man tells me, now he runs this place. Mmm? Yes and loves it, he proclaims. Thank heaven for this place. He's glad to sit quiet, to have his leisure, glad to be done driving truck. His wife and daughters are up in Burnie. Them, they want no part of Arthur River. Faaagh. Burnie, that's another bloody world. His wife has changed suddenly, he volunteers. "Changed"—his word. An ominous little word. It seems an utter black-box mystery to him, her change. I shift my weight and glance at the door and can't help but wonder what the wife has got going for herself. This poor guy, this angry unpleasant soul, has been ravaged by time and fate. He's only about fifty, but he's old and defeated, bitter, irredeemably crimped. His daughters ignore him, he says. Faaagh. He can understand now, he tells me, when you read of a man who has shotgunned his whole family. Beg pardon? I think. It sounded as though you said "shotgunned," but I believe I won't ask. Yah, Red repeats, he can understand how a man could be driven to it. At this earnestly shared confidence, I dip my head. Mmm-*mm*. A polite dip but noncommittal. Then I break for daylight with my canned beef-and-noodles in hand. Thanks, have a nice day. I put a few miles between myself and the red-headed trucker before making camp.

After dinner and then an early-evening siesta, I rise in darkness for a night's work. I drive farther on into the hill country along a lonely dirt road toward a certain geographical feature that shows on my map as a tiny pen squiggle: Tiger Creek. What better spot to go scouting for Tasmanian tigers, I figure, than an obscure drainage of the Arthur River country called Tiger Creek?

It takes me a half-hour to drive out there, and on the way I'm mesmerized by a call-in program on Australian national radio, relayed from some distant city. I've already discovered this cheery aspect of the Tasmanian outback: that radio reception is good, especially at night. With Hobart not very distant (in crow-flight miles, at least) from any part of the island and Melbourne almost a flat shot across the Bass Strait, the signals are strong. Thanks to the Australian Broadcasting Company, the programming is good too. One night a Jimmy Dorsey retrospective, the next night a London opera. Tonight it's the call-in show, this segment devoted to recent interpretive work on the Dead Sea scrolls—a dandy topic, about which virtually anyone can offer a passionately ignorant opinion—and featuring a biblical scholar who is full of wild and dangerous theories. The Essene cult, radical monasticism, John the Baptist's real role, and the identity of the "wicked priest" mentioned in those texts. Could it be that Jesus, a cunning mortal, managed to survive the crucifixion? This vein of thought isn't new, but apparently there is some newly translated evidence.

I park at the roadside near Tiger Creek, where a dirt track leads up into the bush. I'd like to linger and hear more about the scrolls. Instead, pushed along by my own little purpose, I snap off the radio.

The night is warm. The moon is still low. Even by starlight, though, I can see more or less where I'm going as I follow the pale sandy track through dark vegetation. I've got a headlamp and a strong flashlight, but I'll use them only when necessary; walking quietly in total darkness is a better way to jump animals. For some reason, I choose this moment to consider the question of whether Tasmania harbors any dangerous snakes. Come to think of it, isn't there an extremely venomous species called the tiger snake, famous for its bad disposition? A notorious killer, dark as a racer, therefore totally invisible amid shadowy undergrowth? On the Australian mainland, at least,

I'm sure there is: the tiger snake, *Notechis scutatus*, one of the world's deadliest serpents. But my memory for biogeographical minutiae has failed me and I can't recall whether *N. scutatus* made it across that Pleistocene land bridge. If so, gulp, maybe the naming of Tiger Creek has nothing to do with thylacines.

I place my feet carefully. I place them as carefully, anyway, as a person can who's foolish enough to be hiking through the outback at night in cheap running shoes while wearing a headlamp that isn't turned on.

The track is powdery soft under my feet. It curves up along the shoulder of one hill, down through a swale, up another hill, another, then makes a switchback or two and comes to a fork. I choose the right branch, heading away from the car. It winds on, through another set of rises and switchbacks and traverses, to another fork. Wary of getting lost, I begin jotting a record of these forks into my notebook. Again onward. At the crest of a hill, I'm treated to a sudden panorama out over the dark landscape, mile after mile of rolling terrain thick with tea tree scrub and forest, unmarred by a single dot of electricity. The moon is up now. Showing itself intermittently from behind patchy wet clouds, it's large but not full. I'd call it a waning gibbous moon, except that here in the Antipodes with the sky upside down and backwards it may actually be waxing. Can't recall how that works. Far off, I see moonglow on a bend of the Arthur River, glistening in the darkness like a channel of molten lead. When the clouds close again, the moon disappears and the air fills with a warm misty drizzle. I hike on. I follow the track toward nowhere. I listen.

I've been in Tasmania just long enough to have learned some of the noises of nature. The thumpy gait of a running wallaby, for instance, sounds like two people killing ants with baseball bats. The vocalization of the brushtail possum is a low, guttural chuff. The other common possum, the ringtail, gives a high, giddy twitter resembling the call of a loon. Both the brushtail and the ringtail lurk in trees, setting up a commotion of branches and leaves when they feel intruded upon by a human. It's like clearing their throats at stagy volume, I suppose. And besides the audible signals, I've picked up a few that are visual. The wombat is a round-shouldered hulk of dark fur (like a young bear but more mopey) that hesitates and then quietly reverses direction when encountered at night on a path. The Tasmanian devil, also a traveler on dirt tracks after dark, leaves a distinctive hind footprint: a rectangular pad with the four toes aligned across its front edge. The devil is smaller and quicker than the wombat, and there's a white, yokelike marking across its black chest, clearly conspicuous in a face-to-face standoff. Met on a trail at night, a devil acts guilty and decisive, whereas a wombat seems merely embarrassed. The footprint of the thylacine is well known from museum specimens. The silhouette is well known. But the call of the thylacine is indeterminable; it's lost amid the echoes of history.

A howl? No. A bark? No. A song? Not in any conventional sense. Some sources describe the thylacine vocalization as a cough. Some mention a nasal sort of yapping. When irritated, a thylacine uttered (or utters?) a low growl, according to one expert. When excited, it gave (gives?) vent to a harsh hiss. Thylacines in captivity were usually mute. In the wild they were (are?) slightly more vocal. And in the human imagination, they make all manner of eerie racket. For five hours tonight I walk among the possums and wombats and wallabies and devils, hearing those animals, seeing them, reading their tracks by flashlight. I think about the silent, invisible thylacine.

I think about all the people before me who have chased it. Some were bounty hunters, but many others were driven by appreciative curiosity. In early 1937, less than a year after that last captive animal had died in the Hobart Zoo, a small expedition was dispatched by the Fauna Board, the government body then charged with protecting species, to search for survivors of *T. cynocephalus* in the wild. This expeditionary party consisted of a Sergeant Summers, from the Tasmanian police, along with one other policeman and an experienced tracker. They came here to the Arthur River country. They found good evidence that the species still existed, if only sparsely: tracks, recent sighting reports. Sergeant Summers recommended establishing a sanctuary in the area, but he was ignored. Later that year and again the next, other search parties were sent to the southwest, into some of Tasmania's wildest bush, and they also found tracks. The 1938 group echoed Summers's recommendation about a sanctuary, suggesting a different piece of land. This recommendation was likewise ignored. Then came the war. Hitler and Japan were far more pressing concerns than the population status of a predatory marsupial, and another expedition wasn't launched until after VJ Day. In November 1945 a naturalist named David Fleay traveled down from the mainland to lead a party back into the southwest. Months of searching produced no thylacine evidence except a few ambiguous footprints. Fleay's big steel-frame traps, baited with bacon and live birds, caught only devils and other unwanted marsupials. He dragged slabs of meat through the bush but failed to attract a thylacine. One night his party heard a strange scream, possibly the cry of a thylacine, possibly not. In early 1946 Fleay gave up. He returned to the mainland, still convinced that the species survived but with less proof in his hands than Summers had gathered in 1937. From the time of Fleay's expedition to the present, the trail seems to have gotten colder, the evidence more wispy and questionable.

It may be that the 1930s and 1940s was a critical period, during which the last remnants of the thylacine population—scattered among pockets of wild landscape, isolated from each other, each remnant too small to be demographically and genetically robust—came to their collective moment of terminal rarity and then died out completely. We don't know. Later research, on the thylacine's status in particular and on small populations generally, makes that guess seem likely.

In 1957 newspapers around the world announced a dramatic bit of news. A strange animal, suspected to be a thylacine, had been spotted by two men in a helicopter along a remote stretch of beach on the southwestern coast. The men had circled this animal and gotten a photograph. But the photo was unpersuasive, the animal was dark with a bushy tail, and a subsequent investigation (*not* bruited by newspapers around the world) turned up the fact that a fisherman had lost his dog in that area several weeks earlier.

Beginning in late 1963, the zoologist and historian Eric Guiler led a search on the west coast, under sponsorship of the Fauna Board and with direct financial support from the Tasmanian premier, who evidently thought that a thylacine rediscovery might be good for his state's economy or prestige or something. Guiler, as an author of scientific studies and a participant in previous searches, was probably Tasmania's foremost thylacinologist. His expedition worked for eight months, bashing along the bush roads by Land Rover, covering many miles of rough country on foot, setting more than four thousand snares and keeping those snares tended. By Guiler's own figuring, he and his team conducted 126,000 snare-nights' worth of trapping. They caught a lot of native animals, including 338 pademelons, 132 devils, 70 wallabies, 46 wombats, 13 spotted-tailed quolls, 9 brushtail possums, 2 echidnas, 2 crows, and a sea eagle. But no thylacines. Guiler has confessed that "we felt we deserved better luck for all the effort we put into the project." It wasn't so much an act of science, his project, as an act of faith. And despite the empirical discouragement, Guiler's faith held, as true faith generally does. That's what faith is: belief that transcends data.

Faith comforts, but data persuade. Aspiring to persuasion, Guiler continued to hope that physical evidence of the thylacine would eventually reappear. Throughout the 1960s and 1970s there were more searches, by him and others—private quests and official expeditions, some sponsored modestly by conservation organizations or thrill-hungry newspapers, some financed by the searchers themselves, most begun with a peal of publicity and terminated more quietly, none yielding anything solid. Thirty years after the last definite traces of *T. cynocephalus* had faded away, the great thylacine hunt had become almost a mythic activity, a half-bonkers but prideful Tasmanian cultural ritual intermittently renewed for the amusement of the newspaper-reading public. The thylacine took on a status shared elsewhere by Santa Claus, Ned Kelly, UFOs, the Kennedy assassination conspiracy, the Loch Ness monster, and the parachuting skyjacker D. B. Cooper: a gaudy idea, bigger than truth, that helps keep people entertained.

In 1966 hair samples were collected from a suspected thylacine lair. According to one expert, these samples matched thylacine hair as it was known from preserved specimens. According to another expert, the samples didn't match. In 1972 automatic cameras with flash units were set up at two dozen sites, rigged with tripwires and baited with food. They photographed some very startled wallabies. In 1973 an American naturalist spent a few nights in a tree, blowing solos on a small unmusical instrument known as a varmint call. Another effort was mounted in 1979, when the World Wildlife Fund put up $55,000 for a rigorous, two-pronged assault on the thylacine question. One prong was Eric Guiler, using automatically triggered movie cameras. The other prong, consisting of historical research, analysis of previous sightings, and auto-triggered still cameras, was supervised by a second zoologist. Guiler's films revealed "a quite unsuspected abundance" of possums. The still cameras established that the Tasmanian devil is a thriving species and that even pademelons are sometimes attracted by meat. Again Guiler's faith held, despite a lack of evidence. The other zoologist compiled a thorough and clear-eyed report, concluding that the thylacine is "probably extinct" but that then again maybe, just maybe, it isn't.

My nocturnal walk in the Arthur River country leaves time for musing on all of this. I think about Guiler's efforts, his faith, his disappointments. I think about the folks of the International Society of Cryptozoology, an innocent group of impassioned hobbyists who like to travel and dream and make elaborately careful collections of anecdotal evidence regarding improbable beasts, and among whom the thylacine is an icon. I think about Nessie and Sasquatch. I think about the romantic imagination that burns within *Homo sapiens*, about how there's a touch of the cryptozoologist in each of us. I think about the Dead Sea scrolls. I think about the logical form of the proposition "*Thylacinus cynocephalus* no longer exists" and the fact that you can't prove a negative. I think about rarity, the problem of inbreeding, and the various accidental setbacks that a tiny population of thylacines would inevitably have suffered over the past sixty years. Is it plausible that any population could hover so close to zero for so long without succumbing to a fatal downward fluctuation? I wonder. I doubt. All these thoughts simply constitute a full night's disordered reverie, not a syllogism or a coherent opinion. Meanwhile I walk about ten miles. Through the deep hours, in the hills above Tiger Creek, I find time for plenty of thinking because I'm not distracted even once by the sight of a thylacine.

When I get back to the car, around three A.M., the radio call-in show is over. If there were any shocking but persuasive revelations about the identity of the wicked priest, I've missed them. The ABC is playing rock-and-roll.

Next morning I drive back up to Marrawah. I stop at the tavern for coffee. The woman behind the bar would rather not be bothered; having delivered a cup of brownish water, she leaves me alone. In the clarity of a Monday morning, I take another glance at the diamond-shaped highway sign with its TASMANIAN TIGER, NEXT ? KM alert. Just beside that one is a smaller sign, which I didn't notice on Saturday night. EVERYONE NEEDS SOMETHING TO BELIEVE IN, it says. Closer inspection shows me a foam-topped mug, and the fine print: "I believe I'll have another beer."

88

SINCE we're considering the connection between rarity and extinction, I should say something about the passenger pigeon, *Ectopistes migratorius*, now extinct despite having once been probably the *least* rare bird on Earth.

Just two centuries ago the passenger pigeon was almost unimaginably numerous. It wasn't an island species, God knows; it was a phenomenon of continental scope. It inhabited the eastern half of North America, ranging across the great deciduous forests from coastal Massachusetts to the Great Plains and from northern Mississippi up into Nova Scotia. Within a short span of time, its population went from roughly three billion to exactly zero. While it thrived, it was a stupendous biological success. When it perished, it perished quickly and totally. The first of its problems, and the most obvious, was the excessive and greedy slaughter practiced against it by humans. The last of its problems was something else. At a hasty glance, the case of the passenger pigeon seems to be one in which a species passed from extraordinary abundance to extinction without lingering in—or being especially affected by—the condition of rarity. That impression is misleading. People made *Ectopistes migratorius* rare. Rarity then made it vulnerable to a complex of other troubles.

Rarity can be relative. Five thousand thylacines in Tasmania may have been sufficiently many to avert certain problems associated with small population size, but five million passenger pigeons in North America may have been too few. Why? Because social structure and its ecological correlates impose different thresholds of population stability on different species. *E. migratorius* had its own threshold, an exceptionally high one, which was partly a function of its unique social and ecological traits. Direct killing by humans brought it down below that threshold, after which the dynamics of its doom became more subtle and complicated.

Passenger pigeons were fanatically gregarious creatures. We don't know much about their habits, but we do know that. Evolution had committed them to a strategy of exaggerated togetherness. They found comfort, safety from surprise attack by predators, and agreeable conditions for mating within huge and tightly clumped colonies of a sort that for other species would constitute intolerable crowding. They lived on foods that occurred in large concentrations, not uniformly available across whole regions but patchily clustered in space and time. Big crops of acorns and beechnuts and other such mast were important, but also chokecherries, mulberries, elderberries, the seeds of maples and of elms, and (as European settlement transformed the landscape) the grain fields of unlucky farmers. The birds traveled in vast flocks, which put millions of pairs of eyes cooperatively on the lookout for those far-flung bonanzas of food. They nested in vast rookeries spanning as much as two hundred square miles. When they took flight, searching as a group for food or for a roosting site, they literally darkened the sky.

In Virginia, around 1614, one man reported pigeons "beyond number or imagination," adding that "my selfe have seene three or four houres together flockes in the Aire, so thicke that even they have shadowed the Skie from us." A Dutch settler on Manhattan wrote in 1625, "The Birds most common are wild Pigeons; these are so numerous that they shut out the sunshine." In Pennsylvania, a poem from 1729 included the lines "Here in the fall, large flocks of pigeons fly, / So num'rous, that they darken all the sky." Repeated again and again in the historical records, it comes to sound like a received formula, until a new eyewitness gives it the fresh force of observed reality. A French explorer named Bossu, reaching Illinois around 1760, was impressed by the sight of "clouds of doves, a kind of wood or wild pigeon. A thing that may perhaps appear incredible is that the sun is obscured by them." Bossu also reported that a hunter could kill eighty with one shot. Around 1810, the ornithologist Alexander Wilson made a meticulous but dizzying estimate of 2,230,272,000 birds in a single flock. (Daily food intake: about seventeen million bushels of acorns.) If Wilson had rounded downward and said "two billion birds," we would assume he was exaggerating grossly. At about the same time, John James Audubon described a flock that he saw, and saw, and saw, while he was crossing Kentucky. Audubon's flock took three days to pass overhead.

The decline from abundance to rarity seems to have happened fast, possibly during the 1880s. At the start of that decade, *E. migratorius* was still nesting in conclaves of many millions, but by 1888 a sighting of just 175 birds was noteworthy. The sky never again darkened with passenger pigeons. The last bird confirmed in the wild was shot dead in Sargents, Ohio, on March 24, 1900. After that, the species was seen only in zoos. The last known passenger pigeon (famed in the lore of extinctions under its own individual name, Martha) survived until 1914 at the Cincinnati Zoo. She and the Archduke Franz Ferdinand died the same summer. Martha's body was frozen in a block of ice and shipped to the Smithsonian for study. Billions of passenger pigeons had already been killed, with almost no attention given to the bird's anatomy or behavior or ecology. Now suddenly Martha's little carcass was precious, in accord with the scarcity theory of value. Her skin was stuffed and put on display. She stood for a notable fact: The one species that had seemed inextinguishably numerous was extinct.

People are still wondering how it happened. Over the past century, a variety of wild and less wild notions have been offered. There was the report, for instance, that a French Canadian priest had laid a curse on *E. migratorius* for causing nuisance to his parish. "Almost every conceivable cause has been given for the extinction of the pigeon," wrote a historian named A. W. Schorger, who added: "One of the most popular explanations was drowning en masse in the Gulf of Mexico or the Atlantic Ocean. Pigeons were reported as having been washed ashore even on the coast of Russia. Another theory was that, under persecution, the pigeons migrated to Chile and Peru. As late as 1939, they were thought to have been seen in Bolivia." Who knows how much longer they may have survived, hiding out down there with Mengele and Bormann.

Schorger's book, *The Passenger Pigeon: Its Natural History and Extinction*, a historically thorough (but now scientifically outdated) treatment published in 1955, dismissed such self-exonerating human notions as the mass drowning, the Bolivian getaway, the direct intervention of God. Schorger mentioned a few other factors that had been suggested as causes of the pigeon's disappearance: forest fires, epidemic diseases, timber operations that destroyed its mast-bearing deciduous habitat, and unfavorable climatic conditions as those habitat losses drove the surviving birds farther north. Schorger also documented the magnitude of the direct slaughter.

The same gregariousness that had helped protect them from their natural predators made passenger pigeons especially vulnerable to humans. Their instinct for social cohesion was so strong that thousands of birds could be killed before the others would spook. The massacre came to its peak after the Civil War, with a phalanx of professional hunters chasing the big flocks. The hunters were aided by telegraph communications (for passing word about nesting sites) and by railroads (for fast transport of the product). Local people, not just the menfolk but whole families, also took part. The birds were gathered like apples from an orchard, with hundreds left to rot on the ground. "Hunting" is actually too worthy a word for such witless, methodical killing, and though "harvest" seems faintly euphemistic as applied to an animal species, it's probably more accurate. The harvest was accomplished with nets, traps, sulfuric fumigation, clubs, and long poles, as well as with guns. One family in Massachusetts, knocking the birds off their roosts with poles, killed twelve hundred in a night. Some traps could take a thousand at once. Barrels full of salted or iced pigeons were sent back to eastern cities, sold there for as little as a nickel per bird. When an occasional glut hit the urban markets, the price fell even farther and passenger pigeons were given away or fed to hogs.

As late as 1878, at least 1.1 million were shipped to market from a nesting near Petoskey, Michigan. Another nesting in Michigan was remembered in the 1930s by an elderly woman, Etta S. Wilson, who had seen the event as a young girl:

> Day and night the horrible business continued. Bird lime covered everything and lay deep on the ground. Pots burning sulphur vomited their lethal fumes here and there suffocating the birds. Gnomes in the forms of men wearing old, tattered clothing, heads covered with burlap and feet encased in old shoes or rubber boots went about with sticks and clubs knocking off the birds' nests while others were chopping down trees and breaking off the over-laden limbs to gather the squabs.

In addition to the sights and the smells, she recalled sounds:

> Pigs turned into the roost to fatten on the fallen birds added their squeals to the general clamor when stepped on or kicked out of the way, while the high, cackling notes of the terrified Pigeons, a bit husky and hesitant as though short of breath, combined into a peculiar roar, unlike any other known sound, and which could be heard at least a mile away.

It was the song of *E. migratorius* in the throes of extermination. Wilson added:

> Of the countless thousands of birds bruised, broken and fallen, a comparatively few could be salvaged yet wagon loads were being driven out in an almost unbroken procession, leaving the ground still covered with living, dying, dead and rotting birds.

This had occurred in Leelanau County, just west of Grand Traverse Bay, sometime around 1870. Etta Wilson delivered her recollections at a gathering of the American Ornithological Union sixty years later, and there was a confessional vein in her paper: "In my youth I was identified with all the processes of the traffic in Pigeons, first as a retriever, then as a sorter and packer, then as one of those busy persons who plucked the birds, cleaned and cooked them, and finally ate them." But she noted too that her own missionary grandfather had heard and absorbed a warning, from local Indians, about the logical end of such slaughter: "After awhile there will be no more Pigeons." Her grandfather in turn tried to convince the market hunters, but the hunters preferred to believe otherwise: "There'll be Pigeons as long as the world lasts."

The percipience of those Indians and that missionary grandfather wasn't generally shared. Several states had passed laws regulating the pigeon trade, but the laws were halfhearted, seldom enforced, and probably enacted too late. Twenty years earlier, before the war and the spread of railroad lines, real statutory protection might still have saved *E. migratorius*. But a typical judgment back then was the one issued, in 1857, by a committee of the Ohio legislature: "The passenger pigeon needs no protection. Wonderfully prolific, having the vast forests of the North as its breeding grounds, travelling hundreds of miles in search of food, it is here to-day, and elsewhere to-morrow, and no ordinary destruction can lessen them or be missed from the myriads that are yearly produced." Here today, elsewhere tomorrow. But what might become of the species after that—in five years or a decade or one human lifetime—was hard to foresee, and it's not easy to comprehend even by hindsight.

The final cause or causes of the pigeon's extinction remain obscure. Is it really possible that hunters destroyed three billion birds? Is it really possible that humans could be so relentlessly efficient? A. W. Schorger thought it was. He held that "persecution was unremitting until the last wild bird disappeared." But Schorger's view seems to contradict the principle of diminishing returns. As the species became rare, as the flocks became small and the nestings more sparsely scattered, the mass-harvest techniques must have lost their effectiveness and the profit-to-effort ratio must have fallen. The shotgun blasts and the traps and the nets must have yielded far fewer birds per try. Eventually, *E. migratorius* must have been scarcely more attractive as commercial game than any other species of palatable, moderately rare bird. What happened then?

There is almost no scientific evidence. There are only informed guesses. One guess was offered in 1980 by a biologist named T. R. Halliday. Unpersuaded by Schorger's view, Halliday wrote: "The puzzling aspect of the passenger pigeon's demise lies in the fact that during the last years of its existence the species continued to decline at a rate that seems too great to be accounted for simply by hunting." He noted that only thirty-six years had passed between the big nesting at Petoskey and the death of Martha. It was too fast, Halliday thought, for commercial harvest to have accomplished the job of eradication. It made no allowance for diminishing returns. Nor was habitat destruction a satisfactory answer, since sizable areas of beech and oak forest remained uncut when the pigeon population crashed.

Using the methods of theoretical ecology (a graph, a hypothetical curve, terms such as "critical colony size" and "critical fledging rate"), Halliday tried to explain the crash. With its evolved dependence on the benefits of crowding, he argued, this bird had to live in vast numbers or not at all. To find its food, to guard against its natural enemies, it needed millions of pairs of eyes. Incubating its eggs, fledging its young, it needed the womblike security of multitudes. When the size of the flocks fell below a certain threshold, the food-seeking flights became unproductive, the rhythms of mating and nesting were disrupted, the flocks were no longer self-sustaining units, and the species as a whole began to fail. Halliday suggested that "social factors, namely colony size and reproductive success, were related in such a way that, though the species was apparently still quite common, its breeding rate was insufficient to offset mortality." The word "apparently" is there to remind us that appearances are often misleading and rarity can be relative.

With a remnant population in the low millions, scattered among small flocks, the bird might have seemed numerous enough to survive, but it wasn't. It had been deprived of an essential ecological asset: its abundance. If we accept Halliday's view, the passenger pigeon went extinct because hunting had made it rare—relatively rare, anyway—and because that level of rarity was incompatible with the social ecology of the species.

89

THE STORY of the passenger pigeon, though familiar in outline and enlivened by gross human barbarity, is actually a distraction from larger trends. The pigeon, after all, was a mainlander. The case of Hawaii's birds is more representative.

According to Jared Diamond's "Rosetta Stone" tally, mentioned earlier, mainland species have accounted for less than a tenth (16 of 171) of the bird extinctions within recent history. Only eight of Diamond's extinct birds lived and died on the North American continent, whereas three times that many extinctions occurred in the Hawaiian Islands alone. The Mascarene Islands also surpassed North America in extinctions, as did both New Zealand and the West Indies, though even those doom-laden places couldn't match Hawaii, the world's foremost site of bird extinctions since the age of European conquest began.

Distance and size are among the contributing factors. Three thousand miles from the nearest continent, two thousand from any other substantial island, the Hawaiian archipelago is surpassingly remote. It includes eight large and small islands in a cluster, plus a dozen or so tiny ones trailing off to the northwest like the tail of a comet. Its clustered islands (Oahu, Maui, Kauai, Molokai, Lanai, Kahoolawe, Niihau, and the one known as Hawaii or, less confusingly, as the Big Island) are separated by stretches of water just wide enough to foster archipelago speciation. Oahu, Maui, Kauai, and the Big Island are individually larger than most volcanic islands, and the archipelago as a whole encompasses more land area than most remote archipelagos—twice as much as the Galápagos. The large size of the individual islands allows for greater topographic relief and climatic variety, both of which contribute toward greater diversity of habitat, which contributes in turn toward additional speciation. Thanks to these conditions, the Hawaiian archipelago was extraordinarily rich in species—among the *Drosophila*, the honeycreepers, and other groups—at the moment when humans arrived.

Because it was so isolated, and had been for maybe ten million years, it was also extraordinarily vulnerable to disruption. The first human settlers were Polynesians, who came by canoe around A.D. 500. Captain Cook landed in 1778. Soon after him appeared European colonists and missionaries, with their livestock and their guns and their Bibles and their axes and their plows. Nothing could be more predictable, given such circumstances, than that the archipelago would lose dozens of species quickly. What makes Hawaii's case interesting are the peculiar variations on that predictable theme.

One variation involved the arrival of a certain exotic animal, the *Culex* mosquito. If scientific suspicions are correct, this creature was

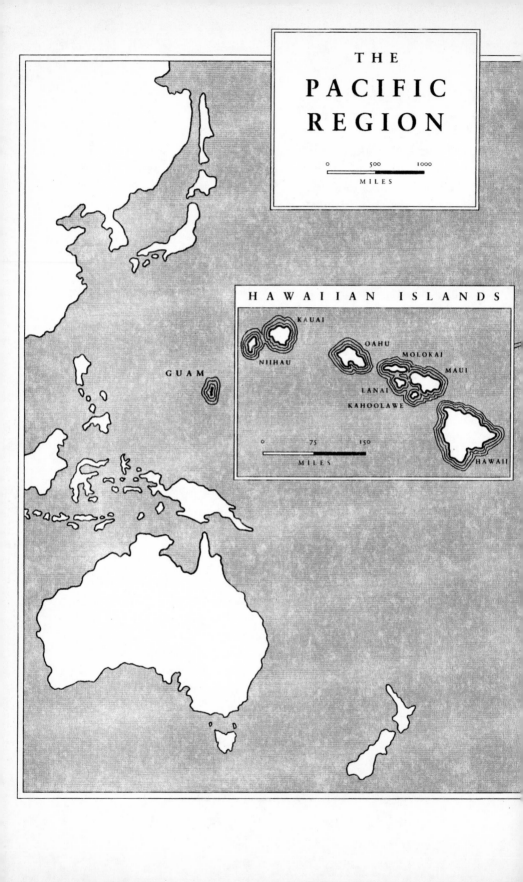

THE
PACIFIC
REGION

0   500   1000
MILES

HAWAIIAN ISLANDS

KAUAI

NIIHAU

OAHU

MOLOKAI

MAUI

GUAM

LANAI

KAHOOLAWE

HAWAII

0   75   150
MILES

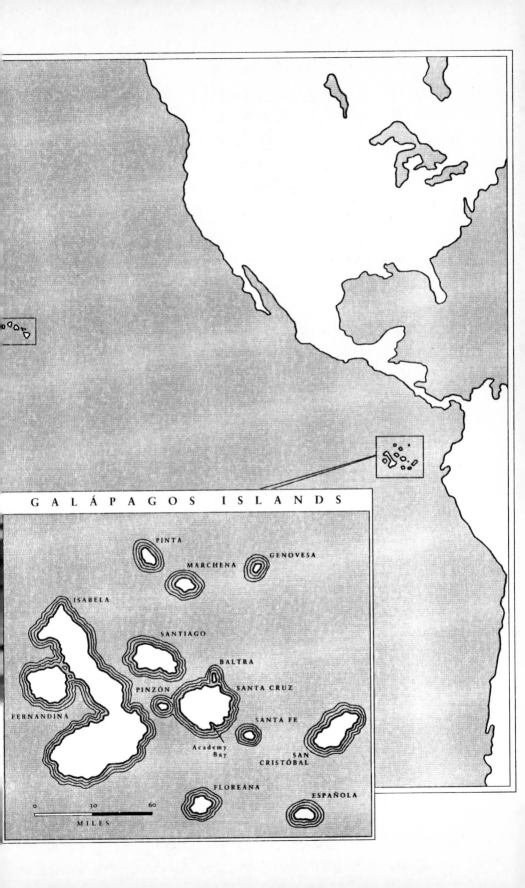

GALÁPAGOS ISLANDS

PINTA

MARCHENA

GENOVESA

ISABELA

SANTIAGO

BALTRA

PINZÓN

SANTA CRUZ

FERNANDINA

SANTA FE

Academy
Bay

SAN
CRISTÓBAL

FLOREÁNA

ESPAÑOLA

0   30   60

MILES

more damaging than other invaders on other islands—more damaging than the dingo, than the pig, than the rat, than the crab-eating macaque, almost more damaging even than *Homo sapiens*. "There are no mosquitoes here," said a cheerful report from two missionaries who reached Hawaii in 1822. But that was about to change.

A later historical source records a moment of ominous recognition: "Dr. Judd was called upon to treat a hitherto unknown kind of itch, inflicted by a new kind of *nalo* (fly) described as 'singing in the ear.' The itch had first been reported early in 1827 by Hawaiians who lived near pools of standing water and along streams back of Lahaina." A little local sleuthing revealed that the British ship *Wellington*, just out from tropical Mexico, had stopped at Lahaina (on the west coast of Maui) the year before, and that a party of men had gone ashore to get drinking water. Before filling their casks, they had emptied some dregs into the stream. The old cask water had been swarming with larvae of *Culex pipiens fatigans*, a night-flying mosquito found along Mexico's west coast. By a nasty coincidence, that particular subspecies naturally favors exactly such habitat as the stream near Lahaina. Within a decade, mosquitoes were well established on Oahu and Kauai as well as Maui, and eventually they spread to all the main islands. Although their presence was annoying for humans, it was worse than annoying for birds, since *Culex pipiens fatigans* is the principal vector of avian malaria.

The *Culex* mosquito is implicated also in carrying encephalitis to humans, heartworm to dogs, and a viral disease known as birdpox to domestic poultry. The only good news was that, as a subspecies adapted to tropical climates, *C. p. fatigans* couldn't colonize the chilly Hawaiian highlands. But in the lowlands, where many of the native birds lived, the mosquito thrived.

Avian malaria is similar to human malaria. It's caused by a parasite of the genus *Plasmodium*, carried from victim to victim in the saliva of mosquitoes. Also like human malaria, it can be especially deadly when it strikes a population with no prior history of exposure (and therefore no acquired resistance) to the disease.

Probably the *Plasmodium* parasite has been intermittently present in Hawaii for thousands of years, brought there in the blood of infected shorebirds and ducks that migrated out from the Americas. But without a resident mosquito to transfer *Plasmodium* from one bird to another, the disease wasn't communicable, so the endemic Hawaiian birds were safe. The birdpox virus, also potentially fatal but not so fast-acting as malaria, seems to have come ashore with the colonists' poultry. It posed no threat to the native birds either—not until mosquitoes arrived to complete the linkage. Then the *Wellington* dumped its larva-infested water, and the players were suddenly all present.

Long before that moment came, Hawaii's native species were already suffering other forms of human-caused disruption. Captain Cook himself had put ashore goats and pigs. Another early explorer, Captain George Vancouver, had introduced cattle and sheep. Within a few decades, all four of those domestic species had gone feral, grazing heavily in the forests, affecting the physical structure of the habitat and the balance among plant species. Meanwhile settlers harvested native trees for timber. They clear-cut and burned, opening large lowland areas to sugar cane fields. Cane was followed by pineapple. Even where the forest was left uncleared, exotic weeds made inroads against the native flora. Alien predators, such as the mongoose and the domestic cat, established feral populations on several islands. Some horses and donkeys also went feral. Axis deer from India were introduced, multiplying so prolifically and browsing so destructively that professional hunters had to be hired to cull them. The Hawaiian ecosystem was benign and accommodating enough to welcome all manner of exotic species. A pair of wallabies escaped from confinement on Oahu in 1916, and even they founded a feral population that reportedly still survives in the mountains behind Honolulu.

For the avian fauna, another threat was more directly lethal: hunting. Colorful feathers were the prime attraction for commercial hunters, and scientific collectors also trapped and shot an unseemly share of native birds. As populations were reduced and hunting began to yield diminishing returns, some species became only more precious. Demand rose in proportion to rarity.

All these consequences of European settlement were only phase two of a process that had begun about a thousand years earlier. The original Polynesian settlers had brought their own menagerie of alien species: dogs, chickens, a Polynesian race of small-bodied pig, and (presumably as a stowaway in their ocean-going canoes) the tropical rat *Rattus exulans*. Those earlier settlers cleared land to plant taro and sweet potatoes. They used fire against the forests. They committed other forms of major environmental impact. Eventually their descendants, the indigenous Hawaiian people, would develop a sense of close human anchoring to the landscape, as intimated in the respectful word applied to that landscape, *aina*. But *aina* wasn't immune to abuse. The oligarchs of the culture were known as *alii*, among whom the preeminent boss on any one island was *alii-ai-moku*, or "the chief who eats the island," suggesting a role that was rather more aggrandizing than holy stewardship.

The indigenous people also took their toll on the native birds. The spectacular feathered capes and helmets that became uniforms of rank among Hawaiian chiefs, and the ceremonial feather-garments known as *kahilis* (loosely translated, "fly flaps"), cost thousands of avian lives. During early periods of Hawaiian history, commoners were even obliged to pay tribute to their *alii* in the currency of feathers. Shiny black plumage from one species, scarlet from another, green from another. Yellow was the most valued color, a stroke of bad luck for species like *Drepanis pacifica*, the Hawaii *mamo*, with its bright yellow rump highlighted against a starling-black body. The annals of Hawaiian fashion tell us that one chief, Kamehameha the Great, possessed a resplendent yellow cape containing the feathers of eighty thousand *mamo*.

Despite the two waves of human invaders, Polynesian and European, a rich assemblage of bird species managed to hang on. Some of the lowland habitat on Oahu and Hawaii remained undisturbed. Some of the smaller islands, such as Lanai, were only sparsely settled. There was no hint of any devastating loss of species throughout most of the nineteenth century. But then, in the 1890s, new signals appeared. Field naturalists on Oahu and in the Kona region of Hawaii started noticing "grievous afflictions" on some birds: lesions and tumors around the face, on the legs, on the feet. These were symptoms of birdpox. At the same time, naturalists saw far fewer birds generally, even in undisturbed forest. If the birds were still numerous, they didn't show themselves. They didn't sing. The woods of Hawaii had gone quiet.

An ornithologist named H. W. Henshaw described those woods in 1902:

> One may spend hours in them and not hear the note of a single native bird. Yet a few years ago these same areas were abundantly supplied with native birds, and the notes of the oo, amakihi, iiwi, akakani, omao, elepaio and others might have been heard on all sides.

What especially perplexed Henshaw was that the habitat hadn't changed. He knew the plant species upon which these birds depended—even knew them, as he knew the birds, by their Hawaiian names—and he was surprised that resources were standing ignored:

> The ohia blossoms as freely as it used to and secretes abundant nectar for the iiwi, akakani and amakihi. The ieie still fruits, and offers its crimson spike of seeds, as of old, to the ou. So far as the human eye can see, their old home offers to the birds practically all that it used to, but the birds themselves are no longer there.

If destruction of habitat didn't explain the birds' absence, neither did hunting. Some of the missing species had never been much persecuted for the feather trade.

On the island of Oahu, the losses were particularly abrupt. Six species of endemic perching birds seemed to have vanished by 1900, from a total of eleven species inhabiting Oahu not long before. One of the vanished birds was the *ou*, a thick-billed little honeycreeper of the genus *Psittirostra*. Just after the turn of the century Henshaw wrote:

> The cause of the extinction of the ou upon Oahu seems to be very obscure. The fruit of the ieie vine is the particular food of the bird, and there are considerable tracts of timber on the mountains of the island where this vine still abounds. So too, there are sections where the introduced guava and the mamaki are still plentiful, and the ou is very fond of the fruit and berries. There apparently being no scarcity of food and shelter, why should the ou have disappeared from Oahu, and yet persist upon other islands where the timbered areas are even more restricted?

A possible explanation, which Henshaw apparently overlooked, is that Oahu is a low island compared with Maui, Hawaii, and Kauai. Both Maui and Hawaii contain large highland areas with peaks rising above ten thousand feet, and even Kauai contains a fair bit of terrain above three thousand feet. Oahu, by contrast, has only a few modest peaks and little highland area. Translating from topography to ecology: Almost every part of Oahu is accessible to a tropical lowland mosquito. The island offers no sizable refuge against *C. p. fatigans*.

A dozen years after Henshaw's alarm, another ornithologist noted:

"Oahu can make the melancholy boast that it has a greater list of extinct birds, in proportion to the total number of species known from the island, than any other like area in the world."

The rest of the archipelago suffered losses too, though not so drastically or so quickly as Oahu. The island of Lanai, low like Oahu and considerably smaller, but relatively neglected by early European settlement, seems to have been spared avian extinctions until its own boom era began. As late as 1920, Lanai's bird populations were healthy. A bird man named George Munro carefully tracked their status. Then in 1923 the pineapple industry took hold, with a base at Lanai City. Along with the pineapple workers came chickens. Along with the chickens came trouble. Munro speculated that the workers "brought bird diseases with their poultry and these, evidently carried by mosquitoes, were fatal to the native bird populations."

Munro was among the first to propose this hypothesis. More recently, Richard E. Warner has published his own comprehensive account, combining experimental data with historical records. The evidence on birdpox comes from field sightings of infected birds, dating back to the time when the populations collapsed. The evidence on avian malaria is more circumstantial, but still persuasive. For instance, the *Culex* mosquito is limited to habitat below 2,000 feet in elevation, while most of the surviving native birds are limited to habitat above that level. Why have Hawaii's last honeycreepers gone to high ground? Warner hypothesized that it's because of lowland malaria—and experimental data reinforced his hypothesis. Warner exposed a few caged highland birds to mosquito attack in the lowlands, where they promptly fell sick with malaria. He concluded: "Any protracted visit to the lowland mosquito belt would mean immediate death from infection of *Plasmodium*. Even if some survived the malaria, onset of birdpox would complete the extermination."

Highland species, of which there weren't many, held on in their mountain refuges. A few other species, broadly adapted, could shift uphill into those safer habitats. But for many species and subspecies of endemic Hawaiian birds, including the most spectacular of the honeycreepers, altitudinal migration was no option. They couldn't go to the highlands, so they went extinct. *Drepanis pacifica*, the Hawaii *mamo*, was last seen in 1899. *Drepanis funerea*, the black *mamo*, disappeared about 1907. *Ciridops anna*, a finchlike little honeycreeper colored in patches of red, black, and gray, has been missing about as long. The native people called that one *ula-ai-hawane*, a name that flows like a warble. The Laysan honeycreeper is extinct. The lesser koa-finch is extinct. So are the greater koa-finch and the Kona grosbeak and the Hawaiian rail. The greater *amakihi* is extinct, as are the Oahu *nukupuu* and the *amaui* and the *kioea* and the Oahu *oo* (not to be confused with the *ou*) and at least three out of four subspecies of *akialoa*. The fourth *akialoa* subspecies, last seen thirty years ago in a swamp on the island of Kauai, has now evidently died out too. The *kakawahie* is most likely extinct, and the *olomau* has been extirpated from Maui and Lanai. You don't have to speak Hawaiian to comprehend the scale of these losses, though I suspect it would help.

90

ON THE ISLAND of Guam, four thousand miles west of Hawaii, the same sort of generalized collapse happened more recently. The forest went silent. The native bird populations came suddenly to the brink of extinction, before anyone could figure out just what was troubling them. As late as 1960 they had seemed fine; by 1983 they had suffered an invisible massacre. I heard about it from a newspaper article in 1985, when this was still a breaking story.

Where had the birds gone? What was killing them? Had they been devastated by an exotic disease, as in Hawaii? Had they been poisoned by cumulative doses of DDT? Had they been eaten by feral cats and tree-climbing pigs and Japanese soldiers who refused to surrender? Had someone set off a neutron bomb? The newspaper article, by a reporter in Philadelphia named Mark Jaffe, was headlined "Cracking an Ecological Murder Mystery."

The article quoted Larry Shelton, curator of birds at the Philadelphia Zoo, who had become involved in the Guam situation by way of efforts to rescue some of the species through captive breeding. "We've never seen anything like this before," Shelton said. No one had—not since H. W. Henshaw on Oahu, anyway. "The complete crash of an avifauna in this manner is unparalleled," he added. Although that statement seemed to overlook the case of Hawaii, Shelton's alarm was understandable.

Guam is the biggest and southernmost of the Marianas, a chain of small islands in the western Pacific that lie almost due south of Tokyo on a line toward New Guinea. Some people like to think of the Mariana Islands as the planet's tallest mountains, since they rise more than seven miles from the pit of the Marianas Trench. As manifested above sea level, though, their peaks are mere hills. Guam itself is only about twenty miles long and eight miles wide. Narrow at the middle, broader at each end, the island in outline resembles a leaping salmon, thrashing in midair with its tail pointed north.

Unlike some other oceanic islands, Guam isn't purely volcanic. The northern end is a porous limestone plateau. It consists of calcium carbonate left behind, over the epochs, by dead reef animals; accreted, compacted, the whole slab was eventually jacked up into daylight by pressure from beneath. Some parts of the northern plateau are thickly forested—those parts that the U.S. Air Force, the main landholder up there, hasn't claimed for runways and ammunition depots and suburbs to house its personnel. The forest is verdant and tropical, but no streams flow through it and no lakes dot its surface, since the limestone absorbs rainfall like a sponge—a petrified sponge, that is. The ground is hard and ragged and sharp. The limy rock tears at your shoes when you hike over it. It slices the knees of your trousers and the palms of your hands, if you have occasion to crawl. The southern half of the island is different, a brownish cow-pie of old lava, and since this volcanic rock is easily crumbled by erosion, the soil is softer. The landscape is open. In the south, at least on the uplands not yet completely expropriated for human housing and commerce and war, the vegetation is grassy. Along the gulches that drain to the southern coast, there are thin strips of ravine forest. While the Air Force controls the north, the U.S. Navy is an important landholder in the south. The birds died away in a mystifying wave moving northward from the savannas of the south.

Until recently, the island supported native populations of eleven different forest-dwelling birds. Five of those were endemic to Guam at the level of species or subspecies. The Guam flycatcher (*Myiagra freycineti*) was found nowhere else in the world. The bridled white-eye (*Zosterops conspicillatus*) was endemic to the Marianas, and its Guam population constituted an endemic subspecies. The rufous fantail (*Rhipidura rufifrons*) and the Micronesian kingfisher (*Halcyon cinnamomina*) were also both endemic at the subspecies level. The Guam rail (*Rallus owstoni*), stranded by its flightlessness, was an endemic species. In addition, the Mariana crow (*Corvus kubaryi*) occurred only on Guam and a small nearby island called Rota.

Sometime after the end of World War II these six birds and almost all of the others began suffering population declines. For a while no one noticed. The trend became obvious during the late 1960s. By that time, the ravine forests of the south were virtually empty of birds. It was spooky. Southern Guam had been quietly purged of its native avifauna, and no one could see what was responsible. Throughout the 1970s the invisible agent of death continued sweeping northward across the island. The remnant woodlands in the north consisted mostly of second growth that had burgeoned after wartime deforestation, but there was a small strip of mature tropical forest beneath the cliff face at the plateau's northern point. By 1983, that strip of forest was virtually the last place where an intact community of Guam's native birds—containing all the expected species at their normal levels of abundance—could be found. On the island as a whole, the range of each species had shrunk almost to nothing. Their populations had become drastically small and therefore drastically vulnerable.

Then the Guam flycatcher went extinct. Blip. The bridled white-eye disappeared. Blip. The rufous fantail, the white-throated ground-dove, the Mariana fruit-dove, the Micronesian honeyeater: blip blip blip blip. Of the eleven forest-dwelling species, these six all vanished from Guam during the mid-1980s.

Surviving individuals of two other species, the Guam rail and the Micronesian kingfisher, were rounded up and evacuated before it could happen to them. They were sent off to live out their time as captive breeders in mainland zoos, under the care of conservation-minded curators such as Larry Shelton in Philadelphia. This emergency airlift gave prospect of preserving small samples of genetic diversity from the larger gene pools, but it was also a concession of ecological defeat. If not utterly extinct, the rail and the kingfisher were now gone (or soon to be gone, when the last uncaptured kingfishers died away) from the wild. Whatever owned the forests of Guam, it wasn't the native bird fauna.

In the meantime, another spooky phenomenon appeared in those forests: an ungodly abundance of spiders. Many of them were orb weavers of the family Araneidae. Whatever was fatal for the birds seemed to be highly beneficial for the araneids. They gathered in huge communal webs, glittering three-dimensional structures suggesting crystal chandeliers. Each web contained dozens of araneids, some full-grown and some tiny. They spun their elaborate lattices, filling gaps between trees, and then sat waiting for food. They closed passages. They blockaded lines of escape. Their silk clotted the understory like bronchial phlegm. They lurked, as only spiders can lurk: eight-eyed and still and vigilant, listening with their feet.

But the spiders themselves hadn't eaten Guam's birds. Their abundance was only a visible symptom of some larger ecological trauma.

91

THE MYSTERY was slow to be solved. In the early 1980s reputable ornithologists were still discussing at least five different hypothetical explanations for the disappearance of Guam's birds, and the paucity of scientific data made it difficult to choose among them: pesticide poisoning, introduced predators, introduced disease, habitat destruction by humans, and typhoons. The range among these hypotheses reflects how thoroughly people were puzzled.

The main pesticide under suspicion was DDT, which had been applied heavily on Guam both by the U.S. military and by local farmers. Disease too seemed a good possibility, based on the paradigm offered by Warner's study of malaria and birdpox in Hawaii. Typhoons were suggested as a secondary cause. Although Guam usually endures several each year, an exceptionally fierce typhoon strikes about once per decade; Typhoon Pamela, of the exceptional sort, had hit Guam in 1976 with two-hundred-mile-an-hour winds and hammered some bird populations when they were already reduced and vulnerable. As for introduced predators, the roster of ecological rabble in that category included monitor lizards, four species of rat, one species of shrew, feral cats, feral dogs, feral pigs, and a species of snake, *Boiga irregularis*. Although most of those aliens had no redeeming merit, the snake was sometimes spoken of kindly, by well-meaning naturalists, for the beneficial role it supposedly played in helping control the rats.

Timing seemed to cast doubt on the introduced-predator hypothesis, since the four rat species, the monitor lizard, and the feral mammals had all been present on Guam since the 1890s, whereas the bird declines became apparent a full lifetime later. But the snake was a more recent invader, having colonized the island in the late 1940s, presumably from a few stowaways on a military ship or plane returning from some part of southeastern Asia after World War II. The snake had, however, been incorrectly identified. Ornithologists were conducting the inquest, and ornithologists are not trained to make fine herpetological distinctions. They were calling *Boiga irregularis* the "Philippine rat snake." Finally, in 1983, someone showed a photo of the snake to a herpetologist. Oh, *that* one, said the expert. That's not a Philippine rat snake, you featherbrains. That's a bird-eating tree snake from the Solomon Islands.

The correct common name for *B. irregularis*, it emerged, is the brown tree snake. With the different name came a different set of life-history attributes and ecological implications. The species is native to New Guinea, tropical Australia, Indonesia as far westward as Sulawesi, the Bismarck Archipelago, and the Solomon Islands. It's a venomous snake with a fierce disposition when harassed, though usually it's not very dangerous to humans. With its fangs at the rear of the mouth, it lacks the ability to deliver a lightning-quick poisonous stab, like a rattlesnake; instead it enwraps its prey and injects its venom methodically in the act of chewing. Graceful and stealthy, it moves slowly through the forest canopy, hunts nocturnally, and lives on a varied diet including birds and bird eggs. Its body is drastically slender and its muscle tone is superb, giving it great horizontal reach between branches. It plays hell on new fledglings, on unfledged young in the nest, and on adult birds as they sit incubating eggs against the cool night air. As a bird predator, *B. irregularis* is more formidable than anything the native Guam species had ever faced.

But even the herpetologist who corrected the identification was reluctant to believe that this was the solution to Guam's mystery. His ecological intuition made him doubt that one snake species could destroy an entire avifauna. Predators generally don't cause the extinction of populations of their prey. More typically, a predator population condemns itself to famine as it overexploits its supply of victims. At a point when its prey population is severely reduced, the starving predator suffers its own population collapse, which allows the prey population to recover. This gives the predator-prey relationship a wobbly sort of equilibrium. If the Guam collapse was really a case of extinctions caused by predation, it was exceptional.

Meanwhile, during the summer of 1981 the U.S. Fish and Wildlife Service in collaboration with Guam's Division of Aquatic and Wildlife Resources had initiated a pair of modest studies on the island—one to assess the problem of pesticides, the other to census the bird populations. A young zoologist named Julie Savidge was hired by the Guam DAWR to work for the summer with the USFWS scientists. Having been accepted into a doctoral program in ecology at the

University of Illinois, she was searching for a field project that might yield her a dissertation. The island caught her fancy in the course of that summer, and the mystery of the disappearing birds stuck in her mind. A year later Savidge returned to Guam, joining the DAWR on a somewhat more permanent basis, and from that point she became the leading researcher on the question of what was causing the losses. To test the disease hypothesis, she collected the carcasses of dead birds and sent them off frozen to a wildlife laboratory in Wisconsin. The lab found no evidence of a Guam epidemic. Savidge also conducted experiments of the sort Warner had used to detect malaria in Hawaii, exposing caged birds to potential disease vectors, such as mosquitoes, and monitoring their condition. Did the exposed birds fall dead of strange fevers? Did they grow tumors on their heads and lesions on their feet? Did other birds, kept under parallel conditions but protected by mosquito netting, stay healthy? Among the captive individuals that Savidge used in her disease study were some bridled white-eyes taken from one of the other Mariana islands, where they were still relatively common. These were delicate green-and-yellow birds with a white ring highlighting each eye. The Guam subspecies was the tiniest of all Guam's forest birds, and for some reason it had also been the most vulnerable—the first to disappear from each habitat zone during the current crisis. One night a small specimen of *B. irregularis*, the snake, forced its way into Savidge's laboratory through an air conditioner. By morning, it had eaten three white-eyes and killed a fourth.

With this hint and others to guide her, Savidge focused increasingly on the snake. From field records kept by the Guam DAWR, from published accounts, and from questionnaires filled out by Guam citizens, she documented that the population of *B. irregularis* had reached high levels on Guam during the period when the bird populations began to decline. Furthermore, the snake population had swept over the island in a wave that correlated geographically with the disappearance of birds—from the southern savannas in the 1950s, through the midlands in the 1960s, up into the northern forests by 1980. Setting traps baited with quail, Savidge established that a live bird left vulnerable overnight faced a high probability of being eaten by a snake. *B. irregularis* had become a familiar nuisance on the island, invading chicken coops, entering houses and stores, turning up in light fixtures and cabinets and toilets and beds, climbing on fences and guy wires and electrical lines, causing power outages, occasionally getting tangled with an infant in a crib long enough to give the baby a serious bite. Some of the larger snakes had killed piglets and caged rabbits, even German shepherd puppies. Guamanians of all stripe were now coping with *B. irregularis*. Savidge began collecting snakes and dissecting them. Inside some she found birds and bird eggs. Her dissertation was finished in 1986, by which time she had established persuasively that the culprit of Guam's ecological murder mystery was the snake.

Julie Savidge was just one of many scientists involved during this period. Another was Bob Beck, a lifelong birder as well as a biologist, who had joined the Guam DAWR about the same time as Savidge. Beck's original mandate was to monitor the status of the declining populations and, if possible, to find some way of stemming the declines. As Savidge clarified the cause of the problem, Beck's role became clearer too: rescuing the birds from the snake, either by protective measures in the wild or by evacuation. Working with him on the various species were a few other biologists who held overlapping responsibilities within the division. They were too late to save the white-eye, the fantail, or the flycatcher. But they rounded up the last of the Guam rails for captive breeding in a set of aviaries on Guam and in zoos on the mainland; and they captured thirty Micronesian kingfishers, nearly all of Guam's surviving population, for transfer to the mainland. Beck himself in 1984 accompanied the first six pairs of kingfishers on a sequence of long airline flights from Guam to New York, nursemaiding them in the passenger cabin with a supply of live geckos for food. Back in Guam again, he and his colleagues contrived physical barriers (such as a swath of sticky goo painted around a tree's trunk) to prevent snakes from getting to the nests of the Mariana crow. They kept vigil over other native birds that by now were badly endangered: the Micronesian starling, the Vanikoro swiftlet, the Guam subspecies of moorhen. They confirmed that even the Philippine turtle dove (an exotic species, originally introduced during Guam's Spanish era) was suffering a high rate of nest predation by the snake. They wrote papers and intradepartmental reports, they did what they could in the field—but it was all desperate and provisional so long as *B. irregularis* flourished on the island.

Still another important player was Tom Fritts, an experienced field herpetologist with the U.S. Fish and Wildlife Service, who arrived from the mainland in 1984. Fritts was initially skeptical that a single snake species could reach such abundance or cause such trouble. He had heard about Guam's supposed plague of snakes and had decided to come see for himself. On his first morning he found a snake dead on the pavement outside his hotel. That was an interesting sign, though possibly it meant nothing. With a bit of searching he found many more snakes, dead and alive, in the urban areas and the forests and virtually everywhere else. That meant something—but what?

For a professional herpetologist like Fritts, the population explosion of B. irregularis was an imposing phenomenon. How many snakes did the island contain? Another herpetologist had already guessed casually that there were at least a million. Fritts, more cautious, made his own estimate on the basis of trapping data: maybe thirteen thousand snakes per square mile. Since the total area of Guam is about two hundred square miles, a straight arithmetical calculation would leave room for almost three million snakes. Admittedly, the snake density figure that Fritts offered was applicable mainly to forest habitat, not to Guam's cities. Still, the message from both estimates was clear: Guam harbored a staggering profusion of B. irregularis. If they had already eaten up most of the native birds, they must have later discovered other food sources that would sustain them—rodents, exotic birds, lizards, and no one knew exactly what else. Even with other foods available, though, the sheer abundance of snakes was impressive. Like most predators, snakes normally can't afford to live among high concentrations of their own kind. But this situation apparently wasn't normal.

There was one other standard for judging the population density of B. irregularis, mentioned by Tom Fritts in an official report: How many specimens could a veteran snake collector gather in a single night? It turned out that Fritts and his colleagues found eight times as many snakes on Guam as in a rainforest at the Amazon headwaters. That rainforest region, in eastern Ecuador, supported fifty-one different species of snake. The Guam snake fauna, in contrast, consisted solely of B. irregularis and a tiny blind snake. The blind snake, a native species, was negligible to Fritts's specimens-per-night data. Therefore B. irregularis itself, on the landscape of Guam, under the scrutiny of an experienced collector, appeared four hundred times more abundant than an average snake species in a representative patch of the Amazon.

It's not easy to fathom such a giddy ecological imbalance. Eight times, four hundred times. I see the numbers for an experienced collector, and I wonder: What about an amateur?

92

My PLANE arrives in the evening. Gordon Rodda, a young herpetologist working with Tom Fritts on the USFWS team, meets me. Most of the year Gordon lives in southern Arizona, where a reptile lover can know deep happiness. The Guam situation, from his perspective, is unfortunate but scientifically interesting—a good gig for a consulting herpetologist. My contact with Ted Case among others has persuaded me that herpetologists are a congenial lot, and Gordon too turns out to be energetic, informal, simpatico. C'mon, he says, we'll go snaking. Grab them like this, put them in here. He issues me a headlamp and a battery pack. For himself, there's also a flimsy gardening glove to protect his right hand. "I don't like being bit," he explains. Coming from a professional, it sounds almost apologetic.

We drive past a guard post and onto the grounds of the Naval Air Station, a hermetic little world of tract houses, grassy yards, cyclone fencing lined with scrub vegetation, the military approximation of suburban tranquility. Just outside the fence is second-growth tropical forest; inside the fence is a banal terrain of mowers and rakes and never-ending warfare against crabgrass. Most of the houses are dark. Either everyone is asleep or they're down at the officers' club playing bridge. Going off-road, we drive quietly through their backyards, over their grass, over their hoses, past their lawn chairs and their patios, panning our beams from the car windows, policing the area for snakes. You couldn't do this in the real world. But it's the right time of night and Gordon knows the good places. The armed guards, the fence, are no barrier against *B. irregularis*, and they're no barrier against Gordon Rodda either. He has a government I.D. He has clearance. I see fifteen snakes before I see my motel room.

It might seem peculiar, Gordon tells me, that the Guam situation progressed to such an extreme before anyone put the blame on the snake. "But really, nobody would imagine that there *were* that many snakes on the island. You just don't see them. Unless you go out at night with a light."

For two hours we prowl the backyards of naval officers. Gordon entertains me with running commentary. "This is the famous power pole #84," he says at one point. It's a huge concrete pylon, stout as a Douglas fir, supporting two high-voltage lines. The upper line carries 115,000 volts, the lower line only a fraction as much. Occasionally an unfortunate snake reaches the lower line, momentarily bridging the gap between that line and a grounded guy wire. The result is flash-frizzled snake and an industrial-strength short circuit that takes out the line. "Sometimes it shorts out with a big ball of fire," Gordon says enthusiastically. "You can see it miles away. And sometimes that ball of fire takes out the upper line too. That one's the biggest power line in Guam. When that one goes out, the whole island goes dark." During the past decade, snakes have caused more than five hundred electrical outages. Pole #84, here in the snake-ridden wilds of the naval officers' backyards, has been hit seventeen times. The Guam Power Authority, for whom these outages are a costly embarrassment amounting to millions of dollars, finally decided to do something.

After a bit of wrongheaded deliberation, the GPA fitted its most vulnerable high-voltage poles with girdles—snug sheaths of metal, like the tin girdles you see around birdfeeder posts to keep squirrels from clawing their way up. The power-pole girdles were six-foot-high cylinders of gleaming and slippery stainless steel. It was a forceful response, also expensive, Gordon notes, "but nobody bothered to ask a biologist whether snakes even climb concrete power poles. They don't. They go up the guy wires." The GPA should have had its own consulting herpetologist, who could have pointed out that tree snakes, unlike squirrels, don't have claws; that they ascend by enwrapping, and that a major power pole is just too fat to be enwrapped. Alerted belatedly, the company fitted its guy wires too with snake barriers—metal disks, like the rat guards on docking lines to an old ship. The disks might be more useful than the girdles, according to Gordon, if it weren't for the fact that their size and their radial spokes still allow a large snake to find purchase enough to climb around one. Meanwhile the Guam Power Authority has devised a still better method of protecting that 115,000-volt line: by shutting it down between dusk and dawn. A concession that nighttime belongs to the enemy.

Although the power outages are annoying and costly, Gordon says, they don't deliver quite the same emotional jolt as some other manifestations of the snake. "The thing that really gets people excited is getting bit on the privates while they're sitting on the toilet. Or cardiac arrest in a six-month-old child, like that one this week." It has just hit the local newspaper: The baby daughter of an Air Force staff sergeant was rushed to a hospital after a five-foot-long snake tried to swallow her arm. The girl suffered a scary respiratory shutdown, but she was saved. She's still under care. No one on Guam has died from snakebite, though this infant and a handful of others have come close. In objective terms, Gordon says, the danger to humans isn't sizable. Bee stings account for more medical emergencies than *B. irregularis.* "But the *idea* of snakes makes it worse."

Our talk is interrupted occasionally by the sight of a snake. Whenever it happens, Gordon tromps on the brake and leaps out. He pulls on the gardening glove. He runs to the cyclone fence and, like Willie Mays in deep center, he makes the grab. Using his free hand and his teeth, he unties the knotted neck of a cloth sack, into which he deftly jounces the snake. This is one of those special little skills that a professional herpetologist develops: dropping a poisonous snake into a sack, then spinning the sack closed before that snake or another can come lurching back out. In his field notebook, Gordon records the length of the snake and a few other facts. Night after night, month after month, he has been gathering similar data. It's a boy's dream of bliss: collecting snakes on a tropical island under contract to the U.S. government.

As we drive again, I incite Gordon to tell me about his Ph.D. work on alligators. The night is quiet and we've got time to squander. He did the field research in Florida. The question he addressed was how alligators guide their travels. It turns out that they have an elaborate navigational sense, Gordon says, though crocodiles are even better. What's the mechanism? he wondered. Do they possess some sort of internalized compass that detects geomagnetic fields? Do they use smell? How important is memory? The real trick in navigation, for alligators or humans or whomever, Gordon explains, is knowing where you are. That's the first requirement: knowing your own position accurately at a given moment. Saying this, he drives up over a curb, across a gently banked lawn, past a few bedroom windows, down a steep grassy hill, and around another power pole, scanning for snakes as he goes.

Have you ever plowed the car into one of these poles while you're watching the fence? I ask.

"No. But I've driven it into a hole."

So far we've collected fourteen snakes. They ride in the back seat, docile within their cloth sacks. It's not an exceptional night's haul. With the onset of the dry season, and with the vegetation along the fence having been pruned back by the Navy groundskeepers, Gordon has lately been finding not quite so many *B. irregularis* here as in earlier months. He's cautious about guessing what the lower numbers might mean. He's obliged to be cautious—he's a scientist. My own obligations are different, and in the privacy of a dark car I have no qualms about pushing him into the speculative mood. Well, Gordon allows, the lower numbers could be significant but then again maybe not. It's possible that the snake, like the western Pacific weather, might be subject to natural cycles of extremity: peaks and depressions of population, or peaks and depressions of obtrusiveness, reflecting how many snakes are nocturnally active in areas where Gordon can collect them. Maybe the snake density here on the Naval Air Station has fallen, whereas the density up in the northern forest might still be increasing. Or maybe snakes all over the island are starving for lack of birds, now that the native avifauna has largely been gobbled up. That seems unlikely, since *B. irregularis* is a generalized predator that also eats lizards and rats. But maybe the lizard and rat populations have been depleted too. If so, then starvation might indeed cause the snake population to decline. At a density of thirteen thousand snakes per square mile, the species would burn through a hell of a lot of food.

Back in the snake's native habitat, circumstances would be very different. In the forests of New Guinea, for instance, *B. irregularis* is just one species of tree-climbing snake among many, harried by its own predators and competitors, afflicted with its own parasites and diseases; any drastic increases or decreases in its population would tend to be dampered amid the welter of ecological interactions. In reaching Guam, though, it escaped from most if not all of its customary predators, competitors, parasites, and diseases. Like a felon on the lam, with a new name and a new haircut, it freed itself from the constraints of context and reinvented its own role. Here it can fluctuate through a broader range of possibility: from boom to bust. Right now, the population might be in a bust phase. Does that mean *B. irregularis* has dropped below its own threshold of precarious rarity? Probably not. That it will disappear altogether from Guam? No such luck. The thing to remember about boom-and-bust cycles is that they are cycles. After a period in decline, the snake density might come back up to thirteen thousand per square mile. The snake will have plenty of food so long as its alternative prey species—the rats, the skinks, the anoles, the geckos—don't follow the Guam flycatcher to extinction. But none of this is certain. It's not science, it's just plausible sciency talk. The natural history of *B. irregularis* and its population dynamics on Guam, not to mention the long-term implications for the ecosystem, are still largely unknown. What's certain is that there are thousands and thousands of snakes on the island, maybe millions and millions, and that Gordon Rodda needs to get his hands on as many as possible.

"There's one!" He brakes. He leaps out. He pulls on the gardening glove and runs toward the fence.

93

OF MY TWO weeks on Guam, I spend a large portion in the company of Gordon Rodda and Tom Fritts, walking through the forest, patrolling the fence lines, strolling at night across the fairways of a golf course: searching for *B. irregularis*.

Gordon and Tom need data. More concretely, they need snakes. Their ultimate mission is to discover a means of bringing the snake population under control so that the native species of wildlife have a chance to survive. They have some ideas about how the snake's pestilential abundance might be reduced, but they are well aware of the sorry truth that *B. irregularis* can probably never be completely extirpated from Guam. Having arrived and taken hold, it's now an ineradicable part of the local biota. The island is too large and intricate, the snake is too cryptic and hardy and fertile, for there to be hope of complete extermination. Biologists would have to destroy the place in order to save it and that didn't even work in Vietnam. But maybe the snake's victory needn't be total. At present *B. irregularis* is a scourge; with a little luck, with a little human ingenuity, it might somehow be reduced to a pest. Before any such reduction can happen, though, the snake must be better understood. Its reproductive biology, its growth rate, its sensory physiology, its demographics, its hunting behavior, its cycles of activity and dormancy, its preferences for one type of prey over another, its susceptibility to trapping, its responses to different sorts of bait—all these must be described and charted. That's the job of Rodda and Fritts. While most Guamanians despise the snake, Gordon and Tom study it with something strangely akin to loving attention.

It's not an evil animal, after all. It's just an amoral and earnestly stupid creature in the wrong place. What it has done here in Guam is precisely what *Homo sapiens* has done all over the planet: succeeded extravagantly at the expense of other species. Encountered in New

Guinea, encountered in Queensland, encountered in Sulawesi or Guadalcanal, *B. irregularis* is a handsome and sleek native reptile, constrained by the natural boundaries of its station in life. Tom Fritts and Gordon Rodda are mindful of that. To them it's a regrettably translocated species of wild animal, not a demon from ecological hell. They catch snakes, kill snakes, embalm snakes, dissect snakes, plot warfare against a whole population of snakes, not in loathing or wrath but with a sober sense of scientific duty.

On a typical morning, I walk the trap line with them.

The trap line is actually a rectangular grid, laid out by Gordon in an area of second-growth forest in the north of the island. It lies inside the greater perimeter of Anderson Air Force Base. Access is restricted and there's not much disturbance, aside from B-52s roaring low overhead and an occasional helicopter. Within an area of two hectares, Gordon has placed eighty wire-mesh traps, specially designed for the accommodation of snakes. The traps are arranged in twenty rows of four, their sites tagged with the same sort of plastic surveyor's tape that Ted Case has used on the *bajada* of Angel de la Guarda. The difference here is that the forest is thick and the individual trap sites connect via a machete-cut trail. Fan palms and pandanus and tangantangan trees arch low over the trail, closing it down into a tunnel. Dark swallowtail butterflies flit through the broken light. Termite nests hang like massive brown goiters on the air-roots of the pandanus. Monstrous spiderwebs interrupt the trail at the height of a person's head. Each web is a venture in cooperative predation, holding a dozen or so major spiders and many more apprentices, some scab-red and corpulent, some black marked with yellow, some solid black. Among the black-bodied spiders is one especially conspicuous type, with long graceful legs and an abdomen the size of a prune.

Parenthetically, here, a confession: I happen to be scared of spiders. They give me the willies, all right? Of course there's no rational grounds for this feeling, it's just a matter of psyche and taste. The bastards unnerve me: too many eyes, too many legs, and that demeanor of slaver-jawed stealth. My worst nightmares feature tarantulas the size of badgers. I burden you with this personal revelation only by way of explaining that my sensitivity to spider population levels, as I straggle through the undergrowth behind Gordon and Tom, is acute.

I try to stay alert for hazards. The terrain underfoot is that rough coralline limestone—jagged, brittle, furrowed hereabouts by old trenches and craters, strewn with Coke bottles and rusty war debris.

Back when Guam was retaken by the U.S. military, in 1944, the last of the fierce Japanese resistance occurred not far from this spot, according to Tom. Bombs, mortars, grenades flying every which way. Some of these ditches may have been dug by the Japanese to store ammo. So watch out for unexploded ordnance, Tom suggests. Watch out for unexploded ordnance and big black spiders, I think.

Tom has noticed the spiders too but he's blessed with sang-froid. "Some days I carry a stick," he says, "and spend half my time knocking webs out of the way." He suspects that the proliferation of spiders is a secondary consequence of *B. irregularis*. It might be a two- or three-channel effect, entailing both reduced predation and reduced competition as the snake has taken its toll on the bird and the lizard populations. Delivered from their enemies, the spiders flourish. They multiply wildly. And the snake itself doesn't bother them, because the snake prefers larger mouthfuls.

Tom is only guessing, but it does seem evident that some sort of unusual outbreak has occurred. I've never seen such a concentration of arachnids in any other tropical forest. They fill the understory with high-volume webs, many of which punctuate the trap line. So we dip and duck around cities of silk, each harboring whole gangs of spiders. I wouldn't be all that surprised to see the wrapped husk of an unwary herpetologist dangling in one of them.

Gordon fiddles carefully with each trap. Is there a captured snake? Not in that one. Not in this one. Not today. Are the door flaps still functional? He pokes, he tests. Has the bait animal gone moribund from exposure? The traps are baited with live geckos, and the snakes prefer vigorous prey, so Gordon carries a fresh supply. Has the bait animal disappeared? Sometimes a snake eats and escapes. Gordon knows that the current trap-and-bait system should be improved; part of his present task is to determine how. I shadow him from trap to trap, helping with the chores, dodging the spiders.

As we work, Gordon tells me something about geckos. The species he's chiefly using for snake bait is *Hemidactylus frenatus*, also known as the house gecko because of its affinity for human-made structures. *Hemidactylus frenatus* is a notorious tramp species throughout the western Pacific, having established populations on any number of islands within recent centuries, evidently by stowing away on ocean-going vessels. It reached Guam among other places and it seems to be thriving here, despite some predation by the snake. *Gehyra oceanica* is a big-bodied native species, eight times as hefty as *H. frenatus* and more restricted to the forest. It's a nice meaty lizard whose size and habitat preference make it an ideal prey item for the snake. Once abundant on Guam, *Gehyra oceanica* is now rare.

Four other native geckos have also suffered declines that may be attributable to the snake. *Lepidodactylus lugubris* is a small-bodied species, formerly common, now less so on Guam than on snake-free islands nearby. *Gehyra mutilata* is a medium-sized species, famous for the trick of peeling free of its own skin, coming out pink and naked and raw, to escape the grasp of a predator. But the trick isn't foolproof and may not be effective against brown tree snakes. The population of *Gehyra mutilata* has been severely reduced. *Nactus pelagicus*, a slightly larger species, is missing entirely. *Perochirus ateles* also seems to be gone, not having been sighted within the past decade. So while newspapers on the far side of the planet have been raising alarms about bird extinctions, lizards too have been threatened, devastated, eliminated. "But it's very hard to get people concerned about geckos," Gordon says. A species of gecko doesn't have the same popular appeal as a little green bird.

The skinks, another family of lizards, have been affected too. Skinks differ from geckos in a few obvious ways—their bodies are smoother and more rounded, their feet are less suited for climbing, they aren't so insistently vocal, and they are generally diurnal. Several species have disappeared from the Guam forest, either totally or mostly, within about the same period as the geckos and birds. *Emoia slevini* is evidently gone. *Emoia caeruleocauda* has become scarce. *Emoia atrocostata* is abundant on a tiny islet just offshore but suspiciously absent from Guam. Has the snake eaten their populations down to nothing? Gordon doesn't care to make reckless claims. There are no data. Slow down and spell everything, I plead. Is *caeruleocauda* one word or two? Which species has the breakaway skin?

It's complicated, as ecological matters always are. By eliminating entire populations of native birds and lizards, *B. irregularis* has committed a violent sort of simplification—but it's still complicated. A drastic ecological disruption tends to ramify in all directions, affecting patterns of competition, habitat use, population size, and predation. Consequences ricochet like billiard balls.

Consider, for instance, Guam's Micronesian kingfisher. It was a predatory bird that probably ate its own share of skinks. On that subject I pester Gordon for some details. Did it favor one kind of skink over another? Did it like small delicate mouthfuls or hefty hunks?

Did it prey on the tree-climbing species more heavily than on the ground-lurking species? Don't ask me, says Gordon. To hear what a kingfisher does, he notes gently, you'll want to consult an ornithologist. To hear what a kingfisher likes or dislikes, he implies, you'll want to consult a kingfisher. Suddenly I'm hungry for behavioral observations about Guam's subspecies of Micronesian kingfisher—observations that may never be gathered, now that the subspecies survives only in zoos. How did it feed in the wild? Did it dive through the tree branches or did it catch its prey in the open? Did it eat smaller skink species with delicate scales, in preference to big skinks with thick scales? If the kingfisher exerted that sort of differential predation pressure, it may have played a significant role in maintaining the balance among populations of competing skinks.

In the kingfisher's absence, by coincidence or otherwise, at least one species of skink has become far more abundant. The population of *Carlia fusca*, introduced from New Guinea or elsewhere during the 1960s, is booming.

What are the actual relationships among all these facts? Hard to say. What happens next? Impossible to predict. It's conceivable that the booming population of *Carlia fusca* might eat the last few individuals of some rare and endemic species of insect. Ecological field research is a seemingly infinite task that must be practiced slowly and painstakingly, the way Gordon Rodda and Tom Fritts practice it, using methods that are hopelessly finite and leaving the wild speculation to visiting writers.

Eighty traps seemed like a lot when we started walking. But at the end of two hours, Gordon and Tom and I haven't collected a single snake. Maybe it's the drought. Maybe the snake population, here in the north of the island as in the south, has passed the peak of its cycle and declined. Maybe the trap design is no good, or possibly we're using the wrong bait. Temporarily, for whatever reason, *B. irregularis* has turned invisible. But the consequences of its presence surround us like a web.

94

I'M CURIOUS about those spiders. A tangled path of questions and referrals leads me to a small, cluttered laboratory inside a building labeled ENTOMOLOGY on the campus of the University of Guam. Mounted specimens, pins, microscopes, dust, cocoons, jars full of guano, stray pupae, stacks of technical literature, and broken-off bits of insect anatomy decorate the premises. The air itself is powdery and humid and sour, like the air of a chicken house. Two entomologists named Don Nafus and Ilse Schreiner work here, elbow to elbow amid the disorder, a pair of focused and unpretentious scientists whose passion for their own field of study seems almost infectious. Then again, if anything in this room *is* infectious, I'm pretty sure I don't want to catch it.

Without explaining what's on my mind, I ask Nafus and Schreiner: Have they seen any recent invasions by exotic arthropods, or any dramatic population outbreaks among native ones? I inquire about arthropods rather than insects because it's a broader category, inclusive also of such charming non-insect invertebrates as ticks, centipedes, millipedes, and spiders. Asking a professional entomologist about arthropods (and not about, say, bugs) is a way of signaling at least some familiarity with the subject. Still, Nafus rocks backward in his chair, rebounding from the dumbness of my question.

"We have invasions and outbreaks *all the time*."

Among arthropods as among reptiles and birds, it seems, Guam is an ecological battleground on which the native species fight unending defensive warfare against the exotics. Nafus and Schreiner can cite a long list of particulars. The citrus spiny whitefly is an exotic. So is the Philippine lady beetle. The Oriental fruit fly is another invader, as is the Chinese rose beetle, the Formosan termite, and the coconut red scale. There's also the agromyzid vegetable leafminer, the poinciana looper, the banana skipper, the Egyptian hibiscus mealybug, the chrysomelid cucumber beetle, several species of eumenid wasp, a nititulid beetle, a magarodid, a psyllid, a bagworm—even the names are strange. Among mosquitoes alone, fourteen different exotic species have come ashore since 1945. Have there been any invasions, what a laugh.

I try a different slant. Extinctions, then? Have you seen any extinctions among the native arthropods?

This question turns out to be less risible. Nafus gives me a cautious yes. Some evidence suggests that there were "a number of butterflies going extinct" at about the same time as the loss of the birds. The general theme of these cases, as for much of Nafus and Schreiner's work, is the conflict between endemic species and exotics.

The swallowtail *Papilio xuthus*, for one, has declined badly since the

1950s. Formerly it was common. Nowadays it's rare, if not extinct. *Euploea eleutho* is another butterfly, endemic to the Marianas, that hasn't been sighted on Guam in decades. Schreiner shows me its picture on a page of the manuscript of an insect guide that she and Nafus are writing. *Vagrans egistina* is still another, endemic to the Marianas but now missing from Guam itself. Again, Schreiner offers its picture. Then the talk turns to a rare lepidopteran that may never have been very common—a species known informally as the guano moth.

The guano moth is graced with that name because its larvae are found only in slag piles of feces left by the Vanikoro swiftlet, a swallowlike bird. The Vanikoro swiftlet rears its young inside caves, gluing its nests to the ceilings with saliva. Like the other native birds, it stands in peril of extinction. Despite the wariness of its nesting behavior, some combination of factors (possibly including the snake) has reduced the swiftlet to just a few hundred individuals, most of which now nest colonially in a single cave. That cave lies in the backcountry of southern Guam, some miles from the nearest public road. There, in the dark of an underground rookery, guano moth larvae can be found squirming merrily on the dung heap beneath the birds' nests. It's a peculiar niche, a narrow niche, but theirs. Although the site is a secret well kept by Guam's wildlife officials for fear that human disturbance will further stress the swiftlets, I've been taken there by one trusting biologist. Don't step in the guano, he told me, you might hurt the rare moths.

At my mention of the guano moth, Ilse Schreiner perks up. "Here they are," she says proudly, and dumps a jarful of grayish black powder onto the table. It's a precious sample of swiftlet guano, brought back to her and Nafus by the same biologist who showed me the cave. Schreiner spreads the stuff out across a clean sheet of computer paper and, with a pair of scissors as the handiest tool, pokes through it avidly, paying minute attention to certain small lumps. "Swiftlet guano is really interesting," she says, "when you look at it under the microscope. Nothing but tiny pieces of insect. Heads and legs and other body parts." She points the scissors at a brown morsel, no bigger than a sunflower kernel, and explains that this is some sort of shell-like case containing a guano moth larva. But what sort of case? She doesn't know. It can't be a cocoon or a chrysalis, enclosing the transitional stage between larva and adult, because she and Nafus have seen cases in a whole range of sizes. So it must be a portable shelter that the larva drags through life, exchanging one size for another as it grows. What's the case made of? Silk and debris, probably. How does the larva construct it? Unknown. The habits of this insect are a matter of guesswork for Schreiner and Nafus. Even its identity is indeterminate, "guano moth" being only a loose way of labeling the thing until further notice. Is it confined to Guam? Maybe. Confined to that single cave? Possibly. Nafus suspects that the species might be new to science, never before named or described.

"There's the adult," Schreiner says, jabbing her scissors at a little carcass. "We'll send it off to a guy in Hawaii who specializes in cave fauna. He may be able to identify it. But we'll be lucky if we get it down as far as genus." Having its precise taxonomic identity ascertained—or establishing an identity, if it's a new species—is a task that will have to wait, however. Schreiner and Nafus are up to their ears in other work, mostly involving exotic pests with major economic implications. Agriculture is what generally pays the bills for entomology, so it's no wonder that the citrus spiny whitefly and the banana skipper take priority over an unassuming guano moth that may or may not be endemic to Guam.

I broach the subject of spiders. Aren't they unusually abundant? Is it possible that their populations have exploded as a secondary result of the snakes having eaten up so many lizards? It's possible, yes—but Nafus and Schreiner can't say more, since they haven't actually studied that question. No one on Guam has. But a few years ago there was an experiment done in the Caribbean. Two scientists removed all the lizards from enclosed patches of habitat on an island. In the aftermath, they found, spider populations increased dramatically.

Can Nafus and Schreiner tell me the name of the black-bodied spider as big as a prune? No, they can't; spiders are a bit outside their bailiwick. But they can steer me toward a fellow who might know. Alex Kerr is his name—only an undergraduate, but crazy for spiders and rather knowledgeable. He works over at the marine lab. I scribble the information into my notebook.

Don Nafus, in the meantime, has become slightly preoccupied. While holding up his end of the conversation, he watches a praying mantis crawl across his hand. The mantis is a fresh hatchling, tiny and delicate as an ant. Because mantises undergo no drastic metamorphosis as they grow (the way butterflies and beetles do), this baby is a miniature version of a mantis adult, standing bold and fierce on its four hind legs, torso erect, front legs held high like a boxer's fists. A full-grown mantis is a formidable predator, with jaws that can chew into a person's finger, but the hatchling strolls innocently across a broad landscape of knuckles, discovering the world. Nafus leans over and transfers it onto my hand, as though I had asked to dandle it.

I admire the baby mantis politely. I let it walk on me while I ask a few more questions. I do my best not to startle it. After what seems an appropriate interval, I invite the little creature to step back onto Nafus's knee. It does.

Nafus pinches it up between two fingers, suddenly, and flicks the corpse toward a wastebasket.

"Exotic," he says.

95

SEVERAL DAYS later, in a careless misstep while walking the trap line with Gordon Rodda, I bury my face in one of the chandelier webs.

I freeze. My pores pucker like sphincters. I feel the strong tiny threads, sticky with some sort of glandular arachnoid exudate, stretched taut against my forehead and cheeks. With false calm, I back my way out. Doubled over, I brush at my hair, swat at my ears, scoop at my collar. The silk is like cotton candy. I peel away what I can. Then I look up to see a long-legged black spider at eye level just in front of me. From this perspective the prune simile seems inadequate; her abdomen looks as big as an eggplant.

Poised in readiness on the far side of the web, she doesn't move. Maybe she's paralyzed with alarm. But if her pulse has gone rapid, like mine, she doesn't show it. She waits forbearingly for me to remove my head from her web. Faintly I realize that she bears me no malice. I don't fit the profile of her accustomed enemies or prey. I'm an anomalous cataclysm, like Godzilla.

I won't claim this moment as a life-changing epiphany. It's just one of two memorable spider events during the last days of my stay on Guam. The second occurs that afternoon. After enough legwork to aggravate Philip Marlowe, I finally make contact with Alex Kerr, the undergraduate arachnologist mentioned by Nafus and Schreiner.

The effort of tracking him hasn't been wasted. When I describe the big prune-bodied spider of the three-dimensional webs, Alex knows exactly what I mean. The species is *Cyrtophora moluccensis*, he tells me. It belongs to the Araneidae family, the orb-weavers. It's found on islands throughout the Pacific. I mention my impression that it seems extraordinarily abundant in the forests of Guam.

Although his response isn't grounded in the methodology or language of science, it has a ring of authority. "Yeah," Alex says cheerily. "I kinda thought there might be a spider explosion."

96

I'M CURIOUS about those spiders for a reason. The reason is more sober than arachnophobia and larger than Guam. *Cyrtophora moluccensis* brings to focus a phenomenon that Jared Diamond and others have called *trophic cascades*, and we haven't begun to comprehend extinction until we comprehend that.

The term refers to cascading disruptions that can pass between trophic levels—that is, between different categories of interrelated organisms in the hierarchy of energy transfer within an ecosystem. The ecosystem itself is not just a landscape full of plant and animal species; it's an intricate network of relationships, including those between predators and their prey, between flowering plants and their pollinators, between fruiting plants and the animals that disperse their seeds. Each such relationship constitutes a link between trophic levels. Trophic cascades, as defined by Diamond in his "Rosetta Stone" paper, are the secondary effects that can ramify from level to level in consequence of a single extinction.

His words: "Since species abundances depend on each other in numerous ways, disappearance of one species is likely to produce cascading effects on abundances of species that use it as prey, pollinator, or fruit disperser." At the low extreme of abundance, of course, a species faces rarity unto extinction.

As an example, Diamond cited a plant group that seems to have suffered—secondarily—from malaria. "All five species of the endemic Hawaiian plant genus *Hibiscadelphus* are extinct or nearly extinct due to the disappearance of their pollinators, Hawaiian honeycreepers, whose long curved bills match the plants' narrow tubular curved flowers." The extreme shape of the flowers implies an exclusive relationship between plant and pollinator. A lengthy history of mutual adaptation seems to have fitted the bills of certain honeycreeper species to the flowers of certain *Hibiscadelphus* species, leaving other nectar-gathering animals (such as bees, moths, bats) excluded. If this is so, then with those honeycreepers eliminated by malaria, the *Hibiscadelphus* species can't be pollinated. As older plants die, no new plants germinate to replace them. The eventual result is extinction. The plants disappear as a secondary result of the birds' disappearance. The mechanism is trophic cascade.

Imagine another set of species and relationships in another tropical forest. This forest is hypothetical, like our island of owls and mice, but the particulars again are all plausible. For continuity, I'll even hypothesize another owl. The full set includes a species of peccary, a species of frog, a species of mosquito, a microbial parasite, a species of monkey, a species of mango tree, a species of leafcutter ant, a species of lizard, a species of wasp, and our species of owl—which in this case is a pygmy, adapted to nesting in small cavities. These species interact with each other in diverse and incalculable ways, but a few of the interactions can be noted. The peccary roots for food along the soft banks of forest creeks, and its rooting creates muddy puddles. The frog and the mosquito both lay eggs in those puddles. As the tadpoles hatch and grow, they feed on the larvae of the mosquito. The mosquito, as an adult, carries the microbial parasite in its saliva, and that parasite causes disease in the monkey. Fortunately for the monkey, the mosquito population is small—and so the incidence of the disease remains low. The monkey eats a wide variety of plant material, but it prefers items that are juicy and sweet. Whenever the mango tree comes into fruit, the monkey gorges on that. The mangos are small enough that sometimes the monkey swallows them whole, and the woody mango seeds pass through the monkey's gut. The leafcutter ant depends on the same tree for its forage. The ant is so specialized, in fact, that it relies on that tree and no other. The lizard is slightly less specialized, feeding on ants, flies, and an occasional wasp. The wasp species builds its nests in the small tree holes originally excavated by woodpeckers. The pygmy owl, like the wasp, nests in abandoned woodpecker holes.

Now let's disturb the ecosystem with a single extinction and postulate what might happen. The peccary population is killed off, say, by hunters.

With the peccary gone, the creeks flow clear, the soft banks remain undisturbed, and after a period of time there are no muddy puddles. So the frog, deprived of egg-laying sites, follows the peccary to extinction. But the mosquito, less specialized than the frog, survives. It adjusts to the absence of peccary wallows by laying its eggs in the tiny pools of rainwater that accumulate on fallen leaves. In fact, the mosquito does better than simply survive; relieved of predation by tadpoles, its population soars. Hungry mosquitoes swarm thickly through the forest, tormenting all warm-blooded animals and infecting every monkey with the microbial disease. This disease outbreak comes as a final blow to the monkey, which (like the peccary) has already been reduced by hunting. So the monkey goes extinct. The mosquito scarcely misses the monkey, shifting its bloodsucking attacks to a variety of small mammals and birds. But the mango tree does suffer from loss of the monkey, which was an irreplaceable partner. Without monkeys to swallow its fruits, to process its seeds by digestion, to shit those seeds onto the ground along with little legacies of fertilizer, the tree can't reproduce. So in the years following the monkey's extinction, the tree sets no saplings. After two centuries of helpless sterility, two centuries without offspring, the last old mango tree dies.

Our tally so far: peccary extinct, frog extinct, monkey extinct, tree extinct. The mosquito is thriving. The microbial parasite, like the tree, has followed the monkey to extinction—it was insufficiently versatile to establish new reservoirs of infection in other mammals preyed on by the mosquito.

The cascading effects continue. The leafcutter ant, bereft of its favored tree species, goes extinct. In the absence of the ant, the lizard population declines sharply. The lizard's decline is a boon to the wasp population, which finds itself no longer harried by lizards. Some few lizards do continue to prey on some few unwary wasps, but far fewer than ever, and wasps become vastly more numerous. They seize every abandoned woodpecker hole, building their nests, raising their offspring, producing ever more wasps, claiming ever more holes. The pygmy owl is the loser in that competition. Faced with a wasp's territorial fervor, the owl is meek. It does not inherit the Earth. Denied access to breeding holes, it goes extinct.

The ecosystem has been transformed by trophic cascades.

97

IN THE SAME paper Jared Diamond cited two other notable cases of what he took to be trophic cascades. One of those two—concerning bird extinctions on the island of Barro Colorado, within the impoundment for the Panama Canal—is probably valid. The other case—concerning a tree species, *Calvaria major*, on Mauritius—is probably fallacious. The inclusion of the *Calvaria* case suggests that even an eminent theorist like Diamond is susceptible to being misled by faulty reports from the field. Fallacious or not, that case is interesting because it involves the dodo. But before getting into it, let's consider Diamond's valid example.

Barro Colorado is a land-bridge island, like Tasmania and Bali, with one crucial difference: its age. Whereas most land bridge islands became insularized about twelve thousand years ago, at the end of the Pleistocene, Barro Colorado became insularized much more recently. It's almost as young as Tom Lovejoy's reserves in the Amazon. At the turn of this century, what now constitutes Barro Colorado was a forested hilltop in the interior of Panama, not far from the Chagres River. Then the canal was cut through the peninsula and the Chagres was dammed to form Gatun Lake, which would help keep the new channel fed with water. As the lake waters rose, the hilltop became an island.

It's a tiny plot of land, only six square miles in area, but it has received an extraordinary amount of scientific attention. Back in 1923, when rainforests were still jungles and few people viewed them as a precious resource, Barro Colorado was declared a biological reserve. Ornithologists came to study its birds. Botanists came to study its plants. Since 1946 the island has been managed as a research site by the Smithsonian Institution. Over the course of seven decades it has been one of the most closely scrutinized patches of landscape in the American tropics. This too makes it crucially different from Tasmania, from Bali, from virtually all other land-bridge islands, of which the early ecological conditions can be known only by surmise: Barro Colorado's early conditions are known by direct observation. For most of its insular history, biologists have been on it like flies on a picnic buffet.

The biologists have seen changes. Back before the reserve was established, some parcels of forest had been cleared for timber or farming. After 1923, the cleared parcels were allowed to revert toward second-growth forest. Windstorms occasionally toppled trees in the mature forest, and those treefalls offered new habitat to certain animals. In 1959, landslides disrupted the vegetation in a few zones of steep terrain. The storms, the landslides, the regrowth of cut forest all carried significance for the larger ecological community. But the most dramatic change was faunal extinctions. Soon after the island was isolated, certain mammal and bird populations disappeared.

The puma was among the first to go. It was a big carnivore with big needs, and Barro Colorado was too small. The jaguar vanished also, presumably for the same reason. Nobody knows what became of the last jaguar—whether it vacated that hilltop before the water came up, or swam away afterward, or found itself marooned there and died lonely—but in any case the result was the same. Even the harpy eagle no longer nested on Barro Colorado. The loss of these large-bodied and far-ranging predators was just the beginning.

By 1970, forty-five species of bird had disappeared. That's a severe degradation for a small patch of forest—severe, in fact, for any patch of forest. "Although Barro Colorado has proven to be a valuable natural laboratory," one researcher has written, "it is clearly a failure as a faunal preserve. The pace of extinctions is unacceptably high."

What made these birds go extinct? The causes are hard to identify, but ecologists have a list of possibilities: competition, predation, lack of sufficient area, lack of appropriate habitat, lack of rainfall, and the random sorts of misfortune that can finish off a small population. Most of the lost species were birds that inhabited meadows, forest edges, and early second-growth vegetation, so the regrowth toward mature forest may actually have deprived them of their habitat. But more than a dozen were species that did require mature forest. Habitat changes don't account for what happened to them.

Among the mature-forest species, most were birds of the understory that did their nesting or foraging on the ground. The great curassow, for instance. The marbled wood-quail. The rufous-vented ground-cuckoo. The black-faced antthrush. They all vanished. Why? One persuasive (though unproven) explanation was offered in 1980 by two researchers named John Terborgh and Blair Winter.

The Terborgh-Winter hypothesis involved medium-sized mammals, some of which occupy a trophic level midway between the big cats and the birds. The roster of those medium-sized mammals on Barro Colorado includes peccaries, possums, armadillos, coatimundis, pacas, agoutis, howler monkeys, and sloths. The peccaries, the possums, the coatimundis, the pacas, and the agoutis all forage on the ground; under normal conditions, they are all subject to predation by pumas and jaguars. With the big predators gone, the medium-sized predators have flourished. Their populations have increased. By how much? Terborgh and Winter referred only generally to "excessive densities," but in a later paper, Terborgh reported that coatimundis, pacas, and agoutis were more than ten times as abundant on Barro Colorado as at similar mainland sites. Possum and armadillo populations were also unusually large—at least double, at most ten times, what Terborgh found elsewhere. Peccaries and monkeys have likewise proliferated, though for them Terborgh gave no numbers.

These increases were bound to be consequential. Coatimundis are nimble omnivores that forage in small bands, snooping ravenously through the understory to eat whatever they can catch or turn up, bird eggs and nestlings included. A tenfold increase in coatimundis might well constitute a fatal disaster for ground-nesting birds. Peccaries and possums also take nestlings and eggs. The Terborgh-Winter hypothesis, echoed by Jared Diamond to exemplify trophic cascades, was that the ground birds of Barro Colorado have been exterminated by an untoward abundance of medium-sized predators.

It makes for a neat scenario: The curassow and the wood-quail and the ground-cuckoo disappeared from this small patch of forest—why?—because all their eggs and their nestlings got eaten—why?—because the coatimundis and their ilk became extraordinarily numerous—why?—because the pumas and the jaguars disappeared—why?—because the pumas and the jaguars were insupportably rare within the small patch of forest and then that small patch became isolated—why?—because a lake rose and turned the place into an island. Insularization was the first cause, after which extinction cascaded from one trophic level to another. Despite the neatness, it may well be true.

98

THE STORY of the Calvaria tree and the dodo is equally neat, though far more dubious. It was told to the world in the pages of *Science*, back in 1977, by an ecologist named Stanley Temple.

Temple had spent several years on Mauritius during the early 1970s studying endangered populations of birds. Plant ecology was a sidelight. In the case of *Calvaria major* as presented by Temple, this tree species had mysteriously lost its ability to reproduce and was therefore approaching extinction. The question, again, was why. The answer was simple but poignant—as poignant as plant ecology gets, anyway. *C. major* was languishing toward a slow death in the absence of its obligate ecological partner, the dodo.

The tree is an endemic species. According to historical records, it had once been common in upland Mauritian forests and was often exploited for lumber. "However, by 1973," Temple wrote, "only 13 old, overmature, and dying trees were known to survive in the remnant native forests of the island." Experienced foresters had given Temple to believe that each of those thirteen trees was three hundred years old. No younger specimens were known to him, and he made the assumption that no younger specimens were known to anybody. The old trees still produced fruits that looked like they might be fertile, but the seeds from those fruits weren't germinating, according to Temple. The current absence (or apparent absence) of young plants and the current failure (or apparent failure) of germination led him toward a risky deductive leap across an epistemological chasm: that there had been "no germination of *Calvaria* seeds for hundreds of years." These bold categorical statements—no younger specimens, none of the seeds presently germinate, no germination for centuries—were all necessary to Temple's argument. Since the dodo itself was long gone, its relationship to another species could be established only by the logic of negatives.

C. major belongs to the Sapotaceae, a family of tropical and subtropical trees that produce globular fruits. Its own fruit is roughly the size of a peach, with a stony endocarp like an oversized peach pit. Protected within the pit is a single seed. The pit is thick-walled and very hard—so hard that the breakout and growth of a tender shoot might seem problematic. It seemed problematic to Stanley Temple. *C. major* seeds were failing to germinate, he speculated, because they found themselves trapped within their pits. But why should such trapping be a problem nowadays if it hadn't been a problem in the past? Because something in their surroundings had changed. They had lost an indispensable helper, Temple guessed.

"The temporal coincidence between the extinction of the dodo 300 years ago and the last evidence of natural germination of *Calvaria* seeds," he wrote, "led me to hypothesize the following mutualistic relationship between *Calvaria* and the dodo." The relationship, as Temple saw it, was one of "obligatory mutualism." He explained:

> In response to intense exploitation of its fruits by dodos, *Calvaria* evolved an extremely thick endocarp as a protec-

tion for its seeds; seeds surrounded by thin-walled pits would have been destroyed in the dodo's gizzard. These specialized, thick-walled pits could withstand ingestion by dodos, but the seeds within were unable to germinate without first being abraded and scarified in the gizzard of a dodo.

A nifty idea, but is it true?

Temple's talk about "obligatory mutualism" is somewhat misleading, even within the context of his own argument. The phrase implies an interdependence that is crucial for each partner. But there is no evidence that the dodo depended crucially on the fruit of *C. major*. Even Temple didn't claim that it had. His actual point was narrower: that *C. major*, for its part, had depended crucially on the dodo. This was a reasonable hypothesis, formulated in fair-minded scientific terms, since it was capable of being tested and disproved.

Temple neglected to mention that the hypothesis didn't originate with him. Back in 1941, two Mauritius-based botanists had published a paper in the *Journal of Ecology*, describing the structure and development of the island's upland forest and suggesting that dodos might have played a role in the life cycle of *C. major*. "The germination and distribution of these remarkable woody seeds were probably assisted by their passage through the alimentary canal of the Dodo," they wrote, "and young seeds of this species have been unearthed with Dodo remains." One of the two botanists was a quiet, careful man named R. E. Vaughan. To Vaughan and his co-author, P. O. Wiehe, the possibility of a dodo-*Calvaria* relationship seemed worth considering, though they made no claim that such a relationship would have had to be "obligatory." Having mentioned the idea in that single sentence, they let it drop.

Temple's short and dramatic paper in *Science* gave the idea new currency. As presented by him, it sounded persuasive, with a sad touch of poetic irony. What had once been adaptive—the sturdy endocarp, protecting the seed from destruction in a dodo's gizzard—had become, in the post-dodo era, a lethal imprisonment.

Although there was no direct proof, Temple did offer some supporting information, which lent his argument a flavor of precision. Using experimental results from other birds, he estimated the force generated by a dodo's gizzard at 11,300 kilograms per square meter. He tested *C. major* pits, in some sort of smooshing machine, to determine their crush-resistant strength. Based on data derived from hickory nuts, he made a guess at the relative power of, on the one hand, his smooshing machine and, on the other hand, the forces applied by a bird's gizzard. From that result, he concluded that *C. major* pits could have withstood a cycle through the gizzard of a dodo. He also fed some *C. major* pits to turkeys. A sizable fraction of the pits were destroyed, but ten did emerge—abraded and scarified, yet intact— from one end or the other of a turkey.

Temple planted the ten pits. Three of them opened. It made a triumphal climax for his study. "These may well have been the first *Calvaria* seeds to germinate in more than 300 years," he squawked.

Temple's paper attracted broader attention than a *Science* paper normally does. In London, the *Sunday Times* took note. Stephen Jay Gould recounted Temple's hypothesis in the pages of *Natural History*. It was embraced by other biologists as a textbook example of mutualism and a cautionary case of the ramifications of species extinction. Jared Diamond cited it in his discussion of trophic cascades. On scant evidence and intuitive appeal, the dodo-*Calvaria* story became famous across great distances. Temple, by the time his paper appeared, had left Mauritius and was at the University of Wisconsin.

Clunking around Mauritius myself, not knowing a *Calvaria* from a fig, I'm curious for a different perspective. So I ask Carl Jones, the sarcastic Welshman who runs the bird-rescue project in the Black River Gorges, about Temple's paper.

The dodo-*Calvaria* thing? What a bollix, Jones says in his amiably brusque way. He tells me, "Of all Mascarene ecologists, no one accepts Temple's version."

99

JONES HAS good reason to disbelieve: His own ladyfriend, the formidable Dr. Wendy Strahm, is a dissenting expert.

Strahm, who knows the plant ecology of Mauritius more thoroughly than any other living human, disputes Temple's basic claim about the present population status of *C. major*. The trees have continued to procreate, she says, though their saplings are neither abundant nor conspicuous. The situation that Temple's hypothesis purports to explain—that situation doesn't exist.

Although Temple was able to find only those "13 old, overmature, and dying trees" in the native forests of the island, Strahm puts the population in the hundreds. Many are younger trees—far younger than three centuries. Therefore they did manage to germinate in the post-dodo era. That fact alone would disprove Temple's notion of obligatory mutualism. The younger specimens of *C. major* could easily be overlooked by a nonspecialist, since they are confusingly similar to related species. Stanley Temple, a bird man, was a nonspecialist when it came to the Sapotaceae. Strahm on the other hand has studied them for years, with the benefit of a venerable mentor: the late R. E. Vaughan, co-originator of the dodo-*Calvaria* hypothesis back in that 1941 paper.

Temple's neat story has also been challenged by Anthony S. Cheke, another expert on Mascarene ecology. The claim that *Calvaria* seeds no longer germinate isn't true, according to Cheke. Evidence from horticultural efforts over the past fifty years confutes it. Each pit bears a circumferential suture line, and that's where the pit will break open if it's destined to break open at all. Some do. Most don't. The germination rate has been low in recent centuries, but not necessarily too low to maintain population stability in a long-lived species of tree. And even a low rate of germination is sufficient to falsify Temple's categorical claim that the seeds can't germinate without help from a dodo.

But if Temple's hypothesis is untenable, what replaces it? Something or other has been pushing *C. major* toward extinction. Even by Strahm's count, the species is perilously rare. The real cause of its decline may be less simple and less dramatic than a disrupted mutualistic relationship with the world's most famous dead bird.

Carl Jones has some ideas on this. At the mention of *C. major* he embarks on a Jonesian discourse. The proper way to think about that particular tree, he suggests, is not as an isolated species with a single mutualistic partner but as one among many similar tree species that had adapted themselves to the ecosystem of Mauritius as it existed before humans started mucking the place up.

Quite a few of those native trees produce seeds that are protected by hard pits. "And this is, I assume, to avoid parrot predation," says Jones. He's just speculating, but his speculation draws on a longer acquaintance with the ecosystem than Stanley Temple's. Before human settlement, Jones says, the island contained perhaps four species of parrot, of which only one (the echo parakeet, a small-bodied bird that's now desperately rare) has survived. Among the lost parrots was *Lophopsittacus mauritianus*, a hefty animal, maybe the largest parrot

ever known. And besides the parrots, Jones adds, there were two big vegetarian bats of the genus *Pteropus*. One of the bat species is now extinct,
the other severely reduced. Unless they were very different from their
surviving relatives, those extinct parrots and bats ate fruit. So each
fruit-bearing tree, Jones explains, must have coped with the threat
posed by a whole guild of sharp-billed and sharp-toothed frugivores.

If the pits of any given tree could withstand a fair bit of nibbling
and gnawing, then the tree would have benefited from the attentions
of parrots or bats as those animals carried its seeds to new habitat.
The ecological purpose of fruit, after all, is exactly that: to attract frugivores who might help disperse seeds. But if its pits were too sensitive to nibbling and gnawing, then the tree was in trouble. Thwarted
reproductively, it would have suffered a population decline. Across
the evolutionary scale of time, in response to such pressure from frugivores, Jones suspects, more than one Mauritian tree species developed extremely hard-pitted fruit. *C. major* falls within this group.

"It gets fed on by the fruit bat. And by the echo parakeet," says
Jones. Now he's describing present-tense realities of *Calvaria* ecology,
not hypotheticals of the past. How does he know about this frugivory? "Because we've seen it." And that's why the species has hard
pits, he posits confidently. "It has nothing to do with dodos. It has to
do with parrots."

Besides the parrots and bats, there are still other factors to consider: seed mortality from fungal infestation, competition from alien
plants, seed predation by rats, browsing on saplings by deer, disruption by pigs, maybe even a share of blame for the infernal monkeys.
Any one of these, or all together, might account for the decline of *C.
major*. Jones dislikes the Temple hypothesis because, in addition to
controverting known facts, it's too pat. "I think the actual truth is
probably far more interesting," he says.

But the actual truth—the intricate relationships among native
trees, fruit bats, extinct giant parrots, seed-killing fungi, exotic plants,
and exotic animals—is dauntingly complex, and not susceptible to
concision into a two-page report in *Science*. It's the work of three lifetimes for a dozen methodical ecologists like Wendy Strahm. In the
meantime, a dramatic but errant hypothesis has become another tantalizing scientific legend.

For every neat legend told and retold, a bit of messy but significant
reality is ignored. We saw that with Darwin's finches, we see it again
with the dodo and *Calvaria*. Furthermore, while some legends have

their uses, they all have their costs. Ignoring reality can be hurtful. Probably neither the dodo-*Calvaria* notion nor the tale of the finches is quite as costly, in that sense, as another legend in the lore of extinct populations: the legend of the last Tasmanian Aborigine.

100

ACCORDING TO this legend, the last survivor of the Tasmanian Aborigine people was a fierce little old woman named Truganini. She died in 1876.

Several photos of Truganini have come down to us, grainy portraits from which she seems to radiate anger and lonely strength. Late in her life she was a familiar character on the streets of Hobart in her red turban and her serge dress. If she had borne any children, they didn't survive. She kept dogs. She was reportedly terrified that her body would be cut up, after she died, by ghoulishly curious white men who called themselves scientists. Her fear on that count turned out to be well founded. Her portrait now hangs in the Tasmanian Museum and Art Gallery. For some years, her skeleton itself was there on display. Her death, on May 8, 1876, at two in the afternoon, has been taken as the terminal event in the twelve-thousand-year saga of the Tasmanian Aborigines. It's an important story, full of pathos and edification, though with a bogus ending.

Truganini had seen a long lifetime of change, war, displacement, persecution, destructive paternalism, and institutionalized abuse that amounted to genocide. She had seen her kin killed off by disease, white people, and despair. The legendary version of these racial travails—which seems sympathetic on its face but which embodies a brutal dismissiveness—holds that the final consequence was total eradication of the native Tasmanians, marked by Truganini's own death. That version has dominated Tasmanian historiography for the past century. For instance: "With no defences but cunning and the most primitive weapons, the natives were no match for the sophisticated individualists of knife and gun. By 1876 the last of them was dead. So perished a whole people." Another instance, from a respected scholar: "The Tasmanian aborigines are an extinct people." This legend offers several sorts of inherent appeal: It's sad, it's dramatic, it's simple, it's final, and although it demands pious regret, it's beyond repair. It does not oblige white Tasmanians to negotiate land-restitution arrangements with survivors, since putatively there aren't any.

But the legendary version has recently become subject to revision. No one doubts the date of Truganini's death or the fact that she was a Tasmanian Aborigine. The questionable point is whether she was the last of her people.

It's a point highly germane to the study of local extinctions because, although the Tasmanian Aborigines were not a discrete *species* of hominid, they were indeed a discrete population—distinct even from the Aborigines of the Australian mainland. Their history, in either its legendary or its revisionist form, represents a classic case of the biological perils of insularization.

IOI

CONFINED ON their island for twelve thousand years, the Tasmanian Aborigines diverged both genetically and culturally. Their hair was more woolly than the hair of mainland Aborigines. Their skin was more reddish brown than black. At the time when Europeans first arrived, the Tasmanians had no metalworking skills, no stone tools with ground edges or wooden hafts, no wooden bowls, as the mainlanders did. They had no boomerangs. Their rock art was rudimentary, far less imposing than the elaborate pictographs at mainland sites like Ubirr, with its x-ray-style turtles and its thylacine. In observance of some inscrutable taboo or preference that arose about four thousand years ago, the Tasmanian Aborigines declined to eat fish. Archaeologists are still wondering why. They pomaded their hair with grease and ocher. They carried fire when it was available, but they don't seem to have known how to make it. They used simple baskets, digging sticks, spears without stone points, and a sort of lightweight club known as a waddy. They built canoes of bundled bark, adequate for paddling to the nearer offshore islands but not for adventurous transoceanic voyages. They lived on a varied hunter-gatherer diet in which kangaroo meat was especially important. They had no written language. Their metaphysical beliefs and their ceremonial practices are preserved only in fragmentary hearsay. But they were notable for the fervor of their occasional dancing and singing.

Their total population was small. In the years just preceding European contact, they may have numbered only three or four thousand. Either their birth rate was naturally low or else the land, as they lived on it, didn't feed many mouths.

The most reliable chronicler of the Tasmanian Aborigine people is probably Lyndall Ryan, whose careful scholarship was presented in her book *The Aboriginal Tasmanians*, published in 1981. Before the European conquest, Ryan explained, the island was divided among nine territorial tribes. The basic social unit for day-to-day life was the band, of which there were roughly a half-dozen within each tribe. A band encompassed about forty people, who foraged together, camped together, traveled together, and knew themselves by a collective name. To our ears those names sound ungainly: the Tommeginer band, the Pennemukeer band, the Luggermairrernerpairrer band, the Trawlwoolway band. Up near Oyster Bay, on the east coast, lived the Loontitetermairrelehoinner band. Truganini was born into the Lyluequonny band of the South East tribe, whose territory lay along the southeast coast near the estuary of the Huon River. Her father was chief of the band. She seems to have inherited his force of character.

By the time of her birth, in 1812, the invasion and expropriation of her people's homeland was well under way. British seal hunters had set themselves up on the Bass Strait islands, clubbing fur seals and harvesting their skins at a rate of thousands per year. The first British penal settlement had been founded in 1803, on the east bank of the Derwent River not far from the site that became Hobart. The number of convicts and of the military personnel who guarded them rose quickly as more men were shipped down from Sydney and two further penal colonies were founded. Some of the convicts escaped, going feral in the hinterlands as bushrangers. In 1807 free settlers seeking land for farming and stock raising also began to arrive by the hundreds. Growing so fast, the European population of Tasmania matched and then surpassed the Aboriginal population sometime during Truganini's early childhood. The area of land under cultivation, the area of settlement along the river valleys, the populations of cattle and of sheep—all these had increased sharply, and they would continue increasing. Conflict between Europeans and Aborigines was inevitable.

The British sealers were men of free spirit and raunchy character who had made contact on their own terms with the tribes of the northern coast, purchasing (in some cases kidnapping) Aborigine women to serve as their concubines and seal-skinning assistants. The

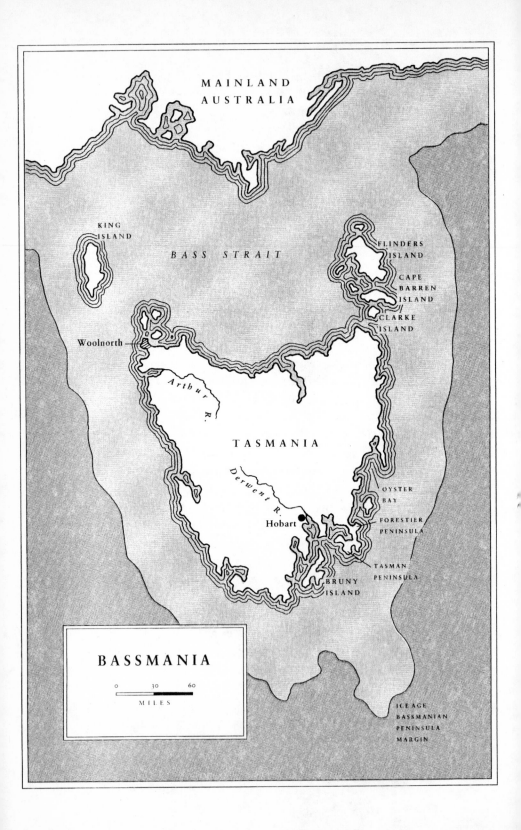

MAINLAND
AUSTRALIA

KING
ISLAND

BASS STRAIT

FLINDERS
ISLAND

CAPE
BARREN
ISLAND

CLARKE
ISLAND

Woolnorth

Arthur R.

TASMANIA

Derwent R.

Hobart

OYSTER
BAY

FORESTIER
PENINSULA

TASMAN
PENINSULA

BRUNY
ISLAND

ICE AGE
BASSMANIAN
PENINSULA
MARGIN

BASSMANIA

0   30   60

MILES

military and civil authorities of the settlements were more benignly
indifferent toward the native people—or so they pretended, at least
initially. The commander at Hobart had orders "to open an intercourse with the natives and to conciliate their goodwill," as well as to
require that the colonists under his authority "live in amity and kindness" with those natives. Despite such adjurations, the British government didn't consider the Tasmanian Aborigines a "civilized people"
and therefore conceded them no legal status, not even the status of a
conquered nation. "They were now British subjects without the
rights of British citizenship," according to Lyndall Ryan. "Thus the
Aborigines had no rights to original land ownership, and any attempts
they would make to defend their land could only be defined in British
law as criminal in intent." They did make attempts. The first European was killed in 1807. Within a year, by Ryan's estimate, about
twenty Europeans and a hundred Aborigines had died in violent skirmishes.

The issue of land ownership is one way of conceptualizing this
conflict, though land ownership in the European sense is not a concept that the Aborigines embraced. Another way of seeing it is in
terms of kangaroos. The Aborigines had depended on these animals
for longer than memory, but now the pale-skinned, pain-in-the-ass
strangers were building their villages and their farms on some of the
best hunting grounds and, still worse, greedily overharvesting kangaroos. During the early colonial years, kangaroo meat became the
main source of protein for the settlers. Market hunters made good
money, enough to compensate them for operating throughout wider
areas as kangaroo populations were depleted near the settlements.
Bushrangers also killed kangaroos. This heavy European reliance on
kangaroo meat was opportunistic and flexible; it would change as the
native Tasmanian herbivores were displaced by sheep and cattle. The
Aborigines' reliance on kangaroo meat was not so flexible.

Their sense of land tenure *was* flexible, but that conflicted with the
new, rigid notion of real estate as construed by the settlers. Property
lines increasingly prevented the various bands from following their
traditional routes and rhythms of nomadic foraging. The island was
filling up with white folks. It was being parceled out into farms and
stock runs. Free settlers streamed into Tasmania during the 1810s and
1820s, and any immigrant who was reasonably well connected back in
London (such as a retired military officer or the second son of landed
gentry) brought a letter of recommendation, redeemable for a land

grant and a convict laborer assigned at slave wages to help the settler work that land. By 1830 the European population had risen to 23,500. A million acres of land had been granted away by a colonial administration that didn't rightfully have it to grant. A million sheep were eating grass that had formerly been available to kangaroos.

Sometime in that period, during her girlhood, Truganini suffered an experience that showed her what sort of "amity and kindness" she could expect from whites. She recounted the episode years later, and it turned up in a book titled *Black War*, published a half-century ago by Clive Turnbull. The young Truganini was on a stretch of coastline near the Huon estuary, her own tribal territory, in the company of a certain Paraweena, the Aborigine man she was matched to marry. Along with another native man, they wanted to cross to Bruny Island, a body of land just a few miles offshore that was also part of the South East tribe's territory. Two white men (Watkin Lowe and Paddy Newell, according to Truganini's indelible memory) offered to row them across. Halfway to the island, Lowe and Newell threw the Aborigine men overboard. When Paraweena and the other fellow struggled back to the boat and grabbed the gunwale, Lowe and Newell cut off their hands with hatchets. In the demure language of Turnbull's book: "The mutilated aborigines were left to drown and the Europeans were free to do as they pleased with the girl."

In the early 1820s, relations between natives and settlers entered a new phase. Until that time, by Lyndall Ryan's account, the Europeans had seen the Aborigines as "timid, gentle people who would prove no barrier to the expansion of pastoral settlement." They weren't numerous, these gentle people. They didn't build villages. They drifted from place to place, asking little from the land and making no claims to possessing it. They were almost invisible. White bushrangers seemed a more serious menace to public safety and commerce. Then in 1823 a band called the Laremairremener, of the Oyster Bay tribe along the northeast coast, killed a pair of stock keepers. Four months later the same band struck again, killing two more whites. The Laremairremener leaders who had incited these raids were captured, put on trial, forbidden to testify at their trial because they weren't Christians, and sentenced to hang. When another three Europeans were speared to death in 1826, another two Aborigine men were promptly sentenced to hang. These men were known as Jack and Black Dick.

Jack and Black Dick belonged to a group from the Oyster Bay tribe that had set up a camp at Kangaroo Point, just across the river from

Hobart. The people at Kangaroo Point had become accustomed to handouts from the colony. By one argument, that familiarity might have made them more dangerous than Aborigines of the bush. In other words, the problem wasn't Aborigines per se, it was Aborigines "gone bad" from the corruptions of contact. On the day when Jack and Black Dick were executed, the lieutenant governor (the chief colonial official, who took his orders from London) issued a notice declaring optimistically that the hangings would not only "prevent further atrocities" but also lead somehow to "a conciliatory line of conduct." Dreamy invocation of conciliatory measures was popular among the British colonists, and destined to be more so in the near future. "Conciliation" would become the standard euphemism covering a multitude of oppressions. What it meant was, *We take your land, we give you our religion and our language and our diseases, and then you and your culture melt away like snow on a griddle*. Jack and Black Dick, as they went to the gallows in late 1826, were considered unrepresentative renegades. It was a happy illusion. The colonists didn't yet realize that this was war.

Six more Europeans were killed in the next several months, and the realization began to hit. There were further raids against solitary stock keepers on the fringes of the settled districts. In 1827, thirty Europeans died. Stock keepers in one area organized a vigilante group and went on a reciprocal killing spree against Aborigines. A few voices (including the editorial voice of the *Colonial Times* newspaper) predicted that this war, if fought to an end, would result in the extermination of the Tasmanian native people—an astute bit of foresight for that time, even if it wouldn't quite prove true—and some suggested that exiling all Aborigines to a smaller island might be more humane. Couldn't they be shunted out to one of those windswept mounds in the Bass Strait, presently occupied only by a few dozen sealers and their Aborigine women? The lieutenant governor of the colony was an ambivalent, placative man named George Arthur, who might well have preferred that travesty of humaneness. But even Arthur was clear-eyed enough to tell his bosses in London: "They already complain that the white people have taken possession of their country, encroached upon their hunting grounds, and destroyed their natural food, the kangaroo; and they doubtless would be exasperated to the last degree to be banished altogether from their favourite haunts." The lieutenant governor faced a tough question: If banishment to another island wasn't the answer, what was?

Within the next few years, under Arthur's regime, three measures were instituted. First, he appointed a pious businessman named George Augustus Robinson to run a mission for a small group of tractable Aborigines living on Bruny Island, off the southeast coast. In addition to the blankets, European clothing, and food rations (a daily allowance of biscuit, salt meat, potatoes, bread, tea), already being given, Robinson's mission would dispense unctuous Christian instruction. Lieutenant Governor Arthur hoped that this menu of treats would attract Aborigine people from other areas and other tribes.

Arthur's second measure was a program of bounty payments for captured Aborigines. The program began in 1830, the same year that bounties were first paid for thylacine carcasses. This coincidence of timing reflects the tendency among British settlers to see the Tasmanian Aborigines in the same terms as they saw the thylacine: as a distinct species, inconvenient to current circumstances and requiring removal. The bounty rates for Aborigines were more generous than for thylacines, five pounds per adult and two pounds per child. Vigilante groups could now profit financially from what was called "black catching." The bookkeeping for Aborigine bounties doesn't seem to have been so conscientious as for thylacine bounties, and it's hard to guess just how much impact the program had. Lieutenant Governor Arthur stipulated that no payment would be made for Aborigines captured while interacting amicably with Europeans in the settled districts, and he tried to discourage gratuitous killing.

When members of the Big River and Oyster Bay tribes continued their guerrilla attacks within the region around Hobart, Arthur tried a third measure. He authorized a military operation that became known as the Black Line.

In concept, the Black Line was bold and decisive. A human chain of soldiers, field police, convicts, and citizen volunteers would sweep across the landscape, capturing or killing every Aborigine in its path. The Line would be set into position some forty miles north of Hobart, forming its linkage between mountains to the west and a point on the eastern seacoast. It would march across hundreds of square miles of rural terrain, progressing inexorably southeastward toward a thin neck of land leading onto the Forestier Peninsula, which led in turn, by another thin neck, onto the Tasman Peninsula. Any and all Aborigines who fled the advancing Line could be easily bottled up on the outermost peninsula. Once there, Arthur declared, they would be allotted that blip of land as an Aborigine reserve. By the estimate of

the operation's commander, as many as five hundred Big River and
Oyster Bay people stood to be captured or driven.

The Black Line began moving on October 7, 1830. It included two
thousand men, armed with a thousand guns, plenty of ammunition,
and three hundred sets of handcuffs. They were ready for hot action
and big results. As Ryan described it:

> For three weeks they beat bushes, built defensive huts from
> which to assail their hidden foe, scoured the countryside,
> got lost in pouring rain, and consumed vast quantities of
> government stores. Despite the apparent disorganization,
> the Line dislodged the Oyster Bay people on the east coast
> so that on 24 October two were captured and two others
> shot at Prosser Plains. The remaining six or seven moved
> through the Line to the north-east, realizing they were
> lucky to escape with their lives.

Two captured, two shot—an embarrassingly meager take. The
colonists went home at the end of October and the convicts were evidently withdrawn too, leaving the soldiers and the field police to continue hunting for stragglers, of which they turned up almost none.
The commander's estimate, that the Line would roust five hundred
Aborigines, was wildly wrong. There had been maybe a dozen fugi-
tives in the area, and most of them were deft enough to elude this parade of noisy white men.

The Black Line was derided in the colonial press for its ineffectual-
ness. But in fact it had served its intended purpose, more or less,
putting the Aborigines of that region on notice that their ancestral
lands were lost. The Oyster Bay tribe and the Big River tribe—what
was left of them—made no concerted effort to return. Their guerrilla
campaign petered out. Most of the Oyster Bay people were dead or
dispersed. The last of the Big River people retreated into high country to the west.

Truganini was meanwhile among the pacific group accepting
British-supplied shelter and rations down on Bruny Island. Another
historic personage at that site was Woorraddy, an Aborigine warrior
with a wife and three young children. Woorraddy was a dignified but
pliable man who in coming years would learn to straddle two cultures, making himself useful to the white conquerors without ever
completely giving up his Aborigine ways. When the well-meaning

but purblind George Augustus Robinson arrived to convert the Bruny Island outpost into a mission, he found Woorraddy and his family, Truganini, and just thirteen others—the remnants of the South East tribe, which not long before had included 160 people.

Robinson's plan, as approved by Lieutenant Governor Arthur, was to establish a model village in which Aborigines could be Christianized and civilized. Nomadic foraging would be out, farming would be in; kangaroo robes would be out, trousers and dresses in; and for Truganini and the other women, consorting with lowlife British sealers and whalers would definitely be out, if Robinson had his way. He meant to create rough facsimiles of decent English peasant families. Oddly, one of his first steps toward creating these families was to separate Aborigine children from their parents, so that the children could be indoctrinated more thoroughly.

The social arrangements don't seem to have been salubrious, and the food rations, lower in protein than the Aborigines' traditional diet, weren't either. Exposure to alien European diseases was also a problem. The mission Aborigines promptly got sick, possibly from influenza, which can be fatal to anyone with a weakened or inexperienced immune system. Woorraddy's wife died. Three others died. The surviving adults wandered away, escaping the bad air, the bad associations, the bad mojo of the place. A change of locale was their customary means of coping with mortal illness, as they explained to Robinson, who wasn't pleased. They moved to the southern half of the island, where a person could still hunt and forage. Kangaroos were plentiful down there, which more than made up for the forfeit of biscuit and salt meat from the mission. Truganini and several other women, to Robinson's disgust, spent much of their time at a nearby whaling station. Robinson himself could see that the mission effort was a bust. He didn't despair, however; instead he dreamed his way toward a far more ambitious and fatally dumbheaded scheme. He began imagining how he might use the Bruny Island people to help him "conciliate" Aborigine tribes all over Tasmania.

He would put his mission on the road, like a carnival, with himself as chief carny. He would circumambulate the island, accompanied by his small cadre of half-converted Aborigines, contacting the various bands of the frontier and wilderness regions and persuading them to accept the blessings of Christian civilization. He would bring them in from the darkness and cold. He would transform them into trustworthy, obedient, properly clad, and God-fearing third-class colonial subjects. It was a grandiose plan that, in Robinson's view, would accomplish two good ends: saving the Tasmanian Aborigines from barbarity and saving them from extinction.

Lieutenant Governor Arthur granted official endorsement. On February 3, 1830, Robinson and his party set out. They walked westward into some of Tasmania's wildest country. Woorraddy was a member of the group, serving Robinson as a bush guide and a liaison to the unpacified tribes. Truganini went along. She was an asset to Robinson in her own right, and by now too she had become Woorraddy's new wife.

They traveled for eight months. One encounter, near the Giblin River on the west coast, was typical. On March 25, Robinson met a local band from the South West tribe. Carrying waddies (their clublike weapons, with which they had been hunting) and dressed in kangaroo robes, they were wary of him. Only two men showed themselves, while the rest hung back in the brush. But with a little gentle coaxing he persuaded them to relax, come forward, shake hands. Through his interpreters, Robinson told them that he was no enemy—that in fact his intent was to be helpful. Presumably he explained something, but not much, of what he had in mind. They were welcome to travel onward with him if they wanted; if not, fine. The South West people did join Robinson's party, and that evening in camp they danced and sang until late. They stayed with him four days. Then, in the middle of the night, they disappeared. Thanks but no thanks.

The conciliation business was harder than Robinson had hoped. Along the southwest coast he found people whose Aboriginal culture was still undisturbed and who preferred that culture to anything Robinson could offer. In the north he found militant bands, embittered by their conflicts with sealers, bounty hunters, settlers, and the Van Diemen's Land Company. Robinson ruminated in his journal:

> The children have witnessed the massacre of their parents and their relations carried away into captivity by these merciless invaders, their country has been taken from them, and the kangaroo, their chief subsistence, have been slaughtered wholesale for the sake of paltry lucre. Can we wonder then at the hatred they bear to the white inhabitants?

He added: "We should fly to their relief. We should make some atonement for the misery we have entailed upon the original proprietors of this land." He proposed to do exactly that. He seems to have been a man of great heart, great energy, and not so great historical vision, whose persistent efforts probably had the net long-term effect of making things worse.

The 1830 journey was only his first try. In the course of it, he established friendly contact with a number of bands (who declined to follow him), took a few prisoners (who were later released), learned a bit of Aborigine language, and acquired some bush skills. But his dream was still a dream, not shared by many people of either race. The Aborigines didn't much care to be rescued from their traditional way of life; the sealers, the stock keepers, the farmers didn't much fancy seeing them rescued from anything. Robinson circled the island and arrived back in the settled districts near Hobart in time to be invited to participate in the Black Line.

He stayed clear of that. He was afraid that Woorraddy, Truganini, and the other acculturated Aborigines of his entourage, if he brought them into the Black Line as scouts, might be shot by trigger-happy soldiers or settlers. And besides, he felt that the operation would only exacerbate matters. No doubt he was right. Despite its anticlimactic result, the Black Line seems to have inflamed "extirpationist" sentiment among grumpy white colonists yearning for a final solution to what they saw as the Aborigine problem. Robinson himself, alarmed by the extirpationist mutterings, concluded that Tasmanian Aborigines would no longer be safe in Tasmania.

He now embraced the same notion that a few other voices had suggested earlier: giving the Tasmanian Aborigines their own asylum on an offshore island. In the clarity of hindsight, for "asylum" we can read "terminal exile." But it seemed like a nice idea at the time. Lieutenant Governor Arthur, who shared some of Robinson's tortured benevolence, again concurred.

In March of 1831, Robinson transported fifty-one Aborigines out to Gun Carriage Island, a tiny hump of land between Cape Barren Island and Flinders Island in the Bass Strait, about thirty miles off Tasmania's northeastern point. Their arrival marked the founding of the Aboriginal Establishment, the official reserve of Tasmania's native people, which would be maintained on one island or another until 1847. Less officially, it was a prison colony. Gun Carriage Island was only a few square miles in area, too small to support even these first several dozen Aborigines—let alone the hundreds more that Robinson intended to round up—except in a state of sedentary, ration-dependent subsistence. It allowed them room to exist but not to live real Aborigine lives.

Robinson proceeded with the same misguided program as at Bruny Island two years before: assigning huts and garden plots, expecting nomads to become peasants. Then he went off on an expedition to some of the other islands, trying to rescue Aborigine women from sealers, and when he returned a month later his captive people were already sick and dying. The surgeon assigned to the Establishment wanted to blame these health problems on hard climate, bad shelter, the insufficient supply of fresh water. Lyndall Ryan wrote that "forcible removal, enforced confinement, and the loss of control over their own lives" were important too. Ryan herself may have underestimated the fatal impact of unfamiliar microbial infections. It should be remembered that the Tasmanian Aborigines were an island population, and that they had been separated for thousands of years from the dynamics of disease and immunological adaptation among humans of the wider world.

Within a year, the Aboriginal Establishment was transferred to a slightly better site on Flinders Island, but the dying didn't stop. By early 1832 there were only twenty survivors.

George Augustus Robinson was back on mainland Tasmania during most of the early 1830s, making further expeditions to gather up those Aborigines who remained at large and becoming famed for his efforts as "The Conciliator." This prattle about conciliation was in the truest tradition of Orwellian doublespeak, serving mainly to anesthetize the collective conscience of the conquerors. Meanwhile, Robinson's tactics were already changing from cajolery to coercion. One stage of that change occurred abruptly in August 1832 along the western coast near the Arthur River, not far from where I would take that futile midnight hike in search of the last thylacine.

Coming south along the coast with his party of co-opted Aborigines (including, as always, Truganini), Robinson crossed the Arthur River by raft. Soon afterward, a large group of Aborigine men appeared in his camp. They were the remnants of three different bands, and they carried spears. Robinson offered them bread. He gave them some beads. This didn't get him far. Something about their demeanor made him jumpy, as he later recorded in his journal: "They were shy and sullen, yet bold and full of bravado." The sullen, bold men went off to hunt, and Robinson spent a nervous night, anticipating some sort of ugly surprise. At dawn he quickly packed his gear. He told one of his Aborigine interpreters to communicate that the people could join him or else go on their way, as they wished, since he didn't intend to threaten them. Lyndall Ryan wrote vividly of what happened next:

> Scarcely had he spoken when the strangers surrounded Robinson holding spears. His own Aborigines fled. Somehow Robinson escaped. As he ran through the scrub he overtook Truganini, who was fearful of capture, for she had relatives among these people, and they had a shortage of women. On reaching the river, Robinson fashioned a makeshift raft out of two spars of wood tied together with his garters at first and then, when they broke, with his cravat. He lay on this raft and urged Truganini to push him across, for he could not swim.

By the time they reached the far bank, Robinson's confidence in his conciliatory skills had been doused with doubt. Truganini's esteem for him had presumably gone a bit soggy too. Scared, confused, incapable of comprehending the Aborigines' resistance, "Robinson would in future carry firearms and use belligerent methods to force surrender," according to Ryan.

For almost three more years, he persisted with his oxymoronic mission: "conciliating" Aborigines at gunpoint and shipping them back under guard to join the Establishment on Flinders Island. He captured about three hundred. Most of those died in transit camps without ever reaching Flinders. Others reached Flinders and then died. Despair and bad diet no doubt contributed to the high death rate by increasing the lethality of unfamiliar diseases. Influenza, tuberculosis, and pneumonia may have been the main killers, though diagnosis was casual and no one will ever know. Of the Aborigines that Robinson didn't capture, many were shot and some probably died in the bush of diseases picked up during contact with whites; a number of children were taken in by settler families (to be raised as, essentially, household slaves); a few women continued living with sealers; the rest disappeared. On February 3, 1835, Robinson reported proudly to a colonial official: "The entire Aboriginal population are now removed."

He was nearly correct, but not quite. One family at least was still at large in the Tasmanian outback.

102

GEORGE Augustus Robinson was tired of bush travel and wanted a less arduous role. He accepted the post of commandant of the Aboriginal Establishment and, in lieu of continuing personally to hunt Aborigines, he sent out a search party led by his sons. In November of 1836, the searchers encountered a family of Aborigines near Cradle Mountain, in the high chilly moorlands of the interior.

The family refused to be sent to Flinders Island. Somehow they got away from the posse and weren't seen again for years. Then, in 1842, an Aborigine couple with five sons (possibly it was the Cradle Mountain family, though historical sources aren't definite) turned up near the Arthur River. This time they didn't escape. They were captured by sealers, transported to the north coast, and from there shipped across to Flinders. They were the last Tasmanian Aborigines to be severed from their land and their culture. One of the five sons was a seven-year-old boy who came to be known as William Lanney.

William Lanney's life makes a pathetic, gruesome vignette, second only to Truganini's for its prominence within the larger story. Twenty-five years later, Lanney would be celebrated as the sole surviving Tasmanian Aborigine male. But in 1842 he was just an inconspicuous boy, one among many. His path converged with Truganini's when the boat brought him to Flinders Island.

Truganini herself had just been shipped back to Flinders after achieving infamy on the Australian mainland. Robinson had taken her up there to Port Phillip (a harbor town on the south coast, now known as Melbourne) three years before, when he was appointed "protector" of the Aborigines for southern Australia. What the whites at Port Phillip had been truly concerned about, of course, was protection for themselves. The idea was that Robinson would use his nonpareil skills and his troop of acculturated Tasmanian Aborigines to lure the fractious mainland Aborigines into some sort of prisonlike asylum along the lines of Flinders. In addition to Truganini, Robinson had brought with him Woorraddy and a dozen others, among whom were two men named Timmy and Pevay. Robinson's mission to the mainland was no great success, but Truganini and a few of her fellow Tasmanians made a splash. Sometime in 1841, Timmy and Pevay, along with Truganini and two other women, broke their ties to Robinson and went rebel, turning themselves into a little pack of outlaw marauders who raided shepherds' huts in the area southeast of Port Phillip. This is how we do it in Tasmania, they seemed to be saying. They wounded four stock keepers. Ryan wrote: "Their tactics had all the marks of sustained guerrilla resistance to white settlement." They may have felt panicky, too, over the prospect of being sent back to Flinders. Their rebellion, already serious, became desperate on October 6 when they killed two whalers. Captured, tried, Timmy and Pevay took all the blame for the killings. They were hung. The three women, including Truganini, were acquitted. Having given Port Phillip something to think about, they were promptly deported.

The others of the Tasmanian group, less notorious but still worrisome to the white settlers of the mainland, were packed off too. Robinson stayed behind in his new role, and at least the Tasmanian Aborigine people were rid of him.

But they continued to suffer attrition. Wooraddy died on board the ship from Port Phillip shortly before it docked at Flinders Island. Maybe he was mortally ill, maybe he was just sick and tired. Ryan wrote of Wooraddy: "At the age of fifty he represented the last of the traditional Aborigines. He never learned to read or write, never wore European clothes, and never ate European food. His belief in his Aboriginal identity remained unshaken through the years of dispossession." William Lanney's family, consigned to Flinders about the same time, didn't make an easy adjustment either. Within a few years both parents were dead, plus two of the five brothers. Fifty-nine other Aborigines had died at Flinders during the early years of the Establishment, and the heavy mortality had led Truganini to predict that before long there would be "no blackfellows to live in the new houses." Robinson himself had shared Truganini's pessimism, admitting in one of his reports that "the race in a very short period will be extinct" unless conditions were somehow improved. Conditions only got worse.

Among the sixty or so Aborigines who remained in the mid-1840s, after Truganini's return from Port Phillip, the death rate was higher than the birth rate. Precious few children survived infancy, and most of those were taken away to the Orphan School at Hobart, where their cultural identity was erased. By 1847, fewer than fifty adults and children were left in captivity on Flinders. At that point, for a combination of mainly coldhearted reasons (one of which was the cost of maintaining a separate establishment on Flinders for such a small number of people), colonial officials in Hobart and London decided to move the group back to mainland Tasmania. A prison camp was standing vacant down at Oyster Cove, just twenty miles south of Hobart. The Aborigines could be put there.

For Truganini, it was the completion of a great circle, since Oyster Cove lay on the coastline just opposite Bruny Island, within the original territory of her own South East tribe. She was delighted to return. The other survivors were also happy with the move (on the logic that any place should be better than Flinders Island) and celebrated their arrival with a shindig of ceremonial dancing. Probably they also sang some of the old songs. But the Oyster Cove compound was nothing more than a cluster of rundown wooden buildings on a bleak mudflat, landscaped with puddles during the wet season, abandoned for use as a penal station because it had been judged too unsanitary for convicts. The euphoria of the new occupants didn't last. The remaining children were pulled away to the Orphan School, leaving just thirty-seven adults.

The Oyster Cove site was mucky, windy, cold. The lodgings were vermin-infested and damp. The diet was poor, doled out in daily rations: this much meat, this much tea, sugar, tobacco. Some of the women consorted with sealers, a favorite old form of degradation that relieved the more tedious degradation of sitting around obediently at the compound; with the sealers, they at least had the grim satisfaction of prostituting themselves for alcohol. Some of the men drifted into the pubs of Hobart. Not surprisingly, under these circumstances, the population growth rate remained negative. Few if any of the Oyster Cove captives were bearing and raising children. The history of the Tasmanian race had dwindled down to a matter of small numbers and recognizable names, a drumbeat of obituaries.

By the time George Augustus Robinson stopped in for his farewell visit, in April 1851, thirteen more had died. Among them, a man named Eugene. A woman named Hannah. Louisa, formerly known as Drummernerloonner, was dead too. Nancy and Old Maria were dead. Wild Mary and Neptune and Cranky Dick were dead. David Bruny was dead—his father had been Woorraddy. The young man known as Barnaby Rudge was dead, survived by his brother, William Lanney. Alphonso, the second husband of Truganini, was dead. Truganini herself, ferociously stubborn, held on. Robinson greeted her and the other survivors, noticing astutely that they didn't seem thrilled to see him. He accepted some shell necklaces from a good-natured sixteen-year-old girl named Fanny Cochrane and her sister, whose father had been a sealer. Then Robinson sailed back to England and never laid eyes on another Tasmanian Aborigine.

The drumbeat continued. Alexander (his real name had been Druemerterpunner when he traveled with the Big River tribe) died. Amelia (formerly Kittewer, of the North West tribe) died. Charley (Drunteherniter) died. Frederick (Pallooruc) died. Harriet (Wotte-cowwidyer) died. Washington (Maccamee) died. Moriarty died, leaving no record of his Aborigine name. Fanny Cochrane's mother died. William Lanney's last brother died.

From the colonial government's perspective, the decline at Oyster Cove was a double blessing. The expense of supporting that population decreased steadily as the people died off, and the disappearance of the Aborigine race was taken as an indicator that British Tasmania had approached some sort of imperial maturity. The rations were cut further. Upkeep on the buildings was neglected. By 1859, the Oyster Cove station was a ruin of broken windows, leaky roofs, rooms without furniture. There were fourteen survivors now, most of them unhealthy and waiting to die. Nine females, five males, of whom Ryan gave an accounting: Sophia spent most of her time in tears; Caroline talked only to her dogs; the other women suffered pulmonary ailments that hadn't yet quite killed them; Jack Allen and Augustus (had he taken his white name from The Conciliator?) were alcoholics; Tippo Saib was senile and nearly blind. William Lanney, by contrast, was a strapping twenty-four-year-old whom the women admired as "a fine young man, plenty beard, plenty laugh, very good, that fellow."

Lanney was a bellwether. If he couldn't endure, prevail, perpetuate the lineage, who could?

Photographs show William Lanney as heavyset and round shouldered, with sad gentle eyes. He looks to have been physically strong, amiable, probably shy. He lived a more fortunate life than some of the others. Raised in the forest by traditional parents until he was seven, then on Flinders Island, then at the Orphan School, he was an adaptable boy, capable of picking up the language and meeting the expectations of the white folks who had taken charge of his universe. At age sixteen he went to sea on a whaling ship. He seems to have been a diligent sailor and earned at least benign condescension, if not respect. He made some friends. Later, by one account, he fell to boozing whenever his ship was in port and stayed drunk for as long as he had money; but that doesn't set him apart from other whalers. Callous jokers in the Hobart community eventually stamped Lanney with a nickname that mocked not just his unpretentious demeanor but also the fact that, by elimination, he was the preeminent male of his race: They called him King Billy. He was introduced to the Duke of Edinburgh, during a royal visit, as "king of the Tasmanians." The historian Clive Turnbull judged Lanney to have been "a poor, drunken piece of flotsam," ridiculed by most people who knew him. Lyndall Ryan was more generous. By 1864, she reported, William Lanney was the only living full-blooded Aborigine male, and he accepted that role with a sense of responsibility. When the handful of Aborigine women at Oyster Cove weren't receiving adequate rations, Lanney lodged an official complaint. He had enough sober dignity to declare, "I am the last man of my race and I must look after my people." He never married.

In 1869, during a shore leave in Hobart, struck with some indeterminate disease, Lanney died in his room at the Dog and Partridge Hotel. He was thirty-four.

His body was moved to the morgue at the Colonial Hospital, supposedly to protect it. But the "poor, drunken piece of flotsam" was suddenly a coveted specimen. Two surgeons at the Colonial Hospital were slavering for the chance to dissect the corpse. Their names were Stokell and Crowther. They were respected medical professionals and personal rivals, each with his faction of partisans. Dr. Stokell was prominent within the Royal Society of Tasmania, a club of gentleman scientists; Dr. Crowther belonged to the Royal College of Surgeons. Evidently both men believed that Lanney's corpse would yield some sort of dramatic anatomical data about how Aborigines differed from Europeans. Having lured Dr. Stokell away from the hospital, Dr. Crowther got into the morgue and, with help from a barber, set to cutting. About nine that evening, Dr. Crowther left the morgue with a parcel under his arm. The parcel contained Lanney's skull. To cover his theft, he had plopped a substitute skull into position, wrapped crudely with whatever remained of Lanney's scalp and face. Dr. Stokell returned and found what Dr. Crowther had done. He conferred quickly with his Royal Society colleagues and, not to be totally thwarted, cut off Lanney's hands and feet.

The next day, William Lanney's much abused remains were carried in a coffin to the cemetery. The crowd of mourners was large. It included many of Lanney's shipmates, suggesting that the whaling profession in late-nineteenth-century Hobart was graced with a higher level of humanistic sensibility than the surgical profession. Lanney's coffin was put into its grave and covered with dirt. About midnight Dr. Stokell came to dig it up. A change of mind, evidently; he wasn't satisfied with the hands and feet.

Through the small hours Dr. Stokell and his Royal Society chums poked and slashed at the body until their curiosity was sated. Then they dumped it back in the morgue. According to one account, there was nothing left but "masses of blood and fat all over the floor." Now again it was Dr. Crowther's turn. Getting word that Lanney's corpse was still in play, he hurried to the morgue, which was locked, and broke down the door with an ax—but he found only those last smears and gobbets. It's not clear what had become of the greater bulk. Something of William Lanney was put into a cask, finally, and given a more permanent burial. "Dr. Stokell had a tobacco pouch made out of a portion of the skin," according to Ryan, "and other worthy scientists had possession of the ears, the nose, and a piece of Lanney's arm. The hands and feet were later found in the Royal Society's rooms in Argyle Street, but the head never reappeared."

This is why Truganini lived her last years nervously. At the end, in 1876, she was the only surviving full-blooded Aborigine from all those "conciliated" by George Augustus Robinson. She had witnessed, almost from start to finish, the destruction of her people. At the time of her birth, in 1812, Tasmania had been a glorious green wilderness harboring native tribes, thylacines, and uncountable kangaroos, though already the first Europeans had arrived and the transformation was under way. Within a few years, Truganini's mother had been stabbed by sealers; her sister had been kidnapped by a sealer, then later shot; her first intended mate had been mutilated and murdered by white men. She had seen Robinson come to Bruny Island in 1829. She had trusted him, followed him, helped him, tolerated him, and eventually watched him leave. She had seen her ancestral culture attacked as though it were a curable disease. Both of her husbands had died from strange germs and despair. She had seen her race wither and her friends die. She had seen Oyster Cove become lonely and empty. Her own longevity made her a national curio. At last she was moved up to Hobart and supported there with a stipend. On her deathbed, thinking of William Lanney, she told the doctor, "Don't let them cut me up."

She added, "Bury me behind the mountains."

But she was buried on the near side of the mountains, at the site of an old prison for female convicts, and there only temporarily. Her body, like Lanney's, was exhumed—considered too valuable to be wasted on worms. In lieu of worms, it was the Royal Society of Tasmania that got hold of her. They cleaned up her skeleton and strung it into a museum display. In this form, as a traveling exhibit, she even revisited Melbourne. Then the skeleton came home again, and until 1947 it was on view at the Tasmanian Museum. These bones, arguably more valuable for political than for scientific purposes, were taken to mean something. They were taken to mean: *We the British-descended occupants and current possessors of Tasmania have a culpable past, as do many proud peoples. We have exterminated the race who lived here before us. This creature, Truganini, was the last of the Tasmanian Aborigines. Now they're extinct. A terrible thing, sigh, but it's done.* Her skeleton was precious because it embodied this convenient lie.

103

IN TRUTH, Truganini wasn't the last. Up on Kangaroo Island, just off the south coast of mainland Australia, in a community of sealers and their wives and offspring, a woman named Suke survived. She was a full-blooded Tasmanian Aborigine. Taken from her tribe at an early age, Suke had missed being "conciliated" by George Augustus Robinson and confined with the others at Flinders and Oyster Cove. She lived until 1888, twelve years after Truganini's death.

Fanny Cochrane, the good-natured girl who gave a shell necklace to Robinson just before he departed for England, survived too. She was ignored in the countdown to extinction because her father had been white—though a photograph taken in 1903 shows her as quite strongly and handsomely Aborigine. She herself married a white man, a sawyer named William Smith, with whom she raised six sons and five daughters. For a while, Fanny and Smith ran a boardinghouse in Hobart. She assimilated gracefully, if not wholly, into the colonial culture, becoming well known and respected as "a staunch Methodist, an excellent cook, and a fine singer," by one account. Although she made it possible for white Tasmanians to be comfortable with her existence, she didn't erase from her memory some of the old Aborigine songs that she had learned as a child on Flinders. In 1899, and again in 1903, her singing was recorded by members of the Royal Society of Tasmania, including one Horace Watson (who seems to have been less predaciously scientific than those earlier Royal Society members who stole Lanney's hands and feet). The 1903 photograph shows Fanny Cochrane Smith poised, like the RCA dog, at the wide end of a recorder trumpet, while Horace Watson paints the cylinder with wax. She sang in her own language. One of the songs has been translated:

> Lo! with might runs the man:
> My heel is swift like the fire,
> My heel is indeed swift like the fire.
> Come thou and run like a man;
> A very great man, a great man,
> A man who is a hero!
> Hurrah!

Stripped of its poetry and its melody, this cold English rendering holds roughly the same representational relation to Tasmanian Aborigine music as Truganini's skeleton held to the living race.

Another of the old songs is far richer, at least to my taste, and seems more distinctly Tasmanian. After all, the heroically fast-running male—yea, though his heel be swift as fire—is a dull type that could be tossed up by any culture. I prefer:

> The married woman hunts the kangaroo and wallaby,
> The emu runs in the forest,
> The boomer runs in the forest.
> The young emu, the little kangaroo,
> The little joey, the bandicoot,
> The little kangaroo-rat, the white kangaroo-rat,
> The little opossum, the ring-tailed opossum,
> The big opossum, the tiger-cat,
> The dog-faced opossum . . .

—they all run in the forest. Or they once did, anyway. In its original form, in its original time, this was musical biogeography, celebrating just what made Tasmania Tasmanian. The dog-faced opossum, of course, was the thylacine.

Fanny Cochrane Smith died in 1905. But still other Tasmanian Aborigines survived. They were the mixed-race offspring of Aborigine women who had been kidnapped and otherwise acquired by sealers. They lived as pariahs, set apart from the mainstream colonial culture, in little communities on some of the Bass Strait islands. They were treated as nonentities by the legal and social adjudicators of British (later, Australian) Tasmania, allowed grudgingly to occupy land on the islands without ever being acknowledged as its owners, and referred to dismissively as "half-castes" or, vaguely, as "the Islanders." They retained much of their old Aborigine culture and sense of identity. But they suffered the double whammy of not being accepted as white and not being recognized as "true" Aborigines. Truganini was the last of her race—so went the official line—and now all that remained were those half-castes, the Islanders. If they weren't quite Aborigines, if their skin wasn't as dark and their hair wasn't as woolly as William Lanney's—so went the official illogic—then what claim could they make to ancestral ownership of the Tasmanian mainland? What claim could they make to reparations? What claim could they make, even, to those Bass Strait islands where they lived as squatters without legal deeds? None.

As a gesture of placation, the Tasmanian parliament in 1912 passed a law granting the Islanders use (but not ownership) of a reserve on Cape Barren Island. In 1951 that grant was rescinded. Real estate on Cape Barren was by then coming into demand. White folks from the mainland were discovering that it was a good place to raise cattle. The Islanders who couldn't qualify for individual leases were expected to get out, decamp to the mainland, find themselves niches in a poor neighborhood of Hobart or Launceston, do their best to adapt, disappear. That's what they did.

In the decades since 1951, the situation has scarcely improved. "It is still much easier for white Tasmanians to regard Tasmanian Aborigines as a dead people," according to Lyndall Ryan, "rather than confront the problems of an existing community of Aborigines who are victims of a conscious policy of genocide." A recent census found about seven thousand Tasmanians, scattered throughout the state, who identify themselves as Aborigine. One of them is Denise Gardner, an Aborigine-rights leader with curly reddish brown hair and a fierce intelligence.

Denise Gardner lives in Hobart. When I reach her by telephone she agrees, reluctantly, to a meeting. She's godawful busy but she'll give me half an hour.

I find her in an old house not far from the waterfront. The house holds cluttered offices that serve as the Tasmanian Aboriginal Centre. On one wall hangs a black-and-red-and-yellow Aborigine flag. A lanky woman in jeans and a loose sweater, pulling vehemently on a cigarette, Gardner looks like a 1960s peace activist or an organizer for the Socialist Workers' Party.

"A lot of people think, if you call yourself Aboriginal, you've got to be tribal," she tells me. What she means by that is: You've got to hunt, gather, practice the old folkways in a wild landscape. "We never said we were tribal people. We're urban blacks. We live in houses. We eat the same food other people eat."

As chief functionary of the Centre, she fights daily skirmishes in the areas of legal aid, family support, and substance abuse among the Aborigine community, as well as the long-term battle over land rights. Forceful and angry, she has no patience for hairsplitting talk about "full-blooded" Tasmanian Aborigines versus some other sort.

"I have one black parent and one white," Gardner says. "But I don't even *know* my dad's family. I was raised by my black family. I was raised in that culture." Racial and cultural identity, as she rightly suggests, aren't measured out in quadroon or octoroon fractions. As for blood, hers or anyone's, she notes: "All people's blood is the same. It's all red."

To possess any discernible bit of Tasmanian Aborigine ancestry was once a grave disadvantage. More recently, it's a point of pride. The Tasmanian Aboriginal Centre was founded in the mid-1970s, and since then there has been a reawakening collective sense of identity and a concerted political effort to gain land rights. But the legend of extinction, the legend of Truganini as the last Tasmanian, has hobbled that effort. "For some years, we had to fight the stigma that we didn't exist."

Why did the legend last so long? Why has it persisted not just as a popular notion but even among supposedly serious historians?

"It may be they're writing what they'd like to believe," Gardner says. Or there may be some other, indeterminate explanation. She brushes the question aside, having no time nor patience for historiographic speculation. "It doesn't change the fact that we *are* here."

104

N. J. B. Plomley was one historian who should have known better but persisted as a purveyor of the legend. "The Tasmanian aborigines are an extinct people," he stated flatly in a pamphlet self-published in 1977, just when the Tasmanian Aborigine rights movement was beginning its struggle.

In the same pamphlet, Plomley added: "Many causes contributed to the extinction of the Tasmanians and, although it may seem obvious to say so, the basic cause was that the death rate exceeded the birth rate after European contact." Plomley moved beyond that level of obviousness to more complicated explanations in another work, his mammoth edition of the journals and papers of George Augustus Robinson during the years of "conciliation."

Plomley titled that tome *Friendly Mission*, with only a mild sense of irony, and no doubt considered himself also a friendly chronicler of the Aborigine tragedy. At the end of the book is an appendix, headed "The Causes of the Extinction of the Tasmanians." By Plomley's reckoning there were five:

- exotic diseases
- appropriation of Aborigine hunting-and-gathering grounds by whites
- direct killing and injury by whites
- abduction of Aborigine women for prostitution and slavery
- general disruption of Aborigine life and culture.

Under the fifth category—general disruption—he added some subsidiary items:

- destruction of food supplies
- encroachment of one displaced tribe upon the territory of another
- disruption of family life
- abduction of children
- lowered birth rate

Influenza, tuberculosis, gunpoint conciliation, kidnapping, syphilis, anomie, despair, competition, predation, loss of habitat, the Black Line, and the overharvest of kangaroos—all these are subsumed within Plomley's categories. And they were interconnected, he recognized, with each one exacerbating others.

Forget for a moment that N. J. B. Plomley promoted the legend of the Tasmanians' extinction. Suspend judgment against him for dismissing the Islanders to limbo. Give him some credit for the clarity of his analysis of the factors that can lead to extinction. Plomley himself probably wasn't aware of it, but the case of the Tasmanian Aborigines—as a genetically discrete population, isolated for thousands of years on an island, well adapted to their accustomed circumstances but unable to cope with the new circumstances when a versatile and aggressive population of invaders entered their range—includes many of the hallmarks that typify insular populations (or species) brought near (or to) the point of extinction.

Plomley even noted pointedly that "the Tasmanians were probably never a numerous people." He seems to have grasped another seemingly obvious but, in fact, complicated and far-reaching truth: that the chief factor contributing toward their jeopardy of extinction was, from the start, their rarity.

105

THESE CASES from Tasmania, Guam, Mauritius, and Hawaii are only a sample. Dozens more species, subspecies, and genetically discrete populations could be mentioned, all native to islands, all lost within recent centuries.

On New Zealand, the laughing owl is extinct. So are the New Zealand quail, the North Island bush wren, and the New Zealand grayling. On Jamaica, two species of macaw as well as the Jamaican iguana are extinct. On Cuba, three species of rodent and two species of nesophont (primitive shrewlike insectivores unique to the Caribbean) have disappeared. So has the Cuban yellow bat. Christmas Island, southwest of Java, no longer harbors the bulldog rat or Captain Maclear's rat or the Christmas Island musk shrew, all three of which were endemic. The Falkland Islands once supported a species of canid called the warrah, otherwise known as the Antarctic wolf; Charles Darwin, stopping there with the *Beagle* in 1833, took a few specimens for his collection. The warrah, with a nudge from Darwin, is now extinct.

The Japanese wolf is extinct. The Tasmanian emu is extinct. Two species of Puerto Rican agouti are extinct.

The Mascarene Islands have lost at least fourteen bird species within the past three centuries—as many as the combined losses throughout the Asian, African, and North American mainlands in the same span of time. The West Indies have lost thirty-five mammals—more than the combined losses throughout Africa, Asia, and Europe. Hawaii alone has lost more bird species than were lost from all the continents on Earth. Compared to the roster of historically recent extinctions from the world's islands, the roster from the mainlands has been strikingly small.

On Samoa, the Samoan wood rail is extinct. On Macquarie Island, the Macquarie Island parakeet. On Tristan da Cunha, the Tristan gallinule. On the Cape Verde Islands, the Cape Verde giant skink. On Wake Island, the Wake Island rail. On Guadalupe Island, the Guadalupe flicker. On São Thomé, the São Thomé grosbeak. On Auckland Island, the Auckland Island merganser. On Iwo Jima, the Iwo Jima rail. On the Ryukyu Islands, the Ryukyu kingfisher. On Lord Howe, a small Pacific island with a big share of endemism, the Lord Howe Island pigeon is extinct, as are the Lord Howe Island white-eye, the Lord Howe Island fantail, and the Lord Howe Island flycatcher. On Réunion, not far from Mauritius, a whole passel of endemic birds are extinct. On Hispaniola, a whole passel of mammals. On the Society Islands, the so-called mysterious starling, no less mysterious now that it's gone. On the islands of the northern Atlantic, between Norway and Newfoundland, the great auk. On Stephens Island, the Stephens Island wren.

Stephens Island is a dot of land between the North Island and the South Island of New Zealand, and the extinction of this endemic wren, *Xenicus lyalli*, is just one more instance among many, tiny in its own scope but emblematic. The bird was a small thing with short wings and almost no tail, either totally flightless or nearly so. It made its living by skulking among rocks and feeding on insects. Even during good times, the species was perilously rare. Besides those disadvantages, presumably it suffered from ecological naïveté—being too trusting for its own good. We can't be confident whether or not it *was* totally flightless, nor just how disastrously trusting, because almost no one ever saw it alive.

It was discovered in 1894 by a lighthouse keeper named Lyall—or, to be more precise, it was discovered by one of Lyall's cats. The cat killed a local bird and proudly presented the carcass to its master, as cats will do. Lyall, being something of a birdwatcher, recognized that the stubby-winged creature was unusual and passed it along to an expert on New Zealand's birds. The expert in turn shipped the preserved specimen off to a British bird journal, along with a manuscript describing this new species and naming it, for the lighthouse keeper, *Xenicus lyalli*. With the wren now internationally famous, collectors arrived on Stephens Island. Apparently a combination of factors—collection, habitat destruction as the island's forest was cleared for pasture, and further predation by Lyall's cats—destroyed the entire population within a very short time.

Twelve specimens of *X. lyalli* now exist in museums. Back on Stephens Island itself, the bird has never been seen again.

106

"Except for island cases," according to an ecologist named Michael Soulé, "we know very little about the process of natural extinction." Soulé is entitled to say so, having pondered that process as carefully as anyone in the world.

Early in his career, Michael Soulé studied evolutionary trends among insular lizards in the Gulf of California. He was interested in biogeographical patterns and their genetic bases. He took note of Ernst Mayr's 1954 paper on genetic revolutions and pursued a related study. He did fieldwork on some of the same arid islands that would later be favored by Ted Case. Then, gradually, Soulé became more concerned about the matter of extinctions. Extinctions of whole species, extinctions of local populations, extinctions on islands, extinctions elsewhere—what exactly was being lost, Soulé wondered, and why? The range of his work, and of his influence, widened.

Eventually Soulé would help invent a new mode of addressing the ecological and genetic aspects of extinction, a more scientific approach to the conservation of biological diversity. For that reason he will loom large in a later section of this book. He appears briefly now because, like N. J. B. Plomley, Soulé made a list of the factors that can contribute to the extinction of a population.

Soulé's list was a generalized warning, not a specific postmortem like Plomley's. It formed part of an essay that he contributed to an edited volume, published in 1983, exploring the overlap of genetics and conservation. Soulé's chapter was titled "What Do We Really Know About Extinction?" Not nearly as much as we need to, was his answer, but what little we do know is interesting and important.

We know, for instance, that island forms are particularly vulnerable. We know that ecological isolation—either by seawater or by other sorts of delimitation—correlates strongly with risk of extinction. We know that predation and competition can in some cases extinguish a population, especially if the predator or the competitor is an exotic. And we know that habitat destruction is another highly effective way to annihilate species. Beyond these quanta of sorry knowledge, there's a bit more. Soulé mentioned eighteen factors, any or all of which might contribute toward a given extinction:

1. Rarity (low density)
2. Rarity (small, infrequent patches)
3. Limited dispersal ability
4. Inbreeding
5. Loss of heterozygosity
6. Founder effects
7. Hybridization
8. Successional loss of habitat
9. Environmental variation
10. Long-term environmental trends
11. Catastrophe
12. Extinction or reduction of mutualist populations
13. Competition
14. Predation
15. Disease
16. Hunting and collecting
17. Habitat disturbance
18. Habitat destruction

Relax, this time I'm not going to annotate. I just wanted you to pass your eyes across such a panoply of biological jeopardies.

In the same chapter Soulé voiced a quiet but ominous observation. "Scores of birds, mammals, and flowering plants have suddenly become extinct, particularly on oceanic islands, during the last 100 years. It is as if some epidemic were raging among these species." By this point, you know as much for yourself. "Now the disease appears to be spreading to the continents," he added.

⋯

PRESTON'S

BELL

⋯

107

THE SPECIES-AREA relationship is one of ecology's oldest and most profound generalizations. Exactly how old and exactly how profound are matters still under discussion. The discussion is no idle debate. It pertains to the preservation or loss of biological diversity on our planet, where the total area of natural landscape grows smaller and more fragmented every year.

Among professional ecologists, the discussion dates back before 1922, when Henry Allan Gleason published a paper titled "On the Relation Between Species and Area" in the journal *Ecology*. Gleason's paper came partly in reaction to another, just a year earlier, by a Swedish botanist named Olof Arrhenius. The Arrhenius paper was titled simply "Species and Area," reflecting the fact that the subject was still in its elementary phase.

But even Arrhenius and Gleason had precursors. A researcher named Jaccard had touched on the species-area relationship in 1908, and fifty years before that H. C. Watson had recorded some relevant evidence among the flora of Great Britain. One square mile of countryside in the north of Surrey, Watson had found, contained half as many plant species as Surrey overall. It wasn't a dramatic finding, but it carried implications that a Victorian botanist couldn't have guessed. And even Watson's work was anticipated by the comment, already quoted, from that cantankerous Prussian biogeographer Johann Reinhold Forster, who had gathered his observations while exploring the southern Pacific with Captain Cook: "Islands only produce a greater or less number of species, as their circumference is more or less extensive."

Less circumference encompasses less area. That much is logically implicit. Less area harbors fewer species. That's the empirical reality, which Forster had seen with his own eyes.

Like Gleason and Arrhenius, like Jaccard and Watson, Forster was referring to plant species. Plants lend themselves well to the collection of this sort of data because an individual plant, having taken up residence within a given area, remains stationary and easily counted. Animals, being mobile, are harder to localize. But if Gleason and Arrhenius followed Forster's model by using plant data, they also saw the species-area relationship as a more ubiquitous phenomenon than what Forster had described. Gleason and Arrhenius weren't concerned with the numbers of species merely on islands. They were concerned with the numbers of species on quadrats, wherever those quadrats might be located.

What's a quadrat? It's a square piece of terrain that has been chosen to serve as a representative sample. A quadrat, unlike an island, is not ecologically isolated. Mark off a patch of your backyard, using nothing but tent pegs and string, and you've created a quadrat. If the quadrat is typical of your yard as a whole, you might use it to ascertain that you're harboring seven dandelions and ninety square inches of crabgrass per square foot. Multiply by the yard's total area, and you know your overall holdings in weeds. Though yielding only an estimate, this procedure would save you the trouble of counting your way across the entire goddamn lawn. The units of area on which Gleason and Arrhenius measured species diversity were quadrats in that sense: artificially delineated plots within a larger area.

Gleason collected his data from an aspen-dominated thicket in northern Michigan. Within that thicket he defined 240 plots, each with an area of one square meter. Then he made a census of the plant species on each plot. The average number per plot was just over four species. The entire system, spanning 240 square meters, supported twenty-seven species. It wasn't a very rich vegetational community, but it was rich enough; and its relative paucity made it manageable for Gleason's purposes.

By conceptualizing his plots into variously sized aggregates (that is, clusters of contiguous plots) and counting the number of species within each aggregate, Gleason could arrange and rearrange his results to reveal differences according to scale. A ten-square-meter area, for instance, contained on average about ten species of the total twenty-seven. A twenty-square-meter area contained thirteen species, a forty-square-meter area contained sixteen species, and an eighty-square-meter area contained twenty species, almost three-quarters of the total for the whole system. The particular numbers are noteworthy only as they illustrate a pattern: Larger areas support more different species than smaller areas. This was the species-area relationship made manifest among a community of Michigan plants.

Gleason's fieldwork was neat and his statistics were interesting—at least to a plant ecologist. He brought the pattern vividly into relief. But to him it remained a matter of pure science. Exactly what it might mean in the wider world was left for later debate.

108

WHATEVER IT did mean, the species-area relationship was not just a phenomenon among plants. Philip Darlington returns to us here because, back in 1943, in his paper on flightlessness among beetles, he had mentioned the same pattern: "Limitation of area often limits both number and kind of species of animals in isolated faunas."

Cuba, Hispaniola, Jamaica, and Puerto Rico served as his four diagnostic islands. He had personally collected beetles on each. The four seemed suitable for comparison because, though different in size, they were ecologically and climatically similar. Cuba, the largest, encompassed an area of about 40,000 square miles. How many species of carabid beetle would such a big island support? By Darlington's actual survey, the number was 163. Hispaniola was slightly smaller at about 30,000 square miles, and it was also less rich in its diversity of carabids. Then came Jamaica, smaller still, with still fewer species. Puerto Rico was not quite a tenth the size of Cuba, with less than half as many species of carabids. Darlington noted the ratio between Cuba and Puerto Rico, at his upper and lower extremes, and remarked that "division of area by 10 divides number of species of Carabidae by 2." It was only a throwaway comment in the 1943 paper, nearly unnoticeable amid his analysis of carabid flightlessness. But Darlington restated the same point later, with more emphasis and more data, in the book for which he is mainly remembered.

This is *Zoogeography*, published in 1957. It's a big, broad-ranging volume, an encyclopedic treatise on the distribution of all types of animal across all parts of the planet, and one of the last great landmarks of old-fashioned biogeography in the tradition that stretches forward from Johann Reinhold Forster. It bears a closer resemblance to works such as *Island Life* and *The Geographical Distribution of Animals*, published eighty years earlier by Alfred Russel Wallace, than to the new wave of biogeographical studies that came immediately after it. Darlington's *Zoogeography* differs in several ways from the later works; the most apparent of those differences is that it's a descriptive book written in plain English prose. It contains a few simple diagrams, a few maps, but no outbursts of statistical abracadabrizing, no tirades of hieroglyphic calculus, almost no mathematics at all. So it makes easy reading for those of us who are mathematically impaired. A modest exception within its generally anumerical prose appears on page 484. Here, in a chapter titled "Island

Patterns," Darlington returned to the species-area relationship. He offered, as Henry Allan Gleason had, a numerical table.

Maybe the sudden sight of numerals, looking so lean and sure, is what makes that page seem especially significant. It's the one part of *Zoogeography* that every modern ecologist knows.

As in the earlier paper, Darlington took his data from the Antilles. Now he considered reptile and amphibian species instead of beetles. He had found that Cuba (at roughly 40,000 square miles in area) supported *this* many species of lizards, *this* many of snakes, *this* many of frogs and toads. Never mind the individual tallies—the total for those groups together was about eighty. Puerto Rico (at roughly 4,000 square miles) supported only *that* many lizards, only *that* many snakes, only *that* many frogs and toads. The total for Puerto Rico was forty. Between Puerto Rico and the next-smallest island came a gap in the order-of-magnitude regression, since Darlington possessed no empirical data from any Antillean landmass of roughly 400 square miles. But he could report that the little island of Montserrat (only 40 square miles in area, a hundredth the size of Puerto Rico) had nine species of reptiles and amphibians. Still smaller was the island of Saba, about 4 square miles in area, with just five herpetological species. Darlington saw among these data the same pattern he had seen for beetles, and he repeated his earlier generalization, noting that the "division of area by ten divides the amphibian and reptile fauna by two."

Using this generalization as his rule of thumb, he dared to fill the gap in his data—for the missing island, the one that would rank in size between Puerto Rico and Montserrat—with an extrapolation. A hypothetical island of 400 square miles, which would be a tenth the size of Puerto Rico, should support half as many herpetological species as Puerto Rico—that is, twenty. In making this estimate Darlington put his toe forward into a new sort of enterprise: theoretical ecology.

Coaxing the numbers into the shape of an argument, he produced a sleek, pyramidal table:

Area (SQUARE MILES)	Species (APPROXIMATE)	Species (ACTUAL)
40,000	80	76–84
4,000	40	39–40
400	20	–
40	10	9
4	5	5

It was as boldly schematic an assertion as any biogeographer at that time would make. But it was only a beginning.

One point remained puzzling: What causes this species-area relationship? If limitation of area results in limitation of species diversity, how is that limitation exerted? What's the underlying process that leaves small places so much poorer than large ones?

109

AN INDEPENDENT researcher named Frank Preston had meanwhile gotten interested in an issue that turned out to be closely related: the matter of commonness and rarity. As with species-area studies, it entailed the recognition and interpretation of pattern.

Within any biological community, some species are common, some species are rare, and some species fall into a middling category of modest but not exceptional abundance. How many rare species does a community include? How many common species? Is there a pattern to the way species are distributed among these categories—a pattern that applies generally across different ecosystems? Based on his scrutiny of data describing commonness versus rarity in certain communities of birds and of insects, augmented by some conceptual leaps of his own, Preston deduced that there was. He called this pattern the *canonical distribution* of commonness and rarity, because in his view it held the force of canon. It seemed almost a law of nature.

Preston's canonical pattern was one variant of a more generalized class of statistical distributions, known (for reasons we can ignore) as *lognormal*. Depicted on a standard coordinate graph, a lognormal distribution traces out a characteristic line: rising sharply from the left, hitting an early peak, then falling off gradually toward the right side of the graph—like a profile of the Rock of Gibraltar. Preston preferred to depict his particular lognormal distribution in slightly different graphic terms, by making the routine mathematical adjustment of converting to a logarithmic scale along one axis. Given that conversion, the same set of data points trace out a differently shaped line: swooping up gracefully from the left, arching to a high rounded dome in the center, swooping down again toward the far right—like the silhouette of a bell. Mathematicians call this a bell curve.

Preston's bell curve, like all bell curves, was tidy and symmetrical.

What it represented was a biological community containing some species that are very rare, many species that are fairly abundant, and just a few species that are very abundant. If that distribution of commonness and rarity was canonical, as he asserted, then it would be characteristic not just of some situations but of almost any complete biological community. Preston also stipulated one further point, which is less obvious and logical than it might seem: In a complete biological community, the few most common species (rather than the many species of middling abundance) will account for most of the individuals.

From these premises, grounded in empirical evidence and raised to significance through the magic of math, Preston derived some potent corollaries. For instance: "If what we have said is correct, it is not possible to preserve in a State or National Park, a complete replica on a small scale of the fauna and flora of a much larger area." The crucial words here are "complete" and "on a small scale." It was a prophetic insight. Preston's canonical hypothesis led toward *The Theory of Island Biogeography* by MacArthur and Wilson and beyond. His bell rang as herald to a revolution.

Frank Preston himself was a curious figure in the history of ecology. He was born in England, passed up an Oxford scholarship to go into apprenticeship with a civil engineer, learned enough chemistry and physics on his own to get a degree at the University of London in 1916, and then went to work for an optics firm. Nine years later he returned to the university and quickly took a doctorate, largely on the strength of papers he had already published. To celebrate that academic triumph, he went on a birdwatching trip around the world. Having fetched up in Pennsylvania, Preston started his own engineering firm, which specialized in glass. One recent source says he became "a giant in the industry, writing more than sixty papers for solid-state physics journals and being granted some twenty industrial patents." His bird studies were an avocation. In journal papers derived from those studies, he listed his affiliation as "Preston Laboratories, Butler, Pennsylvania," which sounds as though it might pertain to biological research of some sort; that impression is quietly shattered by a footnote, in one paper, explaining that the former Preston Laboratories is now "American Glass Research, Inc." But if theoretical ecology was a hobby, it was a hobby he practiced with great rigor and penetration.

He published his first paper on the commonness-and-rarity question in 1948. There followed another in 1960, and then a long, two-part article titled "The Canonical Distribution of Commonness and

Rarity," which appeared in successive issues of *Ecology* during spring and summer of 1962. Those were his last prominent contributions to the ecological literature, though he lived until 1989.

Preston's work is thick with ideas and mathematics, but two bits of it stand forth across time. The first is what Preston called "the Arrhenius equation," conferring pedigree on his own formulation by attributing it to one of his forerunners. "This equation is not given by Arrhenius but is implied by his work," Preston explained. It may be odious to behold, but it's inescapable for anyone concerned with later developments in island biogeography. Close your eyes and I'll print it quickly.

$$S = cA^z$$

All right, you can open them.

The equation declares that the number of species (S) within any given area is related to the size of that area (A) according to a mathematical ratio dictated by two strategically applied constants, c and z. Neither of the two constants, however, is constant. Got that? Not constant in the ordinary sense of the word, anyway. The constant c is specific to different taxonomic groups and geographical regions—one value for carabid beetles of the Antilles, another value for birds of the Galápagos. The constant z is also specific to circumstances.

What z denotes is the degree, more or less drastic or slight, to which the number of species on each patch of landscape decreases with decreasing area of the patches. If area is far more strongly determinate among the Galápagos Islands than among plots of thicket in Michigan, then the Galápagos will show a high value for z, whereas the Michigan thicket will show a low value. The exponential z is known as the slope of the species-area relationship. Watch out, here it comes again: $S = cA^z$. This obscure but important equation has been arrived at by induction from real-world data. In other words, it merely describes the species-area relationship that scientists have observed in the field.

When supplied with some actual numbers and plotted on a double logarithmic grid, the equation takes graphic shape as what Preston and others after him have been pleased to refer to as a *species-area curve*. A species-area curve portrays the correlation between species and area within any gradated selection of habitat patches.

Terminology, terminology—but a little careful decipherment of the technical terms is essential to keeping this stuff from seeming opaque. So I should explain further that there is only one species-area *relationship*—namely, the general truth about smaller areas supporting fewer species—but there are any number of possible species-area *curves*. Every field survey of the species-area relationship in a different context yields its own individualized species-area curve. Every set of islands on which the species of reptiles and amphibians have been counted has its own curve; likewise every set of quadrats on which the plant diversity has been surveyed. Preston included a gallery of examples in his 1962 paper: one species-area curve for birds of the East Indies, another curve for land vertebrates on islands in Lake Michigan, another curve for plant diversity on quadrats of grassland in England, and still others. But now here's an additional insult to your common-sense apprehension of language: If the graph has been adjusted appropriately, with conversion to logarithms along both axes, the "curve" in each case is a straight line.

What's the incentive for such logarithmic jiggering in order to make the curves straight? Well, any set of data points forming a line on a graph can help a scientist move from facts to theory. It serves as a reliable pathway toward insights in the realm of the empirically unknowable. And a straight line is preferred because it's simpler to work with than a squiggly line. The value of z in a given species-area system designates the angle of that line as it streaks across its graph—which is why z is known as the slope. A curve that is actually curvaceous will wind its way through a variety of different slopes, but a straight line, conveniently, conforms to just one. The slope of a species-area curve provides guidance toward theoretical deductions, just as the lip of the Big Dipper provides guidance toward Polaris.

Hello, are you still with me? Have you forgiven me for mentioning logarithms?

110

THE OTHER memorable point from Preston's work is his distinction between *samples* and *isolates*. This brings us back to the matter of how many common species and how many rare species are represented within biological systems.

A sample, according to Preston, is a group of species or a parcel of landscape that exists as part of a greater whole. That is, the sample can be taken either as a representative patch of area within some larger area or as a representative list of items within some longer list. Gleason and Arrhenius, with their quadrats demarcated among plant communities on the landscapes of Michigan and Sweden, were studying samples. The string-bounded patch of dandelions and crabgrass in your yard is a sample. The moths that turn up at a porch light during one summer night in a Tennessee forest constitute a sample of the moth fauna from that part of Tennessee. And a hectare of rainforest in the central Amazon, if surrounded by a vast and continuous expanse of similar rainforest, is a sample.

An isolate, on the other hand, is a sequestered group of species or an isolated parcel of landscape—an island, for instance.

The importance of this distinction is that samples maintain one sort of balance between common species and rare species, while isolates maintain a different sort of balance. A sample will be richer in rare species than an isolate. Why? "The explanation," Preston wrote, "is that on isolated islands we must have an approximation to internal equilibrium and presumably to a self-contained canonical distribution and, since an island can hold only a limited number of individuals, the number of species will be very small." What he meant by "internal equilibrium" is that the rare species, within an isolate, are virtually never reinforced by other members of the same species wandering in from outside. The few lonely individuals of each rare species are utterly on their own. Either they survive and procreate among themselves or else that species disappears from the island. "But on the mainland a small area is not in internal equilibrium; it is in equilibrium with areas across its boundaries and is a sample of a vastly larger area." So a sample area on the mainland contains more species represented at the low threshold of rarity. Those locally rare species, within a sample, tend to be continually (though barely) maintained by individuals wandering in from the larger landscape.

Take away the larger landscape—that is, transform the sample into an isolate—and the balance changes. Wandering individuals no longer arrive. Many of the rare species can't maintain themselves independently. So they vanish from the isolate altogether.

"This is a matter," Preston added modestly, "that does not seem to have been considered previously in studies of Species-Area relationships." True enough in 1962, when he wrote those words. But its time was coming.

III

Two THOUGHTS are afloat in my mind as the puttering old boat of Captain Sultan carries Nyoman the Balinese Tailor and me away from the island of Komodo. One is: I'm glad to have escaped from the Loh Sabita bluffs with my eyes wide and my buttocks intact. The other is: $S = cA^z$.

The weather has stayed fine. The sea is still as flat as a pond. If we were to cruise directly eastward, we could reach our port town on the west coast of Flores within four or five hours. We could spend the night at a clean little rooming house, treat ourselves to roast fish and cold beer at a restaurant, and catch our plane in the morning. But that's not the plan. I've broken the news to Nyoman that I'm not ready to go back to Flores. First I want to explore some of the tinier islands in between. I want to witness the relationship between species and area when one of the species is *Varanus komodoensis*.

So we have carried aboard ten bottles of drinking water and a supply of malt biscuits. Sultan has laid in enough rice and coffee to last us a few days. Fresh fish and squid we can buy from the occasional lonely fishing prau that we encounter at sea. An off-duty ranger from Komodo—it's Johannes, one of the friendly crew who welcomed us in the Loh Sabita backcountry—has joined our party, in the spirit of larking away his weekend. As we sail eastward from Komodo, into a realm where the public ferry doesn't go, we're guided only by the whims of my curiosity and the sea map in Sultan's brain.

This region between Komodo and Flores—which even the better atlases portray as nothing more than a small stretch of blue—is actually a world of lesser islands. They loom within sight of one another like a flotilla of icebergs. Obscure mounds of landscape that tourists don't visit and that even most Indonesians have never heard of: Padar, Rinca, Sebayur, Siaba, Papagaran, Pongo, Gilimotang, quite a few others. They divide the sea hereabouts into a labyrinth of riptides and surge lines, deep channels and boat-eating reefs. In size, they range from Rinca—which at 109 square miles is almost as large as Komodo—down to the smallest hump of dry rock that's big enough to be given a name. Nusa Pinda, at a tenth of a square mile, is representative. With the exception of Rinca, they are mostly uninhabited by humans—too arid, too severe, too useless, nobody wants them. Like the islands on which Ted Case does his work in the Gulf of California, they are hospitable chiefly to reptiles and birds. I think of this region as Lesser Komodoland, in honor of the dragon whose distributional range encompasses it. Greater Komodoland, by which I mean the reptile's complete range, includes also the island of Komodo itself and a swath of habitat along the west coast of Flores. It's important to note that within the boundaries of its range, *V. komodoensis* occurs only patchily. Not every island of Lesser Komodoland harbors a population of komodos.

Rinca and Gilimotang do. Most of them don't. Why is that so?

One factor seems to be area. Rinca is relatively big. Gilimotang is smaller but not too small. Sebayur is smaller still and supports no komodos. Nusa Pinda is prohibitively tiny. This gradation of areas, and

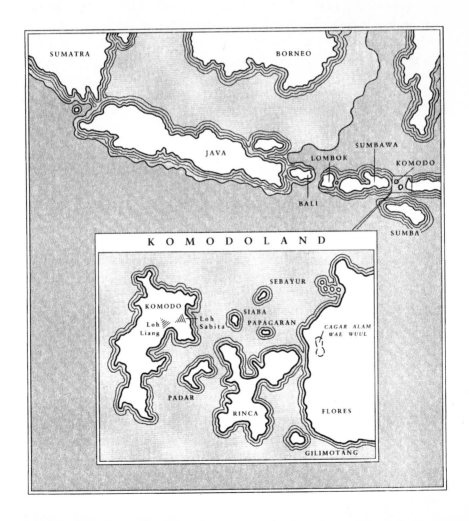

the degree to which it correlates with presence or absence of *V. komo-doensis*, is easy to measure. The other contributing factors are more elusive.

We make our first landing on Padar, wading up onto a white beach while Sultan holds the boat just offshore. Padar is 5.3 square miles in area—as I know from a handy, precise chart in Walter Auffenberg's book, *The Behavioral Ecology of the Komodo Monitor*, which is still riding along in my field bag. Padar's landscape, mostly covered with thick brownish grass, rises into a modest set of volcanic peaks, but the high-lands are not high enough to capture much rain or support much for-est, so it's a dry place with an austere savanna ecosystem. It harbors no human settlements and few large animals. Despite the austerity, Padar was one island on which Auffenberg did find *V. komodoensis*, during the field study that he eventually documented in the book. But that study was done twenty years ago. The population of komodos seems to have been small, and twenty years is a long time in the precarious history of a small population. Sultan has already warned me not to ex-pect much. "*Ora? Tidak ada,*" he has said. Naw, no komodos here. Not now, anyway.

I value Sultan's view on such questions, since he knows his way around Komodoland as few people do and his powers of observation are acute. Johannes agrees with him, offering testimony that reaches me in English through Nyoman: "He say, 1991 was fire. Burned whole island except just here. John came out with people to help fight it. Before that, is deer and komodo. After, is never see them again." The story sounds plausible. A catastrophic fire is exactly the kind of event that could extirpate a small population from an isolated patch of habitat. There is no visible evidence of a fire, not to my eye, but dried-out savanna would burn like paper and grow back fast.

We hike for an hour on Padar. From our landing point on the west side, we climb up a gentle slope toward a saddle between peaks, then down the far slope into a swale on the east side. We find only a few clus-ters of trees punctuating the grassland, which is almost unrelievedly open and sunny—disadvantageous for a lurk-and-lurch predator. Then again, it could be disadvantageous too for large-bodied prey, offering few places to hide. But whether the openness of the terrain might margin-ally favor the prey species over the predator, or vice versa, seems a moot question if both have been killed off or driven out by fire. The grass is waist-high, and we stop to unburden ourselves of ticks.

Moving on, we spook two healthy stags from a small clump of

cover; they go trotting away to higher ground. Johannes mutters a reaction and Nyoman interprets: "Interesting. Last time is people checking, is *no* animals. But now is still two deer. He gonna make a report to the office." Since deer will occasionally blunder off on a reckless sea-crossing swim and arrive safely on a new shore, there is a reasonable possibility that the two stags have recolonized Padar, from Komodo, just since the fire. If they came without female company, then of course they are fated to die celibate, horny, biologically thwarted, leaving the island as empty as ever, unless some females arrive in the meantime to rescue their spirits and help sustain a population. Komodos swim too, occasionally. And so they too might eventually recolonize Padar. For now, though, we find no sign of *V. komodoensis*—no tracks, no komodo shit, no piles of half-chewed bone.

No sign, either, of wild pigs. No sign of goats. The sheer diminutiveness of its area, or some other factor, has left Padar impoverished of large vertebrates. The predominant faunal presence seems to be ticks, of which we pick another dozen off our pants before wading back to the boat. The ticks must be starved, living here in famine conditions with such a shortage of red-blooded hosts. As Sultan poles us out through the shallows, I feel no reluctance about leaving. I'm happy to put the Island of Hungry Ticks in our rearview mirror before the little bastards can muster a massive assault.

Siaba, near the northern edge of Komodoland, is smaller: only 1.4 square miles, according to the chart. We walk ashore, past the mangroves, onto a sandy flat covered sparsely with grass. Twenty years ago, Auffenberg found Siaba empty of komodos, and Johannes now asserts the same: "*Disini, tidak ada ora. Tidak ada rusa. Ada kambing, ya.*" No komodos here. No deer. Only a few goats.

We stroll the flat, looking for signs. The goats of Siaba must live a marginal, though unharried, existence. They even defecate parsimoniously, leaving dainty piles of dry, raisinlike pellets. If there are no komodos, I ask Johannes, then what about *this*? I'm standing over a sinuous tail-drag mark across one of the sand patches. Clawed footprints straddle the tail mark, reflecting a smallish animal but definitely a reptile.

"Ah. *Tidak ora. Varanus salvator*," says Johannes. Not a komodo, no. These prints were left by *Varanus salvator*, the smaller species of monitor lizard. Less burly, less ferocious, known in English as the two-banded monitor, *V. salvator* is the komodo's little cousin.

Farther along the sand flat I find a bleached jawbone and then an-

other set of reptilian tracks. Yes, it's the jaw of a *kambing*, a goat, Johannes confirms. I'm eager to believe that this goat didn't die a natural death. I point to the tracks and, hoping to be contradicted, ask,
"*Ini, salvator?*" The smaller species again? From my own reading and
experience I know that *V. salvator*, which swims well and probably
shows up on most of the islands of Indonesia, is vastly more common-
place than the komodo. I saw it on Krakatau and I might certainly
meet it again here on Siaba. Still, I'd be delighted to discover that *V.
komodoensis* itself has colonized Siaba since Auffenberg's visit and begun feeding on the island's feral goats. But Johannes, with the irk-
some fastidiousness of a person who knows what he's talking about,
declines to support that notion.

"*Tidak ora?*"

"*Tidak,*" he insists. "*Tidak ora.*" No. With just a glance at the
tracks, he's confident they are not from a komodo. His reasoning
comes translated by Nyoman: "Is not possible for komodo to be living here. Because no other animals is living here to eat."

"What about the goats?"

"Goat is belong to somebody. Is putting here."

"Komodos could still eat a goat," I quibble, "even if it belongs to
somebody. Ya?"

"No. If people is know there's komodo here, they're not putting
goat."

That seems suspiciously circular, but I concede anyway. Most likely
Johannes has his facts straight, even if his logic is wobbly. Most likely
Siaba is too small, too harsh, too bereft of big edible mammals and
the ecological conditions they require, to support even a temporary
population of komodos. We return to the boat.

Late in the day we coast up toward the island of Papagaran and
find ourselves on an ebbing tide over a turquoise reef, unable to get
within two hundred yards of land. We're just offshore from a fishing
village—fifty small houses on pilings along the water's edge—but
there's no pier, no channel through the coral. The local fishing people
have no *need* for either a pier or a channel, since they come and go in
their shallow-draft canoes. Determined to make at least a token re-
connaissance of Papagaran, I prepare to swim in. The others say:
You're gonna do *what*? Sultan the mariner believes, like most
mariners, that swimming is for chumps. To save me the effort and
danger, he hails a young boy in a dugout canoe.

The boy consents to paddle me to shore, if I can just manage to avoid capsizing him. There's about one inch of freeboard, but we make it.

Papagaran is a bleak place, a low mound of red rock gouged by small canyons and covered sparsely with brown grassy stubble. The village is tiny, though proud or pious enough to have its own tin-dome mosque. The island is not blessed with a spring, not even a well, so all the fresh water has to be brought in. When that supply falters, who knows what the people do? Drink goat's milk, maybe. Goats seem to be plentiful, thriving in this sun-punished landscape as goats do in sun-punished landscapes everywhere. As I climb up a stony gulch, three or four nannies edge out of my way, grazing on tufts of nothing. With its rocky red soil, its dry heat, its lack of trees, its thirsty village, and its goats, Papagaran combines some of the least felicitous aspects of Galilee and Mars.

The total area is 0.6 square miles. The island was reported by Auffenberg to be devoid of komodos. Undoubtedly it still is.

We cruise southward. Sunset comes early in Komodoland and Sultan has no intention of groping through these dicey waters in darkness, so he suggests that we pass the night at anchor, beside another fishing village along the coast of another island. We can sleep on deck and sail again at dawn.

As we steer for our anchorage, hurrying by twilight, the sky fills with an unworldly spectacle: a great flock of fruit bats, maybe a thousand, each as big as a raven, flying toward their nocturnal feeding grounds on the island of Flores. High and silent, graceful, they come out of the western sky, flapping their huge scalloped wings in slow rhythm as the last light of dusk ignites a peach-colored glow. The moon is a sharp crescent. Against the dusken glow, crossing the moon, the bats silhouette themselves in procession, like a nervous child's dream of Halloween. It's so eerily beautiful that I sketch a bat in my notebook to help persuade myself later that it was real. The islands on the horizon, five or six within sight, become silhouettes too. We anchor.

By kerosene lamp we wolf our rice and fish. Then the lamp is doused. The sea air is just cool enough for comfort. Sultan has produced some old mattresses, and I've got a sarong for a sheet. Now, I anticipate contentedly, the most majestic sunset I've ever witnessed should be followed by a majestically tranquil night's sleep. But the fishing village turns out to be full of roosters and moronically vocal dogs. We're moving again before sunrise.

Sultan has been thinking ahead. He knows that our route down to Gilimotang passes through a difficult narrows, a choke-point between islands where the surge lines and whirlpools are nasty and the bottom is cluttered with wrecked boats. He wants to make that passage at high tide. Having covered the distance in good time, we do clear the narrows safely, riding high, with only the hungry slurp of swirling water to indicate menace.

The moment Gilimotang comes into view, I see that it's dramatically different from Papagaran, Siaba, and Padar: a tall croquette of an island instead of a pancake, not so arid as the others, and far more densely upholstered with vegetation. The area is only 4.0 square miles, which makes it smaller than Padar. But it has tree cover from coastline to peak, with just a few swatches of savanna interrupting the forest. And, as I know from Auffenberg, it harbors populations of deer and wild pig. The deer find food and security on the open savanna, across which they carve their own trails. Such a marbleized mix of savanna and forest entails an abundance of edge between the two types of vegetation, and these are favorable habitat conditions for *V. komodoensis*. Lurking in the forest, three strides from a deer trail that winds out through the savanna, a hungry komodo has the predatory advantage it needs.

A recent census on Gilimotang by Indonesian researchers has produced a count of one hundred komodos. It's not a big number, but on an island this size it amounts to a pretty high density, at least for a species of large-bodied predator. We've barely stepped ashore when we find ourselves standing over another set of tracks: a tail drag on the beach, with clawed footprints astride it.

"Komodo," decrees Johannes.

Wait a minute. Aren't these tracks nearly identical to what we saw on Siaba? I ask how Johannes can be so sure that it's not just a big individual of the smaller species, *V. salvator*.

Nyoman relays my question, listens, and then tells me something or other about the subtlety of proportion by which Johannes discriminates *salvator* tracks from *komodoensis* tracks. The subtlety escapes, eluding either Nyoman's English or my comprehension. Nyoman also explains that Johannes has come to Gilimotang on previous survey expeditions for his department. "Last year he's been check many times. Putting goat here. Always come to eat is komodo." Oh. Okay, then. Watching a couple of big komodos charge out of the forest to rip a goat in half would constitute, I admit, fairly convincing evidence of their presence.

We spend an hour hiking uphill through a long patch of savanna. Cautiously we search the forest edges. We rouse one large komodo from its cover and chase after it, at a safe distance, until that animal disappears into the forest. Through my binoculars I spot a group of four deer climbing slowly across a far slope. For a few minutes I track on the deer, hoping shamelessly to see a komodo spring out and snatch one. When the four have passed beyond sight, I see another, a small doe, and then two more. So there is no shortage of prey on Gilimotang. And the habitat structure seems to be excellent. And the island contains no human settlement. And it's far off the ferry route, seldom invaded by tourists except those few, such as me, whose curiosity and presumption are especially strong. This patch of landscape would seem an ideal sanctuary for *V. komodoensis* if not for one thing: It's too small.

At only 4.0 square miles in area, it offers inadequate long-term security. Its population of one hundred komodos may not be numerous enough to endure the perils of rarity. Accidental factors (like the fire that burned over Padar) and the inherent statistical vulnerability of small populations probably put the Gilimotang komodos in some danger of extirpation.

More on that subject to come, but in the meantime, speaking of danger: Sultan has warned us not to linger too long here on Gilimotang, since we're racing against low tide and we still have to make our return through the narrows.

So when I notice our captain clapping and hollering as we come off the slope, I assume that he's urging us to get our landlubbing, slow-poking asses into gear. But no—now I see what he's pointing at. A five-foot-long komodo stands between us and the boat. Lordy, they're so dense on Gilimotang that they overflow onto the beaches. I try to get close enough for a photo. But this animal jacks itself high on its legs and runs off, moving so sleekly that its tail doesn't leave a mark on the sand.

By now we have visited four lesser islands, all roughly similar in climate and location but different in size. We have seen the species-area relationship laid out in real flesh and turf. The relationship isn't perfectly linear, but it's real. Gilimotang, a relatively large island among this little group, supports species (komodo, deer, pig) that the smaller islands of Siaba and Papagaran don't. Padar too still supports deer, even after the fire. The size threshold for a dragon-haunted island— that is, the minimum area required for supporting a population of komodos—seems to be somewhere between 4.0 (Gilimotang) and 1.4 (Siaba) square miles. It's incautiously precise of me to hazard such numbers, given all the other variables besides area, but the field data are nonetheless suggestive. Philip Darlington, in the spirit of his pyramidal table of species-area figures for the Antilles, would have liked Komodoland too.

No time to loiter. We climb back into the boat. The tide is ebbing and we're not finished. Each of the four islands we've seen is an isolate, in Frank Preston's sense of that word. Now I want to look at a sample. So we sail north toward a place called Wae Wuul, a nature reserve in the backcountry of western Flores, where humans and komodos still share the landscape in a situation of uneasy intimacy and mutual jeopardy, as though they both were endangered species.

I I 2

THE RESERVE is formally known as Cagar Alam Wae Wuul. That translates to the Wae Wuul Nature Conservation Area. It's about eleven square miles in area, twice the size of the island of Padar, but Wae Wuul, unlike Padar, is not demarcated by a boundary of beach and ocean. It's demarcated by an amoeboid outline drawn on a map, specifying the idealized boundary of the reserve. Within that area (according to a recent census) live 129 komodos, give or take a few. Just outside the area, in the forests and savannas that surround it, live still more. There is no fence, no gates of entry and exit. The amoeboid outline is invisible on the ground. It doesn't isolate the Wae Wuul Nature Conservation Area from the greater landscape; it merely delineates a sample.

The region around Wae Wuul is wild country, with only a sparse human presence. It contains hamlets but no towns, trails but no roads. The reserve itself is spread across high valleys and forested slopes about two miles inland, inaccessible by car or by boat. This part of western Flores is a foot-travel region, and not one where hiking is viewed as recreation.

Sultan lands us at another small fishing village on the coast. From there, Nyoman and Johannes and I follow a trail up a small mountain, down again, through a stretch of forest, and finally out across a rolling meadow. The brush is thick, the trail winds through blind turns, and I feel the same nervous vigilance, as I walk, that I'd feel in grizzly bear country. After my scare at the Loh Sabita bluffs, I'm attuned to the reality that a komodo could lunge out of the shrubbery and bite off a person's butt.

Just after noon we reach the Wae Wuul ranger post, where Johannes's colleagues welcome us for water and chat. They assure me of the futility of going out to search for komodos at this time of day, during the afternoon heat, when sensible reptiles as well as sensible people will be resting in the shade. So we kill time. I ask several questions that I've been carrying in my head since I first heard about Wae Wuul. With hamlets on the periphery, do the komodos of the reserve represent a mortal hazard to humans? Is it true that someone was recently attacked?

"Ya, ya." Three years ago there was an incident, they tell me, near a hamlet called Mbuhung. A bad thing, yes, but the fellow survived. When I press for details, one ranger opens a dusty cabinet. From amid the clutter of teacups and binoculars and kerosene lanterns, he brings out a typewritten report.

I push it at Nyoman, who reads and interprets.

The report was composed by the local chief on the day after that latest mauling. It tells of a man named Don Lamu, who strolled out to move his water buffalo onto pasture and "got an attack from a komodo which caused a very serious pain which needed a very long cure." This occurred in September, near the end of the dry season. Don Lamu and a friend were fetching the buffalo from a wallow when they heard a dog barking. They went to look, assuming that the dog had cornered a wild pig. There was no pig. For a moment they were distracted and unwary. "Suddenly," says the report as translated, "a komodo appeared between the two people. One of them run away, but Don Lamu couldn't do anything because the distance between Don Lamu and komodo about fifteen centimeters." Although the fifteen-centimeters business sounds peculiar—too little distance, too finely measured—the essence is that this animal appeared from nowhere and struck at short range. Don Lamu tried to kick, but the komodo "fall upon back of Don Lamu's knee until it serious injured." He kicked with the other leg. "But the animal gave respond by its teeth so that his left leg was also torn by the komodo's teeth." With its second lunge, the komodo had seized hold. Here the report adds a sentence that hints at the same lonely intimacy between predator and victim that I heard in the story of Saugi's mother, back at Loh Sabita:

"Then Don Lamu sat down and held the komodo's mouth while he was shouting asking for help."

Don Lamu's friend returned with a chopping knife and whacked at the animal until its jaws opened. He moved Don Lamu some yards away. But the komodo came at them again. "Don Lamu's friend was angry and a fight happened," says the report. The unnamed friend plays a heroic role, no doubt partly because Don Lamu was in serious condition when the report was composed and the friend, as sole witness, supplied the narrative. With only his chopping knife, the friend stood off the komodo and killed it.

The chief's account ends with a modest but poignant plea. *Demikian hal ini kami sampaikan dihadapan Bapak untuk bersama memikirkanya.* Translated: "That's our report. We hoped that everyone would try to find out the way how to overcome the problem." The chief's unspoken meaning: For the love of Allah, help us rid ourselves of these flesh-eating beasts.

There is no Xerox machine at the Wae Wuul ranger post. The rangers sit patiently, smoking clove cigarettes and bantering with Johannes and Nyoman, while I copy the entire text of the report, in Indonesian, by rote, into my notebook. *Demikian hal ini kami sampaikan dihadapan Bapak untuk bersama memikirkanya.* Then the original is returned to its cabinet.

It's a resonant document, well worth the hike to Wae Wuul. Why? Because the problem alluded to by the chief, with such gentle anguish, is much bigger and more general than the matter of one species of giant lizard that eats dogs, goats, buffalo, deer, horses, men, women, and children in the forests and fields around this little reserve. All over the planet, humanity is at war against other species, against the wildness of wild landscape, against the redness of nature's tooth and claw. Humanity will win. The only point at issue is the final severity of the peace. It's not difficult to predict the future of wild landscape in western Flores. There will be major and irreversible transformations, probably within a few decades, as a result of increasing human population and increasing human needs and wants: forest replaced by pasture, pasture replaced by crops, hamlets and trails converted to towns and roads, habitat lost, normal predator-prey relationships disrupted, komodos killed off with chopping knives whenever they venture outside their reserve and make themselves lethally troublesome. Somewhere nearby, there may eventually be a classy hotel, devoted to ecotourism and accessible by air-conditioned van. Don Lamu, if he avoids further mishaps, could live to see it.

The Wae Wuul Nature Conservation Area, protected by Indonesian law, may not be directly affected by these transformations. Its boundary may hold. The towns and the crops and the roads, all the other forms of disruption, may encircle that amoeboid outline without crossing it. Within the reserve, *V. komodoensis* may survive, at least in a small population, at least for a while.

But as the surrounding landscape changes, Wae Wuul will cease to be part of a continuous ecosystem that is larger and richer than itself. It will no longer be a sample; it will become an isolate. As far as komodos are concerned, another tiny island.

လ

THE

COMING THING

လ

113

In 1967 Princeton University Press published a small, dense, eye-crossingly mathematical book that I've already mentioned, *The Theory of Island Biogeography*. It was the first in a series that Princeton intended to bring out under the collective rubric Monographs in Population Biology, and its premier position in that series hinted at the fact that this drab little volume concerned itself not just with islands and not just with biogeography.

Population biology, at that time, was still a new label for a new combination of scientific concerns and perspectives. It had subsumed a number of disciplines that, throughout earlier decades, earlier centuries, were generally treated as distinct: evolutionary biology, taxonomy, biogeography, ecology, the demography of plant and animal species, genetics. During the 1930s and 1940s, evolutionary biology (as descended from Darwin and Wallace) had at last become informed by and integrated with genetics (as descended from Gregor Mendel, of whose work Darwin and Wallace were ignorant), largely through the efforts of such men as Sewall Wright, R. A. Fisher, J. B. S. Haldane, and Theodosius Dobzhansky. Also during that era, taxonomy had been added to the synthesis, thanks much to Ernst Mayr and George Gaylord Simpson. By the 1950s, a great ecologist and teacher named G. Evelyn Hutchinson, born British but latterly fetched up at Yale, had started to bring a modestly mathematized style of ecology into conjunction with those other elements. Hutchinson's own research specialty was limnology, the study of lakes. It was a canny choice for an ecologist of his bent, in that lakes are neatly bounded ecosystems, relatively small and relatively impoverished, and therefore lend themselves well to relatively thorough ecological investigation. They share that with islands—not surprisingly, since a lake is the reverse image of an island. But the similarity between lacustrine and insular ecosystems isn't what puts G. Evelyn Hutchinson into this story. Among the graduate students who would go on to add luster to Hutchinson's reputation as a teacher—foremost among them, in fact—was a young man named Robert MacArthur, born in 1930.

After finishing a master's degree in mathematics, MacArthur had turned to ecology, arriving at Yale in 1953 for a doctoral program under Hutchinson's guidance. By 1955 he had published a paper in *Ecology*. By 1957 he had produced a provocative theoretical contribution that appeared in the *Proceedings of the National Academy of Sciences*. His dissertation work, a study of community structure and niche partitioning among different species of warbler, also yielded a paper for *Ecology*, which appeared in 1958 and became recognized as a minor classic. Young MacArthur was a hot prospect. He possessed the right combination of talents and ambitions to make a big impact at that particular time on that particular field. He was formidably bright, restless, almighty curious, innovative, and mathematically proficient. He loved the natural world, especially birds, but he had no interest in spending his life as a descriptive field naturalist. He cared about ideas and deep mechanisms, about order and explanation, not just about creatures and landscape. He was eager to change the very character of ecology.

MacArthur felt that the science of ecosystems should venture beyond description. It shouldn't limit itself to collecting and indexing facts. It should find broader patterns in the natural world, and from those patterns it should extract general principles. It should measure and count and perform calculational abstractions that would illuminate the essential amid the contingent. It should construct mathematical models that would function as usefully as a slide rule. It should be vigorous enough and bold enough to make predictions. It should offer theory. Olof Arrhenius, Henry Allan Gleason, and a few other ecological researchers had made modest efforts in that vein. So had Hutchinson himself. And Frank Preston had begun ringing the bell of canonical distribution just as MacArthur reached his scientific adulthood. But ecology as generally practiced was still a loose-jointed, descriptive, nonquantified, nontheoretical enterprise.

MacArthur began to teach. He continued to publish interesting papers. His reputation blossomed. In 1965, after several years at the University of Pennsylvania, he accepted a professorship at Princeton, his role there to include the overall editorship of the Monographs in Population Biology, which offered promise as a forum for the newfangled ecology that he favored. MacArthur would co-author the first monograph himself. By that time he had met and begun sharing ideas with another young scientist, Edward O. Wilson.

Wilson was a myrmecologist, a specialist on the biology of ants. He had grown up in the South, a nature-loving boy who collected insects and kept snakes, spending much of his time alone in the woods and swamps of Alabama and northern Florida. He had become interested in ants by the time he was nine, and soon after that he began doing precociously serious myrmecological fieldwork. Why ants? To some degree it was their inherent fascination, to some degree it was a choice urged by necessity. "I had use of only my left eye," Wilson explains in an autobiographical essay published in midcareer. The right eye "was mostly blinded by a traumatic cataract when I carelessly jerked a fish fin into it at the age of seven." With only monocular vision, he had found it hard to watch birds or mammals in the field. But the left eye was very acute, especially for fine detail at close range. "I am the last to spot a hawk sitting in a tree, but I can examine the hairs and contours of an insect's body without the aid of a magnifying glass." Many young boys collect insects; this one-eyed boy was doing real science.

At age thirteen Edward Wilson began a rigorous study of certain ant groups in the Mobile area and made his first publishable observation. Several years later, with adolescent earnestness, he decided that he should pick an entomological specialty. Based on the challenge and opportunity they offered, he chose flies (the order Diptera), in preference to ants. But it was 1945 then, and World War II had interrupted the supply of insect pins, which came mainly from Czechoslovakia. Since ants are best preserved in small vials of alcohol rather than mounted on pins, Wilson switched back to them—and has never regretted it. "In a sense, the ants gave me everything," he declares. His highest career aspiration at that age was to be a government entomologist, "to ride around in one of those green pickup trucks used by the U.S. Department of Agriculture's extension service" and help farmers cope with their crop pests. But he had a first-class scientific brain, as it turned out, which carried him from the University of Alabama to the University of Tennessee and then to Harvard. The particular attraction of Harvard, as put to him, was that its museum held the world's largest collection of ants. "Get a global view," an older colleague had told him; "don't sell yourself short with entirely local studies." When he met Robert MacArthur, in late 1959, Wilson had recently returned from a long stint of fieldwork in the tropics.

His skills and his interests were complementary to MacArthur's. Whereas MacArthur had shifted from mathematics to mathematical ecology, Wilson was a taxonomist and a zoogeographer. He had made extensive collections in New Guinea and Australia, on the island of New Caledonia, in the New Hebrides, on Fiji. He knew more about the Cerapachyinae and the Ponerinae subfamilies of ants in Melanesia than any man-jack on Earth—what they looked like, how they made their livings, which species lived on which islands. Like MacArthur, though, he was interested in more than the description of natural-history phenomena. As his head and his field notebooks filled with ant data, Wilson had begun to see patterns. For example: The number of ant species on an island tended to correlate closely with island size.

Wilson's brainload of purely descriptive data seems to have helped bring on an existential headache. In 1961, feeling mildly depressed and unsure about his professional direction, he took a sabbatical. During this getaway from Harvard he went to South America. He loved fieldwork as much as ever, and the rainforests of Surinam, to his great pleasure, were rife with previously unstudied ants. He also visited the islands of Trinidad and Tobago, just offshore from Venezuela. Trinidad is large, Tobago is small. On Trinidad he found more species of ant than on Tobago.

Pondering the descriptive sort of work he had been doing, Wilson yearned for something more. "It just didn't seem enough to continue enlarging the natural history and biogeography of ants," he confides in his short memoir. "The challenges were not commensurate to the forces then moving and shaking the biological sciences." His recent acquaintance with Robert MacArthur and some other young mathe-matical ecologists, and his awareness of G. Evelyn Hutchinson's work, had persuaded Wilson that "much of the whole range of popu-lation biology was ripe for synthesis and rapid advance in experimen-tal research; but this could only be accomplished with the aid of imaginative logical reasoning strengthened by mathematical models." He was a mediocre mathematician, by his account, with no training beyond algebra and statistics. So during his brooding summer in Trinidad and Tobago, besides collecting ants, he taught himself some calculus and probability theory out of textbooks. Returning to Har-vard, he enrolled in a mathematics class and sat conscientiously, as an associate professor, in a small desk among a crowd of undergrads. Eventually, after further remedial classes, he had enough math to feel comfortable. Meanwhile he began a collaboration that would take him beyond the descriptive.

"In 1962 Robert H. MacArthur and I, both in our early thirties, de-cided to try something new in biogeography," Wilson relates. "The discipline, which studies the distribution of plants and animals around the world, was ideal for theoretical research. Biogeography was intel-lectually important, replete with poorly organized information, un-derpopulated, and almost devoid of quantitative models. Its borders with ecology and genetics, specialties in which we also felt well prepared, were blank swaths across the map." Wilson told MacArthur that he thought biogeography could be made into a rigorous, analytical science.

There were striking regularities in the welter of data, Wilson said, that no one had explained. The species-area relationship, for instance. More particularly, the recurrent ratio to which Philip Darlington had called attention: With each tenfold decrease in area came roughly a twofold decrease in species diversity. Among the ant species of Asia and the Pacific islands Wilson had noticed another pattern. Newly evolved species seemed to originate on the large landmasses of Asia and Australia and to disperse adventurously from there out to the far-flung islands. As those dispersing species colonized small and remote places such as Fiji, they seemed to supplant the older native species that had gotten there before them. New species were continually arriving, old species were continually going extinct, and the net effect was . . . no gain or loss in number of ant species. It looked to Wilson like some sort of natural balance.

Yes, said MacArthur, an equilibrium.

114

"HE WAS MEDIUM tall and thin, with a handsomely angular face," as Wilson remembers MacArthur. This was written a decade after MacArthur's early death. "He met you with a level gaze supported by an ironic smile and widening of eyes. He spoke with a thin baritone voice in complete sentences and paragraphs, signaling his more important utterances by tilting his face slightly upward and swallowing." Although it may not have been a complete sentence when MacArthur first spoke the word "equilibrium" to Wilson, presumably he tilted up his face.

"He had a calm understated manner," Wilson continues, "which in intellectuals suggests tightly reined power. Because very few professional academics can keep their mouths shut long enough to be sure about anything, MacArthur's restraint gave his conversation an edge of finality he did not intend. In fact he was basically shy and reticent."

To that portrait, subjective and affectionate, Wilson adds a statement that stands as dead accurate: "By general agreement MacArthur was the most important ecologist of his generation."

The two of them brainstormed together during 1961 and 1962 over this notion of a biogeographical balance. They scrutinized Wilson's ant data from Melanesia. They looked at patterns of distribution among bird species in the Philippines, Indonesia, and New Guinea, which they extracted from published work by Ernst Mayr and others. They referred back to Darlington's enumeration of beetle and reptile species on the various Antillean islands, and they looked into the case of Krakatau, for which there were those historical records of recolonization by animal species after the cataclysmic purge. They read Frank Preston's recent paper on the canonical distribution of commonness and rarity, finding it essentially in agreement with their own views. They grew convinced that the species *lost* from an island, during a given span of time under ordinary circumstances, are roughly equal in number to the species *gained* by the same island over the same span of time. Unless the island itself is very recent in origin or has undergone a sudden disruption, the rates of losses and gains tend to cancel each other out. The result is a dynamic stability. The *number* of resident species remains steady while, with one species replacing another, the roster of *identities* changes continually.

Just how are species lost from an island? By local extinction. And how are they gained? Two ways: (1) by speciation, when a single old species splits into a pair of new species, and (2) by immigration, when a species arrives and becomes established. MacArthur and Wilson suspected that the second of those two, immigration, is vastly more frequent than the first. Speciation could be disregarded, then, and the equilibrium they envisioned could be expressed as a balance between immigration and extinction. It's worth repeating that extinction in this context refers usually to the local extinction of a population, not to the global extinction of a species. Mainlands or neighboring islands, having supplied the flow of transoceanic immigrants, supply still more as extinctions occur.

MacArthur and Wilson drew a simple, vivid graph that showed two sloping lines forming an X: the immigration line coming down, from left to right, and the extinction line going up. The downward slope indicated that immigration events tend to *decrease* in frequency as an island becomes crowded with species. The upward slope indicated the converse—that extinction events tend to *increase* in frequency with increasing crowdedness. Each of the lines was curved slightly, into a gentle concave sag, indicating subtle changes in the rate at which immigration decreases and extinction increases on an ever fuller island.

In midgraph, the lines crossed; their crossing point marked equilibrium. The number of species corresponding to that graph point is the island's normal complement of species, remaining roughly constant through time. At least that's what MacArthur and Wilson hypothesized.

They also created an intricate mathematical model—a long sequence of differential equations capable of accepting numerical data at one end and, like the canning line in a pickle factory, emitting a conveniently transmogrified product at the other end. The product of this canning line was predictions. The model foretold particulars of equilibration (how much elapsed time? how many species?) on a given island.

MacArthur and Wilson then co-authored a paper, which the journal *Evolution* published in 1963. Titled "An Equilibrium Theory of Insular Zoogeography," this paper contained the essence of their concept, lucidly presented, but for one reason or another it didn't have the immediate effect of turning the whole science of ecology sideways. That was still to come.

They weren't finished with the theory. They expanded the range of supporting data to include plant species as well as animals. They added some discussion of related topics—the process of colonization, the process through which an island community might resist colonization by new species, and also a bit about evolution on islands. Four years later, their expanded version became volume one of the Princeton monographs, published as *The Theory of Island Biogeography*. The slight change in title, from the article to the book, suggested a major increase in confidence and scope.

The new title even sounded a little presumptuous. *An* equilibrium theory of island biogeography had become *the* theory of island biogeography. It wasn't actually the only such theory; the scientific literature was already lightly peppered with theoretical notions concerning the biogeography of islands. But none of the others was so forceful, none was so detailed, and none ever proved so influential. The title's presumption was vindicated. This was the book that changed things.

115

To SOME people, few but vocal, Edward O. Wilson is a bugbear, infamous as the leading enunciator of a branch of science called sociobiology. To others, he is famous and much admired for the same reason. Yes, it's the same Edward O. Wilson who published *Sociobiology: The New Synthesis* in 1975, daring to suggest that human social behavior might be partly shaped by evolutionary genetics. That phase of his life and career is a story in itself (how he was accused of inventing pseudoscientific justification for a racist and sexist status quo; how he was denounced in print by leftist intellectuals and on the placards of angry young demonstrators; how he achieved the unwelcome distinction of having a pitcher of water dumped on his head, by one of those demonstrators, while addressing the American Association for the Advancement of Science), but it's not the story I'm telling. To still other people, more recently, Edward O. Wilson is a brilliant polymath and an elegant writer, a statesmanlike voice of warning about global losses of biological diversity, a wise elder to the conservation movement. And among his entomological colleagues, he is a towering authority on the taxonomy and behavior of ants.

Aware of all those Wilson personae, I'm interested in still another. To me he's the sole surviving member of the MacArthur-and-Wilson partnership, and therefore a maker of history. He's the one living human best qualified to explain how such an informal and seemingly peripheral field as island biogeography became so formalized and so central to the science of ecology. He's also, as I discover, a generous, mild, and unassuming man. He offers me three hours of intellectual hospitality from the middle of his busy workday as calmly as a good-hearted Methodist minister visiting the sick. "Please call me Ed," he says. "And I'll call you David, if I may."

His office, on an upper floor of Harvard's Museum of Comparative Zoology, is a large, cheerful room decorated in memorabilia and ants. Unlike the ants that come and go through my own office, his are in cages. The cages are neat plastic units arrayed on long lab tables. Some of them, no doubt the ones holding tropical species, are warmed gently by red electric bulbs. An oversized bronze sculpture of an ant, big as a lobster, sits on a pedestal bearing an award inscription. An old Georgia license plate, a nonsense memento of the sort that arrive in the mail as gifts, reads HIANTS. And the skull of a sabertooth cat (or a facsimile cast, anyway) rests on a table, offering a note of mammalian relief to the myrmecological theme.

Far up on the left wall hangs a row of imposing, black-framed photographs: five venerable men. Are they Ed Wilson's distinguished predecessors here at the Museum of Comparative Zoology? That's my guess; and I succeed in identifying one of those faces as William Morton Wheeler, a great American ant maven of the early twentieth century. The rest are strangers to me. Five dignified elders, they glower down on everything. Has the Harvard administration issued these photos as mandatory office decor, like the unctuous shot of the president in every post office? On the opposite wall, ironically contrapuntal in identical black frames, five ants glower back.

Far across the room, inconspicuous beside Wilson's small desk, hangs another photo: the young Robert MacArthur.

Before we get serious about islands, Wilson hosts me to lunch. From his office refrigerator he pulls out turkey sandwiches, bottles of lingonberry juice, and paper-wrapped pieces of baklava, all of which he seems to have shopped for at some local deli himself. He apologizes, this eminent man, this wonderful lunatic of politeness, for the food's not being fancy. But the turkey is fine and the lingonberry juice has come all the way from Finland. As we stuff our faces we talk ramblingly about the politics of scientific funding (especially ecology versus molecular biology, with ecology getting short shrift) and about the sociopolitical challenge of conserving biological diversity. Both those related subjects concern him deeply. But the Finnish juice gets us off on a digression about Scandinavian surnames, and he voices curiosity about mine.

It's Norwegian, I answer. The original version, before Ellis Island—so I've been told, anyway—translates roughly as "cow-herder" or "cow-man." I suppose that could be stretched to "cowboy," if a person liked the ring of the word, which I don't. Besides, I say, in Norway there's no swaggering ethos involving horses and dip tobacco and pointy boots. My prattle on the subject moves Wilson to confide that he has always chafed at the ordinariness of his own name. If it had to be WASP, he says, at least it could have been something more robust. Such as Stonebreaker, he muses. Curling his mouth to a little smile, he stares into space and repeats, Stonebreaker. I imagine it: *Dr. Edward O. Stonebreaker, the renowned ant expert and sociobiologist, announced today at Harvard that he would no longer use the name Wilson and that, furthermore, no Marxist yoyos would ever again dump water on* HIS *head.*

But he doesn't seem to be a man of grudges, or of ponderous self-image. The joke of this jokey name is one that he plays on himself. Yes, he decides, Stonebreaker would be his choice, because it so nicely suggests generation upon generation of big-muscled, stolid ancestors, some of them wearing manacles at the ankle. He chortles.

After the baklava, we carry coffee cups to a seminar room just up the hall from his office. Anticipating the focus of my questions, he has offered to show me some slides. What he means, I discover, is that he'll give me a private lecture on the origins and development of the MacArthur-and-Wilson theory. A projector sits ready. Wilson clicks up the first slide and then, triggered by that image into an outpouring of scientific and personal reminiscence, talks for almost an hour before clicking on to the next.

The slide shows two young men, dressed for tropical fieldwork, on a sun-scorched cay of white sand. One of the two, thin and angular, wears chinos and deck shoes and a red-and-white baseball cap with the brim pulled low, barely revealing a three-day beard.

"MacArthur always hated to be photographed," Wilson tells me.

116

IN THE LATE 1940s and the 1950s, Wilson says, G. Evelyn Hutchinson gathered around himself at Yale a small circle of extraordinary graduate students, to whom he was teaching evolutionary theory and "stressing over and over again that ecology needs an evolutionary twist." It was an unconventional approach back then, when ecology and evolution were still generally treated as separate and the integrated perspective of population biology had only begun to take hold. Hutchinson and these particular students were cutting a little deeper than others, says Wilson, in exploring how evolutionary adaptation played a role in the dynamics of complex communities. Anyway, one of the Hutchinsonians was a young fellow named Larry Slobodkin, who had just written a little ecology textbook expounding the quantitative model-building approach to evolutionary problems.

Wilson himself, up at Harvard, wasn't part of Hutchinson's circle. He had not yet met MacArthur or taken his South American sabbatical. But he knew Larry Slobodkin, and they got along well. Since their interests meshed, he and Slobodkin started planning a joint project.

Slobodkin had some of the mathematical skills that Wilson lacked, as well as a firm grounding in ecological theory. Wilson for his part was strong on myrmecology, on biogeography, and on speciation theory. "Slobodkin and I saw that we could put these two bodies of knowledge and these approaches together, and maybe do a book on population biology," Wilson recalls. They had even discussed it with a publisher. "Then Slobodkin hesitated. He said, 'We really need a third author'"—a specific fellow, a certain mathematician whom Slobodkin had in mind—"'because he's just so damn good. He's so brilliant and he's the coming thing in ecology and full of ideas. I want you to meet Robert MacArthur.'" During a conference in December of 1959, they did meet. "And he and I hit it off immediately," Wilson says.

I gaze at the young man in the red-and-white cap, now famous, now departed, whose image remains frozen on Wilson's screen.

MacArthur was just back from a postdoctoral year at Oxford. He struck Wilson as "a somewhat ethereal, delicate, slightly Anglicized American who clearly was thinking deeply about these subjects." Also, just as clearly, he was ambitious to do something exciting and drastic in population biology. Fine; so was Ed Wilson. They quickly became friends, started a dialogue through the mail, and traded ideas toward that proposed three-author book. Wilson even drafted a few chapters. But then for some reason the friendship between Slobodkin and MacArthur went sour—it was Slobodkin who took a dislike to MacArthur, by Ed Wilson's recollection, and MacArthur who felt wounded—so the book project fell apart. Wilson himself went off on that sabbatical and began his remedial studies in math. When he returned north, his attraction to quantitative methodology was stronger than ever, and he resumed the contact with MacArthur. They met when they could, at conferences, at the University of Pennsylvania when Wilson was invited down there to speak, at MacArthur's summer home in the small town of Marlboro, Vermont. They had "runaway conversations about biology—about the future of population biology and so on." They were young and energetic, dreaming the sort of wild plans that energetic young men often dream. But these two weren't just young and energetic. Though Ed Wilson doesn't say so, they were also disciplined, scientifically sophisticated, and extraordinarily smart.

They hankered to collaborate on some project or another. "And I felt that the main collaborative effort would be in biogeography, because that was my passion." Wilson was mesmerized by the biogeo-

graphic patterns that he had found among his own field data on tropical ants and had read about in the works of Darlington, Mayr, and
others. He was especially intrigued by species-area curves. "I became
increasingly interested in them because to me they represented, possibly, an equilibrium. They represented some kind of law." He and
MacArthur talked a good deal about what they saw as "the balance of
nature."

The old photo of MacArthur still shines before us. Without a good
fan in the projector, this precious slide would have long since suffered
meltdown.

"Okay," says Wilson, "now I'm coming up to the crunch." He has
told the story before, in some of his classes, but never to a smaller or
more interested audience. "Slobodkin drops out," he says. "Mac-
Arthur and I become thick as thieves. We're now enchanted by the
idea that something really important might be done with biogeography. So I keep saying, 'Biogeography, that's the field of the future.'
Balance of nature. Equilibrium and so on. We're looking at those
curves. Then in '62, the summer of '62, after talks in Pennsylvania
and Marlboro, at his summer house, I get through the mail a little
two- or three-page letter with the equilibrium model—the crossed
lines—from MacArthur. And he says, 'I think that's the way the whole
thing might be approached.' And I look at this thing. And I say, 'Yes,
that's it.' "

117

THE EQUILIBRIUM theory of island biogeography is not a piece of
conceptual art. It's a tool. MacArthur and Wilson developed it for two
reasons: to explain and to predict.

Although its mathematical details are egregiously complicated, its
essence is simple. Think of it as analogous to your digital watch. You
don't need to comprehend the circuitry on a silicon chip in order to
read the time. Probably you have even mastered the task (after a trial-
and-error struggle, if you're like me) of setting the little alarm tweeter
and using the stopwatch function. Meanwhile you've preserved your
happy ignorance of the chip. Yes? The deal with this equilibrium theory is similar. MacArthur and Wilson's 1963 paper, in which the theory was first presented, contains a moderate amount of arcane

mathematical circuitry. Their book, from 1967, contains an immoderate amount. You and I are gonna ignore it. We're interested in telling time and in using the tweeter, not in the how-to of electronic digital engineering.

Two patterns of real-world data served as the starting points for the theory, which was devised to account for them jointly.

First pattern: the species-area relationship. The ants that Ed Wilson knew so well, on islands of the western Pacific, showed a nicely regular version of that relationship. There were more ant species on the bigger islands, fewer ant species on the smaller islands. The carabid beetles and the reptiles and amphibians of the Antilles showed other neat versions of the species-area relationship, as Darlington had reported. The land birds of certain Indonesian islands showed still another. In each case, larger islands contained more species than smaller islands, and when the numbers of species were graphed against the sizes of the respective islands, the graph points arrayed themselves (with a little logarithmic jiggering) as a straight line. Despite its straightness, as you'll recall, the line was what scientists call a curve. The slope of each species-area curve could be expressed as a decimal number, which varied from one group of islands to another.

To restate the slope business in plain English: Among some island groups, the big islands contained *many* more species than the small islands, while in other island groups, the big islands contained only marginally more species than the small islands. Frank Preston had said as much in his 1962 paper, as MacArthur and Wilson acknowledged (though Preston had been talking mainly about sample areas, not isolates). The equation offered by MacArthur and Wilson for the species-area relationship was the same one that Preston had credited to Arrhenius: $S = cA^z$. The exponent z represented the slope of the curve—that is, the extremeness of the correlation between species number and island size within each group of islands. That exponent took one value for Antillean beetles, another value for Indonesian birds, still another for ants of the western Pacific. Every set of islands was different, though not *too* different; there did appear to be some inherent consistency in the degree to which big islands contained more species than small islands. The average slope value among the cases presented by MacArthur and Wilson was roughly 0.3—just another number to you and me, but a benchmark to anyone studying the interplay of species diversity and area.

The second pattern underlying MacArthur and Wilson's theory, like the first, had long been familiar to biogeographers: Remote islands support fewer species than less remote islands.

This pattern shows itself in several different ways. An island of some given size, if located near a mainland, generally supports more species than a similar-sized island far offshore. Also, a small island near a large island (for instance, one of the satellite islands around New Guinea) generally supports more species than a small island with no big neighbor. And finally, an island that is part of a small-island archipelago, though far from the nearest mainland or big island, generally supports more species than a solitary island equally far from major neighbors. In each case, the species richness correlates inversely with the degree of isolation.

MacArthur and Wilson's predecessors had commonly explained that pattern in historical terms. Remoteness was an impediment that only eons could overcome. Impoverishment together with remoteness suggested that an island's history had been relatively brief. Colonization of any new oceanic island took time—vast sweeps of time, if the island was remote—and remote islands were generally not ancient enough to have acquired great richness of species. So said the historical hypothesis.

But MacArthur and Wilson suspected that history wasn't the answer. Time was the limiting factor only during the earliest period on a new oceanic island, they believed, and most of the world's island ecosystems had long since come to maturity, to a state of balance, to equilibrium, with the number of species on each a reflection of ongoing processes, not historical circumstances. The ongoing processes that most shaped the balance, they argued, were immigration and extinction.

On page 21 of their book MacArthur and Wilson reprinted the sagging-X graph to illustrate this ahistorical theory. The immigration curve slopes downward from the left. The extinction curve slopes upward to the right. The decrease in immigration rate and the increase in extinction rate are graphed not against elapsed time but against the number of species present on a given island. As an island fills up with species, immigration declines and extinction increases, until they offset each other at an equilibrium level. At that level, the rate of continuing immigration is just canceled by the rate of continuing extinction, and there is no net gain or loss of species. The phenomenon of offsetting increase and decrease—the change of identities on the roster of species—is known as *turnover*. One species of butterfly arrives, another species of butterfly dies out, and in the aftermath the island has the same number of butterfly species as before. Equilibrium with turnover.

This sagging X of crossed curves became the most renowned and provocative graphic image in the ecological literature. It or its variants would eventually show up in countless books and papers. It would be fervently saluted by many ecologists and fervently denounced by quite a few others. If Preston's canonical curve was the bell heralding change, MacArthur and Wilson's equilibrium figure was the flag of revolt.

Their conceptual model does explain the two patterns of real-world data, and its explanatory power is what made it forceful. Small islands harbor fewer species than large islands—why?—because small islands receive fewer immigrants and suffer more extinctions. In MacArthur and Wilson's schema, that's known as the area effect. Remote islands harbor fewer species than near islands—why?—because remote islands receive fewer immigrants and suffer just as many extinctions. That's the distance effect. Area and distance combine their effects to regulate the balance between immigration and extinction. It all fits together ingeniously.

The sagging-X figure as it appeared in the book was just a generalized version of specific equilibrium graphs that can be drawn for specific islands. The curves will be steeper for some islands than for others, because of particular localized circumstances. As the steepness varies, the crossing point shifts downward or up (indicating a lower or higher turnover rate, at equilibrium) and leftward or right (indicating a lower or higher number of species, at equilibrium). When either curve is especially steep—reflecting the fact that immigration decreases especially sharply or extinction increases especially sharply—their crossing point shifts leftward, toward zero. The shift means that, at equilibrium, in this particular set of circumstances, there will be relatively few resident species.

In other words, high extinction and low immigration yield an impoverished ecosystem. To you and me it's just a dot in Cartesian space, but to an island it represents destiny.

118

WEEKS went by. MacArthur had mailed Wilson his little sketch of the crossed lines and Wilson had said eureka; but the full theory, with its supporting arguments and math, was not yet on paper. Then one day they were sitting together near the fireplace in MacArthur's living room, as Wilson recollects, with notes and graphs spread out before them on a coffee table. This was late 1962. They felt satisfied that their equilibrium model gave a good explanation for the two real-world patterns. But that wasn't enough. Both patterns—one reflecting the distance effect, one the area effect—were relatively uncomplicated, and some alternative theory might account for them equally well. MacArthur and Wilson needed further evidence. They needed to show an intricate match between what their theory said and what *was*.

MacArthur suggested a way. Using the mathematical machinery of their theory, they could calculate how quickly a newborn island should approach equilibrium, and how that rate of approach should correspond to the island's eventual rate of turnover. The numbers would be different for each different newborn island. They should pick one and try it. Having made the calculations, based on the actual area and remoteness of their chosen island, they could compare their theory-derived answers against empirical data. The theory would be either supported or contradicted. It was a good suggestion with one practical drawback: Newborn islands are rare. And not many have ever been scientifically studied during the approach to species equilibrium. So there wasn't much empirical data available. Lonely nubs of fresh lava, cooling like slag in midocean, don't often attract ecologists.

At this point Ed Wilson had an idea. Let's look at Krakatau, he said.

Wilson himself had never been there and didn't propose to go, but he knew something about Krakatau from the literature. He knew that the great eruption of 1883 had killed everything, leaving only a cauterized mound, and that the ash had barely settled before spiders and insects and fern spores began arriving to fill the biological vacuum. He knew that the cauterized mound was effectively a newborn island. He recognized that this mound (under its new name, Rakata) could offer a well-documented case of an island's approach to equilibrium.

It was as close to a theory-testing experiment as two eager young bio-geographers could get.

So MacArthur and Wilson began calculating. They focused on birds. Extrapolating from a species-area curve for the avifaunas of other Asian islands, they figured the number of species that Rakata should support at equilibrium. They figured how much time should have been necessary, after the sterilizing explosion, for Rakata to reach that equilibrium. They figured what the turnover rate should be at equilibrium. Baldly put, their projections were:

- equilibrium number: thirty species
- time to equilibrium: forty years
- turnover: one species per year

Then they went to the empirical data.

After the earliest biological expedition, led by Professor Treub back in 1886, Rakata had been visited and surveyed again a number of times, notably in 1908, 1921, and 1934. The data from those surveys had been summarized in a Dutch journal paper of 1948. Consulting it, MacArthur and Wilson found that Rakata in 1908 had supported just thirteen resident species of bird. Recolonization had barely begun. Between 1908 and 1921, they found, the number of bird species rose to twenty-seven. Then it leveled off. Between 1921 and 1934, there was no net change, with just twenty-seven species recorded again. That suggested a form of stability, dynamic or otherwise. Among the twenty-seven bird species identified in 1934, though, five of the old species were gone and five new species had replaced them. That was turnover. Equilibrium had arrived within roughly four decades, it seemed, and the equilibrium number was roughly thirty, as predicted. True, the turnover rate appeared to be lower than their projection, but not so much lower as to discourage MacArthur and Wilson.

"We were very excited," Wilson tells me. "We said, Good Lord, here was the first time that you could come up with a bare-ass model like that, of what an equilibrium would look like, and you actually have a case where you can measure equilibrium and turnover, and it seems to fit."

Three decades later, the fit of the Krakatau data to the MacArthur-and-Wilson model continues to be a point of dispute. Is it a stretch, is it a pinch, or is it a stunningly good match? Ecologists are still making expeditions to Rakata, still surveying the bird species, totaling the number, gauging the turnover; and they are still comparing their findings against the coffee-table calculations of MacArthur and Wilson. Why do those other ecologists bother?

They bother because the equilibrium theory of island biogeography has been so successful at capturing minds. For much of the past thirty years, it has defined an important framework of research and debate within the world of professional ecology.

And the Krakatau case contributed vividly to its success. Wilson's idea about looking at those data had paid off. "But that didn't make me satisfied," he adds. "I said, What we need is a whole system of Krakataus."

119

"THERE'S *nothing* more romantic than biogeography," Ed Wilson says with quiet passion, and in the spell of the moment, by God, I almost believe him.

The aura of romance has been embodied by scientists like Philip Darlington, who was one of Wilson's early role models at Harvard. Darlington, when he wasn't living the cloistered life of a beetle taxonomist and a curator, was "a tough character, a real macho field person," as Wilson recalls him. Phil Darlington would march in a straight line through the thick of a tropical forest—off the trails, up and down the mountains, following only his compass—in order to collect beetles. That was the way to get the real fauna, the deep-woods fauna, he advised Wilson. Another formidable predecessor was William Diller Matthew, a paleontologist based at the American Museum of Natural History, who studied the biogeography of extinct mammals and made field expeditions to places like Java, Mongolia, and Wyoming. Earlier still, before Matthew and Darlington, there were what Wilson describes as the countless heroic, Kiplingesque naturalists of the past. He doesn't mention Alfred Wallace by name and he doesn't need to, since we both know that no single human fits the "heroic, Kiplingesque" description more aptly.

Wilson's paean is leading us toward a point. Yes, biogeography was always romantic, physically challenging, enlivened by adventurous field trips to exotic locales. But intellectually it was a muddle. Amorphous, disorderly. It lacked quantitative rigor. It made no use of the experimental method. But in the early 1960s Wilson had begun to believe that maybe it could be transformed.

While working with MacArthur on their equilibrium theory, he became energized by that prospect: to recast biogeography as an experimental science. Keep the romance, dispose of the muddle. Apply some mathematical rigor. Endow it with a capacity, not just to describe and to explain, but to predict. The potential was enormous. One problem, though. He himself couldn't play a substantive role in any such transformation, Wilson knew, if he never left urban Massachusetts.

"So I said, I've got to get into the field."

He wasn't as young and unfettered as he'd been during his fieldwork in Melanesia. He had more responsibilities: a family, a professorship at Harvard. He couldn't just gallivant off to some tropical archipelago for months that would stretch into years. But he longed to create an important field experiment on the equilibrium aspect of biogeography. He knew it would be possible, he says now, "if I could just find the right place for it. I kept poring over maps of the United States, you know, places I could get to quickly. And there it was: southern Florida."

During the summer vacation of 1965 he took his family to Key West. He got hold of a fourteen-foot boat with an outboard motor. He began exploring the little universe of mangrove islands that speckled the shallow blue waters south of the Everglades.

There were thousands of these mangrove islands, more than any mapmaker had ever bothered to name. Florida Bay was full of them and, still farther south, the Key West area had its own share—tiny tufts of greenery punctuating the saltwater flats out beyond Boca Chica, beyond Sugarloaf Key, offshore from the Snipe Keys and Squirrel Key and Johnston Key, out between the Bob Allen Keys and the Calusa Keys, north of Crawl Key, north of Fiesta Key, all the way back to Key Largo. They bulged up dark against the bright sea horizon like bison grazing a frost-covered prairie. Wilson cruised his boat among them.

He saw that many consisted solely of vegetation, lonely clusters of mangrove standing knee-deep in the shallows. The red mangrove tree, *Rhizophora mangle*, is a water-loving and salt-tolerant species that thrives in such circumstances. It supports itself on a single main trunk and a constantly thickening network of prop roots, broadening outward, stabilizing its stance in the underwater muck, gradually claiming more area. A small mangrove island might comprise just a single tree, or it might include several red mangroves with a black mangrove taking security in their entanglement. The whole cluster might be hardly much bigger than a beach umbrella.

An island so minuscule, lacking dry ground, couldn't support any large-bodied terrestrial animals. It would be devoid of mammals and probably also of reptiles. But there would be tree-dwelling arthropods—mainly insects and spiders, with maybe a few centipedes, millipedes, isopods, scorpions, pseudoscorpions, and mites. In such a tiny and simplified ecosystem, the diversity would be modest and the population sizes would be small. A few dozen different species, a few thousand individuals. One determined scientist could collect and identify virtually every creature.

Wilson was determined.

He returned in 1966 with a young grad student named Dan Simberloff, another in his long series of mathematically proficient collaborators. Together, Wilson and Simberloff designed a project that would become the first experimental test of the equilibrium theory—and, for that matter, one of the first gambits in experimental biogeography of any sort. It would become famous for its logical elegance, for its results, and for its gonzo methods. The idea was to pick a few of the smaller islands and census them completely, identifying every species of resident arthropod; then to empty each island of its animal life, as Rakata had been emptied by the eruption; finally, to monitor what happened next.

Would recolonization occur? Yes, certainly. Wing power and wind would bring adventurous arthropods back, filling the zoological vacuum.

Would it happen quickly or slowly? Would the rate correlate with an island's remoteness? Would the number of species on each island eventually settle at an equilibrium? If it did, would the new equilibrium number be roughly the same as what Wilson and Simberloff had found originally? And would there be turnover? At what rate? Would it all conform to predictions from the theory? Those were the ultimate questions.

The immediate question was, How the hell are we going to empty an island?

They hired National Exterminators, of Miami. The men of National became intrigued with this challenge and rose to it. They erected scaffolding and towers above the muddy-bottomed shallows.

When the initial censusing was finished, they wrapped each island in a huge tent of plasticized nylon. (Although the avant-garde artist Christo later gained fame for wrapping parcels of landscape, including islands, his debt to Ed Wilson, Dan Simberloff, and National Exterminators has not generally been noted by art historians.) Then they filled the tent with Pestmaster Soil Fumigant #1, a pesticide gas consisting mainly of methyl bromide. Methyl bromide is lethal to arthropods, and under some circumstances it will also hurt red mangroves; but Wilson and Simberloff arranged for the fumigations to be done at moderate dosages, administered nocturnally when air temperatures were lower, and with those considerations the trees remained healthy. The gas treatment lasted a few hours. The insects and spiders fell dead, the tent came off, the gas blew away, and the island was suddenly empty of animal species—but otherwise unchanged.

"This is where Dan Simberloff carried the heavy part of the work," Wilson says. "Dealing with the company, being there, supervising most of the fumigation. And beginning the monitoring afterward."

Here in the seminar room, Wilson's slide carousel reawakens. The shot of the camera-shy young man blinks away and new images start to appear. Click: a mangrove island. Click: The island is wrapped with dark plastic. It looks like a giant tomato plant inside a Hefty bag on a night when the gardener expects frost. Click: closeup of a beetle. At least in the visual dimension, we have now moved beyond Robert MacArthur.

This Florida work entailed some peculiar difficulties, Wilson tells me, besides just the problem of fumigation. He and Simberloff got buzzed by helicopters. It was the Coast Guard, checking them out, "because there were a lot of anti-Castro Cubans in the area, and they were using these islands, we were told, as weapons caches." The Coast Guard patrols were alert for anyone out on the flats who didn't look like a fisherman. It's easy to imagine how Wilson and Simberloff, two guys from Harvard wearing field khakis, climbing around in the branches of a mangrove or up to their shins in the shallows, causing giant black tents to occlude entire islands, might have provoked some suspicion. The mud was another thing, says Wilson. It sucked at their ankles like bathtub caulk. They sank, they stuck. Simberloff got a brainstorm and made some large plywood paddles for himself and Wilson to wear like snowshoes. The paddles were drilled full of holes, supposedly to relieve the suction. "But you couldn't move in them," says Wilson. "I jokingly referred to them as simberloffs. I'd say, 'Our simberloffs aren't going to take us anywhere.' Dan didn't seem to care for that joke very much. But he was really giving it his best."

After the fumigations, Wilson traveled back and forth to Harvard but Simberloff stayed in the area for a year, monitoring the recolonization process. He worked beastly hard, according to Wilson. Because the new populations were so small, and because the point was to study living ecosystems in a state of renewal, Simberloff and Wilson couldn't simply collect arthropods and carry them away for identification. They had to make their identifications by sight or take photos and then release each individual arthropod unharmed. What with the tree-climbing and the mud and the climate and the identification problems, the project demanded an acrobatic entomological taxonomist who was immune to sunburn and despair, proficient with a camera, and blessed with big feet. History doesn't record Dan Simberloff's shoe size, but he coped. During the course of the experiment, Wilson says, Simberloff transformed himself from an urban-raised mathematician into a very good field biologist.

And the results were gratifying. "To my absolute delight, and to Dan's delight, because this was going to be his thesis," Wilson says, "we watched the numbers of species rise to what was obviously equilibrium within about a year." That was true on three of the islands, anyway. Each of those three was recolonized quickly. On each, the species diversity rose to a temporary peak, then settled back down and stabilized at a lower level. The stable levels were nearly identical to what each island had supported before fumigation. Here was strong confirmation for the idea of equilibrium.

On a fourth island, meanwhile, recolonization happened more slowly. That island was the most remote. It eventually came to an equilibrium of its own, but its equilibrium number was lower than on the less remote islands. So the distance effect was confirmed too.

After two years, the four islands were censused again. Each island still held about the same number of species as it had after the first year—but some new species had been added, some old species had disappeared. Immigration was continuing, offset by extinction: turnover.

It was a nice bit of science, and it brought Simberloff his doctorate. He and Wilson together published three papers on the project in *Ecology*, for which they received an award from the Ecological Society of America. The work became quietly famous among other ecologists and biogeographers. It lent empirical validation to the equilibrium theory at a time when that validation was crucial. The mangrove-experiment papers—coded in dry scientific format as Wilson and Simberloff (1969), Simberloff and Wilson (1969), and Simberloff and Wilson (1970)—began turning up among the source citations of other scientists' journal articles, along with MacArthur and Wilson (1967). The theory gained credence from its applicability to the mangroves, and its adherents began to see a great breadth of other possible applications, most notably in the realm of conservation planning. This all seems ironic, in retrospect, given the role that Dan Simberloff would later play—as preeminent critic of the application of the equilibrium theory to conservation problems.

120

AFTER THE publication of *The Theory of Island Biogeography*, in 1967, Robert MacArthur had five years to live. He continued teaching at Princeton. He wrote fifteen more papers. He helped found the journal *Theoretical Population Biology*. He played his editorial role in the Princeton monograph series. In 1968 he went down to the Keys and visited Wilson, who had another sabbatical and was spending it there. Wilson took him out to see some of the mangrove islands, in which Dan Simberloff was still climbing around. They talked excitedly, MacArthur and Wilson, about island biogeography as a quantitative paradigm and where it might possibly lead. MacArthur was always a great talker, and conversation with his students and his colleagues was one of the chief ways he exerted his magic influence. Sometimes the conversation involved scribbled equations, groping efforts to put nascent ideas into the crisp language of mathematics. Not many ecologists could match his mathematical sophistication. But mathematics for MacArthur was only a means toward the larger goal: understanding how evolutionary processes and ecological tensions combine to shape biological communities.

One theme underlay most of his work. This theme—it seems almost a truism now, but MacArthur himself considered it worth stating—was the search for patterns. He emphasized patterns and equilibria and ongoing processes, while de-emphasizing the sort of one-time, contingent events that figure in historical explanations.

Where lies the distinction between those two types of explanation, the process-oriented and the historical? A historian pays special attention to the differences between phenomena, because they shed light on historical contingency. "He may ask why the New World tropics have toucans and hummingbirds," MacArthur wrote, "and parts of the Old World have hornbills and sunbirds." The hornbills of Africa and Asia are large-bodied, omnivorous birds with huge beaks, allowing them to fill roughly the same ecological niches as the toucans of tropical America; likewise the sunbirds of Africa and Asia are small-bodied, bright-colored nectar drinkers, filling roughly the same niches as American hummingbirds. The history-minded biogeographer wonders why hummingbirds, not sunbirds, have occupied the suitable niches on a given continent. MacArthur himself was more interested in the similarities among phenomena, because similarities reveal the workings of regular processes. He was more inclined to wonder why hummingbirds and sunbirds, despite their different ancestries and their independent histories in two different regions of the planet, are so similar. I've already quoted MacArthur's statement that to do science "is to search for repeated patterns, not simply to accumulate facts," and the patterns that especially concerned him were the patterns of biogeography.

The geographical distribution of animal and plant species had utterly captured his interest. The subject was full of deeply interesting questions. Why did such-and-such a community contain *these* species but not *those*? Why did such-and-such species live *there* but not *here*? Answers existed, MacArthur knew, and those answers were important, he believed. Of course the best of the earlier biogeographers, Darwin and Wallace, had also searched for repeated patterns and sought answers for the deeply interesting questions those patterns raised; MacArthur was merely continuing their tradition while inventing new methods. In his early career he had been a mathematician and later he labeled himself an ecologist, but during the final years, according to Wilson, he preferred to think of himself as a biogeographer. For most of the twentieth century, biogeography had been descriptive, not theoretical; it hadn't maintained the standard of profound provocation that Wallace and Darwin had set. Robert MacArthur, like his distinguished Victorian predecessors, was a biogeographer with a hungry ambition toward theory. One of his last field projects was a study of niche breadth and total abundance among the birds of Puercos Island, off the south coast of Panama. The birds of Puercos were not surpassingly significant in themselves. But they embodied patterns, and maybe the patterns were telling.

Sometime in 1972, MacArthur's illness was diagnosed. The romance ended.

He had renal cancer. It progressed quickly. He went up to the house in Vermont and, working against time, with no access to libraries, produced a short volume titled *Geographical Ecology: Patterns in the Distribution of Species*, summarizing his own final view of what mattered most in ecological science. He gave special thanks in the preface to Ed Wilson, who "showed me how interesting biogeography could be and wrote the lion's share of a joint book on islands."

By late autumn, MacArthur was home again in Princeton. Although he and Wilson had pursued different projects since their collaboration, the friendship remained strong. Ed Wilson tells me of their last conversation. He had heard that MacArthur was low, so he telephoned. "I knew he was dying. I didn't realize he was within hours of passing on."

"What did you talk about?" I ask.

It was just the same sort of chat as he'd had with MacArthur many times. They discussed ecology. The future prospects, the unanswered questions, the personalities. "We talked about recent developments in the field, and we gossiped about who had his head screwed on right," Wilson says. MacArthur barely mentioned his own illness, ignoring it as though he had a hundred years to live, and he and Wilson rattled on about other matters. The call lasted half an hour. Wilson made notes immediately afterward, thinking not so much about deathbed wisdom from a historic personage as about his own later sentimental interest. He stuffed the notes into a file, he says, and hasn't looked at the file in two decades. He seems to prefer trusting his memory. Memory knows things that notes could never remember.

About those notes: "I took them down as to what we said, and as to what we thought of certain people, and this and that. And that was it. I mean, there was no . . . We didn't inject sentiment." What he seems to mean is that true emotion, of which there was plenty, is often wordless. "Or say farewells. I didn't really expect that it would be that quick, anyway. That was it."

The death of MacArthur as perceived by Wilson—that *can't* be it, I think. There's undoubtedly more. There's surely some little human detail, incidental, piercingly vivid, so poignant that even memory chooses not to remember it. I want to say: Let's look at the notes. Let's see what it is you've forgotten. But of course I don't.

By now I've cored the heart out of Ed Wilson's workday, and it's time to let him escape back into the present. I've caused distraction enough, not least by reminding him of the note file. The file is still around somewhere, presumably in the ant-filled office, where the photo of Robert MacArthur hangs like a private icon. "Someday I'll pull that out and read it," Wilson says, "and may have a different perspective. I might even do that. I think I will do that. I've been meaning to pull it out for some time now," he says.

"It wouldn't hurt to read it," he says, though neither of us is entirely convinced.

I 2 I

MacArthur was dead but the theory lived. It grew steadily more influential among the professional community of ecologists and population biologists. You never caught wind of this at the time, neither did I, no reason we would have, but journals like *Ecology* and *Evolution* and the *American Naturalist* were full of it. The science of ecology was undergoing a revolution, some people said. Invigorated by the new techniques of mathematical modeling, shaken out of its conceptual timidity, liberated from its plodding obsession with merely cataloguing the particulars of natural history, ecology was in upheaval. It was finally becoming a theoretical, predictive science. The preeminent revolutionist had been Robert MacArthur, according to this view, and his most resounding battle whoop was *The Theory of Island Biogeography*. Not every ecologist agreed that the book was ingenious, persuasive, and useful—but those who didn't agree still found MacArthur and Wilson's equilibrium theory hard to ignore. It was the coming thing.

One way to measure its growing influence is from the record of its citation by other scientists. A citation record, like a set of career stats in *The Baseball Encyclopedia*, is a sequence of crabbed, coded entries into which can be read a whole vista of failure or success. Each journal paper in ecology (as in most other branches of science) concludes with a selective bibliographical list, usually labeled "references" or "literature cited." Cited works are any that have been alluded to by author and date in the text of the paper, the implication being that these were especially helpful—or anyway inspirational, or maybe just provocative because of their adamantine dumbness—to the author citing them. Such citations serve to unify and focus the scientific discourse. They specify just which ideas and which sets of data stand as relevant background to the ideas and data in a piece of new work. The sheer frequency with which a scientific book or paper is cited reflects the breadth of its influence. Citation records are therefore a good way of keeping score. And scientists do keep score.

Like those baseball statistics, citation records have their own encyclopedic repository. It's the *Science Citation Index*, a series of big, boring volumes, supplemented annually. Every instance of citation for virtually every scientific book or article published within recent decades is encoded in one of those volumes, which stand collectively ready to serve as a scholarly resource (or as one wall of a hurricane bunker) at your local university library. So from the *Science Citation Index* we can track the influence of MacArthur and Wilson's theory.

First there's the short version, "An Equilibrium Theory of Insular Zoogeography," published as a journal paper in 1963. It attracted little notice. In 1964, only two other scientific papers acknowledged it. Just six citations the following year, just seven the year after. The paper showed no sign of becoming a runaway hit.

In 1967 came the book. In 1968, while the world pondered student revolutions not scientific ones, and Denny McLain won thirty-one games for the Detroit Tigers, *The Theory of Island Biogeography* collected a mere five citations.

But then its spell began to take hold. The numbers climbed, plateaued, climbed again. In 1969, the book was cited more than two dozen times. Most ecological publications don't ever receive such attention. In 1970, twenty-nine citations. It drew around thirty in each of the next few years, then in 1974 it surged up to seventy-nine. Word was out. Ecologists had started muttering to one another about insular equilibrium, immigration versus extinction, the area effect, the distance effect, turnover. In 1976, ninety-four citations. I know the exact number because, peering through the low rim of my glasses at maddeningly tiny print, I have counted them. In 1977, more than a hundred. Still rising in 1978. These were the glory years for MacArthur and Wilson's model. Island biogeography was no longer a subject on the fringe, interesting chiefly to Kiplingesque beetle taxonomists. It had become one of the central paradigms of ecology. In 1979, the book was cited by other scientists 153 times, an exceptional total. Its influence still hadn't peaked.

In 1982, Willie Wilson of the Kansas City Royals (no relation to Ed Wilson of Harvard) batted .332. He led the league. That same year Reggie Jackson, playing for the Angels, hit thirty-nine home runs. He was flaunting his escape from the Yankees. With the Oakland A's, Rickey Henderson stole 130 bases in 1982, more even than Lou Brock at the height of his powers. Also that year, one decade after Robert MacArthur's death, *The Theory of Island Biogeography* was cited 161 times.

122

WHY DID THE book have such impact on ecology and population biology? Not because islands were so important, but because *The Theory of Island Biogeography* brought the island biogeography paradigm to the mainlands.

MacArthur and Wilson had made that point on their first page. They quoted Charles Darwin's early hunch that the zoology of archipelagos "will be well worth examination." They noted that patterns of species distribution on islands had played a major role, from Darwin's time onward, in the development of evolutionary theory. They added an important observation:

> Insularity is moreover a universal feature of biogeography. Many of the principles graphically displayed in the Galápagos Islands and other remote archipelagos apply in lesser or greater degree to all natural habitats.

They meant that literal islands, surrounded by water, are only one sort of insular situation. Also to be considered are virtual islands, surrounded by other kinds of barrier.

"Consider, for example, the insular nature of streams, caves, gallery forest, tide pools, taiga as it breaks up in tundra, and tundra as it breaks up in taiga," wrote MacArthur and Wilson. Taiga is subarctic conifer forest. Tundra is treeless plain. In the far north, they dapple into each other along the border zone, forming paisley and polka-dot patterns. A tree-dwelling species of animal that inhabits a small dot of taiga surrounded by tundra is effectively insularized. Consider also lakes, which are insular for the fish and amphibians that inhabit them.

Consider mountaintops, which are cooler and wetter than the surrounding valleys, and which often support utterly different plants and animals. The equilibrium theory was directed at all these situations.

The new mode of thought began to show itself promptly. In 1968 an imaginative ecologist named Daniel H. Janzen published a short piece in the *American Naturalist* titled "Host Plants as Islands in Evolutionary and Contemporary Time." For an herbivorous insect, he noted, each individual plant represents an island of habitat. Janzen had read and absorbed MacArthur and Wilson more quickly than most of his colleagues.

In 1970 David C. Culver published a paper subtitled "Caves as Islands." Culver brought equilibrium theory into his analysis of the biotas of certain West Virginia caves. Also in 1970 François Vuilleumier published a study of bird species on "páramo islands" of the northern Andes. *Páramo* is a type of meadow-and-scrub vegetation that occurs at around 10,000 feet on some peaks of the Andean chain, isolated there between the snow line and the tree line. Since páramo constitutes a distinct botanical community, with each patch supporting bird species distinct from the birds of the lower-elevation forest, Vuilleumier chose to analyze his páramo data in insular terms: area effect, distance effect, slope of the species-area curve. The slope of that curve, he reported, fell nicely within the narrow range of slope values that MacArthur and Wilson had described in their own collection of examples. About the same time, S. David Webb produced a paleontological study that also paid its respects to MacArthur and Wilson. Webb considered North America itself as a single vast island (fully isolated when the Bering and Panama land bridges were submerged) and found that, over the past ten million years, the diversity of North American land mammals had held roughly at equilibrium, with turnover.

Webb's study, and Culver's, and Vuilleumier's, and Janzen's, served as quiet indicators that the new mode of thought was taking hold. Another study in the same vein attracted wider attention, partly on the strength of contrariety. This one was "Mammals on Mountaintops: Nonequilibrium Insular Biogeography," published by James H. Brown in 1971. Brown announced, politely but firmly, that his research results did *not* fit the theory.

Brown's "islands" were forested mountaintops poking up out of the Great Basin desert. The Great Basin is generally dry, low, austere, and flat, an ocean of sagebrush spread across that huge region of the

American West bordered roughly by Las Vegas, Mount Whitney, Reno, Boise, Pocatello, and Salt Lake City. Protruding above the sage-purple plains are some forest-topped ranges of mountains. At their higher elevations, above 7,500 feet, these mountains get cool temperatures and enough rainfall to support piñon and juniper woodlands. They also support certain species of small mammals that can't live on the plains below. The northern water shrew (*Sorex palustris*), for instance, doesn't inhabit the lowlands of central Nevada; but it does survive as an isolated population in the moist heights of the Toiyabe range. The pika (*Ochotona princeps*) needs a subalpine climate, nutritious highland grasses for food, and slopes of talus for shelter; it's absent from the sage flats, where it can't find those essentials, but it does exist in the Ruby Mountains southeast of Elko. Besides the northern water shrew and the pika, Brown found the ermine, the Uinta chipmunk, the yellowbelly marmot, the bushytail woodrat, and nine other species of small mammals on mountaintops within the Great Basin. These were all boreal species, normally native to cold forests of the north.

Brown outlined seventeen mountaintop islands, most of them in Nevada, each one rising separately from that vast swale of sage between the Sierra Nevada in California and the Rockies in Utah. The Sierra and the Rockies, offering large zones of high-altitude habitat and source populations for the boreal mammals, represent mainlands relative to the insularized mountaintops. For each of his islands, Brown gathered the familiar sorts of data: area, distance from mainland, list of species in residence. Like the other researchers, he made charts. He drew species-area curves. He saw patterns, and from the patterns he deduced meaning. But the patterns and the meaning in this case were different.

These mountaintop mammals, Brown concluded, were not descended from immigrants who had crossed the sage. Instead they were relicts, left behind when their habitat dried up around them.

Their ancestors had arrived sometime in the Pleistocene, maybe fourteen thousand years ago, maybe earlier, during one of those glacial episodes when cooler and wetter climatic conditions had allowed piñon-and-juniper woodland to spread across lower elevations. The low-altitude woodland had formed a network of habitat linking the solitary mountaintops to the Sierra and the Rockies. While the cool conditions endured, the small boreal mammals enjoyed broad distribution throughout that habitat. Then the last ice age ended and the climate changed. The network dissolved into fragments. The boreal mammals were extirpated from the lowlands, but they survived (at least for a while) in their highland enclaves. The pika and the bushytail woodrat and the northern water shrew found themselves stranded—just as the brown bandicoot and the wombat had been stranded, at about the same time, on those land-bridge islands between mainland Australia and Tasmania.

The area effect showed strongly in Brown's data. The larger islands of mountaintop habitat supported distinctly more species than the smaller islands. So Brown's species-area curve was fairly neat. But it seemed unusually steep; the slope of the curve (as derived from that familiar species-area equation) was above the upper end of the range that MacArthur and Wilson had described. This meant that the species-area correlation for Brown's mountaintop islands was not just consistent, it was severe. A big mountaintop held not only more north-country mammal species than a small mountaintop, it held *far* more.

The distance effect, on the other hand, was nonexistent. Proximity of each island to either the Rockies or the Sierra, the habitat mainlands, signified zilch. There was no inverse correlation between distance and species number, as the equilibrium theory said there should be. The Stansbury Mountains, rising just forty miles west of the Rockies (in Utah), held only three species. The Panamints, just fifty miles east of the Sierra (in southeastern California), held only one. Some of the more remote islands, of about the same size, contained more species than those less remote islands. So distance was irrelevant to diversity. What did that mean?

It meant that immigration wasn't contributing to the species richness of *any* of the islands. The little mammals weren't achieving new colonizations across the Great Basin—not occasionally, not rarely, not ever. The pikas and water shrews and chipmunks and woodrats weren't so vagile and adventurous as reptiles and birds often are. With their high metabolic requirements, their tiny footsteps, their vulnerability to stress and predation, they had no chance in hell of crossing a forty-mile stretch of sagebrush. The barriers to dispersal and establishment were unbreachable. Brown said it clearly in his summary: "Apparently the present rate of immigration of boreal mammals to isolated mountains is effectively zero."

MacArthur and Wilson had tried to free ecology from historical explanations, but this was a case in which history couldn't be ignored.

The distribution patterns represented artifacts of historical circumstances rather than manifestations of timeless processes. Colonization
had happened during the Pleistocene—historical fact. Then the climate changed, the habitat split into fragments, and the mountaintop
populations were trapped—historical fact. Immigration came to a
halt—historical fact. During the subsequent ten millennia, Brown
guessed, it had been solely a matter of survival (for some populations)
and (for others) extinction.

Without immigration, there can be no immigration-extinction
equilibrium. That's why Brown's title announced a "nonequilibrium"
case of island biogeography. He was not so much refuting the
MacArthur-and-Wilson model as he was using it to highlight an alternative sort of situation.

Instead of turnover, Brown found only extinction. Slowly but irre-
versibly, one species after another had been lost from each island, and
those losses weren't offset by gains. When the Stansbury Mountains
lost their last ermine, no Uinta chipmunks arrived to compensate.
When the Panamints lost their last yellowbelly marmot, no incoming
pikas made up the difference. The species number for each island
didn't hover at equilibrium; the species number only went down. Insularization, for the Great Basin mountaintop communities, entailed
an inexorable decline in diversity.

Does this sound ominous? Does it sound familiar? The same phenomenon would eventually be known by various labels, one of which
is ecosystem decay.

123

BROWN'S PAPER IN 1971 was an early hint of what would soon become an important trend: using the equilibrium theory to illuminate
matters of habitat fragmentation and species extinction on the continents.

MacArthur and Wilson had presaged that trend with a comment in
their first chapter:

> The same principles apply, and will apply to an accelerating
> extent in the future, to formerly continuous natural habitats
> now being broken up by the encroachment of civilization.

The breaking-up process, they added, was "graphically illustrated by Curtis's maps of the changing woodland of Wisconsin," which maps they reprinted in the book. The Curtis maps gave visual force to MacArthur and Wilson's point that this theory of islands was not just a theory of islands.

John T. Curtis, a professor of botany at the University of Wisconsin, had studied the progressive destruction of native forests and grasslands during the decades since white settlers had invaded the Midwest. His work appeared in 1956 as one chapter of a volume titled *Man's Role in Changing the Face of the Earth*. The book's reach was global, but Curtis himself had focused on a single small unit of land: Cadiz Township of Green County, Wisconsin. It was a square tract measuring six miles on a side. Curtis had depicted its botanical transformation with four dated maps: one each for 1831, 1882, 1902, and 1950.

Back in 1831, before settlers began clearing it, the area had been covered with deciduous forest dominated by basswood, slippery elm, sugar maple, oak, and hickory. What would later be Cadiz Township was in those years a hardwood wilderness that Daniel Boone or John Muir could have loved. By 1882 it had become a patchwork of rectilinear woodlots within a larger matrix of cropland. By 1902, as mapped by Curtis from historical records, the forest patches had grown smaller and fewer. By 1950 they were still further reduced. These last parcels of woodlot in mid-twentieth-century Cadiz Township resembled a sprinkling of cracked pepper on a bare china plate. One sneeze, it seemed, and they'd vanish.

Curtis's maps were arresting. Like time-lapse photography from high overhead, they revealed the relentless diminishment whereby a mainland expanse was chopped into largish islands, then nibbled down into tiny ones. The remnants were too dinky to support much in the way of large-bodied, forest-dependent animals. The maps showed how human enterprise and human population growth had reduced one square of wild landscape to dysfunctional bits. And of course Cadiz Township of Green County, Wisconsin, was only a small model of the world.

124

THE SAME SORT of pattern that Curtis had documented in Wisconsin was envisioned on a broader scope by Jared Diamond. One place he saw it happening was New Guinea, and that caused him special concern.

Diamond, our man of the elephant-eating komodos and the bird-extinction tally, is a dauntless field ecologist who began making ornithological expeditions to New Guinea in 1964. He was just out of graduate school with a doctorate in physiology. The prospect of settling into a one-track career as a lab researcher on membrane physiology, his professional specialty, seemed too narrow. He had other aspirations and talents. Since boyhood, he had been a serious birdwatcher. He was blessed with a strong constitution and a fine ear for birdsong, the two requisites for doing ornithology in the tropics. Just by hiking and listening, matching the bird calls he heard against the aural field guide in his brain, he was capable of surveying the avifauna of a forest—even, he discovered, a forest so dense and difficult as those of New Guinea. During that first visit, the island captured his lifelong devotion.

"New Guinea is my intellectual roots," Diamond says. He feels as though he was born there. "Half of my emotional life goes on in New Guinea. Cut off from New Guinea, I would feel like Rachmaninoff cut off from Russia." Diamond came from a musical family, which might go some way to explain the acuity of his birding ear.

By the early 1970s he had become alarmed over the rate at which humans were destroying tropical rainforest. It was occurring not just in his beloved New Guinea but also in the Amazon, Malaysia, Madagascar, Central America, West Africa, the Philippines, and elsewhere. It took two basic forms, very different in socioeconomic character but similar in result: large-scale timber extraction by corporate entities, and small-scale slash-and-burn agriculture by hungry people. Diamond himself had seen it in both forms.

At the end of an otherwise arcane paper on equilibrium theory as applied to bird diversity, published in 1972, he mentioned the problem. Throughout the tropics, he wrote, rainforest "is being destroyed at a rate such that little will be left in a few decades. Since many rainforest species cannot persist in other habitats, the destruction of the rainforests would destroy many of the earth's species and would permanently alter the course of evolution." He wasn't the first biologist to voice that concern, but Diamond added a gloomy insight, extrapolated from MacArthur and Wilson's work:

The governments of some tropical countries, including New Guinea, are attempting to set aside some rainforest tracts now for conservation purposes. If these plans succeed, the rainforests, instead of disappearing completely, will be broken into "islands" surrounded by a "sea" of open country in which forest species cannot live.

The likely result of establishing those refuges, he warned, was the same as occurs on peninsular bits of landscape when they become isolated as land-bridge islands: Rare species suffer extinction and overall species diversity falls. It happened on Flinders Island and King Island in the Bass Strait, and on Brown's mountaintop islands in the Great Basin. It happened on Barro Colorado in Panama. It could happen, Diamond argued, to any patch of ecosystem that became insularized—for instance, to a nature reserve or a national park.

Although Diamond didn't cite him, Frank Preston had made the same point a decade earlier in his discussion of samples versus isolates. To refresh your memory: "If what we have said is correct, it is not possible to preserve in a State or National Park, a complete replica on a small scale of the fauna and flora of a much larger area."

Diamond's term for the phenomenon was *relaxation to equilibrium.* The sort of equilibrium to which he alluded, of course, was MacArthur and Wilson's, and the relaxation was actually a loss of species, as the species number came to equilibrium again at a new, lower level commensurate with the new, smaller area of the insular patch. It was another way of saying ecosystem decay. Diamond's phrase might be mistaken to imply stoical indifference—*relaxation* does sound languid and soothing—but indifferent is one thing that he wasn't.

125

IN 1974, the upheaval initiated by MacArthur and Wilson entered its next phase. The equilibrium theory by then had been published, widely noticed, and empirically tested. It had been projected onto a variety of natural situations—Janzen's plant islands, Culver's cave islands, Vuilleumier's páramo islands—and had helped to illuminate them. Intellectually, it was a success. The next phase was more practical: applied biogeography. Could the theory be used for problem solving in the real world?

A small handful of scientists, speaking up individually, suggested that it could. In their view, it was pertinent to the fate of natural landscape and wild creatures all over the planet. At a time when humanity was cutting forests and plowing savannas at a rapid pace, when habitat everywhere was becoming fragmented and insularized, the equilibrium theory embodied minatory truths. It was not just an interesting set of ideas—it was goddamned important. If heeded and applied, it might help save species from extinction. Conspicuous amid this chorus of opinion was Dan Simberloff.

Simberloff declared in print that the work of MacArthur and Wilson, along with the obscure papers by Frank Preston, had "revolutionized" biogeography with the suggestion of dynamic equilibrium. "Island biogeography has changed in a decade," Simberloff wrote, "from an idiographic discipline with few organizing principles to a nomothetic science with predictive general laws." He was a bright young man with a recondite vocabulary. "Idiographic" to "nomothetic": He meant that MacArthur and Wilson had succeeded just as they had hoped to, transforming biogeography from a descriptive endeavor into one that could articulate some of nature's governing rules. Yes, the theory brought fresh insight to the ecology of islands; but still more significant, Simberloff added, was that it also addressed insularized habitats on the mainlands. "We can therefore use island biogeographic theory to further our understanding of a variety of evolutionary and ecological phenomena and even to aid in the preservation of the earth's biotic diversity in the face of man's ecological despoliation."

Simberloff's statements appeared in 1974. Jared Diamond was saying much the same thing. That agreement, as you'll see, was a notable and fleeting convergence.

The following year Diamond published a journal article titled "The Island Dilemma: Lessons of Modern Biogeographic Studies for the Design of Natural Reserves." It would become one of his best-known papers. Because it represented a culmination in the development of these ideas, and because it triggered such vehement reaction, "The Island Dilemma" is worth looking at in detail.

Like Simberloff, Diamond believed that MacArthur and Wilson had touched off a "scientific revolution." One aspect of the revolution was a heightened awareness that insularity can occur under natural conditions on the mainlands: a mountaintop, a lake, a tract of woodland surrounded by meadow. As humanity chops the world's land-scape into pieces, those pieces become islands too. A nature reserve, by definition, is an island of protection and relative stability in an ocean of jeopardy and change. So the dynamics of parks and reserves can be described—and predicted—by equilibrium theory. The theory, according to Diamond, yields a handful of serious implications.

On his way toward examining the implications, Diamond revisited some of the theory's logical and empirical underpinnings: the species-area equation and Darlington's tenfold-to-twofold ratio, among others. He invoked Vuilleumier's páramo islands in the Andes, as well as Brown's work on mountaintop mammals of the Great Basin. He nodded in the direction of Krakatau. He described the mangrove experiment by Simberloff and Wilson. He also drew on his own study of "relaxation times" for the satellite islands around New Guinea. He noted the high rate of extinction on Barro Colorado. This body of evidence and theory, as Diamond viewed it, led to certain conclusions about the survival prospects of species in isolated reserves, such as:

- A reserve newly isolated will *temporarily* hold more species than its equilibrium number—but that surplus of species will eventually disappear, as relaxation to equilibrium occurs.
- The rate at which relaxation occurs will be faster for small reserves than for large ones.
- Different species require different minimum areas to support an enduring population.

At the end of the paper he offered a set of "design principles" for a system of nature reserves, including:

- A large reserve can hold more species at equilibrium than a small reserve.
- A reserve located close to other reserves can hold more species than a remote reserve.
- A group of reserves that are tenuously connected to—or at least clustered near—each other will support more species than a group of reserves that are disjunct or arrayed in a line.
- A round reserve will hold more species than an elongated one.

Besides stating them verbally, Diamond presented his design principles graphically, in a figure suggesting tiddlywinks on parade. Circles of various sizes and in various spatial arrangements were shown as a menu of dichotomous options. Each pair was labeled with a letter: principle A, principle B, and on up to principle F. The clear visual message was that some patterns of insularity, in the abstract and in reality, are more damaging than others.

Diamond made two claims about his design principles: that they were applicable to conservation planning and that they derived from MacArthur and Wilson's theory. Although both claims were controversial, the first met especially strong challenges. Some ecologists rejected the whole notion of imposing abstract solutions on such various situations; and some who accepted a few of the principles were unwilling to swallow the whole group. There were quibbles about principle F, reservations about E, qualifications that should be added to C—but none of the others provoked such heated and lasting disagreement as the one Diamond listed as principle B.

Four little tiddlywinks were "worse" than one big tiddlywink, according to Diamond's graphic figure. But were four small reserves, in the real world, necessarily worse than a single large one? That question would be argued for a decade. The argument would grow bitter and personal. It would acquire a handy label: SLOSS. The acronym stood for "single large or several small." During the late 1970s it would become ecology's own genteel version of trench warfare.

126

THE CHORUS of opinion about applied biogeography rose to fullness in 1975. Besides Diamond's paper, there were conspicuous statements by John Terborgh, Robert M. May, and others.

May's appeared in *Nature*, under the title "Island Biogeography and the Design of Wildlife Preserves." He allied himself with Diamond on the single-large-or-several-small question. "In cases where one large area is infeasible, it must be realised that several smaller areas, adding up to the same total as the single large area, are not biogeographically equivalent to it," May wrote. Rather than being equivalent, the small areas "will tend to support a smaller species total."

Terborgh, an old friend and kindred soul of Jared Diamond's, agreed. His scrutiny of the Barro Colorado situation and his field-work in other tropical forests had helped persuade him that the equilibrium theory was both scientifically valid and applicable to conservation planning. He paid homage to MacArthur and Wilson, noting that "the methods and the way of thinking they developed are extensible to a much larger range of situations, including the design of faunal preserves." Terborgh was not so specific as May on the SLOSS question, but he argued that the equilibrium theory implied a need for leaving corridors of habitat as connections between reserves—as Diamond had warned in another of his design principles. Terborgh further noted that if reserve planners hope to save large-bodied predator species, the planners had better make their reserves huge.

Ed Wilson joined the chorus. Along with a collaborator named Edwin O. Willis (whose long-term study of the Barro Colorado birds had made him an important influence on Terborgh's thinking also), Wilson co-authored a paper on applied biogeography, proposing many of the same principles as Diamond. Wilson and Willis's piece was published as the final chapter in a multi-author volume, derived from a somewhat historic symposium held at Princeton in November of 1973.

The choice of Princeton as the venue for that symposium, and the date, were significant. One year had passed since the death of Robert MacArthur, and the meeting was a memorial event. The participants included Jared Diamond, James Brown, Robert May, Edwin Willis, John Terborgh, and a handful of other ecologists of the new mold, as well as G. Evelyn Hutchinson and Ed Wilson, all of whom had cause to pay homage. The book appeared several years later as *Ecology and Evolution of Communities*. It was dedicated to MacArthur.

THE HEDGEHOG

OF THE AMAZON

127

WE'RE A SPECK in the sky above the jungle of jungles. Below, but not far below, is an unbroken expanse of canopy that stretches away vanishingly in every direction: Amazon forest. The ceiling today is low and gray, the terrain beneath us looks sublimely inhospitable, and our noisy little two-engine plane is thumping along gamely at about five hundred feet, a precarious altitude where the air is as curdy as soured milk. We're headed north from the city of Manaus. Occasionally the plane lurches twenty or thirty feet upward. Or it sinks. It buffets like a kite while we try to concentrate our attention on the ground. About an hour of flying under these circumstances, I know from experience, is as much as my stomach will tolerate.

"If the pilot gets lost," Tom Lovejoy shouts above the engine's baritone whine, "we could end up in Venezuela." Then he gives me his chipmunk grin.

From where we sit, dangling above it, the forest seems nothing more than a magnificent abstraction of flatness and chlorophyll—cryptic, monotonous, green. That's at first glance, anyway. But the magnificent abstraction conceals a magnificent wealth of particulars, and with second and third glances I can discern some detail. The greenness resolves into hundreds of different shades, representing hundreds of different tree species. Here and there it's punctuated by the crown of one individual tree blossoming gaudily in yellow or magenta. Steam rises in cottony plumes from some spots, where the wet exhalation of plant metabolism is putting vapor back into the sky. "Fifty percent of the rainfall in Amazonia is generated by the forest itself," Lovejoy tells me. It seems only logical that an ecosystem so grandly intricate, so organismic, should have its own breath.

I crane my face to the window. Scenery reels by like a Bach fugue at 78 rpm. The vomit-alarm glands underneath my jawbone begin tingling. The gentle bucking of our low-flying little plane isn't the problem; the problem is that I've been staring too long at the ground. If I relax and breathe deeply, if I lose myself in the spectacle, I can possibly avoid tossing my heavy Brazilian breakfast onto Dr. Lovejoy's shoes.

I notice five macaws, large and blue, gliding in formation above the trees. I see no other wildlife, only foliage and steam. The forest canopy covers everything like a tent. There don't seem to be any gaps—no glades, no savanna, no ponds. No splits in the green upholstery revealing brown-water or black-water rivers. Not hereabouts, anyway. Only treetops. The great junction of the Rio Negro and the mainstem Amazon has fallen away behind us. Even the clay-red slash of road leading out of Manaus, our first line of orientation from the airport, has gone its own way—and we've gone ours, on a more abstract path. There's nothing much below us except plants and animals in unimaginable profusion. There's nothing much ahead of us, as Lovejoy has said with such dire satisfaction, except more plants and more animals and then Venezuela.

The plane banks. The ground rises toward us from one side. We sweep down, straighten, and pass low, as though on a bombing run. This would be murder on my balky stomach if I weren't distracted by a new set of sights. As I watch, the canopy comes to an abrupt edge. The forest is suddenly gone. Beneath us now, instead, is a different sort of landscape. We have crossed an ecological boundary.

We're over the zone of clear-cut. We're looking at near-naked Amazon clay covered thinly with weeds. The tropical earth is as bare as a shorn ewe.

I see a stubble of tree stumps. A few charred logs lie crisscrossed in slash piles, remnants of the fires that took everything else. The sawyers and burners have done their work thoroughly. They have destroyed much more than trees. Besides removing the rainforest, they have removed the conditions that make rainforest possible: the darkness, the humidity, the shade-loving understory vegetation, the various flying and climbing beasts, the big predators, the small prey, the leaf litter, the soil fauna, the protection from wind and from erosion, the mycorrhizal fungi, the chemical nutrients, the pulsebeat of ecological interactions. The wet breath of the forest is gone too.

Now the plane banks again, tipping my window toward another peculiar sight. Standing up out of the scorched clear-cut is a small patch of intact forest, almost perfectly square. It looks like a piece of shag carpet tossed down on a dirt floor.

"That's a ten-hectare reserve," says Lovejoy.

We're in the airspace above Fazenda Esteio, the same hardscrabble cattle ranch surrounding the reserve where I'll later chase *Pithecia pithecia*—those leaping monkeys marooned in their tiny fragment of forest—with Eleonore Setz. Right now I'm getting a much different perspective on a similar fragment. The ten-hectare patch is known to Lovejoy as Reserve #1202. In the terms more familiar to most of us, it's a twenty-five-acre piece of woefully isolated tropical forest. It constitutes one unit within a mammoth enterprise that Lovejoy himself launched, back in 1979, to explore the dynamics of ecosystem decay. He has brought me up here today for an overview of the world's largest experimental study of the principles and implications of island biogeography.

Reserve #1202 is just one of many. Other such patches of forest lie scattered amid the clear-cuts of Fazenda Esteio, and still others on neighboring ranches, each one peculiarly rectilinear and ranked within an orderly gradation of sizes: one hectare, ten hectares, a hundred hectares, a thousand. This one, at ten hectares, is roughly the size of four city blocks. It's an island of jungle in a sea of man-made pasture. It was isolated in 1980, carefully preserved amid the scorched desolation, and since then it has been intensively studied. Around its perimeter I notice a well-worn trail.

Our plane lifts through another banked turn and back over the area from a different angle. Now into view comes a second island, still smaller. This one is barely more than a tuft of trees. Like the first, it's outlined by a square trail, but the patch itself is no longer square. It's tattered like an old flag. Sunlight and wind have parched and eroded it. Trees have fallen—tall rainforest trees, stout-trunked but shallow-rooted, incapable of enduring such exposure. The canopy is full of gaps. The shade-loving understory vegetation has shriveled.

As the plane dips us down for a close pass, I see movement just outside the patch. Aha, I think, Amazon fauna. I prepare myself eagerly for a glimpse of a monkey, or a tapir, or with great luck maybe even a jaguar.

Wrong. What I've spotted are cows.

They are white as bone. A native rainforest herbivore couldn't afford whiteness—it would need camouflage—but these beasts aren't native. They are oblivious to the local realities of predation and defense. Spooked by our engine noise, they lumber around stupidly.

"That's a one-hectare reserve," says Lovejoy.

Even from this height, I can tell that the one-hectare patch is a husk of its former self. What changed it? Not chain saws, not fire, not cattle—not directly, anyway. Within its surveyed perimeter, it has been carefully protected against those factors. What changed it was sheer insularity.

Beyond the particulars of these two reserves stands a more general question: Is there a quantifiable, and therefore predictable, relationship between insularization and doom?

What is the threshold of ecosystem decay? What is the lower limit of ecological cohesion for a parcel of landscape? If a one-hectare fragment is too small to sustain itself, if a ten-hectare fragment is also too small, then how much is enough? Stated otherwise, in Tom Lovejoy's language: What is the minimum critical size of a piece of Amazon rainforest?

128

DURING THE earlier years, from 1979 onward, Lovejoy and others called it the Minimum Critical Size of Ecosystems Project. In Brazil there was also a Portuguese name, suggesting a less narrow focus: Projeto Dinâmica Biológica de Fragmentos Florestais. That translates, by easy cognates, to the Biological Dynamics of Forest Fragments Project. It could mean many things.

Actually, it has meant many things. Over the course of its history, Lovejoy's enterprise has encompassed a number of major subprojects and smaller studies: on the local snake fauna, on bats, on primates, on rodents, on the solitary bees, on the social spiders, on the distribution and ecology of palms, on such important tree families as the Lecythidaceae and the Sapotaceae, on the reproductive biology of strangler figs, on soil nitrogen, on microclimate and plant-water relations within the reserves, on seedling regeneration along the reserve edges, on lizards, on frogs, on beetle diversity, on symbiotic relationships between ants and plants, on stream dynamics, on the role of fungi in rotting leaves, on various aspects of the bird fauna. Eleonore Setz's work on the little population of *Pithecia pithecia* is another of the many discrete studies. A recent report lists thirty different topics of research. The phrases "ecosystem decay" and "minimum critical size" appear nowhere on that list, though they represent the original, transcending agenda.

In the late 1980s the English name of the project was changed, under some pressure from a special review committee, so as to suggest wider purposes and to achieve semantic congruence with the Brazilian version. Now it's officially the Biological Dynamics of Forest Fragments Project. Lovejoy himself liked the old name, and so do I. The old one was narrow, yes, but also more vivid, and it gave a better sense of the project's theoretical roots. Those roots include the species-area relationship, the equilibrium theory, the early efforts by Jared Diamond and others toward applying island biogeography to the design of nature reserves, and the conundrum that became known as SLOSS.

Single large or several small? That question had been argued, argued again, argued bitterly and to a point of logical impasse, when Lovejoy got the idea of addressing it by means of a giant experiment.

129

"THE FOX KNOWS many things," wrote Archilochus, a Greek poet of roughly 700 B.C., "but the hedgehog knows one big thing."

The hedgehog is more familiar on our side of the world as a porcupine. And the one big thing that it knows, of course, is the value of being prickly. The fox, a predator well armed with teeth and claws and speed and wit, but smallish and gracile, is obliged to be more versatile.

Forty years ago, the intellectual historian Isaiah Berlin wrote a memorable essay titled "The Hedgehog and the Fox," using that pair of animals to illuminate what he considered a bifurcation within the novelist Leo Tolstoy. Berlin suggested that the published works of Tolstoy show a certain two-mindedness, reflecting a tension between the great writer's natural disposition and his conscious convictions. On the one hand, here was a creative genius gifted with wondrously acute powers of observation and invention; on the other hand, a philosopher chasing the ultimate single answer. Tolstoy's particularized, pluralistic view of human history stood opposed to his theoretical, monistic beliefs. As he looked at the world, part of him saw Oneness and part of him saw Many. So Tolstoy himself, according to Isaiah Berlin, using the categories of Archilochus, was at once both a hedgehog and a fox. Besides fitting the categories onto Tolstoy, Berlin also applied them more broadly. The intellectual particularist Balzac, for instance, was a fox. Pushkin, Herodotus, Joyce, also foxes. Single-minded visionaries such as Dante and Dostoyevsky and Nietzsche were hedgehogs. It isn't that hedgehogs don't also see the multifariousness of life and experience; but like Tolstoy in writing *Anna Karenina*, they see it in light of a single big idea.

By this standard, Tom Lovejoy can be considered the hedgehog of tropical ecology.

Lovejoy is a soft-spoken fellow in early middle age, deceptively youthful in appearance, deceptively easygoing in manner. He doesn't come across as a man of grand obsessions. He looks as well scrubbed and ingenuous as a graduate student from Indianapolis, though in fact he's the scion of a wealthy Manhattan family and bears an unmitigated Ivy League pedigree. He is presently counselor to the secretary of the Smithsonian Institution, which makes him an influential science maven with a broad mandate. In the offices and boardrooms of Washington, he wears a bow tie. He claims that dangling neckties collect gravy stains and are more of a nuisance than bow ties. But you and I know that there is simply a bow tie sort of person—quirky but smart, unassailably self-confident, sublimely immune to the vagaries of fashion, rooted in New England—and Lovejoy is one. On the streets of Manaus, during an Amazon downpour, he deploys a collapsible umbrella. But when he hikes into the forest, his khakis are trail-weary and his boots are old. His Portuguese is fluent. His ear for the local birdsong is expert. He knows a forest falcon from a caracara, a hoatzin from a sunbittern, and one species of antshrike from another. He has been commuting to the Amazon since 1965. That was just after he finished college and before he got rolling on his doctorate.

Professional ecology is a small world where paths commonly cross. Lovejoy, like Robert MacArthur, did his Ph.D. under the broad, owlish wing of G. Evelyn Hutchinson at Yale, though Lovejoy was eleven years younger than MacArthur and arrived after he had left. Also like MacArthur, Lovejoy approached theoretical ecology by way of a solid empirical grounding in ornithology. For his dissertation research, Lovejoy spent a few years netting and banding birds in the forest near Belém (formerly Pará, the coastal Brazilian city where Alfred Wallace had established his earliest base), at the mouth of the Amazon. Eventually his notebooks would document seventy thousand individual bird captures. He was looking at patterns of diversity and abundance among bird species within forest communities, a topic that could be traced straight back, beyond MacArthur, to Frank Preston.

There was a further coincidence between Lovejoy's early career and the work of Robert MacArthur. Before he ever visited Brazil and fell in love with the country, Lovejoy had taken part in an ornithological expedition to East Africa, where he noticed that isolated mountain forests seemed to constitute ecological islands above the grassy plains. He had passingly contemplated shaping his dissertation around that. But it wasn't an irresistible idea at the time; it didn't seem hugely resonant. MacArthur and Wilson had not yet transformed the conceptual framework of ecology, and islands weren't the commanding paradigm they would shortly become.

"My first awareness of island biogeography was when I was in Belém, about 1967 or '68," Lovejoy recalls, "and this Harvard graduate student showed up with a copy of the book." By "the book" he means, without need to specify, *The Theory of Island Biogeography*. "Hot off the presses. I'll never forget that. And it was all very interesting. But it didn't particularly apply to anything I was doing." Not at that moment, it didn't. Later it would.

In 1973 Lovejoy became program director of the United States branch of the World Wildlife Fund. Today WWF-US is a vast corporate organization that performs the function, among others, of funneling money to small conservation projects all over the world. Back when Lovejoy signed on, it was still small itself, but even then it played a role in grant disbursal. Lovejoy's job involved reading and assessing the grant proposals. He had the help of one secretary. Funding would flow, or not, on the basis of his judgments. Many of the proposed projects involved statutory set-asides of habitat—that is, the designation of new parks and reserves. So, at just over thirty years old, he found himself making consequential recommendations about which parcels of the planet's richest ecosystems might be protected and which, by default, might not. He remembers two essential questions that had to be answered about such protected parcels. Where should they be? And how *big* should they be? He wanted to base his advice on the best available scientific wisdom. He read the early journal literature on applied biogeography. Then came SLOSS.

"I began to realize," he says, "that there was this rather serious problem relating to how you design a conservation area."

130

THE SLOSS debate went public in 1976, when Dan Simberloff and a colleague named Lawrence Abele published a short paper in *Science*, voicing their concern over the recent vogue of applied biogeography. What made them uneasy was this business of deriving neat principles of reserve design from MacArthur and Wilson's theory. They cited Diamond's paper "The Island Dilemma" and pointed also to the brief but prominent essay by Robert May, "Island Biogeography and the Design of Wildlife Preserves," which had run in *Nature*. They quoted May's statement that several small reserves "will tend to support a smaller species total" than will a single large reserve of equal area. Not so fast, said Simberloff and Abele.

The theory itself hadn't been broadly enough proven, they warned, to justify such confident application. And the most basic principle being offered by May, Diamond, and others—that nature reserves should always consist of the largest possible continuous area—might not be correct. It didn't follow necessarily from the theory. Some of the same evidence cited on behalf of the single-large option could also be adduced to support the several-small option. It was all premature, according to Simberloff and Abele. The complex decisions involved in reserve design weren't reducible to a half dozen sleek principles. Furthermore, given the cost and the irreversibility of ambitious conservation programs, any half-baked application of a three-quarters-baked theory could conceivably do more harm than good.

Simberloff's own stance here is an interesting matter. Back in the late 1960s, as Ed Wilson's graduate student, he had virtually been present at the creation of the equilibrium theory. Sweating his way through the Florida mangroves, he had helped gather experimental data that lent support to that theory. He seemed at the time to believe in the theory's predictive (as well as its descriptive) value. Was he contradicting himself, then, when he criticized its application in 1976? Not necessarily. But if his scientific convictions hadn't changed, his emphasis had. His attitude was subtly but firmly different from just two years earlier, when he had stated in print that island biogeographic theory might indeed "aid in the preservation of the earth's biotic diversity." He wasn't now repudiating the equilibrium theory. He was just cautioning that it might not fit all island ecosystems quite as well as it fit the Florida mangroves, and that it didn't provide any simplistic guidelines for conservation.

To some degree he was reacting against what he saw as the others' careless haste. Another factor was new data. Simberloff and his co-author Abele had each been involved with recent field studies casting doubt on the notion that a bigger reserve is invariably better.

Simberloff had gone back to the mangrove islands—not just to the general area, but to the same individual clumps that he and Wilson had fumigated. In late 1971, with the original experiment finished, he had altered a few of those islands, cutting channels through the root tangle and the canopy. His cuts reduced the total area only slightly, but a new level of fragmentation was introduced. Each mangrove island was now a tiny archipelago of islets. How would that fragmentation affect the total number of arthropod species supported on a given archipelago? Simberloff had waited three more years, allowing the islets again to equilibrate; then in the spring of 1975 he had returned for another census. His results were equivocal; a single large mangrove island did not always support more species than several small ones. He found a case in which four severed fragments combined for a higher total of species than the original island. In another case, two severed fragments supported fewer species than they had as a single intact island. Simberloff announced these mixed results in his 1976 paper with Lawrence Abele.

Abele's own fieldwork on marine ecosystems had turned up a similar situation. His "islands" were heads of coral, and the resident species were marine arthropods dependent for habitat on one coral head or another. Abele had found a consistent pattern: Two small coral heads harbored more arthropod species than one large coral head of equal total area. Abele's field results were intricate and equivocal, like Simberloff's, but their net significance was to challenge the categorical postulate that a big island is always richer than two small ones.

Given the long history of species-area studies and the recent excitement over the equilibrium theory, it seemed heretical. Simberloff and Abele might as well have announced that the pope *isn't* Catholic and the bear *doesn't* always shit in the woods.

What could account for the reversal against expectations? Simberloff and Abele mentioned some factors: (1) size of the mainland pool of species, (2) lack of differences in dispersal ability among the mainland species, and (3) competition between species. Each of those factors could contribute, for complicated reasons, toward a situation in which two small islands would contain more different species than a single large one.

Another potential factor—not mentioned by Simberloff and Abele in their 1976 paper but destined to figure eventually in the SLOSS debate—was habitat diversity. If two small islands offer three different types of habitat between them, and one large island offers only two types of habitat, the pair of small islands might conceivably harbor more species.

Or possibly not. And possibly a single large island might contain more habitat types, not fewer, than two small islands. Possibly the habitat factor and others might determine that, in a given situation, large size and lack of fragmentation do indeed support greater diversity. Simberloff and Abele granted such possibilities. In the real world, islands are various. The principles of reserve design as enunciated by Diamond, on the other hand, assume a certain uniformity. The point that Simberloff and Abele wanted to make, they explained, was not that several small reserves are necessarily preferable to a single large reserve. Their point was that the species-area relationship (as subsumed into MacArthur and Wilson's theory) gives no guidance as to what is preferable, and that each situation should be judged on its own details.

"In sum," Simberloff and Abele concluded, "the broad generalizations that have been reported are based on limited and insufficiently validated theory and on field studies of taxa which may be idiosyncratic." Even within a context of formalized scientific discourse, those words don't sound especially harsh. But the paper's effect was incendiary.

131

"WHEN THAT was published," one biologist has told me, "that was sort of like dropping a match in some very dry tinder. I think it's no overstatement to say there was an explosion of rebuttals to that."

The Simberloff-and-Abele paper appeared in January 1976. In September, *Science* ran a set of responses by six different authors, including Jared Diamond and John Terborgh. Diamond's came first. He noted Simberloff and Abele's accusations about prematurity, insufficiently validated theory, idiosyncratic field data. He acknowledged that Simberloff and Abele had posited circumstances in which several small reserves might seem preferable to a big one. And he wrote: "Their reasoning from their assumptions is correct but minimizes or ignores much more important conservation problems. Because those indifferent to biological conservation may seize on Simberloff and Abele's report as scientific evidence that large refuges are not needed, it is important to understand the flaws in their reasoning." The debate was on.

Among the flaws, Diamond argued, was their emphasis on the fact that two small islands sometimes contain more species than a large one. It might be true in certain circumstances, wrote Diamond, but it was irrelevant. How so? Because the sheer number of species present on some island or group of islands, or within some wildlife reserve or group of reserves, was not the real issue. The real issue was how an island or a reserve might affect extinction or preservation of particular species in the larger context. And in the larger context, not all species are equal.

Why aren't they equal? Because some species live in much greater jeopardy of extinction than others. The rare species, the highly specialized species, the less competitive species, and the species having low aptitude for dispersal and colonization—all these species might be missing from a reserve system if the reserves were numerous but small. Meanwhile the common species, the generalists, the good competitors, and the adventurous species would most likely be present. Diamond acknowledged that several small reserves might very well contain a large number of species, but most or all of them, he argued, would belong to that second category: the good travelers, the feisty competitors, the generalists. Those are precisely the species least in need of protection. They are resourceful and tolerant of disruption. They are abundant in all sorts of landscape. They are resistant to extinction. They might survive even without any system of reserves. The species most needing protection are members of the first category: the specialists, the less competitive creatures, the reluctant travelers. "A refuge system that contained many species like starling and house rat while losing only a few species like ivory-billed woodpecker and timber wolf," Diamond wrote, "would be a disaster."

He rejected Simberloff and Abele's rejection of his design principles. Idiosyncratic, my eye. Premature, my ass. Of course his language was muted, but that was the spirit. Biologists had better alert themselves, Diamond concluded, to this island dilemma.

John Terborgh and the other responding authors made similar arguments. Terborgh even declared that the logic of Simberloff and Abele, if accepted uncritically, "could be detrimental to efforts to protect endangered wildlife." That comment seems to have stung.

In the same issue of *Science*, Simberloff and Abele were allowed the last word. They repeated their charge about inadequate data. They repeated their assertion that the species-area relationship offers no generalized guidance. They alluded, now, to the factor of habitat diversity. Raspberry to you guys and the island dilemma you rode in on, said Simberloff and Abele, though not in precisely those terms.

They also reacted to Terborgh's stinging comment. "We regret being cast as the bêtes noires of conservation," they wrote. But they didn't retreat. "Our conclusion still stands: the species-area relationship of island biogeography is neutral on the matter of whether one large or several small refuges would be better."

The audience for this debate was other biologists. Among them, Tom Lovejoy had reason to be especially interested.

132

"I WOULD READ these papers. I would discuss it with people. The controversy was raging," says Lovejoy. "And it was upsetting, in a way. Because it was intuitively obvious that some species need large areas, and if you're going to protect them, you just have to *have* large areas." Big-bodied predatory mammals, for instance. A population of jaguars or wolves or tigers simply can't sustain itself within the confines of a small area.

On the other hand, Lovejoy admits, Simberloff and Abele were correct in pointing out that island biogeographic theory provides no logical basis for assuming that large reserves are necessarily better. "Because the theory merely treats all species as being equal," he says. It makes numerical assessments, not qualitative ones. It predicts that the equilibrium number of species on a large reserve should be greater than the equilibrium number on a small reserve, but it doesn't distinguish between a large reserve full of common species and a large reserve full of rare species. It doesn't take account of whether a species harbored in one reserve is desperately jeopardized throughout the rest of the world. It knows a big island from a little island but not an ivory-billed woodpecker from a starling.

The SLOSS debate was made doubly complicated, then, by the fact that it entailed two closely linked issues: (1) What is the optimal strategy for designing nature reserves? and (2) What does the equilibrium theory say, if anything, about the optimal strategy for designing nature reserves? One issue was mainly practical, the other mainly intellectual. Lovejoy, a pragmatic conservationist, cared more about the first than the second.

"I used to discuss it and worry about it a lot, with my scientific colleagues," he recalls. He came to believe that conservation planners like himself needed more than just theory and argument. They needed data. Specifically, they needed data on the structure of ecological communities and the pernicious phenomenon that Diamond was calling relaxation. They needed answers to a handful of technical-sounding but crucial questions:

- What exactly happens when a sample of habitat, newly isolated, relaxes to equilibrium? Species are lost, yes, but which species?
- Are the losses random, or are they determined by aspects of the community structure of the ecosystem?
- Will a particular nature reserve give long-term protection to those species in greatest jeopardy? Or will the reserve lose its jeopardized species over a short span of time while maintaining its populations of species that don't badly need protecting?
- Will several small reserves, offering similar habitat, all lose the same species and save the same species? Or will they vary randomly from each other, preserving assemblages of species that are different and complementary?

These questions hadn't yet been answered—not by Simberloff's work in the mangroves, not by Brown's study of Great Basin mountaintops, not by other empirical investigations, and certainly not by the early phase of SLOSS.

Lovejoy himself viewed the SLOSS debate with some wariness. "It may be my political nature, but I've never gotten into the middle of that pissing match," he says. "Simply because I didn't see what could be added to it. I thought it would be better just to go out and try to get data." But how?

Then, just before Christmas in 1976, Lovejoy was invited over to the National Science Foundation for a brainstorming session with a few other biologists, one of whom was Dan Simberloff. "We were sitting around, talking about this problem," Lovejoy recalls. "And all of a sudden, in one little lump, this project leaked into my mind."

133

"THIS WAS one of those brilliantly simple ideas," says Rob Bierre-gaard, a younger biologist who became Lovejoy's first recruit to the Minimum Critical Size of Ecosystems Project.

Bierregaard has known Lovejoy since 1969. They went to the same prep school, some years apart, and were inspired there by the same science teacher, a remarkable man named Frank Trevor. It was on Trevor's recommendation that Bierregaard had met Lovejoy—back when Bierregaard was a freshman at Yale and Lovejoy still a graduate student—and through Lovejoy in turn he was introduced to G. Eve-lyn Hutchinson, the eminent ecologist who had been Robert MacArthur's mentor. During his undergraduate years Bierregaard studied ecology under Hutchinson, and in Bierregaard's recollection that opportunity was "sort of like taking your catechism from God." The linkage among Hutchinson, MacArthur, Lovejoy, and Bierre-gaard, with Frank Trevor hovering beneficently nearby, is more than an artifact of old-boy-network dynamics within professional science; it's a major chain of influence in the history of modern ecology, from pioneering theory to pioneering applications.

Lovejoy was just back from his two years of doctoral fieldwork when he met Bierregaard at Yale, and he hired the undergrad as a student assistant to do the computer chores on his dissertation data. After Bierregaard finished his own doctorate, Lovejoy hired him again—this time at the World Wildlife Fund in Washington, where Bierregaard made five hundred bucks a month and lived rent-free in Lovejoy's basement. The consummate right-hand man, he would eventually serve eight years as field director of Lovejoy's Amazon project. Bierregaard above all other people is entitled to call the MCSE idea simple, since he played such a big role in its complicated day-to-day implementation.

"Somebody had to go in," says Bierregaard, "and find out what re-ally happens with this whole process of ecosystem decay in frag-mented habitats." Lovejoy's brainstorm, in late 1976, offered a way to do it.

The crucial questions by then had come into focus. The difficult part was devising a workable but sufficiently grand experimental structure. It had to be much broader in scope than Simberloff and Wilson's project among the mangroves. It had to consider larger fragments, a richer ecological community, a longer time scale. And it had to incorporate before-and-after surveys of the species that were resident on each fragment, not just guesswork about what species *might* have inhabited some land-bridge island before its isolation, back in the mists of the Pleistocene. The new enterprise would need to be different from any ecological field experiment ever done. It would need to be wildly ambitious but economical. Tom Lovejoy, the pragmatic conservationist with the theoretical education, saw those requirements clearly. He also happened to know about a certain provision in Brazilian law.

This provision (let's call it the fifty percent provision) related to tax structure and land rights in the Amazon region, including that area known as the Manaus Free Zone, which encompassed a handful of ranches such as Fazenda Esteio. Under the Free Zone incentives, big swaths of rainforest on Fazenda Esteio and elsewhere were being cleared to plant pasture for a few scrawny cattle. But the fifty percent provision stipulated that fifty percent of the forest on any given fazenda had to be saved. In effect, it mandated the insularization of scattered rainforest fragments. Lovejoy's idea was to make scientific virtue out of legal necessity. With a little painless cooperation from the ranchers, maybe the mandatory insularization could be shaped into an experiment.

There was nothing so esoteric about that. Like other brilliantly simple ideas, though, it seemed obvious to everyone else only after the first hedgehog had thought of it.

The next step was a long campaign of persuasion. Lovejoy flew down to Manaus and started jawboning. He talked with scientists at Brazil's National Institute for Amazonian Research (INPA, by its Portuguese acronym), in particular Dr. Herbert Schubart, who was intrigued enough to become his collaborator. He talked with Brazilian conservationists, including Maria Tereza Jorge de Padua, at that time in charge of the country's national parks. Schubart and Jorge de Padua helped him with entree to others. He explained his idea to officials at SUFRAMA, the Manaus Free Zone Authority. He also visited some of the ranchers. With his diplomatic charm and his fluent Portuguese, Lovejoy coaxed people into sharing (or at least indulging) his vision. He created a broad partnership. And eventually it was agreed: The destruction of forest, destined to happen anyway, would be arranged so as to incorporate certain patterns. A remnant here, a remnant there, a bigger remnant over there . . . and the rest clear-cut.

BIOLOGICAL
DYNAMICS
OF FOREST
FRAGMENTS
PROJECT

Manaus
(Barra)

Rio Uaupés

Rio Negro

Rio Amazon

Rio Madeira

Rio Tapajós

Santarém

Belém
(Pará)

Rio Tocantins

SOUTH
AMERICA

0   150   300

MILES

Some of the remnants would be square, their edges linear and abrupt. They would be measured out with decimal precision, spanning a progression of sizes: one hectare, ten hectares, a hundred, a thousand. A vast tract of man-made pasture, hospitable only to cows, would surround the patches of untouched jungle. Those patches would be studied, both before and after isolation, by scientists whom Lovejoy, Bierregaard, and Schubart would enlist. There would be multiple replicate patches, for purposes of comparison—so that results from a ten-hectare patch could be matched against those from a one-hectare, from a hundred-hectare, or from another ten-hectare patch in a different location. Across two or three decades of study, the patches would yield information about the phenomenon of ecosystem decay.

Although the conceptual outline was vast, reflecting the size of the issues to be addressed, the execution phase began modestly, reflecting the size of Tom Lovejoy's disposable resources. At first Rob Bierregaard was the only staff person, and his operating budget was minuscule. "When I had the project set to go," Lovejoy remembers, "I brought him down, introduced him to people, and said, 'Here you are, Rob. See what you can do.'"

Bierregaard recounts it in somewhat more vivid terms. "Lovejoy just abandoned me," he says amiably. "We went to Belém, and I met some people there, and he said, 'All right, get on a plane, go back to Manaus, charter a plane, do an overflight of the area where the sites are. And work with Schubart.'" Bierregaard rewrote the project proposal in order to get government approval, and Schubart translated that into Portuguese. They submitted it. "This was the middle of Carnival," says Bierregaard. "The theme in Manaus for Carnival was Disneylandia. Here I was going to Carnival in Brazil, my first time out of the country, I was really gonna get some Latin American culture—and there were Mickey Mouse costumes coming down the Carnival avenue. It was also in the peak of rainy season. So somehow I had to charter a plane, I had to talk to the ranchers, and all this on the Portuguese one has in about two months of working with some tapes. Unfortunately I'd stopped one chapter before the past tense. It was terrible."

But when Carnival was over and the Mickey Mice ended their capering, the problems turned out to be solvable. Official approval was granted a half-year later. Eventually he would be able to reminisce—even in Portuguese, using the past tense—about those hard early days.

Bierregaard established his field presence in September of 1979. He spent $245 that month, and offered a long explanation in his first financial report as to why he had been so profligate. He opened a vest-pocket office in Manaus and, in October, purchased a vehicle. The surveying of reserve sites had begun, but so far there had been almost no clearing of forest. The reserves were like those hilltops on the Bassmanian peninsula during the last low-water episode of the Pleistocene: islands-to-be. But this time, when insularization occurred, a team of ecologists would be watching.

A few months later Lovejoy himself came back down from Washington. He was encouraged by what he found. Bierregaard had adapted fast. Things were happening. It was still just the humble start of what would grow into a vast project—with a $700,000 annual budget and a sizable roster of scientists, support staff, interns (young volunteers from the United States, Brazil, and elsewhere), and Brazilian woodsmen—but Lovejoy could see his idea becoming reality. He and Bierregaard took off together into the forest. They hiked up a long trail to the site of what would later, after the cutting and burning, become one of their insular reserves. "It was beautiful," says Lovejoy. "Big ground birds would fly up from the trail as we came along. Finally we stopped. And he built us a little fire, getting it started in the damp wood by using a bit of old tire for tinder, as they do in the jungle." Bierregaard heated up two cans of *feijoada*, the dark, bean-rich Brazilian stew. Then he pulled out a miniature bottle of wine that he'd saved from his last airline flight. "And sitting there in the jungle," says Lovejoy, "we drank a toast to that biology teacher back at our prep school." Frank Trevor had died some years before. He had always dreamed of a visit to the Amazon.

134

"I HAVE a tendency to start things," says Lovejoy, "and then get other people to do the follow-through." Sensitive to the realities of collaborative effort and national pride, he corrects himself: "—other people to *help with* the follow-through."

It's a few days after our flight over the project aboard the little plane. In the meantime we have gone back out on foot. We have hiked down a path, away from the clear-cuts, deep into an area of undisturbed forest, where I've had the chance to appreciate its magnificent wealth of particulars at close range. We have walked through a wilderness of filtered green light, surrounded by gauzy humidity, buzzing and hooting sounds cast forward against deeper silence, toucans rustling in the canopy, hummingbirds and azure-winged morpho butterflies in the few sunny openings, cotingas, screaming pihas, epiphytes, lianas, leafcutter ants marching in long processions through the understory, discreet lizards, lichens, fungi in bright pastels. We have slept at a field camp, in hammocks, and heard the sibilant roar of howler monkeys filling the night like a choir of Harpies. For dinner and breakfast we have eaten *caldeirada*, an aromatic Brazilian chowder of river fish and cilantro. And on top of the *caldeirada* Lovejoy has fed me as much information as I could swallow. Now there's a respite for digestion.

"As a kid I was known as an instigator," Lovejoy adds. "Always starting things, getting in trouble." He pauses again, surrendering to a reflective smile. "Evidently it's carried over into adult life."

With Bierregaard installed as field director, the Minimum Critical Size of Ecosystems Project grew. Other scientists came to do fieldwork on the site. Lovejoy himself spent most of his time in Washington, at the World Wildlife Fund offices, and from there he provided guidance and support. On the fazendas north of Manaus, more than twenty reserves were laid out in the midst of what was still forest. Simple field camps were established. Ecological surveying began. The first question that had to be answered was: What species of animals and plants inhabit these demarcated squares before any isolation occurs? To find out, the researchers deployed mist nets for birds, traps for small mammals, binoculars for primates, and all manner of other techniques for collecting and identifying fauna and flora. The project's budget increased, so did support staff, and more woodsmen and interns were sent out to help gather data. The first isolates were created during the dry season of 1980: a one-hectare patch and a ten-hectare.

The understory, outside the patch boundaries, was cleared by machete. Fallers with chain saws took down the trees. The jackstraw debris of timber and slash was left to go crisp for a few months; then it was torched. The fire raged and died. The two patches remained, rectilinear islands of green in an ocean of charcoal. One of those patches was Reserve #1202, the ten-hectare square that Lovejoy showed me from the plane. I wasn't aware at the time, with my stomach gone burbly, that I was looking down on such a swatch of scientific history.

Phase two of the fieldwork started at once. Bierregaard and others went back out with their mist nets and their traps, this time to study the consequences of insularization. Some early results were published in 1983 and 1984, under the joint authorship of Lovejoy, Bierregaard, and a few colleagues. Those results were dramatic but not surprising.

First and most obviously, both reserves had lost their large predators. The jaguar, the puma, and the margay cat were no longer present. Why not? In addition to the noxious commotion of chain saws and flames, which had probably helped scare them away, there was an ecological reason: The reserves were too small to provide adequate cover and food for such big-bodied predators.

The paca, the deer, the white-lipped peccary, and the other sizable prey were likewise gone. Inadequate forage, inadequate security, inadequate something or other. The precise ecological details are unknown; what is known is that those species hadn't found what they needed.

The primate populations were also affected. The ten-hectare reserve had contained a band of golden-handed tamarins (*Saguinus midas*) before isolation. Not long after, they came down from the trees and hightailed it off across the clear-cut. Two bearded sakis (*Chiropotes satanus*) also found themselves isolated within the ten-hectare patch. Unlike the tamarins, the sakis were canopy-dwelling creatures unaccustomed to ground travel. Their lonely confinement on such a small island must have been hard for these two individuals, since bearded sakis ordinarily live in large groups and travel wide circuits through the forest, feeding on fruits and seeds. The two trapped sakis couldn't do that. Within the reserve, only one tree was in fruit, and before long they had eaten it clean. Then they disappeared, first one, then the other. Maybe they died of starvation. Maybe, pushed to extremes, they came down and made a run for it. No one knows.

A third primate species within the ten-hectare area was the red howler monkey (*Alouatta seniculus*), which had the advantage of being folivorous. Leaf eaters, like cows, can usually find something to eat. Still, the howlers had trouble. At first there were eight. By 1983, they had been reduced to about five, and the skeleton of a young male had turned up on the forest floor.

Among birds, a peculiar trend had appeared just after isolation: a higher capture rate in the mist nets, suggesting increased population density within the reserves. The likely explanation was that these patches of forest had assumed short-term significance as lifeboats, artificially crowded with birds displaced from the clear-cut terrain. If that was so, the increase in bird density wasn't expected to last, since the habitat of the reserves couldn't support it. And it didn't last. Within a year, bird density had fallen back to normal in the ten-hectare reserve and was lower than ever in the one-hectare.

The avifauna of each reserve continued declining. The declines were notable not just numerically but also in terms of which species were disappearing—the ant-following birds particularly. These are species of the understory that eat insects spooked up by columns of army ants on the march. Some ant followers are versatile enough to find food by various methods, but others (such as the white-plumed antbird, *Pithys albifrons*, and the rufous-throated antbird, *Gymnopithys rufigula*) are specialists, depending utterly on the armies they follow. No ants, the antbirds don't eat. Both the white-plumed and the rufous-throated antbirds seem to require a largish feeding area—large enough that they share it with at least a couple of ant colonies putting their armies on the march daily. And even a single colony of these ants requires a modestly big area of its own. With a half-million mouths to feed, one hectare isn't sufficient. After six months of isolation, therefore, army ants were gone from the one-hectare reserve, and the white-plumed and rufous-throated antbirds were gone too. By the end of a year, three other species of ant-following birds had been lost. They had followed the ants to oblivion.

Among butterflies, the shade-loving species of the deep forest became rare, some disappearing altogether. Meanwhile the number of light-loving species (such as those flashy blue morphos, adapted to forest gaps and edges) increased. This represented turnover, in MacArthur and Wilson's sense. Even with no net loss of species, turnover among the butterfly fauna was a hint of corrosive changes.

Some of those changes involved the vegetation. Trees along the perimeters of the reserves had received unaccustomed exposure to sunlight, drying, and winds. With their leaves scorched, with their inherently weak root structures subjected to extraordinary stresses, they began dying and falling. Sometimes a falling tree took another tree with it, and every new gap offered another channel for the intrusion of sunlight and winds. Light-tolerant species, including weeds from the pasture, invaded. Hot breezes reduced humidity and raised temperatures within the reserves, drying leaf litter on the forest floor. Near the center of the ten-hectare reserve, the number of standing dead trees increased dramatically. And beyond being stressed, some of the plants were downright confused: A mature tree of the species *Scleronema micranthum* came into flower six months out of sync with the rest of its species. Vulnerable in a season of scarcity, it lost a large part of its seed crop to seed predators. The forest had commenced to unravel.

The ecosystem was in decay. Species were vanishing, relationships were disrupted, even climatic conditions had gone bad. The lesson of these early data was unmistakable. A tiny patch of Amazon rainforest—one hectare, say, or even ten—could not maintain its existence in isolation. It just wasn't big enough to sustain itself.

Then what *was* the minimum critical size? If that question had any meaningful answer, it wouldn't come quickly or easily.

135

WHILE Lovejoy was launching his experiment, the SLOSS debate continued, with Jared Diamond conspicuous on one side and Dan Simberloff on the other.

Throughout the decade that followed their 1976 exchange in *Science*, the familiar arguments were repeated and some fresh ones were added. Each side gathered partisans. At issue was not just one question of reserve design strategy, single large versus several small, but also such broad matters as whether the species-area relationship was profoundly instructive or not, whether the equilibrium theory was scientifically valid or not, and whether either of those two concepts had any relevance to conservation.

Journal papers appeared in quick succession. Even as cited in the usual shorthand format whereby professional ecologists know and remember them, they pile up into a longish list: Gilpin and Diamond (1976), Abele and Patton (1976), Brown and Kodric-Brown (1977), Simberloff (1978), Diamond (1978), Abele and Connor (1979), Connor and Simberloff (1979), Gilpin and Diamond (1980), Simberloff and Connor (1981), and dozens more. Each of those papers was full of elaborate logic and conviction. Many were besmeared with math. Some also contained facts. Scientists characteristically fight out their battles in the journal literature—lobbing papers back and forth like flaming arrows—and that's how these scientists fought SLOSS. It didn't go on as long as the War of the Roses, it didn't spill as much blood, but it was something more heated than a cordial discussion among colleagues.

The specifics were intricate and tendentious. I recommend them as a cure for insomnia. A few salient themes did arise from the welter of voices, and those few are worth mentioning. Habitat diversity was one.

Representatives of the Simberloff camp argued that habitat diversity is far more significant than sheer area in determining how many species can exist in a given reserve. A handful of different habitat types will support many more species than a large zone of homogeneous habitat, and the best way to preserve a handful of habitats is with a handful of smallish reserves.

Representatives of the Diamond camp disagreed—but they disagreed obliquely, not diametrically. They didn't deny the significance of habitat diversity. In fact, Diamond himself had emphasized habitat diversity just a few years before, in a study of bird species turnover on the Channel Islands. But now Diamond and others found it reasonable to propose that, for practical purposes, the laborious inventorying of habitat diversity can be ignored. In the real world of speedy ecological destruction caused by humanity on the march, they suggested, a shortcut conservation method is justifiable, even necessary. They were thinking of planners like Tom Lovejoy, obliged to make consequential decisions on the basis of incomplete ecological data. Under those circumstances, according to the Diamondites, the sheer area of a proposed reserve is a highly significant datum. It can be taken as a rough gauge of both species abundance and habitat diversity.

No it can't, said the Simberlovians. Presume to make such an equation, they said, and you're closing your eyes to the texture of ecological reality.

The fine points of this controversy filled dozens of pages in the *American Naturalist*, *Biological Conservation*, the *Journal of Biogeography*, and other journals. Which side was right? It's anybody's call. In caricatured form (and angry debates do tend toward caricature), the opposing views became known as the "habitat-diversity hypothesis" and the "area-per-se hypothesis," despite the fact that few Diamondites would have agreed that "area-per-se" was an accurate label for their position.

Taking up another issue, the Simberlovians pointed out that several small reserves would offer insurance against certain forms of catastrophe. If a typhoon or a forest fire struck one small reserve among a scattered system of several, the other reserves would probably be spared, whereas the same typhoon or fire might sweep straight across a single large reserve and eradicate whole populations of plants and animals. If a virulent disease took hold in a large reserve, it might likewise spread widely, while several small reserves would offer the protection of quarantine. An exotic predator, like the tree snake on Guam, might also cause more damage within a single large reserve. The same predator would need to accomplish several successful colonizations, each one against high odds, in order to terrorize several small reserves. Don't put all your eggs in one basket, the Simberlovians were in essence arguing.

The Diamondites, for their part, noted that a single large reserve would offer greater security within its interior, while several small reserves would suffer proportionately more disturbance along their edges. A large reserve, too, would be necessary to save large-bodied and wide-ranging predators, such as the jaguars and pumas of the Amazon. And a large reserve might provide a margin of safety against climate change; if one region of the reserve became too dry or too hot for a given species, that species might be able to move to a wetter or cooler region without being stopped by a boundary. A small reserve would afford no such migrational latitude.

So it went: pro and con, argument counterbalancing argument. Small reserves might be more cost-effective for saving tropical insects and rare plants, which tend to be localized within small pockets of habitat. A careful selection of small reserves might harbor more plant and animal species than one large reserve, and those scattered reserves might even cost less, in total, than a single continuous parcel of land. On the other hand, many plants and animals within small reserves might represent common and widespread species not really needing protection. One large reserve might contain fewer species overall but more species that are rare and jeopardized. On the other hand, small reserves might . . . On the other hand, a large reserve might . . . And so on.

All these assertions are summarized in the subjunctive—things that *might* happen, rather than things that had happened or indisputably would—because the SLOSS debate by its nature was hypothetical. And the deep text beneath much of the hypothetical arguing was the equilibrium theory of MacArthur and Wilson.

A large reserve might receive more immigrations and endure fewer extinctions, according to Diamondite logic. Why? A large reserve presents a bigger target to dispersing individuals—hence a higher immigration rate. A large reserve supports large populations, thereby buffering each species against the dangers of rarity—hence a lower extinction rate. Of course a higher immigration rate combined with a lower extinction rate results in a larger number of species at equilibrium. Q.E.D., said the Diamondites.

B.S., said the Simberlovians. There you go again with MacArthur-and-Wilson.

136

THEY FOUGHT to a standoff. As the conflict continued and the lines of alliance and opposition became clear, the arguments grew progressively more rancorous and abstruse. But despite all its vehemence, not every aspect of what Tom Lovejoy calls "that pissing match" was relevant to our interests here. One portion that *was* relevant involved a biologist named Michael Gilpin.

Gilpin, based at UC–San Diego, is a pal of Ted Case, the swashbuckling herpetologist of the Gulf of California, with whom he shares a certain robust personal style. You wouldn't mistake Gilpin for a corporate lawyer; more likely you'd take him for a high school track coach. Like Case, he's an aging athlete who sees no reason why stiffening joints and cholesterol and a forty-seventh birthday should consign him to sober-sided adulthood. His laugh, sounded often, is a full-hearted, high-pitched bray. He reads widely, plays hard, thinks fast but deeply, and his world remains larger than science, though his scientific world is large. My own intermittent multiyear interview with Mike Gilpin has been conducted in kayaks, on ski lifts, while jogging along city streets, in the cab of an old truck crossing Nevada by night, at campsites in the Montana backcountry, and over more than a few glasses of beer. He also competes in triathlons, but for that foolishness he's on his own.

Gilpin was trained originally as a physicist and worked in the 1960s for Hughes Aircraft, "developing lasers to blind Viet Cong," as he puts it cheerily. That life was unappealing, so he dropped out for the Peace Corps. After a few years in the Mideast he returned to the United States and found himself diverted toward the science of ecology by watching Paul Ehrlich one night on the *Tonight Show*. Ehrlich was talking about human population growth, the extinction of species, and what amounted to ecosystem decay. Gilpin took it to heart. He went to grad school in ecology and finished a doctorate quickly.

Reborn as an ecologist, he showed a special talent for computer modeling. He also performed well in collaborative matchups with other ecologists. He did some abstruse theoretical work in population genetics, he studied the dynamics of predator-prey interactions and competition between similar species, but then he came into contact with "two real good island biogeographers" who helped shift his interests in that direction. One was Ted Case. The other was Jared Diamond. Strong in math, masterly as a computer programmer, Gilpin during the SLOSS debate made a good partner for Diamond and a worthy opponent for Dan Simberloff.

His work with Diamond, contra Simberloff, came to focus eventually on something called the "null hypothesis." This was a hot little battle fought out parenthetically to the larger war.

The null-hypothesis controversy was somewhat more purely scientific than the matter of SLOSS. It concerned deductive procedures and their philosophical underpinnings. But it did contain some practical implications of its own. Its starting point was a study, published by Diamond in 1975, of bird distributions and community structure on fifty islands of the Bismarck Archipelago, just east of New Guinea.

The Bismarck bird distributions were irregular, a messy patchwork of avian rosters differing from one island to the next. Amid the messiness, though, was a hint of order. Certain species or combinations of species appeared to be mutually exclusive. Diamond considered that significant. Among the factors determining these distributions, he concluded, was interspecific competition. Some of the bird species just couldn't coexist with some others. Competition made it impossible for them to live sympatrically. They had prevented each other from colonizing their respective islands.

Or maybe not. In a 1978 paper Simberloff attacked both Diamond's methods and his conclusion. It would be more parsimonious, Simberloff argued, to entertain a null hypothesis before seizing on the hypothesis of competition. What was that null hypothesis? Randomness. Maybe the Bismarck distributions, which seemed to reflect competition, in reality reflected nothing but chance. The human mind is always eager to see meaningful pattern, after all, even where meaningful pattern doesn't exist. We compose the stars of the night sky into hokey, connect-the-dots constellations. We imagine that we've caught sight of animal shapes in the clouds, flying saucers in swamp gas, a conspiracy behind every witless assassination, and the future in tea leaves. Simberloff suspected Diamond of making the same sort of credulous leap. He urged skepticism. If sheer chance was sufficient to explain those bird distributions, Simberloff argued, then Diamond's claim that the data reflected a more causal process (namely, competition) was superfluous and logically unacceptable.

From that start, the null-hypothesis dispute rolled forward through time and across journal pages, with Mike Gilpin joining Diamond's side and a biologist named Edward F. Connor co-authoring papers with Simberloff.

Diamond, the bird ecologist, had collected much of the Bismarck data during his arduous field expeditions in the New Guinea region. He himself had done the original, seat-of-the-pants analysis. But with Gilpin's collaboration he could go further. Gilpin, the math jock and wizard programmer, was capable of holing up in his house and writing a thousand lines of ingenious computer code before he emerged again into daylight. Using an early-model Vax computer, he pushed the analysis of Diamond's Bismarck distributional data to new levels of statistical sophistication.

Simberloff and Connor had their own durable, synergistic partnership, which gave a certain two-against-two symmetry to the flaming-arrow battle of the papers. A Connor-and-Simberloff paper would appear, offering new and more trenchant criticisms of the original Diamond work; that paper would be answered by a Diamond-and-Gilpin paper, to be answered in turn by a Simberloff-and-Connor, which a Gilpin-and-Diamond would then rebut. It got testy. They derided each other's mathematical competence. They used exclamation points, which are rare! and drastic! in scientific literature. They accused each other of malpractice. It's slapdash and illogical, claimed Simberloff and Connor, to embrace the competition hypothesis without first having disproved the null hypothesis. No it isn't, argued Gilpin and Diamond. And furthermore, they said, your "null" hypothesis ain't so perfectly null. It contains hidden assumptions that introduce ecological bias. Does not, said Connor and Simberloff.

There was a bit of repetitive bickering, as well as formidable displays of ecological ratiocination and computer-aided statistical hot-doggery. Translate the whole thing into theological terms and it looks roughly like this: Diamond and Gilpin were rational Unitarians, Simberloff and Connor were devoutly agnostic.

Do you hunger for a detailed account of who said what and how the other guys answered? My own working hypothesis is that you don't. But I wanted at least to mention this business for two reasons: because Mike Gilpin will reappear later, in a context more directly germane to the extinction of insularized populations, and because some knowledge of the null-hypothesis controversy is prerequisite to appreciating what confronts me as I knock on Dan Simberloff's door.

137

SIMBERLOFF'S office is in a nondescript building on the campus of Florida State University, in Tallahassee. It stands near the end of a cinderblock corridor. Though the surroundings are drab, the door itself is festooned with newspaper clippings. One headline catches my eye:

IMAGE ON TORTILLA DRAWS CROWDS TO HIDALGO HOME

The copy reads: "They come alone or in groups to view a tortilla in a foil-lined cup above a home altar. Candles flicker while people wait in a darkened parlor for a chance to see for themselves what all the fuss is about." Hidalgo is a small town on the Mexico side of the Rio Grande. The fuss, it seems, is about a provocative pattern. Whether that pattern might be meaningful is a question the newspaper reporter has left to others. "They want to decide first-hand if the story is true, if an image of Jesus Christ did miraculously appear on the last tortilla made by Paula Rivera, a Hidalgo housewife, on February 28." Some academics are content with Gary Larson cartoons on the office door. Dan Simberloff's humor seems a little more dry and pointed. What this clipping represents, of course, is the continuation of the null-hypothesis argument by other means.

There are a half-dozen more in the same vein. JESUS' IMAGE DRAWS CROWDS AT FOSTORIA, reads the headline above a report about dark smeary shapes on the side of a soybean-oil storage tank in Ohio. VIRGIN MARY'S "IMAGE" DRAWS FAITHFUL, CURIOUS, reads another, announcing the case of an Arizona woman who found Our Lady of Guadalupe in a yucca plant. WOMAN MOVING G.E. JESUS is a good one, involving a Tennessee lady who declared she was on a mission from God. The mission was to move her General Electric freezer, upon which she had discovered a facsimile of Christ's visage, to a site where the public could more conveniently venerate it. "Arlene Gardner insists that God chose her appliance to reveal the face of Jesus, but some of the 3,000 viewers who have seen it in recent weeks say it looks more like country singer Willie Nelson." That last part seems especially implausible. Why would God trouble himself to make Willie Nelson's face appear on a freezer?

I suppose it's worth stating that I'm not making this up. We're dealing strictly with facts—or, to be slightly more accurate, with real-life newspaper journalism. We're dealing with the data on Dan Simberloff's door. Maybe the best of his headlines is MIRACULOUS FACE APPEARS ON SCROTUM! "IT'S A MIRACLE!" SAY EXPERTS, about which the details can be left to your fertile imagination. Simberloff has also posted a story about a Moslem man in England who found Allah's name written inside an eggplant. Somewhere in western Asia, no doubt, the face of Mohammed has shown up in a chapati, though Simberloff so far doesn't seem to have gotten wind of that one. But there is another account of Jesus revealing himself in his favorite medium: "Maria Rubio was rolling a tortilla for her husband's dinner last fall when she noticed that the skillet burns on the tortilla resembled the mournful face of Jesus Christ, crowned with a wreath of thorns. Since then, 8,000 curious pilgrims, most of them Mexican-Americans, have trekked to the Rubios' small stucco house in rural Lake Arthur, N.M., to view what they consider a sacred icon."

Standing at Simberloff's door, my head swims with questions. For instance: corn or flour? I would love to read my way through the full dossier, but I'm already late for my appointment and I don't want him to catch me lurking outside. So I knock.

138

A FRIED tortilla is data. An oval shape of oil-darkened splotches represents pattern. To assume that the pattern is meaningful—that it's a face, that in fact it's the face of Jesus made manifest as a miraculous message—is to adopt a theory about the pattern within the data. Dan Simberloff has strong convictions on the general subject of theory.

"Maybe I haven't been blunt enough," he tells me.

He's alluding to his writings on SLOSS, his role in the null-hypothesis controversy, and his published criticisms of what he considers the overzealous application of MacArthur and Wilson's theory. "It's hard for me to believe that I *haven't* been blunt enough," he adds. "But theory can cause a lot of trouble."

It has led to some disastrous mistakes, Simberloff says. And besides the actual disasters, there have been some near misses. He mentions a handful of cases in which bad conservation decisions were barely averted when the prevailing theory was successfully challenged. Officials in Costa Rica, for instance, were going to give up on jaguars and harpy eagles because they had read somewhere that rarity to a certain precise degree (fewer than fifty individuals) dooms a species to extinction. Their jaguars and eagles were already that rare, so they were inclined to abandon all hope. Simberloff knows of similar situations in Australia, where state governments were prepared to eliminate protection for certain species, based on theoretical notions suggesting that the species were irremediably rare.

Another notable triumph for wrongheaded theory, in Simberloff's view, occurred when the United Nations Environmental Programme, the International Union for the Conservation of Nature, and the World Wildlife Fund adopted—in their joint *World Conservation Strategy* document of 1980—the reserve-design principles that Jared Diamond had promoted in "The Island Dilemma."

And there's the case of Israel. "I've done some work in Israel, on leaf-mining insects," says Simberloff. "When I was there in '85 or so, a man from the Nature Reserves Authority came to see me. He was desperate. And here's what he was desperate about. Israel has a wonderful system of nature reserves in a country where it's very hard to *keep* nature reserves." They have more than two hundred reserves, he says, though the country itself is smaller than Maryland. Furthermore, a large portion of the Israeli citizenry seems to care fervently about nature. Like the English, Simberloff explains, Israelis have a strong proprietary affection for the remnants of their own native fauna and flora. Most of those reserves are adjacent to bodies of water, where the ecological communities are rich. Many of them were established decades ago, back when Israel itself had just gained statehood. For years there had been squabbling between the Nature Reserves Authority and two interests that wanted the reserves *dis*established: agriculture and the military. Agriculture coveted the sites for their water, and the military wanted them for strategic reasons. But the political pressures had been resisted; the reserves kept their protection. Then, just about the time of Simberloff's visit, a new factor threatened to tip the balance.

"They read, both the military and the agricultural interests, that small reserves are not able to preserve viable populations. And almost all of the reserves in Israel are very small. So at a cabinet-level meeting they said that the weight of scientific evidence shows that these are worthless anyway, and what's the good of them? And they wanted me to go and look at their reserves, and then to provide some scientific support. You know, it was utter bullshit," he says.

The Israelis had caught wind of SLOSS. They were swayed by the argument that a number of small reserves are necessarily inferior to a few large ones—the same argument that Simberloff dismisses as "a cocktail-party idea having the trappings of science."

I'm inclined to agree with him on one point, among others: It seems unlikely that he hasn't been sufficiently blunt.

"I could give you other examples," he says. "The theory that doesn't really have strong empirical support can be very dangerous." In the dangerous class he includes not just Jared Diamond's reserve-design principles but also the equilibrium theory itself. Theory in any branch of science entails a risk of detachment from reality, but especially so in a science as multifarious as ecology. If the theory is applied in decision making that affects how humanity treats portions of the world's landscape, the risks and the consequences are still greater. "It's not like, at worst it's neutral. At worst it's *bad*."

The proper way to bring science into conservation planning, Simberloff says, is with detailed ecological studies of particular species in particular places, not by applying grandiose theories. The term for what he favors is *autecology*, embracing an imperative to learn the ways of the creature itself, and the immediate relationships that connect it to its place, before drawing conclusions about the overall structure of the community to which the creature belongs. This contrasts with *synecology*, attending more to the community dimension and the sort of organizing principles that Diamond saw among the Bismarck Archipelago birds. Although the difference between the two is only a matter of emphasis, it's a clear enough difference to be significant. Autecology, Simberloff's preferred approach, tends to be more descriptive than theoretical, and descriptive ecology is out of fashion. Some people are still doing that sort of work, he says, but they don't get enough credit from their colleagues and they don't exert enough influence on conservation policy. Too many other ecologists are in a hurry to generalize. Too much attention goes to the hotshot theories.

"It's sad. There's an element of physics envy in all of this," Simberloff tells me. "That's what is really at issue here. Conservationists and conservation scientists feel that unless they can point to a theory—and the more quantitative it is, the better—they won't be able to get people to respect their views and ideas." The people whose respect is so crucially at issue are government bureaucrats, conservation administrators, and politicians, as well as scientific colleagues. In order to be heeded in the councils of policy, some ecologists believe, they've got to deliver deductive conclusions in concise mathematical form. "And it's too bad, but ecology isn't that kind of science."

I ask: Is this the legacy of Robert MacArthur?

No, Simberloff says. He finds that explanation too simple. MacArthur happened to be a very, very smart guy who came along at the right time, but the yearning to mathematize ecology was already in the air. If it hadn't been MacArthur, it would have been someone else a couple years later.

How did you yourself become disenchanted with the equilibrium theory? I ask.

"Well, you know, it goes back to, I don't know . . ." And he diverges from there into a roundabout answer, leaving me to continue wondering: back to where?

To the very beginning, presumably. Simberloff after all is the fallen archangel who had stood at the right hand of Ed Wilson while Wilson and MacArthur were creating this particular universe. That's one way of seeing it, anyway, admittedly a little melodramatic. Dan Simberloff himself is not inclined to melodrama. His mind makes razor-clean, ruler-straight cuts. His nostalgia, if any, is not on display to every visiting journalist. Having cast his memory backward, he starts over.

"I'm sort of viewed as a professional crank," he says truly. "I'm disenchanted with lots of community ecology. And my disenchantment to some extent with the equilibrium theory stemmed from the same source. That is: Modeling is fun, sometimes it produces elegant structures, but there is a tendency to reify models. To take them as nature, when really all they are is proposed abstractions of nature. I'm concerned when a literature begins to develop on the models themselves, rather than on nature." There's a long history of such airy theorizing in the journals of ecology, he says. Too much conceptual scat-singing, too little observed data. Far too little experiment. "And the equilibrium theory, it seemed to me, was increasingly that sort of beast. That is, Ed and I did that experiment to try to test it directly." They went to the mangroves. They censused real arthropods on real islands under the real Florida sun. They not only tested theory against reality; they did it with data from a controlled, focused, carefully manipulated situation. They found a way to convert small bits of functioning ecosystem into a rigorous experiment. "Not many other people did."

This brings us around to Tom Lovejoy's experiment in the Amazon, for which Simberloff professes a guarded admiration. The original idea was a great one, he says. The execution has been mixed. Too bad that the experimental conditions aren't more precisely controlled. Too bad that there aren't more replicate patches. "But hell, no one had done anything like that before. So it was a neat thing to do." And it has produced some good work. He mentions a herpetologist named Barbara Zimmerman, who happens to be one of his own doctoral students. Zimmerman's study of frog species on the reserve sites, he says, tends to show that habitat diversity is more important than sheer area. Based on those findings and others, she's a skeptic. She thinks that the application of equilibrium theory to conservation, and the notion that a single large reserve will always be better than several small ones, are hooey. But her results have come out of Lovejoy's big experiment, and that's good, Simberloff says.

We talk for two hours about these matters and others. He loses me occasionally with his brisk, concise explanations of intricate scientific ideas, and at one point he shows impatience with my obtuseness. Like Robert MacArthur, Dan Simberloff is clearly a very, very smart guy. And he knows it. He is perfectly civil and cooperative but, unlike Ed Wilson, he doesn't squander his charm on strangers. It would be easy to make a faulty judgment of the man. He possesses a dangerous critical intelligence, yes. He's blunt, yes. He would gladly cut the throat of an ill-conceived idea and swat down an innocent misstatement like a fly. He seems to care zip about what most people think of him. He despises ecological theorizing when it's done irresponsibly, which in his view is most of the time. He values fieldwork, autecology, and experiment. He seems curt and unemotional. It would be easy to conclude, as some among his peers evidently have, that he cares far more for sleek scientific rigor than he does for the natural world. I might have reached that conclusion myself. But somehow our conversation swings back to his work on the mangrove islands. He has been down there a few times since the 1970s, he tells me. It's not so far from Tallahassee. He still has some ideas that he would like to pursue in that experimental context. But he hasn't been to the site now for five years.

"I'll tell you why I stopped working there," he says, though I haven't asked.

With each trip, he would notice another diner or trailer park or some other touch of human encroachment upon the natural landscape. "It became very depressing to work down there, because I could see the Florida Keys going to hell around me," he says. Then, casting back again to the early days of his work with Ed Wilson, he shares a story.

"After we figured out roughly what I was going to do, I guess in '65 or '66, I went down and spent quite a while in the Florida Keys. Getting to know the system, scouting out islands that we might be able to use, et cetera. There was an island at that time near Seven Mile Bridge—it was the third island to the west of Seven Mile Bridge, the first substantial island—called Ohio Key, for some reason, I don't know. It had a lot of forest. No one lived on it. And fringing mangrove swamp. Off that swamp there were two islands that in our notes would be called E4 and E5, if I remember correctly. And I practiced censusing those islands. We may have practiced fumigating one of them, I'm not absolutely sure." He and Wilson also made a collection of snails from the two islands. Ed still has the snail shells in a box, says Simberloff, and uses them to illustrate certain points when he teaches evolution. In a purely symbolic, sentimental way, these two tiny islands off the edge of Ohio Key were special. Simberloff came to know them intimately. "I mean, they were my first mangrove islands. They were about a third the size of the ones we ended up fumigating. They each had about twenty-five species." They stood just beside Ohio Key, no farther away than from here to the door, he says, gesturing. That proximity made Ohio Key itself seem a bit special. Relative to E4 and E5, it was the mainland.

"And then in the mid-1970s," he says, "I don't remember the exact year, I was driving down to one of the field sites for those later experiments. I came off Seven Mile Bridge, passed over the first key, it was called Little Duck Key, and the next key, which was Missouri Key. Normally, from Missouri Key I would see the trees of Ohio Key. Instead of Ohio Key I didn't see *anything*. Then, when I got to the end of Missouri Key, I could see that the reason was, there were no trees there. The entire key, which would have been in the range of four acres, had been leveled and cleared. It was now all crushed coral. It had been turned into a trailer park."

It was renamed Holiday Key, Simberloff says drily. "And it was too new even for there to be trailers there. So, there was this one central place, a trailer that was the store, and then there were all these stanchions, you know? For hooking trailers. And off this gleaming coral monstrosity were my two islands, E4 and E5. They were still sitting there."

By now I see where the story is headed and, despite my wariness, I've begun to like Dan Simberloff. He's not just smart as a cobra. He's also complicated. It would be easy to assume that his sleek scientific rigor and his adamantine empiricism are incompatible with a deep love of nature. But that, after all, would be just another theory.

"I was so . . ." He pauses.

He starts again. "I drove right over it into the next key, which is Bahia Honda, where the state park is. And I pulled over, and I cried. I couldn't handle it. It was just so sad. And it so epitomized what was happening in the Keys."

Simberloff knows, better than most of us, that the same sort of thing is happening on islands and mainlands all over the world. But sometimes the most personalized losses seem the most dire.

"That's why I stopped working there," he says.

139

Barbara Zimmerman's article on Amazon frogs, which Simberloff mentioned as we talked, was co-authored by Lovejoy's own field director, Rob Bierregaard. It was published in 1986. It's a provocative paper on several counts, and my own copy shows multiple layers of chicken-scratch notations from the half-dozen times I've read it over the years. Although it does carry Bierregaard's name, Zimmerman is the first author, and the frog data derive from her work. Really it's two papers in one: an analytical section on habitat requirements among thirty-nine species of frog, and a section devoted to arguing that the equilibrium theory holds no relevance whatsoever to conservation.

Like her mentor Simberloff, Zimmerman is a confirmed autecologist. She believes that the best route toward conserving frogs is the laborious, incremental, muddy-footed study of—what? yes, of course—frogs. She disbelieves ardently in theoretical shortcuts. Nearly three years of fieldwork convinced her that the survival of frog populations within the Amazon reserves depends on contingent details of available breeding habitat, not on reserve size. She identified three types of critical habitat, each of which supports different frog species: large streams, small permanent pools, temporary pools. The supply of those habitat types within the various reserves, she found, does not correlate with reserve area. In other words, ironic as it may seem, Zimmerman's involvement with the Minimum Critical Size of Ecosystems Project persuaded her that the very concept of minimum critical size (at least as an abstraction detached from particularities of habitat on a given site) is vacuous.

Maybe that's not so ironic, in fact. Maybe it's just science properly done, wherein conclusions based on empirical data don't always conform to initial assumptions.

Forget about sheer area, she argued. Forget the surveyor's boundaries. A hundred-hectare reserve containing a rich supply of breeding habitat will have more conservation value than a five-hundred-hectare reserve containing little. If you want to protect amphibians, she said, then ignore all those species-area curves, ignore $S = cA^z$, ignore the notion that only a big reserve is a good reserve, and focus your protection around streams and puddles.

This was the lesson of the frogs. In the paper's other section Zimmerman and Bierregaard made a broader claim. Cast in regretful tones, it sounded almost elegiac. "Because of the similarity between islands and nature reserves, there were great hopes that the equilibrium theory of island biogeography would produce guidelines for the design of nature reserves." What became of those hopes? Punctured, deflated, squashed. What became of applied biogeography? Nothing. What happened to the equilibrium theory itself? Ecologists got tired of reading about it. "The inescapable conclusion," according to Zimmerman and Bierregaard, is that MacArthur and Wilson's theory "has taught us little that can be of real value planning real reserves in real places."

Dan Simberloff must have been gratified when he read this. Jared Diamond, presumably not. Tom Lovejoy stood apart from the pissing match, as always, and tried to stay dry. His project was still young and more data would come. The lesson of the frogs wasn't necessarily the lesson of the birds or the monkeys or the beetles. But because its narrow assertions are based solidly on experimental results, and because its broader argument is cogently stated, the Zimmerman-and-Bierregaard paper may have seemed, to some biologists, almost terminally persuasive. Was the SLOSS debate settled, then?

No. Probably SLOSS will never be settled, only eclipsed by newer frameworks of dispute. In the meantime, anyway, there was another interesting voice to be heard. A young man named William Newmark had entered the discussion.

140

NEWMARK wasn't associated with the Lovejoy project. His scientific ideas hadn't been shaped among Florida mangroves, New Guinea birds, or Amazon frogs. He worked in a chillier setting.

In the mid-1970s, Newmark came out of the University of Colorado with a degree in political science. Even as an undergraduate doing a poli sci major he had taken a smattering of ecology, and in one of those courses he had gotten his first exposure to the notion of ecological insularity. "This was in '74," he says. "I can distinctly remember the lectures, talking about island biogeography and saying that there may be some application to nature reserve design." MacArthur and Wilson's equilibrium theory was mentioned at least passingly in the course. Soon afterward, Newmark switched to biology, finished another bachelor's degree in that, and then started a graduate program at Michigan. While researching his master's thesis, a modest project concerning data collection within the Yellowstone ecosystem, he discovered that the archives of Yellowstone National Park contained faunal sighting records dating back through earlier decades of the park's history. The records told which species of animal had been seen when. They also told, implicitly, when certain species had stopped being seen. It was a potentially revealing body of information.

Newmark learned that the same sort of information existed for other parks too. Historical sighting records had found their way piecemeal into a wide range of chronicles, reports, and checklists. From those sources, Newmark realized, a patient researcher could compile a fuller set of checklists, indicating which animal species had been present in a given park when the park was founded and at various times since. In 1976, for instance, it was a matter of record that

Yellowstone harbored moose and elk and bison and two species of deer, ravens and bald eagles and Clark's nutcrackers, mountain lions and lynx, bobcats and wolverines, black bears and grizzly bears, badgers and coyotes and ermine, among other species. How closely did that list match the list from, say, sixty years earlier? In 1916 there were moose and elk and bison and . . . virtually the same list, with a single addition: wolves. The gray wolf, *Canis lupus*, had been present in the old days but was now gone.

The sorry tale of wolf eradication in Yellowstone has little pertinence to island biogeography (and much pertinence to a benighted brand of wildlife management known as "predator control," once practiced by even the National Park Service), but it exemplifies the more basic fact that rosters of fauna within American national parks have been subject to change. Some species, once present, are later absent. And campaigns of active extermination aren't the sole factor responsible for such disappearances. Insularity is another. That wasn't lost on William Newmark.

Having finished his master's thesis, he was shopping for a doctoral topic. He remembered the sighting records. "I don't think I would have probably chosen the dissertation topic if I hadn't by chance run across these checklists when I was working in Yellowstone." He remembered also what he had heard, in the ecology course back at Colorado, about MacArthur and Wilson's theory. And he was aware that Jared Diamond, among others, had noted a parallel between nature reserves and islands. Newmark in response felt the same sort of empirical hankering as Tom Lovejoy: It would be nice to get some data on the relationship between insularization and ecological change. He decided to use faunal sighting records, from Yellowstone and other parks, for a study of that relationship.

His working premise was that national parks are analogous to land-bridge islands. They represent bounded areas of natural landscape, formerly connected to much larger areas, that have become (or are in the process of becoming) insularized within an ocean of human impacts. They may be younger than Tasmania, they may be larger than Barro Colorado, they may be surrounded by wheat fields and fenced pastures and towns and highways instead of by water, but ecologically the situation is similar. If his premise was correct, Newmark guessed, certain patterns might appear in the sighting records.

He left Ann Arbor in autumn of 1983, headed west. "I was driving a small Toyota station wagon and I was sleeping in the back of it," he says.

"I would basically drive into a park, introduce myself to the park scientists, the park superintendent, and request permission to go through their files and develop a checklist, or use their checklists as a starting point, and gather this information on every species." Over the course of several months, he made a circuit through the western United States and up into two Canadian provinces, visiting twenty-four different parks and park clusters. Yellowstone and Grand Teton together constituted one cluster. Glacier National Park and Waterton Lakes, contiguous on the international border, were another cluster. He also went to Zion and Bryce Canyon (separate parks, not clustered) in Utah, Yosemite in California, Crater Lake in Oregon, Olympic and Mount Rainier in Washington, and the huge Banff-Jasper cluster along the Canadian Rockies. At each stop, with the indulgence of the superintendent and other park officials, he combed the files. He focused especially on the larger mammals, because sighting records for them were more thorough and reliable than for less conspicuous animals. If a given species was present within the park, there would be documentary evidence. If a given species was formerly present and latterly seemed to be missing, then he asked when it had last been sighted. Back in Ann Arbor later that winter, he organized his data, filling in the gaps with further research by mail. He compiled uniform mammal checklists for all the parks and park clusters. He mailed his checklists back to the park scientists for review and correction. He turned up some facts that, individually, seemed unfortunate but not greatly troubling. Collectively, their implication *was* troubling.

The red fox was missing from Bryce Canyon National Park. According to Newmark's research, it hadn't been seen there since 1961. The spotted skunk and the white-tailed jackrabbit, by his tally, were likewise missing from Bryce Canyon. Mount Rainier National Park had lost the lynx, the fisher (a small predator similar to a weasel), and the striped skunk. Crater Lake National Park had lost the river otter, the ermine, the mink, and the spotted skunk. Sequoia and Kings Canyon National Parks, standing adjacent to each other in central California, had also lost their populations of red fox and river otter.

Newmark found more than forty such cases. None of these local extinctions seemed to be the result of direct human activity—not hunting, not trapping or poisoning. Whatever had caused them was invisible.

Besides compiling his species checklists, Newmark gathered quantitative information concerning habitat diversity within each park. He

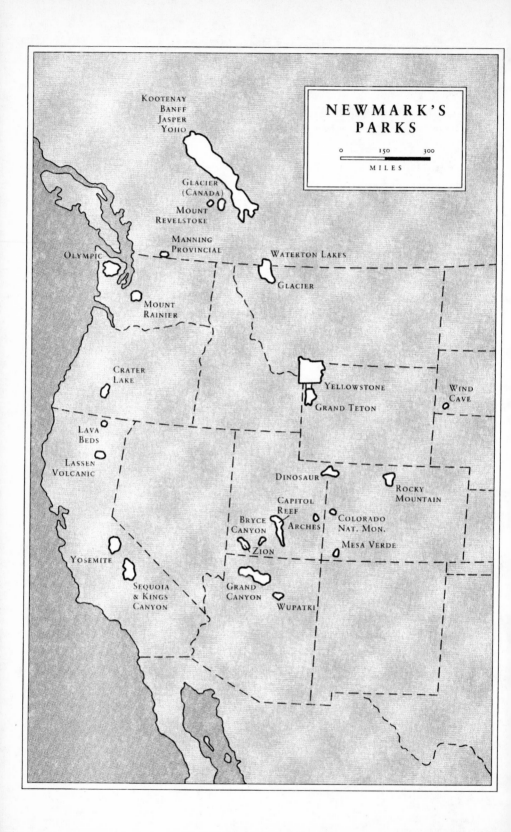

also took note of the size of each park and the year it was established. Species diversity, habitat diversity, area, age—on this body of data he performed a battery of mathematical analyses. Then he looked for larger patterns, and found some. Eventually he wrote a doctoral dissertation that turned heads in the scientific community, which not many doctoral dissertations do.

In early 1987 he published a summary of his results in *Nature*. Because *Nature* is an august journal and science writers watch it for leads, Newmark's findings made news in the wider world. The *New York Times* ran a story under the headline SPECIES VANISHING FROM MANY PARKS. The *Times* reporter, without troubling his readers with chatter about island biogeography, announced that mammal species were disappearing from North America's national parks "solely because the parks—even those covering hundreds of thousands of acres—are too small to support them."

The full dissertation had meanwhile achieved some circulation in photocopy form. From the first set of photocopies were taken further photocopies, grainy and bad, like home dubs from a bootleg recording of Bob Dylan performing half-drunk at a local bar. Word spread that this newcomer named William Newmark, whoever he was, had done an important bit of work. He had discovered evidence of ecosystem decay among America's own treasured national parks.

In the dissertation, Newmark had written:

> Several predictions follow from the hypothesis that nature reserves are analogous to land-bridge islands. These are: (1) total number of extinctions should exceed total number of colonizations within a reserve; (2) number of extinctions should be inversely related to reserve size and; (3) number of extinctions should be directly related to reserve age.

The three predictions came right out of Frank Preston's discussion of samples versus isolates, as further developed in MacArthur and Wilson's theory and as interpreted for application in Diamond's "The Island Dilemma."

A nature reserve, by its essence, is a sample of landscape that is destined to become an isolate. The total number of extinctions "should exceed total number of colonizations" because immigration will be impeded. The number of extinctions "should be inversely related to reserve size" because (a) reserve size will determine population size for each species and (b) small populations face special jeopardies. The thrust of these first two predictions is that a newly delineated reserve will temporarily hold more species than its equilibrium number but that the surplus will progressively be lost, and that the losses will be comparatively great if the reserve is comparatively small. As time passes, species will continue being lost until the reserve has "relaxed" to a new equilibrium. The reserve age will be relevant as a rough estimate of how much time has elapsed since the sample became an isolate and, therefore, how far its loss of species has progressed.

Do the parks of North America conform to those predictions? Scrutinizing the fourteen western parks and park clusters for which records were most complete, Newmark found that they did. Bryce Canyon National Park, Lassen Volcanic National Park, and Zion National Park were the three smallest on the list. According to his research, they had each lost close to forty percent of their larger mammal species, either through direct human persecution or through insularization. Newmark factored out the species lost by direct human persecution—such as the gray wolf, which had been extirpated intentionally from Bryce Canyon, as it had from Yellowstone—and showed the subtotals of those extinctions with no apparent cause except insularization. Of this sort, Zion had suffered five. Lassen Volcanic, six. Bryce Canyon, with the smallest area, had suffered not quite so many extinctions—four—but Bryce Canyon is also a younger park than the other two. Maybe its time would come.

The largest park clusters had suffered the fewest losses. Yellowstone and Grand Teton together, twenty times as big as Zion, had lost nothing but the wolf.

Newmark's mathematical tests showed that the inverse correlation between extinction and area was statistically significant. In plain words, large parks had consistently lost fewer species than small ones. And young parks had also lost fewer species than old parks (though the correlation of those variables wasn't quite so strong). He concluded that the fourteen parks and park clusters had experienced "a mammalian faunal collapse," most likely caused by insularization. That was the essence of what he published in *Nature*, and that's what the *New York Times* carried.

Newmark had reaffirmed Preston's warning, given twenty-five years earlier: "If what we have said is correct, it is not possible to preserve in a State or National Park, a complete replica on a small scale of the fauna and flora of a much larger area."

The dissertation itself made another remarkable point, concerning the issue of habitat diversity versus sheer area. In one phase of his data analysis, Newmark had totaled up the checklists for each park and park cluster, and against those totals he had set a question: Were the differences in mammalian richness best explained by sheer area or by habitat diversity? Area is an easily measured quantity. Habitat diversity is less easily measured, since it involves complex and blurry gradations, but Newmark tried to touch at it indirectly by way of some other parameters—latitude, elevational range, plant diversity. Among those three, plant diversity might seem the most pertinent, since vegetation is such an important component of animal habitat. But he found that plant diversity didn't correlate significantly with the number of mammal species within each park. Latitude didn't correlate either. These results are notable. They tend toward refuting the habitat-diversity hypothesis, propounded by the Simberlovians during the SLOSS war. An argument in rebuttal can be made: that Newmark's three parameters missed the essence of habitat diversity. Still, it's interesting that, of the three, only elevational range showed a significant correlation with species number. And the strongest correlation that Newmark found was between species and sheer area.

The large parks contained more mammal species than the small parks—that much wasn't unexpected. What was unexpected, at least by some biologists, was that the best predictor of species diversity was sheer area, regardless of park-to-park differences in elevational range, latitude, or vegetation. This gave new support to the area-per-se hypothesis, which some Simberlovians had so persuasively reviled. It set Newmark's North American park mammals in counterbalance to Zimmerman's Amazon frogs. For conservation planners like Lovejoy, it had serious implications. But it never made the *New York Times*.

141

"THERE IS ONE idealized will-o'-the-wisp goal, which we may or may not achieve," Lovejoy says over coffee at a hotel in Manaus. "Which I think is the ultimate goal for conservation anyway. Which is to have intact, fully diverse ecosystems persisting in reserves." He chooses his words carefully, even when his syntax is loose, and the most crucial of several crucial words in this statement is "persisting."

He's not talking merely about the miniaturized reserves of his Amazon project. He's talking about real nature reserves throughout the world. It's one thing to mark off an isolate of some ecosystem, hang a sign saying MINUSCULE NATIONAL PARK, post wardens to guard the boundaries from encroachment, and call it protected. Whether the species and relationships within that isolate persist over time, after the landscape all around it has been trashed, is another thing.

Take this central Amazon ecosystem as a test case, Lovejoy says. Consider the species diversity, just among trees. Assume that any ten-hectare sample amid the continuous forest supports about three hundred tree species. "I want ultimately to be able to define a national park size so that, a thousand years from now, somebody can go in and sample ten hectares and *still* come up with three hundred species of trees. And draped around all that is the invertebrate diversity and everything else"—ants, ant-following birds, frogs, peccaries, butterflies, monkeys, jaguars. "I'm not interested just in the preservation of individual species. But of the whole system. And that's . . . you know, I may not achieve that. But that's what I'm thinking about."

He admits that the project has had setbacks. Fewer reserves have been isolated than he had hoped. Changes in government policy have eliminated the fiscal incentives for cattle ranching in the Manaus Free Zone, so the ranchers are no longer clearing much land for new pasture—which is good for the forest but bad for Lovejoy's experiment. Second-growth vegetation has sprung up in some of the clear-cuts, blurring the line (in both a physical and a conceptual sense) between isolated reserves and continuous forest. And there have been the routine hassles inherent to tropical field biology. Equipment has been hard to maintain. Funding has been insecure. Research permit renewals have been subject to bureaucratic delay. Leishmaniasis, a nasty disease carried by tiny crepuscular flies, has taken a toll on fieldworkers.

And the research results have proven more complicated than Lovejoy foresaw. The area effect, the distance effect, habitat diversity, edge effects, small-scale changes in climate, relationships between species, and the vagaries of chance are all implicated in the biological dynamics of forest fragments. Even after more than a decade, there's no simple message. Although the enterprise has yielded insights on the problem of ecosystem decay, ideally it would have yielded more. Still, Lovejoy doesn't seem discouraged.

"In a funny kind of way, we could have been sitting there in the forest drinking *cachaça*" for all these years, he tells me, "and the project would have had some real benefit." *Cachaça* is that old sugar cane moonshine used by Alfred Wallace for pickling specimens. What Lovejoy means is that the emblematic value stands to some degree independent of the empirical results. The Minimum Critical Size of Ecosystems Project has been recognized broadly throughout the scientific and conservation communities as a single visionary idea, not just a collection of loosely related studies. "Its existence has influenced people to think about the problem, around the world, and respond by setting aside bigger areas."

142

LEAVING Lovejoy in Manaus, I return to the forest. I want to see it again from a perspective less Jovian than the view from a low-flying plane. I've arranged to spend a few days at one of the field camps, looking over the shoulders of two bird-banding interns named Peggy and Summer.

The self-mocking label by which these interns refer to themselves, Peggy and Summer and the others, is "bird slaves." They are generally young people with a bit of scientific training, or at least serious birdwatching skills, who have signed on to trade their labor for a tiny stipend and a chance to spend time in the Amazon. Peggy has a degree in earth sciences and environmental studies from UC–Santa Cruz. She has worked previously for the Peregrine Fund. Summer is a nonscientist with a streak of New Age idealism. Her previous experience includes eight years with a collective that ran a natural-foods bakery in Berkeley. Ever since she was a girl, Summer tells me, she wanted to come to the Amazon. "I have these throwback Neanderthal genes," she says, meaning it literally for all I can tell, and she feels them urging her to escape civilization for a more primitive life in the wilderness. She was inspired to volunteer for this bird-slaving tour of duty by her concern over rainforest destruction. But when she finally arrived here and began work, she admits with admirable candor, she found the forest terrifying. "I was scared shitless. It was just kind of a mental attitude. Then, after three months, I started to relax a little. Now I'm getting better." In response to my questioning, both Peggy and Summer declare that they have no thoughts whatsoever about the theory of island biogeography. As far as they are concerned, this project pertains to the ecology and preservation of the Amazon. Isn't that enough?

Well, yes. Okay. On one scale of judgment, it certainly is. Whether that scale of judgment encompasses the real significance of Tom Lovejoy's enterprise is a question that, for the moment, can be set aside.

Before dawn on the first day, we start walking. I follow them up a dark trail and out into a clear-cut, where the second-growth vegetation is short, thick, and weedy. Our destination is a hundred-hectare reserve. The two women wear headlamps powered by large battery packs and I carry a dippy little flashlight. We scuff along, ankle-deep in fallen *Cecropia* leaves as big as busted umbrellas. We step across dead logs, from the soggy flanks of which sprout lewd fungal growths in unearthly colors. The trail is slippery and narrow, a machete-carved tunnel through the weed zone. Then we enter the reserve itself, where the canopy is high and the understory is more open.

Lianas dangle down like a galleon's rigging. Capuchin monkeys yelp in the distance. Toucans and macaws shift from perch to perch overhead. The tree trunks are patterned with lichens and termite tracks. Epiphytes sit high in the limb crotches. From somewhere nearby comes the wolf-whistle call of a screaming piha. Although I catch glimpses by lamplight, my hearing tells me more than my vision. Memory supplies some ecological detail, and imagination adds a few beasts of its own.

I place my steps cautiously. I've heard Peggy mention that she surprised a sizable bushmaster (the world's largest species of viper, *Lachesis mutus*, famed for its propensity to strike without warning) at twilight on one of the trails. "Yeah, you could hardly miss it," she said. "A two-and-a-half-meter bushmaster, you're not likely to step on it." A snake so big and lethal that you couldn't set your foot on it in the dark? This leap of logic marks Peggy for an optimist, in my view, and as a pessimist I'm glad to be following ten paces behind while she walks point.

Beating the dawn is crucial for tropical bird banders. Their mist nets must be opened and ready by first light, when the birds begin moving. But working before dawn also exposes the banders to the crepuscular flies that carry Leishmaniasis. The disease is curable though ugly, I've been told, and the cure is ugly too: a series of injections of antimony. For protection I'm relying on long sleeves, pants tucked into boots, drugstore insect repellent, and fool's luck.

Peggy and Summer stop walking and I stride up beside them, still unaware that we've reached the net line. The fine nylon mesh is invisible until I've nearly smoodged my face into it. I flinch, from the same arachnophobic instinct that made me edgy on Guam. The central Amazon harbors at least one species of tarantula big and cocky enough to prey on birds. It may be the same species that Alfred Wallace knew and collected (another measure of his heroic fortitude) for its charm as a specimen. Rob Bierregaard too has seen one, in a mist net just like this, where it had killed a tangled bird before he could intervene. Mercifully, there's no sign today of the bird-eating spider. If there were, I'd probably push my head into its clutches and get wrapped up like a sulfur-rumped flycatcher.

The net line transects the reserve like a stutter of hyphens across a page. The nets are strung permanently on poles, so that each morning they need only be opened. Peggy and Summer and I set to it. Within a few minutes, we have deployed the whole line, a diaphanous barricade stretching three hundred meters through the forest. Then we vacate the area, crossing a sumpy blackwater stream and climbing a low slope. There, at the far side of the reserve, Peggy sits on a log. Get comfortable, she advises me.

The headlamps and batteries come off. The first hint of daylight leaks down through the broken canopy and then, soon, beams of sunshine. We're in the edge zone, where direct light and sun-loving species have begun penetrating the forest. A hummingbird appears before us to hover and gawk.

The hummingbird, a shiny little creature, has two tail feathers as long as toothbrushes. *Phaethornis superciliosus*, says Peggy, who knows a thousand things by their names.

After an hour we return to the nets. Peggy gently untangles the first bird. She soothes it, examines it, measures it. At a glance she and Summer both know it's a cinnamon-crested spadebill, *Platyrinchus saturatus*, of the Tyrannidae family. Its wide bill is adapted for scarfing small insects. This individual is a recapture, previously banded and released. Back at the office in Manaus, it's already on the computer. Now the fact of its lingering presence, and the details of its current condition, will be added to the file.

Peggy handles it deftly. She puffs feathers aside with a tickle of breath, checking the skin for parasites. She dictates to Summer: sternum configuration, percentage of fat, amount of body molt, wing length, tail length. Summer records the data. So do I, in my own notebook, without knowing why. My gut tells me only that there's something important, something preciously real, about such specificity. The tail length is twenty-five millimeters. The wing length, fifty-three millimeters. Body molt, none. Band number, 24998. The bird weighs ten grams.

Peggy opens her hand. For a moment the dainty brown spadebill sits dazed on her palm, heart pumping furiously, and I have time to wonder about its future. How long can *Platyrinchus saturatus* maintain itself in a hundred-hectare reserve? Will this species eventually disappear, like the margay cat and the golden-handed tamarin? Will it follow the birds that follow the ants? Will it prove to be ecologically desolate in the absence of solitary bees? Will it suffer terminally from the rupture of some other mutualistic relationship? Will it die back to a state of precarious rarity, then damage itself with inbreeding, lose its adaptive vigor, and be finished off by a minor accident? Or is one hundred hectares of insularized habitat all the universe that a population of cinnamon-crested spadebills will ever need?

And furthermore: Will the Lovejoy experiment answer these questions?

The bird flies suddenly, vanishing into the forest. "*Ciao*," Peggy says. Yes, I think. Goodbye and good luck.

Ṇ

THE SONG
OF THE INDRI

Ṇ

143

THE PEOPLE of Madagascar know something about minimum critical size. They have an adage: *Ny hazo tokano tsy mba ala.* It means: One tree doesn't make a forest.

But if not one, then how many?

The years of debate over SLOSS, the years of empirical work at Lovejoy's Amazon site, the decades of research and discussion inspired by MacArthur and Wilson's theory—all that time and effort have yielded no simple answer. Instead of a simple answer, there has been complicated progress. Around 1980, the very terms of discourse began to change. Since then, island biogeography has split open like a chrysalis and a metamorphosed creature has emerged: the concept of population viability.

If the people of Madagascar have an adage that captures the essence of that concept, I haven't heard it. Like the rest of us, they're faced with the need to absorb a new way of thinking.

144

THE FOREST reserve of Analamazaotra lies on the rugged eastern slope of Madagascar, sixty miles from Antananarivo. The road is bad and the rail line is sinuous, winding down off the plateau and then traversing the canyon-gouged face of the escarpment. Most visitors to Analamazaotra come down from the capital by train, which takes about five hours. It's a fast and comfortable trip by the standards of Madagascar (where travel can be difficult), notwithstanding two dozen stops in villages with unpronounceable names. About halfway along, the landscape changes from high savanna and rice paddies to rainforest. The forest is patchy, interrupted in more than a few places by swaths of naked earth where the hardwoods and tree ferns have been felled and burned. Hill rice grows sparsely on some of those charred slopes. Traveller's palm, a weedy Madagascan representative of the banana family, thrives on the edges of the disturbed ground. Eucalyptus plantations cover sizable stretches of terrain. Exotic to Madagascar, these eucalypti harbor virtually no wildlife and are good only for firewood and cheap timber. To ride the train from Antananarivo to Analamazaotra, then, is to view a diorama of ecological wreckage. The little forest reserve itself is surrounded by jeopardized, changing landscape. The rail line continues onward, down off the escarpment and onto the eastern coastal plain, where the topography is more suitable for dense human settlement and the native rainforest has long since been almost completely destroyed. The daily train pauses only briefly, near Analamazaotra, before heading on down. The station stop is a village known as Perinet.

It's a French name, left from the colonial period. The newer name that appears on train schedules is more authentically Malagasy, but the old one lingers because this is a famous place and Perinet is the name under which it's famous. Perinet is where people come, from all over the world, for a glimpse of the indri.

The Hôtel-Buffet de la Gare, true to its name, sits beside the tracks. Its back deck is the station platform. The hotel is another colonial vestige, a large building with a wide rosewood staircase leading up to the second-floor rooms and a spacious dining hall that serves also as lobby and bar. Half a mile from the hotel—a short stroll out the front door and down a narrow lane—stands the gate of the Analamazaotra reserve. The reserve isn't fenced and a person could walk into it anywhere, but dutiful visitors present themselves and their permits-of-entry at the gate. A sign there reads:

EAUX ET FORÊTS
RÉSERVE D'INDRI
BABAKOTO

Eaux et Forêts is the Department of Waters and Forests, charged with overseeing this reserve and others. *Babakoto* is the Malagasy name for the indri.

By one account, *babakoto* translates as "grandfather." By another account it's "little father," and by still another, "the ancestor." Madagascar has its own subtle protocols of denomination, its own shifty epistemological rules, and most things that are knowable here are known in multiple variants. That's especially true at Perinet.

The Analamazaotra reserve was established in 1970, but it belongs to a much older tradition. As early as 1881, in the twilight of the indigenous monarchy (which had been established by the powerful Merina tribe and lasted until conquest by the French), Queen Ranavalona II had seen fit to decree that the forest "may not be

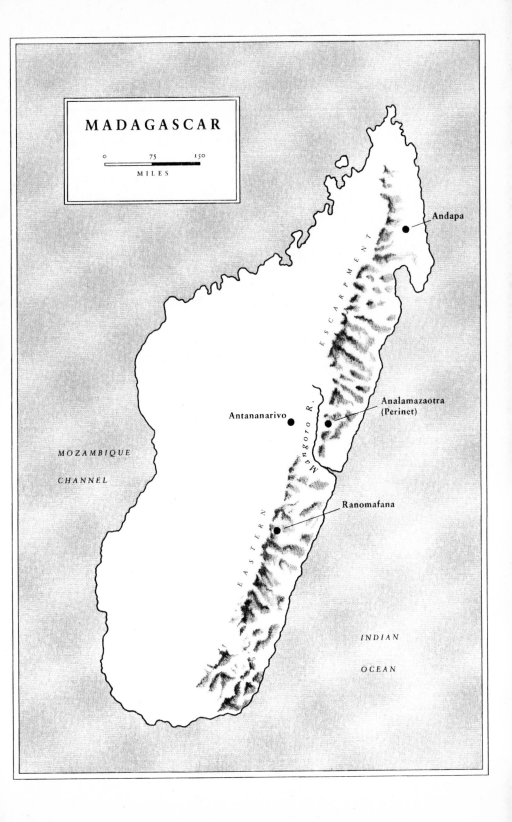

MADAGASCAR

0   75   150
MILES

Andapa

ESCARPMENT

Analamazaotra
(Perinet)

Antananarivo

MOZAMBIQUE

CHANNEL

Mangoro R.

Ranomafana

EASTERN

INDIAN

OCEAN

cleared by fire with the aim of establishing fields of rice, maize or any other crops; only those parts previously cleared and burnt may be cultivated; any persons effecting new clearings by fire or extending those already in existence shall be clapped in irons for five years." A generous view of the queen's decree would be that it reflected a traditional Malagasy reverence for the forest as "the robe of the ancestors" who were buried there. A less generous view would be that the decree was mostly a flexing of regal muscle. Notwithstanding the ancestors' robe and the queen's muscle, a total ban on forest clearing couldn't have been enforced, even if the monarchy had endured, since the Malagasy people resented such imperious edicts and were already exerting irresistible demographic pressure against the limits of available land. The French occupation began in 1895, and in 1903 the colonial authorities announced their own set of sanctions against burning forest for rice or pasture. This prohibition, like the one from Ranavalona II, seems to have been more about asserting control over the country's economic life and resources than about preserving natural landscape. Again it was a futile policy as well as a heartless one, tantamount to enacting a law against having babies and feeding them.

Neither a Merina monarchy nor a colonial government could have hoped to protect every hectare of forest, but beginning in 1927 the French tried a more modest approach—this time, it seems, with a genuinely foresightful concern for preserving some of the island's natural wonders. They established a system of nature reserves. At first there were ten, scattered throughout the country, each representing a unique ecological community. Two others were added during the late stages of French rule, and still others after the country achieved independence in 1960. Nowadays there are six categories of protected area recognized under Malagasy law, of which three are crucial to the conservation of species: strict nature reserves, national parks, and special reserves. The strict nature reserves are closed to everyone except Eaux et Forêts officials and approved researchers—or theoretically closed, anyway, with enforcement irregular. The national parks are big parcels of wondrous landscape in which anyone is welcome; but there aren't many. Montagne d'Ambre National Park, in the far north, protects upland rainforest. Isalo National Park, in the southwest, contains a weird labyrinth of sandstone gorges and spires and an unusual dry-forest community. Ranomafana National Park encompasses one of the last great expanses of eastern rainforest, as well as those three sympatric species of bamboo-eating lemurs. The special reserves, most of which are tiny compared to Isalo or Ranomafana, were each established for a narrower purpose—generally, to protect a particular species of animal or plant. Analamazaotra is a special reserve, and what makes it special is the indri.

The indri, largest of all surviving lemurs, is also the most spectacularly peculiar. Its neck is long, its limbs are lanky, its eyes glow yellow brown in a gawky black jackal-like face. Its ears are smallish and round, like a koala's. Although the dark-and-light pattern of its body fur suggests a giant panda, its shape is more gracile and humanoid. It appears to be built for basketball. Its legs are strong, its hands and feet are quick. It has almost no tail, just a nub, and that nub-tailed condition is unique among lemurs, unusual for any arboreal primate. Its scientific name, *Indri indri*, suggests nothing whatever except that the species is one of a kind. It won't live in captivity. It flees from humans and doesn't tolerate disturbance of its habitat. You could call it the ultimate wild animal.

It grows to the size of a small baboon but it moves through the forest without touching the ground—by making broad jumps from the trunk of one tree to another, sometimes twenty or twenty-five feet across gaps. This mode of locomotion is astonishing. With a great thrust from the legs, the indri leaps out; sailing horizontally through space, it turns; catching hold of the next tree trunk, it shoves that one aside and ricochets onward to the next, and the next, and the next, moving faster than a person can run. Perinet tourists of the more energetic variety sometimes try to keep up, stumbling along through the understory. Hopeless. But eventually the indri stops. It settles into a high fork of limbs and becomes virtually invisible, although with persistence, with binoculars, with help from a keen guide, a visitor can spot it again. And if timing is right and the visitor is lucky, the indri will sing.

The song of the indri is an unearthly sound. It carries through the forest for more than a mile. It rings in the air back at the Hôtel-Buffet de la Gare. It has been said to be one of the loudest noises made by any living creature. It's a sliding howl, eerie but beautiful, like a cross between the call of a humpback whale and a saxophone riff by Charlie Parker. Biologists speculate that it may be useful for territorial spacing between adjacent indri groups (the basic social units, each consisting of a mated pair with one or two offspring), and possibly also for passing other sorts of information (about breeding availability of young adults, for example) across greater distances. It could be a threat, then, or a warning cry, or a way of advertising for new genes. The significance of the song is an unsolved mystery, one of many in connection with the indri. The most thorough study within recent decades was done by a young biologist named Jon I. Pollock, but even Pollock didn't claim certainty for his findings.

Pollock's study and others have supplied a bare outline of indri ecology and behavior. The species survives only in remnants of northeastern rainforest, from around the town of Andapa southward as far as the Mangoro River. (The forest at Ranomafana, which looks like good habitat but lies south of the Mangoro River, has no indri.) Before the centuries of human disturbance, it ranged wider. It feeds on leaves and fruit from many different tree species, but despite the dietary versatility it doesn't attain high population density in any given region, as some versatile herbivores do. It reproduces slowly: a single birth every three years, sexual maturity at about age six. Not many mammals, not many animals of any sort, have such low reproductive potential. And the song itself, according to that hypothesis about territorial spacing, may be a tool for ensuring that indri groups are few and far between. By mandates of its own ecology, the species remains rare, with all the jeopardy that rareness entails.

The mated pairs seem to be monogamous, though extended fidelity is unproven. An indri infant rides aboard its mother for much of the first year, as if stuck to her pelt with Velcro. Females are the dominant gender. They take precedence over males for feeding and grooming, and they don't tolerate being bullied in the matter of mating. Besides the song, there's also another sort of vocalization: a short, sharp bark (*coup de klaxon*, in the nice phrase of one French biologist) that functions as a warning against predators and noisome human observers. It seems that an indri's hearing is exceptionally acute, at least for detecting the song of another indri. Each animal possesses "a large membranous laryngeal sac" that, by Pollock's best guess, might be used in producing the song. Otherwise the indri's anatomy and physiology are poorly known; few scientists have laid hands on a dead specimen, let alone a living one. It *is* clear, though, that the indri's slow rate of reproduction and its low population density make the species especially vulnerable to extinction. Probably its body size does too, since larger animals tend to have larger needs. The dozen or so species of lemur that have gone extinct since humans colonized Madagascar were all large-bodied animals, generally much bigger than the lemur species that have survived. If that trend holds, the indri will be next.

Jon Pollock did his fieldwork at Analamazaotra and two other sites. For slightly more than a year he followed certain family groups, gathering data about their population structure and their behavior, charting their dietary preferences, mapping their territories, recording their strange music. That was two decades ago. Eventually he published a handful of papers, which today account for much of what is known about the species, since almost no one else has studied it. Pollock may still be the world's leading authority. I can think of only one other person in recent years who knew the ways of the indri intimately, and that person died young without leaving any of his knowledge in print.

"Forests inhabited by *Indri* resound each day to loud modulating calls emitted by two to four members of each group either spontaneously or in response to other groups' calls," Pollock wrote. "The song of an *Indri* group, which consists of a discrete succession of these loud calls, lasts for between 40 sec and 4 min, occurs usually during the morning, is emitted at least once on most days, and can be detected by the human ear up to 2 km from its source under optimal conditions." Pollock examined the acoustical structure of several recorded songs, using a Kay sonograph and a real-time sound-spectrum analyzer, whatever those are. His report on the acoustical work, lumpishly titled "The Song of the Indris (*Indri indri*; Primates: Lemuroidea): Natural History, Form, and Function," was published in the *International Journal of Primatology*. "The song consists of a series of calls, each lasting from 1 to 4 sec, whose main energy lies from 500 to 6000 Hz, separated by short pauses of up to 3 sec," he wrote. "Pure tones of the call appear with up to four harmonics and may be modulated by as much as 2000 Hz within each call." This captures some of the physics but none of the soul.

145

ON THE FIRST evening of my first visit to Perinet, several years back, I heard about a Malagasy teenager named Joseph, recommended by the hotelier as the finest of the local guides. Joseph knows the Analamazaotra reserve better than anyone, I was told. Yes yes, Joseph knows the ways of the indri. Intimately. Joseph grew up in the forest, he loves all the animals, even the plants, he has made himself *expert.*

The hotelier himself was a gracious Malagasy of early middle age whose unassuming dignity inspired trust, and I was happy to take the recommendation. Later that evening I was introduced to his protégé, his star guide, this Joseph.

A laconic and serious young man. He looked about eighteen. Spoke a little English. Seemed to possess a steady, internalized confidence. He wasn't shy, he wasn't brash, merely focused. A professional who cherished his work. It turned out that Joseph was just his French name, convenient for tourists to remember. Like the indri, he had more labels than one, and his comfortable name, his Malagasy name, was Bedo.

I asked Bedo whether he'd be willing to lead me on a long quiet hike through the forest. Yes, of course. When? I asked. Whenever. I had already spent an afternoon in the reserve with another guide, a precociously cynical youngster who presumed that I'd tip him well for manhandling the boa constrictors and chameleons into photogenic poses. Bedo seemed promisingly different. For one thing, he would gladly take me through Analamazaotra by night, when the nocturnal lemurs were at large.

The indri itself is diurnal. It lays up at night and feeds by day; it does most of its singing (though not all of it, I would discover) in the morning. I knew that much and, although *Indri indri* is what mainly had brought me to Perinet, my time there was short and I didn't want to waste a good evening. I'd go looking for the indri the next day, I thought. Meanwhile, tonight there was a chance to see such shy little spooks as the greater dwarf lemur.

We started walking an hour after dark. Bedo carried a powerful battery beam and I had my dinky flashlight. The forest rustled and hooted with invisible life. I was comfortable in the knowledge that it's a benign forest: no jaguars, no tigers, no rhinos, no poisonous snakes that I'd heard about and, at least on this evening, no leeches or mosquitoes. We hiked for about four hours, crisscrossing the reserve on a network of steep trails. It was an overcast night with hardly a trace of moon, and the footing was tricky. Bedo set a brisk pace. He knew these jungle paths like an American kid knows the shortcut to the neighborhood ball diamond. After the first hour I put away my light, preferring to walk in his tracks and peer into the trees along his beam. Too bad I couldn't see with his eyes. There, he would say. Huh, what, where? *There there,* and he would point with the light. Ninety feet away in a high crotch of limbs, a pair of orange dots would glimmer back. His vision was impressive, but his effectiveness as a guide involved much more than just sharp eyesight; it involved knowledge and intuition and the intensity of his attention. He showed me the greater dwarf lemur, yes, and the brown mouse lemur, little animals squatting in high branches where I never would have noticed them even in daylight. He found four different species of chameleon, some miniature and some giant, all of them slow-moving, swivel-eyed, and camouflaged—but not sufficiently camouflaged to elude Bedo. His attunement to the ecosystem seemed preternatural. To me he paid little notice, as long as I kept the pace and appreciated the animals after he spotted them. He was no cynic, pandering to the predictable tourist impulses, angling for tips. He was a naturalist, inexhaustibly intrigued by his subject and tolerant of the innocent twit who chose to tag along. At one point, while we paused on a hilltop, waiting for my breath to catch up with us, from far in the distance came a high, mournful wail. Another. Sliding notes, harmonies, graceful razor slashes of sound.

Indri.

Bedo said nothing. He listened. This was something he didn't need to identify, not even to such an ignoramus as myself. His face lit in a special way. His face lit, I think, with love.

Finally, on a trail that had looped over several ridges and left me disoriented, he brought us down. We emerged from the reserve along one of its far borders and trudged back to the hotel on a village path. It was nearly midnight. I was sweaty and tired and I had missed dinner. While a grumpy waiter went foraging to the kitchen on my behalf, Bedo sat on a bar stool paging idly through a looseleaf binder. Are you hungry, Bedo? He shrugged.

Over his shoulder, I saw that the binder held mimeographed pages of typewriter print and rough illustrations. It was a quick-and-dirty draft of a taxonomic key to Madagascar's chameleons. Some visiting herpetologist had taken notice of Bedo's interest and left a copy of this treasure in his hands. Here, Bedo said, showing me *Chamaeleo brevicornis*. Pen sketch of a wonderfully ugly reptile with funnel-shaped eyelids and a rhinoceros nose. And here, *Chamaeleo parsoni*, the big one. And here, *Chamaeleo nasutus*. Bedo browsed on through the binder. He seemed immune to hunger and weariness. The pages were dog-eared from an excess of careful fondling.

Later, in my notebook, I referred tersely to this "amazing guide." If I had known more of his history and had foreseen his future, I would have scribbled a thorough portrait. But I didn't expect to revisit him as a subject or a person. I was simply struck with a strong sense that if there was justice in life, then this young man might be discovered and nurtured by someone from the larger world. I didn't know that he already had been. I didn't know that Bedo of Perinet was the same teenage naturalist who had helped Pat Wright track and photograph the golden bamboo lemur at Ranomafana.

146

MADAGASCAR's special reserves, the group that includes Analamazao-tra, are more numerous and generally smaller than the two other categories of protected area. Currently there are about two dozen special reserves, most of them established in the 1950s and 1960s, when no one in Madagascar or anywhere else was discussing the notion of minimum critical size. The special reserve of Nosy Mangabe encompasses only 520 hectares, not much more than one of Tom Lovejoy's isolated patches of forest. The Mangerivola reserve is slightly bigger, at 800 hectares, and the Cap Sainte Marie special reserve runs to all of 1,750 hectares. These are not immense parcels of landscape. For a standard of comparison from the United States, consider that Bryce Canyon National Park—which, according to William Newmark's work, seems too small to support a population of red fox—covers 14,400 hectares. The special reserve of Analamazaotra, at 810 hectares, is one of the smallest. You could walk its perimeter in a morning.

The perimeter is crisply demarcated. You could walk it in street shoes, and you wouldn't need a machete. Hell, you could rollerskate a good part of the way.

Along one side of the Analamazaotra reserve, from the hotel to the entrance gate, runs a blacktop road. The road isn't broad or busy, but it represents a significant boundary. The gap in the forest canopy is wide enough to prevent, say, an indri from jumping across. Along another side runs a bigger road, a national highway, recently built as an aid project by the Chinese. No chance of an indri crossing that. A third stretch of perimeter is bounded by the village of Perinet and the rail line—again, an impassable zone for an oversized lemur with an arboreal style of travel and a fear of human ruckus. The rest of the boundary of Analamazaotra is marked only by a path. But outside the path is unprotected forest, degraded by woodcutting and clearing for hill rice. There's also a eucalyptus plantation, which can't support lemurs of any kind, let alone indri.

My point is this: The Analamazaotra reserve is no sample within a vast tract of wilderness. It's much closer to being an isolate. If it isn't already insularized, so far as lemurs are concerned, it probably soon will be.

Under current conditions, a recklessly venturesome indri might still wander in or out, crossing through the damaged forest or the inhospitable eucalyptus. When the eucalyptus has been harvested, when the damaged forest has been leveled for rice, the possibility of such crossings will disappear. Birds and bats will still enter and leave. But to its population of indri, the Analamazaotra reserve will be a separate and circumscribed world. Whether that population remains viable over time will depend on a number of factors, most of which are more complicated than a sheer-area measurement of 810 hectares.

The list of factors starts with: How big is the Analamazaotra population of indri?

Pat Wright, who worked there for a season before shifting her focus to Ranomafana, suspects that the number may be as high as two hundred. Probably she's being optimistic. She admits her uncertainty, noting that no careful census has ever been done at Analamazaotra. Nor can any decent estimate be made of indri numbers throughout the whole country, she adds. Although everyone agrees that the species has declined as its habitat has been fragmented and destroyed, no one can claim to know how many indri remain. "There may be more than we think, there may be less," Wright says. "What we think we know is just a dream." Jon Pollock, focusing his own fieldwork on a few family groups, apparently never tried to count the full population at Analamazaotra. But his published data on average group size and population density make possible an informed estimate. Assuming that Pollock's numbers were accurate, Analamazaotra at the time of his study contained about eighty indri.

Two decades later, the figure may be less. The population has more likely shrunk than grown. Let's make the hopeful assumption that it has stayed about constant. If so—if the population has held constant at eighty for the past twenty years, and if it continues to hold at that level into the immediate future—the indri of Analamazaotra are in trouble. Virtually all recent work on the concept of population viability tells us that eighty animals just aren't enough.

147

IN THE LATE 1970s, a few ecologists began using a pregnant new term: *minimum viable population*. It could be traced back to a journal paper that two researchers in Australia had published, in 1971, on the ecological requirements of certain species of Macropodidae, the marsupial family that includes wallabies and kangaroos. The researchers, A. R. Main and M. Yadav, had studied macropod distribution on small land-bridge islands offshore from western Australia. For each island under scrutiny, they noted the presence or absence of various macropod species, the island's size, and its diversity of vegetation. Main and Yadav were proceeding in parallel to the fashionable new school of island biogeography, rather than as part of it. They didn't mention the species-area relationship—not by that label, anyway— although it overshadowed their whole study. They didn't mention the equilibrium theory. The sources they drew on, and the context in which they wrote, were almost completely Australian. They referred to their study animals with informal Aussie names, some of which sound as fantastic as creatures from Tolkien: quokka, tammar, euro, boodie. Main and Yadav were interested in macropod conservation, not theory.

They addressed a practical question: How much area was necessary to "adequately satisfy the ecological requirements" of each of these species? Ecological requirements would include sufficient habitat not just for one animal or a few but for a self-sustaining population. How much landscape was needed to support a population of quokka? How much landscape for a population of euro? Could the minimum area requirement of each species be deduced from its presence or absence on those variously sized coastal islands? Some of the smaller islands held only a single macropod species, the tammar or the rock wallaby. Larger islands supported more species and bigger-bodied species. The biggest macropod, the red kangaroo, didn't exist on any of the islands. Main and Yadav gathered the sort of data that would have allowed them to draw a species-area curve for their collection of islands and to explain the irregularities of that curve in terms of habitat diversity. Just as marsupials had evolved in isolation from other mammals, though, their work seems to have evolved in isolation from the influence of MacArthur and Wilson.

They took a further analytical step that made their paper unusual for its time. They calculated the density at which each macropod species populated its habitat. For instance, one animal per hectare for the quokka; one animal per ten hectares for the (bigger-bodied) euro. Then, by multiplying the density of a given species by the area of the smallest island on which that species survived, they produced an interesting set of numbers. "When these data are extrapolated to small island situations," they wrote, "it is possible roughly to estimate minimum viable population sizes."

That seems to have been the first use of the phrase "minimum viable population." Actually it's more than a phrase; it's an important idea. Although several earlier biologists had written vaguely about "minimal populations" and "viable populations," Main and Yadav had said something new. But there wasn't much immediate response.

Main and Yadav also implied but didn't emphasize a point that would later be recognized as significant—that the minimum viable population might differ from one species to another. Still, their results for the tammar, the euro, and the red kangaroo all fell within a narrow range. In the last sentence of their paper, they added: "The minimum viable population (defined as the population size likely to persist indefinitely) appears to lie between two and three hundred animals."

148

SEVEN YEARS later, a graduate student in the Department of Forestry and Environmental Studies at Duke University submitted his dissertation. It was an analysis of the task of determining minimum viable population sizes, as illustrated with the case of the grizzly bear. The student's name was Mark L. Shaffer.

Shaffer at that time had never set eyes on a grizzly in the wild. He had grown up in western Pennsylvania, attended college there, and gone off to the University of Pennsylvania for a master's program in land-use planning. But he had found himself more interested in how wildlife uses landscape than in how humans do. He transferred to Duke and started fresh toward a different kind of degree. He took some classes in computer modeling. Also, like so many other biologists of his generation, he became fascinated with island biogeography. When the equilibrium theory began being applied to conservation issues by Diamond, Terborgh, Wilson, and others, Shaffer followed that work in the literature. He read the SLOSS arguments. He was concerned about the worldwide destruction of habitat, the fragmentation of what remained, and the loss of species. He formulated his concern in terms that rang faintly of his earlier work in land-use planning.

He recalls thinking: "We've got a finite amount of land. We're in competition for it with nature. So the question is always, How much do you set aside?" It was the same question that Tom Lovejoy was asking, but Shaffer's route toward an answer was different.

He knew that inside the question *how much?* lurked the question *how many?* That is, he knew that the minimum critical size of an ecosystem was a less fundamental measurement than the minimum critical population of each constituent species. When any species becomes too rare within an isolated patch of habitat, that species can't survive. It isn't viable. It goes extinct. But how rare is too rare?

This question leads to a battery of others. What's the numerical threshold of viability? Is there a single magic number for all species, or does the viability threshold differ for different species? Does it differ for the same species in different contexts? If so, what accounts for the differences? Is the determination of viability threshold a purely scientific task, or does it also involve sociocultural values? Can the sociocultural element be quantified? Shaffer had picked a doctoral topic with some meat.

He chose the grizzly bear as his exemplary case for two reasons: because it was a large-bodied animal at the top of its food chain, and because field biologists had already studied it closely. Of the first reason, he says, "If you focus on things on top of the food chain, and save enough land for them, then probably you're saving enough for the whole food chain." The second reason was relevant because his analysis would depend on types of data available only for an exceptionally well-studied species. Shaffer would need information about longevity, causes of death, age at first reproduction, average litter size, sex ratios within various age classes of the population, social structure, and breeding system. That sort of information didn't exist for wolves or mountain lions. For grizzlies, thanks largely to a pair of biologist brothers named Frank and John Craighead, it did. The Craigheads had studied the grizzlies of Yellowstone National Park between 1959 and 1970, and had published much of their data.

The Yellowstone grizzlies offered one other advantage. The greater

Yellowstone ecosystem is a vast area of woodland and meadow and mountain slopes and river drainages encompassing not just Yellowstone National Park and Grand Teton National Park but also contiguous portions of seven national forests, several wildlife refuges, part of the Wind River Indian Reservation, and some Bureau of Land Management holdings, as well as bits of private and state land, most of it still wild enough to be hospitable to grizzly bears. Despite the patchwork of ownership status, these various pieces constitute a single ecological whole. Because the ecosystem is surrounded by developed terrain (including farms, ranches, barbed wire, towns, suburbs, highways, railroad tracks, irrigation canals, power lines, airports, golf courses, guardrails, trailer parks, malls, lumber mills, movie theaters, gas stations, gun shops, pizza parlors, parking lots, picket fences, barking dogs, traffic lights, stop signs, and concrete lawn ornaments), terrain that *isn't* so hospitable to them, the grizzlies of greater Yellowstone are effectively insularized. They stand discrete as a population, on a discrete ecological fragment. So the Yellowstone grizzlies became Shaffer's empirical paradigm.

The first chore he faced, before considering the specifics of that population, was to define the term announced in his title: "Determining Minimum Viable Population Sizes: A Case Study of the Grizzly Bear (*Ursus arctos* L.)." What exactly is meant by "minimum viable population"? Main and Yadav had defined it as "the population size likely to persist indefinitely," but that was vague. How probable is "likely" and how long is "indefinitely"? Main and Yadav hadn't committed themselves to specificity on these parameters, nor in the meantime had anyone else. Shaffer did. Admitting that his choices were arbitrary, he specified a ninety-five percent likelihood and a time period of one hundred years. If a population of some given size had a ninety-five percent chance of surviving for as long as a century, then by his proposed standard it could be called viable.

Inserting precise numerical values in place of the vaguely worded expectations allowed him to construct a framework of assessment. Another important service that Mark Shaffer performed, before launching into his prognosis for the Yellowstone grizzlies, was to clarify the matter of what generally brings about extinction among small populations.

The causes are multifarious but they can be divided into the two types I mentioned earlier, deterministic (human-generated) factors and stochastic (accidental) factors—or, in Shaffer's preferred language, "systematic pressures" and "stochastic perturbations." Systematic pressures are those that can be predicted and controlled, such as sport hunting, bounties, pesticide applications, destruction of habitat. Stochastic perturbations are those that elude human prediction and control, either because they are genuinely random or because they result from nonhuman causes so intricate and obscure as to *seem* random. Stochastic perturbations introduce uncertainty into the fate of a population—and the smaller the population, the greater the uncertainty. So the study of uncertainty and its consequences is crucial to the conservation of rare species.

Systematic pressures, important as they may be, were outside the purview of Shaffer's project, he explained. He wasn't writing a polemic against bear hunting. What he was concerned with were those stochastic perturbations. How would they affect small populations? At what population threshold would the effects become fatal?

"In general, there are four sources of uncertainty to which a population may be subject," Shaffer wrote. He listed them: demographic stochasticity, environmental stochasticity, natural catastrophes, genetic stochasticity. Behind the jawbreaker terminology were some easily digestible ideas.

Demographic stochasticity means accidental variations in birth rate, death rate, and the ratio of the sexes. Say you have an extremely rare species—make it another hypothetical beast, call it the white-footed ferret. Only three individuals survive, a female and two males. The female breeds with one male, gives birth to a litter of five, then dies. By an unfortunate chance, the newborn ferrets are all males. Demographic stochasticity. Now you have seven white-footed ferrets but no females, and extinction is inevitable.

Environmental stochasticity means fluctuations in weather, in food supply, and in the population levels of predators, competitors, parasites, and disease organisms with which your jeopardized species must cope. Say you have eighteen white-footed ferrets, with a balanced sex ratio, but the prairie dog colony on which they depend for food and shelter is being killed off by a virus. Your ferrets are not susceptible to the virus. Still, without enough prairie dogs to eat, they begin starving. They die back to a mere handful. A three-year drought makes their lives miserable, each long dusty summer adding stress to their situation; then comes a ferociously hard winter. Environmental stochasticity. Hungry, inadequately sheltered, the ferrets go extinct.

Neither of these two types of uncertainty could destroy a large population of animals. There's safety in numbers. But a small population is vulnerable to a gentle nudge.

Natural catastrophes (floods and fires, typhoons and hurricanes, earthquakes and volcanic eruptions) can deliver a nudge that isn't gentle. Such catastrophes aren't totally random, in that they do have physical causes; but the causes are so complex as to be virtually unplumbable, the timing is unpredictable, and therefore these events loom as another sort of uncertainty. Your small population of ferrets might endure the hunger, the droughts, and the hard winter, building themselves steadily back up to a population of three or four dozen— only to be eradicated, blotto, when some catastrophic flood puts their home under water.

Then there's genetic stochasticity. This refers to the vagaries by which certain alleles become more common or more rare within a gene pool, irrespective of the influence of natural selection. The vagaries can be costly in two ways. First, helpful alleles can become so rare—by that random process I described earlier, genetic drift—that they disappear accidentally. Second, harmful alleles that are recessive, rare, and (because of their rarity) ordinarily carried only in the heterozygous situation can become just common enough within a small population that they occur homozygously. Inbreeding increases the chance of homozygosity, by pairing family-carried alleles with themselves. And when recessive alleles occur in the homozygous situation, they achieve expression. If they're harmful recessives, what they express is harm.

The sum total of harmful recessive alleles, within any given population, is known as the genetic load. In a large population, the load can be carried almost without consequence; a few recessive alleles, shuffled among many dominant alleles, produce no effects. In a small population, the load can be more burdensome. Since small populations are often forced toward inbreeding, they frequently suffer from the expression of those harmful recessive alleles. The result is called inbreeding depression. If a given population carries a substantial genetic load and that population suffers an abrupt reduction in size, inbreeding will become more likely, and inbreeding depression may cause trouble.

Another form of genetic stochasticity that affects small populations is the founder effect. Remember the founder effect? Remember the flamingo pink socks? When a small number of individuals become isolated from a larger population (either because those individuals have founded a new colony or because they are sole survivors, the larger population having died away), the small population will contain only a meager sample of the larger population's genetic diversity. To some degree that genetic sample will be random. Rare alleles—whether they're harmful, neutral, or helpful—will have been lost. Even losing the neutral alleles, let alone the helpful ones, may eventually have negative consequences. How? Consider the sock drawer again. When you pack hastily for a trip, groggy in the early morning darkness and grabbing socks at random, you're likely to miss the one flamingo pink pair. But what if your plane makes an unscheduled stop in Las Vegas on Halloween? Of course you'll wish you had them. The founder effect deprives small populations of rare and seemingly useless alleles that might later, under changed circumstances, turn out to be useful.

Genetic drift compounds the founder-effect problem, stripping a small population of the genetic variation that it needs to continue evolving. Without that variation, the population stiffens toward uniformity. It becomes less capable of adaptive response. There may be no manifest disadvantage in uniformity so long as environmental circumstances remain stable; but when circumstances are disrupted, the population won't be capable of evolutionary adjustment. If the disruption is drastic, the population may go extinct.

Environmental stochasticity can deliver just such disruption. So can a natural catastrophe. In fact, all four of Shaffer's stochastic perturbations can interact in a dire feedback cycle by which a small population spirals down to extinction.

Imagine, again, your population of white-footed ferrets. Imagine you have a slightly more reassuring abundance—say, eighty. They live as resident predators within two separate prairie dog colonies, one on a gentle plateau and one beside a river. Comes a catastrophic flood, a once-in-a-century event, unpredictable and unavoidable. It covers your riverside colony and drowns every animal. You still have forty other ferrets on the plateau, but two droughts and then a tough winter reduce that number to twenty. Suddenly you're worried. And you should be. Several of the remaining females, for lack of other options, breed with their sons. Several males breed with their sisters. As a result of inbreeding, some of the offspring are born sterile, and as those animals reach adulthood the overall birth rate goes down. Other offspring of incestuous matings are born without resistance to a bacterial disease; that disease hits the colony, and it kills them. Still others, also victims of inbreeding depression, are born with a general lack of vigor. Nothing is specifically wrong with these ferrets, nothing you can point to, but they don't have the robustness of a heterozygous individual. Another harsh winter, and they die. Now you're down to seven ferrets, and by the sorry breaks of the game only two of your seven are female. One female is too old to give birth. The other female produces a litter of five healthy young, then a coyote eats her. The five young are all males. Final tally? You have eleven white-footed ferrets, consisting of ten males and an elderly female. Your species is history. Take a picture and kiss it goodbye.

149

WHAT DESTROYED your population of ferrets? Well, let's suppose that poisoning by government agents and loss of habitat to wheat farming were partly responsible. But those systematic pressures didn't deliver the final blow. Poison and habitat loss merely reduced a once abundant species to a last beleaguered eighty. After that, the four kinds of uncertainty were enough to carry your ferret to its doom.

Mark Shaffer's dissertation, like William Newmark's, turned out to be more than a journeyman's academic exercise. Shaffer provided the first orderly discussion of those four kinds of uncertainty, and he took the significant step of offering a meaningful (though unabashedly tentative) definition of minimum viable population:

> A minimum viable population for any given species in any given habitat is the smallest population having at least a 95% chance of remaining extant for 100 years despite the foreseeable effects of demographic, environmental, and genetic stochasticity, and natural catastrophes.

In a later publication Shaffer would propose a revised definition with higher numbers, based on his feeling that a five percent chance of extinction was too much and that a century of security was too little. He would also add the important insight that creating any such definition, upon any numerical standard, involves society's values as well as biology's powers of prediction. He would recognize that cultural and political questions, as well as scientific ones, will always be part of the process. How much do we care? How badly do we want to preserve white-footed ferrets? What degree of security do we demand?

Species all over the world, reduced to small populations in fragments of habitat, are jeopardized by the four types of uncertainty. Your ferrets might be anywhere. They might be this animal or that plant. They might be the grizzlies of Yellowstone, the snow leopards of Annapurna, the mountain gorillas of central Africa. Or imagine them on a tropical island. Imagine them with the face of a jackal, yellow brown eyes, black-and-white fur like a giant panda's. Imagine you know them as *babakoto*. They sing. Imagine that the last eighty live in a little forest reserve called Analamazaotra.

150

PAT Wright came to Analamazaotra in 1984. It was during her first visit to Madagascar, an exploratory trip for scouting the prospects of doing lemur research at various field sites. She and another primate ecologist decided to go down to that famous village, Perinet, for a glimpse of that famous big lemur, the indri. "Being tropical biologists, of course, we didn't think we needed a guide," she remembers. "As we were searching for the animals, suddenly this little boy appeared on the trail. And he said, 'They're over here.'" The boy's name was Bedo. "He was bright-eyed and cheerful, and he spoke a little French. We both liked him very much," Wright says. "At the very beginning, he was an extremely charismatic personality." And he led them, as promised, to a group of indri. Later, when Wright and her colleague set out on a longer walk, they took the boy along. "We could see right away that he really knew the forest well. He could point out various forest animals, birds and things, along the way. He had very good eyes." Wright delivers this much in a flush of easy recollection.

"So," she adds, with sadness and love in her voice, "that was the first time I ever met him."

She returned to Analamazaotra a year later with one of her students from Duke University, David Meyers, and together they began making field observations on *Hapalemur griseus*, the close relative of her still undiscovered golden bamboo lemur. Little Bedo became their regular guide. He worked with them each day in the reserve, he shared their meals, he spent most of his waking hours with them. They lived like a family. Bedo was about fourteen then, Wright recalls. "He was really enthusiastic over everything about nature. There was nothing that didn't interest him. He loved the snakes, he loved the smallest insects and the chameleons. And he was really good at the lemurs."

He had grown up in a house just across the road, one child among many of a fisheries manager employed at the reserve by Eaux et Forêts, and so Analamazaotra had been his childhood playground. His forest skills derived from devotion, experience, empathy, and another attribute. "He had incredible eyes," Wright repeats. "I've never seen anybody better."

He was ingenuous but resourceful. Wright tells a story about Bedo's efforts on behalf of a primatologist who needed fecal samples from the species of lemur he was studying. Bedo absconded with his mother's cooking pot, stood under the animals for as long as it took to catch them defecating, and returned proudly with the pot full of lemur shit. He had character strength and integrity that was remarkable in a youngster, Wright says. It showed in his attitude toward nature and in his pious adherence to the traditional strictures of his tribe. Wright and David Meyers conversed with him mainly in French but began teaching him English. He was bright. When her work on the bamboo-eating lemurs took her away from Perinet—down to Ranomafana—Wright wanted Bedo to join her there and help with the field survey. But first he had to finish his school year. She left him money for the bus and explained how he should travel to Ranomafana, a daunting journey for a young village boy who had never been anywhere. She knew there was some uncertainty about this proposition. Would he decide to stay home, would he show up at Ranomafana, or would he set out on the trip and then get lost? She remembers the exact date when that question was answered, in late June, because it coincided with a national holiday. She had been out of camp for a long stretch of hours, tromping across the hillsides, bushwhacking away from the trails, following a group of lemurs. Suddenly, there was Bedo. "He not only had found Ranomafana, but he had found us, in the forest," she says. He carried a sleeping bag over his shoulder, and a small pack. He was ready to work.

Eventually he returned home to Perinet, left school, and started making good money (at least by Malagasy village standards) as the best and most-requested nature guide at the Analamazaotra reserve. Pat Wright still visited the village occasionally and stayed close to him as he grew through adolescence.

Once she brought him to the capital. He had never before seen Antananarivo. They took an elevator up to the roof of the Hilton Hotel and gawked at the view. "We were twelve stories high," says Wright. "Higher than any indri had ever been. It was just amazing to see him. He just couldn't believe it." Then he went back to the elevator and rode up and down, up and down, until he was sated with that thrill. Wright also showed him the spectacle of a large grocery store, and they strolled through the Zoma, the open-air market, on its busiest day. At first Bedo seemed enchanted by the whole urban spectacle. But then, Wright says, the enchantment paled. "I remember walking along the street, and I said, 'Well, Bedo, how do you like the big city?' He looked at me and said, 'There's no animals. I don't like it very much. There's no animals anywhere.' He said, 'You can't hear them, you can't see them. They're not here.' And it was just with such sadness."

About the same time, Bedo met a world-class nature photographer named Frans Lanting, who had come to Madagascar on assignment for *National Geographic*. Lanting recognized Bedo's skills and hired him as an assistant, first for work at Analamazaotra, later for some of the trips Lanting made to other parts of Madagascar. Intermittently over the course of a year they traveled together, finding and photographing wildlife. "I crisscrossed the island from east to west and from north to south," Lanting says, "and I never met another individual like him. He was exceptional. He moved through the forest like a lemur." While Lanting and a second helper wrestled with equipment, Bedo would track lemurs or scout ahead for other creatures. He would climb trees. He would notice chameleons. He would discover rare birds on their nests. Because of his agility, his forest sense, and his affinity for the animals, Lanting began calling him by a nickname: *Babakoto*. Little father, ancestor. Indri.

One aspect of the work was especially hard for Bedo. Although Lanting is a conscientious man with his own passionate devotion to untrammeled nature, even he found a small bit of trammeling to be professionally necessary. With some of the more elusive species, he resorted sometimes to capturing animals in order to photograph them. Later the captives would be released; nonetheless, Bedo didn't like it. He chafed. When an animal was caged, he suffered an empathic sense of spiritual imprisonment. "That's what made him one of the wild creatures of the forest," says Lanting. Mostly they photographed the animals as they found them, under circumstances that demanded more patience but less compromise—at large in the forest. "That's where he was happiest."

Lanting too recognized that Bedo was bright, and that his extraordinary love of the natural world was an important resource in a country so troubled and precious as Madagascar. Clearly, such a gifted young man deserved to be cultivated toward some larger possibility than simply guiding tourists through the Analamazaotra reserve for quick cash. Bedo would never be a scholar, he lacked the disposition for long years of training as a scientist, but he might make a wonderful educator, Lanting thought. Maybe Bedo could share his enthusiasm and his knowledge with impressionable Malagasy children, not just with affluent Western visitors. So Lanting tried to help, involving himself in the boy's education. "And that's where things became frustrating," he says. Bedo was an adolescent, after all. In his case the eternal adolescent afflictions of confusion and headstrong impatience were exacerbated by his life at the Hôtel-Buffet de la Gare. Elsewhere he was just another poor village boy, but at the hotel he was a celebrated guide, fawned upon and well paid by foreigners. His schoolboy existence, in contrast, seemed dull and pointless. He became truant. He started to flunk out. Lanting wanted to help pay for Bedo's schooling, as did Pat Wright, but the effort was futile if Bedo didn't attend. Lanting visited the school principal, hoping to arrange some framework of support and accountability that would encourage Bedo to persevere. Couldn't there be, Lanting wondered, a structure of expectations and incentives? Couldn't the principal send him some sort of modest, regular report as to how Bedo was doing? But structure, incentives—these were alien concepts that could be translated only loosely. Establishing a framework of accountability proved impossible. "In that sense," Lanting says, "Madagascar's like quicksand."

Not long after Bedo left school, I arrived at Perinet for my own first visit, becoming another of the foreigners who fawned on him, paid him well, and made his work as a guide seem more promising than further education. On the night he led me through Analamazaotra, I wasn't aware of his connections with Wright and Lanting. I had no inkling that he was already Madagascar's most famous young naturalist, admired not just by tourists but by a number of eminent foreign biologists who had either done fieldwork at Analamazaotra or at least stopped there to see and hear the indri. I knew nothing about the tension in his life between school and forest. I knew nothing about his travels to Ranomafana and elsewhere. I certainly didn't appreciate that his quiet charm, his confidence among English- and French-speaking clients at the hotel, his expertise, his earning power, his emergence as a minor celebrity, might make him resented in his own village. I didn't begin learning any of this until later, back in the United States, when Pat Wright told me by telephone that Bedo had been murdered.

The details were still indeterminate, Wright said, but there was no doubting the basic fact. His body had been found. In a voice dry and heavy with sorrow, she reminded herself and me, "He had the best eyes."

151

MARK Shaffer's dissertation, discussing the four kinds of uncertainty that jeopardize small populations, was submitted in 1978 but not published. His first journal article on the subject didn't appear until 1981. Meanwhile two other scientists, Ian Franklin and Michael Soulé, working independently, tackled the same question: How rare is too rare?

Although neither Franklin nor Soulé used precisely the same phrase—minimum viable population—as Shaffer had, they each analyzed some of the factors that determine such a threshold. They even dared to suggest certain tentative numbers. Whereas Shaffer had looked mainly at the demographic aspect of uncertainty (because that was what presented itself in the grizzly bear data he was using), Franklin and Soulé both considered the genetic aspect. They published their analyses, in 1980, as separate chapters of a volume titled *Conservation Biology: An Evolutionary-Ecological Perspective*. Soulé himself was one of the book's editors, and he had invited Franklin to contribute. Their individual treatments converged on one remarkably similar conclusion, which gave a sense of reliability to what they both said.

Franklin was a quantitative geneticist. He discussed how genetic variation is lost in a small population (by the founder effect and genetic drift) and how the consequences of that loss (including inbreeding depression and loss of adaptability) affect both the short-term and the long-term prospects for survival of the population. In the short term, the main concern is inbreeding. "There will be a minimum population size, depending on the species, at which the population will be able to cope with the inbreeding effects," he wrote. Below that minimum population size, inbreeding depression will be problematic. Harmful recessive alleles will tend to turn up homozygous (like snake eyes in craps) among the offspring, and the population will become genetically maimed. Intrepidly, Franklin offered a generalized estimate of this minimum size in relation to inbreeding. Based on the collective experience of domestic-animal breeders and on some other evidence, he suggested the number fifty. If the population barely exceeds that minimum, if it escapes the dangers of inbreeding and manages short-term survival, then the gradual loss of genetic variation over the long term becomes a concern.

The long-term prospects depend on a balance between two counterpoised factors: mutation and genetic drift. What one giveth, the other taketh away. In any population, genetic drift tends to eliminate the rarest alleles from being passed down between generations. The mutation process, in the meantime, adds a sprinkling of new variation to the gene pool.

Some mutations will be potentially harmful, some will be neutral, some will be helpful—and even those that are neutral under present ecological conditions may turn out to be helpful when conditions change. So the phenomenon of mutation shouldn't be thought of as necessarily bad. It supplies fresh possibilities for adaptation and evolution.

In a large population, the rate of mutation will more than compensate for the rate of loss, and genetic drift will be no issue. But a smaller population will experience fewer mutations. In a very small population, Franklin explained, the losses by drift will be greater than the gains by mutation, and the richness of genetic potentiality will leak away. How small is too small to maintain a positive balance? Franklin again offered a bold estimate. Based on laboratory work on *Drosophila*, he placed the balance point at five hundred. Below that number, a population will have less adaptability and diminished prospects of long-term survival with each passing generation.

These estimates became famous as "the 50/500 rule": for short-term avoidance of inbreeding, fifty; for maintaining long-term adaptability, five hundred. Being specific, the numbers were destined to attract criticism. Some biologists considered them deplorably simplistic and misleading. The 50/500 rule also caught the attention of park planners and wildlife managers, including those Costa Rican officials who (as Dan Simberloff told me) felt obliged to give up on their harpy eagles and jaguars, since the populations at issue were below fifty. Why did the numbers carry such force? Because they were plausible, because they were based on evidence (though only a small amount) as well as on sound theoretical argument, and because people were hungry for answers. This isn't to imply that the 50/500 rule was a pernicious mirage. Even if Franklin's numbers weren't accurate, they certainly helped to sharpen the focus of the discussion.

Another valuable service of Franklin's chapter was that it clarified the crucial distinction between *census population* and *effective population*. The census population is the sheer number of living individuals. The effective population is a number derived mathematically, reflecting the patterns of breeding participation, gene flow, and loss of genetic variation. Among sexually reproducing animals and plants, the effective population will almost inevitably be smaller than the census population. But how much smaller? That depends, Franklin explained, on how many females within the population are reproductively active, how many males win a chance to mate with those females, how much difference in fecundity distinguishes one female from another, how much imbalance exists in the ratio of effective females to effective males, and how much the population size fluctuates from one generation to the next. In a herd of elk, where a few great bulls gather harems and the younger males get little or no chance to mate, the effective population will reflect that inequity of opportunity. Franklin offered a few simple equations relating the various relevant factors to effective population size. The general message of his algebra is that, depending on reproductive habits and demographic history, the effective population may be drastically smaller than a mere head-count of individuals would suggest.

Why is this important? Because any minimum population threshold pertains to the effective population, not the census population. In other words, the precautionary wisdom of population genetics tells us that eighty indri aren't actually eighty indri.

152

MICHAEL Soulé, in his own chapter of *Conservation Biology*, divided the genetic issue into three concerns: first, inbreeding depression over the short term; second, loss of adaptability over the longer term; third, the danger that significant evolutionary change among vertebrate species was "coming to a screeching halt" throughout a large part of the planet. He discussed the first and the third concerns, pointing back to Franklin's chapter for a consideration of the second.

Like Franklin, Soulé had been reading the animal-breeding literature as well as the records of experimental genetics. He described a brothers-and-sisters mating experiment that had been conducted with a breed of swine known as Poland China, yielding various symptoms of inbreeding depression after just two generations. The average number of Poland China piglets per litter fell sharply; the survival rate among newborn piglets fell sharply; and the sex ratio of newborns shifted toward males, resulting in a shortage of young sows. Soulé had also investigated cases involving guinea pigs, poultry, mice, and Japanese quail. He cited obscure agronomic papers such as "Inbreeding as a Tool for Poultry Improvement" and "Influence of Inbreeding on Egg Production in the Domestic Fowl." Pragmatic animal breeders, he reported, had discovered by trial and error just how much inbreeding is tolerable to a line of domestic stock before the poor creatures start going weird. "Their rule of thumb is that the per generation rate of inbreeding should not be higher than two or three percent," Soulé wrote.

Among wild creatures the situation would be somewhat different. A line of domestic animals, with its centuries-long history of manipulated breeding by humans, is generally less vulnerable to inbreeding depression than most wild species, because in a domestic line many of the harmful recessive alleles have already been bred out. So Soulé proposed a slightly lower limit as the safe standard for conserving wild populations: one percent inbreeding per generation. At that modest rate, a small population in the wild would probably be spared from inbreeding depression over the short term—for five or ten generations. How does it translate into population size? For the inbreeding rate to be held to one percent, Soulé calculated, the effective population size must be at least fifty. So, by his own path, he had come to the same numerical conclusion as Franklin. A rock-bottom

minimum of fifty, as effective population, should be considered necessary for the avoidance of inbreeding depression over the short term.

Soulé labeled it "the basic rule." The word "rule" sounds more ur-
gently persuasive than "recommendation" or even "warning," and
probably Soulé's label is what encouraged other biologists later to
speak of "the 50/500 rule" in reference to his work and Franklin's. Of
course, both these numbers represented early and tentative attempts,
by a pair of daring thinkers, to give some provisional quantification to
a concept that was still in the process of being defined.

Soulé's third concern was bigger and more worrisome.

"This century will see the end of significant evolution of large
plants and terrestrial vertebrates in the tropics," he wrote. The end of
evolution? Yes—although what he meant, more precisely, was the end
of speciation. Existing species of trees and of vertebrate animals
might continue to evolve incrementally within a single lineage, but
they would no longer split into new and distinct lineages. This is a
damned serious claim, since such splitting constitutes one of the main
sources of biological diversity. Soulé's logic was based on the worldwide attrition of wild landscape as well as on the biogeography and
genetics of speciation. Because of habitat destruction and fragmentation throughout the tropics, he argued, and because the nature reserves that humanity grudgingly sets aside will be too small, we
should expect a time in the near future when vertebrate and tree speciation virtually stop. Each species will have barely enough habitat (at
best) to maintain itself as a viable population, but not enough to allow
it to split into several divergent populations. Each reserve will offer
inadequate area, inadequate topographic relief, inadequate pockets of
geographical isolation, to foster allopatric speciation among large
creatures. As species are lost to extinction, the losses won't be counterbalanced by new species, and Earth will grow gradually impoverished of large-bodied animals and plants. The invertebrate animals
and simpler plants—those creatures that are generally much smaller
and more numerous—may not be so sorely affected. But for the overall diversity of large vertebrates and trees, if Soulé is correct, the consequences will be severe. Species of rare ape will disappear, and no
new ape species will arise. Species of feline will go extinct, and no new
felines will evolve. Species of dipterocarps, the great hardwoods of
Asian rainforests, will be lost and not gained. The world will be an
emptier, lonelier place. Still, we humans can probably look forward to
sharing the future with a fair number of beetles, tapeworms, fungi,

tarweeds, mollusks, and mites. Dandelions and silverfish are also a good bet.

The whole idea is profoundly gloomy, important, and persuasive. Even with that weighty message, though, Soulé's individual chapter in *Conservation Biology* was a less significant contribution than what he had done toward making the book itself happen. And behind the book was something larger still: a new branch of biological science.

153

DURING THE LATE 1970s Michael Soulé was teaching biology at UC–San Diego. His early research specialty had been the genetic basis of morphological variation among reptiles, but he had grown steadily more worried about the larger problems of habitat destruction, habitat fragmentation, genetic deterioration of small populations, and extinction. He saw that many ecologists and population biologists were beginning to address these problems but that their efforts were mostly narrow and scattered. There was no coordination and not much cumulative effect. There was no common forum for scientists concerned with extinction and how to prevent it. Soulé felt that something had to be done. In 1978 he and a graduate student named Bruce Wilcox convened a gathering in San Diego. They called it, grandiosely and hopefully, the First International Conference on Conservation Biology.

Papers were presented by about twenty biologists, including Jared Diamond, John Terborgh, and the padrone of conservation-minded biologists, Paul Ehrlich. Soulé and Wilcox edited the papers into that compendious volume, *Conservation Biology: An Evolutionary-Ecological Perspective*, published two years later. The book's cover featured a brown-tinted negative image of African antelope, nicely suggesting the threat of negation by extinction. The brown tint turned out to be useful as a marker, since the phrase "conservation biology" would repeat itself on other publications and the subtitle, though descriptive, didn't roll off anyone's tongue. This volume became known, among cognoscenti within the field, as the Brown Book. Eventually there would be a rainbow of others.

It wasn't the first book to address the subject. Raymond Dasmann had published *Environmental Conservation* back in 1968, and David

Ehrenfeld's *Biological Conservation* had appeared in 1970. But the Brown Book was something different: a collaborative effort by a broad group of scientists, many of whom had spent their professional lives under the influence of *The Theory of Island Biogeography*, all of whom agreed that the biological world was falling to pieces and that action had to be taken. These scientists were united by a bone-deep sense of ongoing loss. With each further hectare of ecosystem destroyed and with each additional species extinguished, they realized, the biological universe to which they had devoted themselves intellectually (and emotionally) became smaller. If the trend continued, the near future would see ecologists and field biologists increasingly supplanted by museum curators, paleontologists, and historians—people who could remind the public that once upon a time our planet had been graced with living elephants, bears, and lemurs.

In the book's introductory chapter Soulé and Wilcox defined the enterprise. "Conservation biology is a mission-oriented discipline comprising both pure and applied science," they wrote. On the pure side, it encompassed ecology, evolutionary biology, island biogeography, genetics, molecular biology, statistics, and a handful of other disciplines in the background, such as biochemistry, endocrinology, and cytology. On the applied side, it included economics, natural-resources planning, education, conflict-resolution techniques, and everything that island biogeography could teach regarding the design of nature reserves. Until recently, according to Soulé and Wilcox, academic snobbery had left many biologists unwilling to touch the subject, because those academics saw "applied" biology as the traditional province of lesser intellects trained in wildlife management and forestry. "But academic snobbery is no longer a viable strategy, if it ever was." The posture of superior detachment was a fatuous luxury that just couldn't be afforded anymore. "There is no escaping the conclusion that in our lifetimes, this planet will see a suspension, if not an end, to many ecological and evolutionary processes which have been uninterrupted since the beginnings of paleontological time." Personally, wrote Soulé and Wilcox, they hoped that it might be only a suspension. The purpose of the book, the purpose of conservation biology, was to prevent such an interruption from being permanent.

At least some of the anger and clarity in this statement came from Michael Soulé himself, an interesting man whose own career has tracked the historical progression of these scientific disciplines—from evolutionary biology to island biogeography to viable-population theory and conservation biology—more closely than any other person's.

Soulé grew up in San Diego during the 1940s and early 1950s, when the city was still surrounded by small canyons full of chaparral vegetation and a boy could harvest abalone and lobster along the tide line below Sunset Cliffs. He played in the canyons; he collected butterflies; he looked at pond water through a microscope. He had a tolerant mother who let him keep snakes in his bedroom. Forty years later he's a professor at Santa Cruz, where I visit him in his office.

In his mid-teens, Soulé tells me, he belonged to a club of junior naturalists associated with the San Diego Natural History Museum. "It was sort of an informal group of nature-freak kids who would be called nerds, I suppose, these days," he says. "But we were all nature fanatics, and we'd go on field trips into the mountains and out to the desert, and go collecting marine invertebrates along the coast." They went down into Baja on one trip and even got out to some of the islands in the Gulf of California. It foreshadowed research expeditions that he would make later, to many of those same islands, while he was a grad student under Paul Ehrlich up at Stanford. That was the early 1960s, and by then Soulé had fallen under the spell of an exciting new theory, just published by two young rebels named MacArthur and Wilson.

During his years of doctoral work, he spent some time on the island of Angel de la Guarda and collected lizards at the Cañon de las Palmas site, where Ted Case more recently has been chasing chuckwallas. One of Soulé's earliest publications was a paper on reptile biogeography among the Gulf of California islands—a paper that owed much to MacArthur and Wilson. He and his co-author charted the distributional patterns and discussed the species-area relationship and the distance effect. Later, Soulé would do island work also in the Adriatic and the Caribbean. Still later, he would publish a paper describing ecosystem decay on insular fragments of chaparral habitat in San Diego County. The equilibrium theory was to him a stunningly useful tool, like the microscope through which he had once peered at pond water.

I ask him when he became aware of that theory.

"The day it was published," he says.

After finishing his doctorate and teaching for several years at a university in Malawi, in southeastern Africa, he came home and took a position at UC–San Diego. He continued his research on reptiles.

Meanwhile he felt increasing concern about the conflict between human population growth and wild landscape. He had been sensitized to the population problem by his graduate advisor, Dr. Ehrlich (as had many Americans who knew Ehrlich only distantly, as author of *The Population Bomb*), and in Malawi he had seen some of its consequences, both in terms of human travail and in terms of toll on the landscape. The surviving wildlife of Malawi wasn't nearly so abundant as the great game herds up in Kenya or Tanzania, and it was mainly restricted to a few small national parks. The changing landscape of southern California gave Soulé further data, reminding him that habitat loss isn't solely accountable to starving Africans with machetes and plows.

"It was clear that San Diego was being destroyed," he says. The chaparral canyons were succumbing to urban sprawl. "They were paving paradise and turning it into a parking lot, as the song says. And when you see a place where you were born and raised being destroyed, it's very poignant. That probably creates a deeper—in *me* it created—a deeper degree of angst and concern than anything else. Yet, to be in a place long enough to see it go from either pristine to not pristine, or from tolerable to intolerable, before you are really moved to action . . ." His voice trails off. Michael Soulé was certainly moved to action, but apparently not as early as he wishes he'd been.

"Forgot what you asked me," he says.

"When you first became concerned about extinctions."

Oh yes. Well, it was gradual. It was a slow dawning. First there was that simpler worry about human population growth overwhelming the landscape. He was also aware of what Diamond and others were saying in the early 1970s about the implications of the equilibrium theory for small nature reserves: loss of species, relaxation to equilibrium, lower diversity on isolates than on samples. Then in 1974 he went on sabbatical to Australia. While he was there, an eminent wheat geneticist named Otto Frankel asked Soulé to come down and give a seminar in Canberra. He was interested in Soulé's work on islands, Frankel said, because he thought it might be relevant to a concern of his own, genetical conservation. "I'd never heard of genetical conservation before," Soulé recalls. And he knew beans about wheat. But he said sure, he'd be glad to give a seminar. The contact with Frankel led to a major collaboration that stretched across some years and yielded a co-authored volume, *Conservation and Evolution*. It's another of the milestones in the field.

Conservation and Evolution was published in 1981 with a green cover. The title isn't catchy, so it became known as the Green Book.

154

SOULÉ'S OFFICE IS tiny, and we share the space with his bicycle. Something about California ecologists, I suppose—Ted Case also parks a ten-speed in his office. On the wall beside Soulé is a print of zebras. On another wall, a photo of the volcanic fires of Kilauea, in Hawaii. There's a file cabinet, a greaseboard for sketching ideas, a Christmas cactus in bloom on the desk, a Macintosh computer, and a cast of the skull of a sabertooth cat. I remember seeing the sabertooth skull in Ed Wilson's office and wonder passingly whether this is a runic token of membership in the elite of island biogeographers. Soulé is a lean middle-aged man in gym shoes, with a graying goatee. As we talk, he eats lunch: instant lentil soup from a cardboard cup. His life seems serious and compact, in a way that I find admirable, and he answers all my scientific questions and most of my biographical ones quite obligingly. Only a spirit of nosy mischief compels me to ask about his years at the Institute of Transcultural Studies, a Zen center in Los Angeles.

No mystery, he says. He was tired of the university scene. Burned out. The personal environment had come to seem stultifying. This was 1978, same year as the first conservation biology conference. He had had it with academic values and academic scuffles. So he resigned his tenured position at San Diego, went up to the center, and for a time functioned as its director. Helped with the medical clinic. Started a program of Buddhist studies. But he also continued with biology during that period. No, I shouldn't assume that he had despaired of studying nature or laboring to conserve it. He had just gotten a bellyful of the professorial life. At the Zen center he finished his work on two collaborative volumes—the Brown Book, from the conference, and the Green Book, with Frankel—and he wrote several papers. Did some international consulting on conservation genetics. Also, there was that essay titled "What Do We Really Know About Extinction?" with his list of eighteen contributing factors, which appeared in a volume called *Genetics and Conservation* (later to be known, by chromatically inclined insiders, as the Gray Book). It was a reasonably productive time, the Zen period, even in scientific terms, Soulé explains.

Yes, evidently so. I had pictured five years of ethereal meditation in a saffron toga. As I turn the conversation back to other topics, it occurs to me that considering the significance of those publications as well as their sheer quantity, the Institute of Transcultural Studies must have offered a work environment that other biologists might do well to consider.

On the strength of those early writings, Soulé emerged as one of the pioneer theorists on population viability. Then, in 1982, the ambit of theory intersected the ambit of practice when he got a call from a man named Hal Salwasser, who held the position of national wildlife ecologist for the U.S. Forest Service.

Salwasser had a problem that he hoped Dr. Soulé might help solve. Back in 1976, Congress had passed the National Forest Management Act, amending the directives by which America's national forests were to be managed. Prior to that, a law dating from 1897 had dictated that managers should "protect and improve the forest," which was a sweet notion but too vague, leaving latitude for decades of excessive and ecologically damaging timber harvests. "Improving" the forest could be construed, and often had been, in a shamelessly reductionist sense. To carve logging roads through an old-growth wilderness, or to clearcut a mixed stand of trees and then plant seedlings of one fast-growing species, might both be taken as improvements; if biological diversity was reduced in the process, well, tough splinters, that wasn't the Forest Service's mandated concern. But in 1976 the mandate had changed.

The new law made plain that increasing the standing stock of harvestable board-feet did not alone constitute protecting and improving a forest, and that managers were obliged also to "provide for diversity of plant and animal communities" within the forest boundaries. Furthermore, in the planning regulations associated with the new act, Salwasser found this command: "Fish and wildlife habitat shall be managed to maintain viable populations of existing native and desired non-native vertebrate species in the planning area." Viable populations of the native vertebrates? Fine, Salwasser was happy to try. But what exactly did it mean? What constitutes viable? He was calling with the hope that Soulé and Soulé's academic colleagues might share their thoughts toward an operational definition.

"I think it was a seminal event," Soulé says.

Soulé and Salwasser convened a workshop, under Forest Service sponsorship, that brought a small selection of viable-population theorists together with some of the agency people. It was held in Nevada City, California, in the foothills of the Sierra. Among those attending was Mike Gilpin, the aging triathlete with the nimble touch for computer programming, who remembers it vividly.

155

"And so I get invited to this meeting," Gilpin tells me during one of our many talks. A very small meeting, he explains, funded by the Forest Service and hosted by this fellow Hal Salwasser, an ecologist, who was charged with developing a new set of planning regulations. In addition to Salwasser and Soulé and Gilpin himself, the group included just three other Forest Service biologists, two other academics, and Mark Shaffer, who by that time held a job with the U.S. Fish and Wildlife Service.

"Now, I didn't know much of the background," says Gilpin, "but when Salwasser finally got around to closely reading his own National Forest Management Act of 1976, he saw the line in it that said, 'Each forest superintendent shall maintain viable populations of each vertebrate species in each national forest.' " It's a loose quote, rendered from memory, but there's no blurriness to Gilpin's recollection of the crucial two words. "So. One had to decide what a 'viable population' was."

Soulé had come up with the number fifty, Gilpin reminds me, as a rough approximation of the minimum population necessary for short-term avoidance of inbreeding depression. Ian Franklin had generated the other number that was currently in the air, five hundred, for long-term survival and adaptation. But how short was short-term? How long was long-term? How should a bunch of federal land managers construe that sticky phrase, "viable populations," which had somehow found its way into their planning regulations? If they went to the standard ecology textbooks, says Gilpin, they wouldn't find much help. The whole body of thought was too new. The ideas were just being invented.

"Anyway, we show up at this meeting," Gilpin recalls, "and, sort of the typical thing, you're sent some material in advance of the meeting, and you never read it. You figure you'll read it the night before, in the motel." That's what Gilpin had figured. But with circumstance and lassitude intervening, he had figured wrong. "So I got to this meeting not really knowing what they were talking about. I'd never read the National Forest Management Act and had not been interacting with any sort of real-world biologists for a long time." The Forest Service personnel turned out to be bright people who were eager to nudge their agency toward doing the right thing, if only some reputable scientific specialists would help them to say what that was. "We had this question read to us, and we sat around for two days dealing with the problem of population viability." Gilpin remembers the frustration that hit him—confident guy, agile mind, broad training—when he realized that he couldn't solve it.

"I was somewhat humiliated," he says. "I was also struck by the *reality* of this question."

He and his fellow theorists were being asked to make what seemed like a basic, uncomplicated forecast. "There's the species. You have some information on it—you don't have *enough* information, but you do have some information. Now," says Gilpin, casting himself in the managers' role, "tell me whether that species will be here, or not, fifty or a hundred years from now." It was a binary question: yes or no. If this branch of science had acquired any predictive capability whatsoever in the years since MacArthur and Wilson, here was a chance to deliver. "You didn't have to say the numbers, you didn't have to say what the genes were going to do. You just had to make the simplest possible prediction. Yes or no. Fifty years." Would the species survive for another half century or wouldn't it? When he realized that none of them—not himself or Soulé or the others—could offer a firm answer, Gilpin knew that their work hadn't come as far or made itself as useful as he had wanted to believe. Not yet, anyway. There was still too much uncertainty standing between population viability as an abstraction and the capricious empirical events of any particular case.

After the brush with reality at Nevada City, Mike Gilpin began giving the problem more thought.

156

THEORY-HEADED wizards such as Gilpin and Soulé were unaccustomed to being asked for guidelines that a huge federal agency might actually put into practice. It was gratifying but also perplexing. It was probably a little scary.

The earlier rule-of-thumb numbers, fifty and five hundred, were obviously too general and absolute. A population of grizzlies in the greater Yellowstone ecosystem would not suffer precisely the same risks, nor react to them in precisely the same ways, as a population of spotted owls in western Oregon or a population of snail darters in Tennessee. The threshold of viability and the factors that determine it would vary from case to case. "The moral of this tale," Soulé wrote later, "is that the closer we approach reality, the more complex the problem appears."

Soulé himself had meanwhile emerged from the Zen center and reentered the academic world. Whether that carried him closer to reality or farther away is a matter of opinion about which even he seems to harbor ambivalence.

He took a position at the University of Michigan. Among the first things he did there was organize another small workshop on population viability. It was held at Ann Arbor in October 1984, with sponsorship by both the Forest Service and the Fish and Wildlife Service as well as support from several private organizations such as the New York Zoological Society. Mike Gilpin, Hal Salwasser, and Mark Shaffer participated again, joined now by more than a dozen select population biologists, evolutionary geneticists, and statisticians. This was the meeting, as Gilpin remembers it, that "definitely put all the pieces out on the table." From the table, those pieces went into a book. Soulé again served as editor, and the ideas brought to focus at Ann Arbor appeared eventually in *Viable Populations for Conservation*. The cover of that one is bright blue.

The Blue Book is thin in pages, dense in content. Math and computer modeling figure prominently. In their individual chapters, Gilpin and the others examined various implications of demography, genetics, effective-population size, spatial relationships among multiple small populations, and probability theory. They also discussed some practical realities: the genetic impoverishment of the California condor, the need for institutional cooperation in managing American grizzly bears, the cost-to-benefit ratio of forestalling the extinction of the Sumatran rhino, among others. The comprehensive overview, the voice of wisdom and synthesis, came from Soulé.

In an introduction and an epilogue he subsumed the technicalities and offered perspective. For instance:

> The problem that we address in this book is "How much is enough?" Put more concretely, it is: What are the minimum conditions for the long-term persistence and adaptation of a species or population in a given place? This is one of the most difficult and challenging intellectual problems in conservation biology. Arguably, it is the quintessential issue in population biology, because it requires a prediction based on a synthesis of all the biotic and abiotic factors in the spatial-temporal continuum.

Another instance, rather more pungent:

> There are no hopeless cases, only people without hope and expensive cases.

Soulé was the graybeard who could speak in these tones. He was the one who could trace the history of this scientific enterprise, which he and others now called conservation biology, backward through time—not just to David Ehrenfeld and Raymond Dasmann and Paul Ehrlich, not just to MacArthur and Wilson, not just to Frank Preston, but beyond them to a broader and longer intellectual and ethical tradition encompassing Rachel Carson and Aldo Leopold and Charles Elton and Gifford Pinchot and Theodore Roosevelt and Thoreau and Darwin and Saint Francis and Lao Tzu and a less familiar figure named Ashoka.

"Who was Ashoka?" I ask him when he has finished his soup.

The first Buddhist king of India, he tells me. "There was a period where Buddhism was the dominant religion—a short period of time—before the Mongol invasions of India. And Ashoka was a Buddhist king who issued a number of edicts called the Pillar Edicts. They called that because they were written on pillars and put around the countryside. Like some of the prophets and kings of Israel, he recognized that it was possible to abuse and overexploit natural resources. So he issued rules about forestry and agriculture and the conservation of natural resources. They were one of the first sets of laws ever issued on conservation."

"What was the time period?"

"This was before Christ." Maybe about 100 B.C., Soulé says. But he's just guessing, he adds, don't hold him to it. He claims to have a bad memory. Keeping the dates of such-and-such a conference and such-and-such a workshop and so-and-so's important journal paper all straight in his head—that's hard enough, let alone ancient history like his boyhood in San Diego or the reign of Ashoka.

Our rambling conversation has already used up most of the time slot he promised me, so I let Ashoka drop. Dates can always be checked and corrected. There's another subject that interests me more. Since Michael Soulé has been such a pivotal figure—arguably, *the* pivotal figure—in the development of conservation biology and viable-population theory from its origins in island biogeography and elsewhere, I want to ask him at least one further question: about SLOSS. What I'm hoping, of course, is that he'll spill some gossip.

"Is it my imagination," I say, "or was that debate a little more heated than the average scientific disagreement?"

Soulé answers. His answer is discreet, statesmanlike, generous to all parties, and not for quotation.

So I back off, and come at the same subject again with a different approach. This approach is less gossipy and more decorous, unfortunately, but it's the best I can do. I ask Soulé about a curious journal paper that he published, in 1986, with Dan Simberloff as his co-author. I've read the thing three or four times, and though its scientific content is straightforward, I remain fascinated by its invisible political dimension. The pairing of authors seems odd. The SLOSS debate had long since grown truculent by 1986, with two factions fighting it out in the journals, and that truculence had been aggravated still further by the null-hypothesis controversy. Soulé was a close colleague of Mike Gilpin's, and his own record of work tended to identify him with the Diamond faction. Simberloff was the archenemy of that faction. If SLOSS was trench warfare, this paper seemed the equivalent of Lloyd George in a bear hug with Kaiser Wilhelm. What happened?

"I never met Dan Simberloff until I was in Moscow one time," Soulé says. They were both there as guests of an eminent Soviet biologist and they found themselves staying at the same hotel. They had dinner together one night and then did some sightseeing, in the course of which Soulé discovered that, God help him, he *liked* Simberloff. The guy was good company in a foreign land. "He's one of the most urbane and sophisticated people I've ever met. Extremely cultured and knowledgeable about music and art. It was wonderful for me to be able to hang out with him. And at the time, I had a manuscript with me . . ." Soulé's famously imperfect memory fails him again as he tries to recall how the manuscript had arisen in conversation. Maybe he had discussed it in a seminar, though he isn't sure. Anyway, the subject was on his mind and the manuscript was among his papers.

"It was about trying to resolve this dispute between Diamond and Simberloff. Those were the two key antagonists," he tells me unnecessarily. Then he corrects himself to a more emollient wording: "—*pro*tagonists."

Before getting to know Simberloff, he says, "I considered myself more on the side of Diamond." His manuscript treated all the same issues of nature-reserve design that had troubled Tom Lovejoy and others, that had driven the SLOSS debate, and that had led to the entrenchment of the two factions. Soulé was now attempting to get past the old, polarized view of those issues by way of the more recent insights in viable-population theory. He was also hoping to make peace. There were so many powerful forces against which conservation biologists had to fight, he believed, that they just couldn't afford to fight against each other. "I gathered up my courage and I said, 'Dan, I want you to read this and see what you think of it.' "

Simberloff read the manuscript and thought enough of it to offer constructive criticism. He suggested some revisions and additions. Eventually, at Soulé's invitation, he signed on as co-author. The paper appeared in a good British journal under the title "What Do Genetics and Ecology Tell Us About the Design of Nature Reserves?" The word "Us" carried a bit of special significance, given the pair of authors: What points can we *all* agree on, Diamondites and Simberlovians both? But there weren't necessarily many. In the first sentence of the abstract, Soulé and Simberloff declared: "The SLOSS (single large or several small) debate is no longer an issue in the discussion about the optimal size of nature reserves." No longer an issue? Was that wishful thinking? I ask Soulé.

"Yes," he says.

Yes, but not entirely. Although some people continued fighting the old war, belaboring the old SLOSS dichotomy, others were moving into the new mode of thought. The more forward thinkers concerned themselves with population viability, not just with small areas versus large ones, and explored the notion of keystone species, meaning those plants and animals that might be especially crucial to maintaining the cohesion of an entire ecological community over time. Soulé and Simberloff emphasized keystone species and population viability as essential to the task of designing nature reserves. And they opined that if SLOSS wasn't a dead issue, it probably should be. With or without a little wishful thinking, the paper was a worthy effort.

I ask him about its effect. Did it help at all toward coaxing people out of the trenches?

"Who knows. I mean, my impression is that it did. But I don't have any direct evidence of that."

Soulé, the mediator, the synthesizer, is slow to claim credit and cautious about making leaps from fact to hypothesis. He's a good population biologist with a salubrious grounding in Zen and a strong sense of the probabilistic component of events. He knows that so much is always uncertain.

After two hours I thank him and prepare to clear out. While Soulé goes to find me a reprint of one of his more obscure articles (a gracious gesture that will spare me some library legwork), I notice an interesting small touch of his office decor. It's a photocopy of a newspaper clipping, taped inconspicuously to the side of a bookshelf. The headline says FACE OF CHRIST APPEARS IN TORTILLA AS IT FRIES.

When Soulé returns, I remark on it. Dan Simberloff's office door is upholstered with similar stories, I say. Was it Simberloff who sent this clip? Soulé can't recall. Possibly it was Dan, yes. In pen, at the bottom, someone has written: "Tickets on sale now!"

"He has a wild sense of humor," Soulé says.

I try to imagine that side of the complicated guy whom Soulé got to know in Moscow. Simberloff as incorrigible cutup. Simberloff as manic goofball, sending found comedy through the mail for the yuk of it. Um, okay, maybe. I still have a sense, though, that the tortilla thing represents more than pure silliness. I mention my own hypothesis about Simberloff pursuing the null-hypothesis argument by other means.

"Yes. He's a professional skeptic." Soulé gives me a tiny, dry smile. It's not as though the connection hadn't occurred to him. "I'm an amateur skeptic. But I do what I can."

157

MARK Shaffer's contribution to the Blue Book was titled "Minimum Viable Populations: Coping with Uncertainty." It carried forward the line of thought that he had presented in his dissertation nine years earlier. Although systematic pressures inflicted by willful human activity can drive a species down to the margin of viability, Shaffer wrote, the terminal event—extinction—will generally involve an element of chance. Uncertainty is part of life, yes, but for small populations, uncertainty is the enemy of survival.

To that truth there are two corollaries, he added. "First, the smaller the population, the more likely extinction is in any given period of time. Second, the longer the period of time, the more likely extinction is for a population of any given size." The concept of minimum viable population is merely a formalized version of those corollaries, he explained, an effort to quantify the risk of extinction for any animal or plant population when it's sorely reduced in size. This formalized concept gives us the prerogative of deciding how much of such risk we, human society, are willing to accept. How badly do we want rhinos in Sumatra? How badly do we want elephants in Africa? How badly do we want grizzlies in Yellowstone? "Does a 95% probability of persistence for 100 years make extinction sufficiently remote," Shaffer asked, "or all too imminent?" He didn't supply an answer. He knew that it's a political question, not a scientific one.

But he made one dour prediction. His own rough estimates of viable population size led him to conclude that "the size and number of current nature reserves are likely insufficient to provide a high level of long-term security for at least some mammalian species, especially those that are large or rare."

He wasn't alone in this pessimism. Soulé and others were singing the same song.

158

Two MONTHS after Bedo's murder I return to Perinet, hoping to gather the facts. But there don't seem to be any facts, aside from the stark one that Bedo is dead. In lieu of facts, there is testimony. There are voices. The voices speak at me in French and in Malagasy and in English. Some of them are angry voices, some grief-weary, some defensive. I don't discover truth; I collect versions. It's *Citizen Kane* and *Absalom, Absalom!* all over again.

I call on Bedo's father, in the house just across from the Analamazaotra reserve. His name is Jaosolo. Because I am accompanied by Pat Wright and David Meyers, he welcomes me and begins talking about his son. Yes, Bedo loved the lemurs. The reptiles too, yes, all of the animals. Sometimes people would go to the forest and search for a rare species, and they wouldn't find it. When they went with Bedo, ah, they always found it. The grand professors from Europe would come, they would look up into the trees at a creature, and they would say: What's that? They wouldn't know. Bedo would look up, he would know. He would tell them the species. Always people would come, they would ask for Bedo. Where is little Bedo? Lately a Japanese group came to the hotel, just last month. They had arranged earlier that Bedo would guide them. They asked, Where is Bedo? He wasn't there. It wasn't his fault, but he wasn't there. I couldn't tell them why, says Jaosolo, his voice sounding tired and unsteady. "*Tous les visiteurs, professeurs ou autres, demandait toujours lui.*" Bedo, he was the best.

Jaosolo has taken the loss hard. Barefoot, a small dignified man in the beret and uniform of Eaux et Forêts, he goes to a wooden cabinet and rummages there for mementos. The house is unlit and the windows, blocked with blinds, admit almost no sun. He puts on his delicate gold spectacles. He shows me a document stamped with a red seal, his official complaint over the death. He shows me a copy of *National Geographic* full of Frans Lanting's work as assisted by Bedo. He shows me snapshots. "*Ceci, c'est la dernière photo.*" In this last picture, a strapping young Bedo stands surrounded by forest, wearing a T-shirt printed with lemurs.

I speak with the only eyewitness, a young friend of Bedo's named Solo. I speak with the gracious hotelier, Joseph Andriajaka, who watched Bedo grow up and encouraged his development as a guide, who recommended his services to so many Perinet visitors, including me. I sit near the railroad bridge across the little river that flows through the village, questioning Bedo's brother and sister. This is the murder site. Details are confused, but no one disputes the point that, whatever happened, it happened right here. The body was recovered, eventually, from the river. Bedo's brother has little to say. His sister speaks fiercely and vividly, telling me the tale. I stare at the muddy water. I ask heartless questions. Where exactly? How many days? What made the body so hard to find? Why did he go toward the river when he was attacked?

Later I talk with Alison Jolly, another eminent lemur biologist, who was in Perinet on the day of the killing. I talk with Jean-Marie de la Beaujardière, an honorary consul of Madagascar, who arrived the day after. Pat Wright and Frans Lanting and David Meyers share with me what they know—or, more accurately, what they have heard. I listen to accounts of the murder ten or eleven times, in three languages, with and without help from a translator. Each time it's different. In Madagascar, as Alison Jolly warns me, "things immediately move into the realm of myth. Usually within about a day."

I am variously told:

Bedo was killed with an ax. He was not killed with an ax. He was killed with a rock, thrown hard, which hit him in the temple. He drowned. His dead body was dropped into the river. He jumped into the river while trying to escape. No, his body could not have been in the river until just before it was found, days later, because the searchers wouldn't have missed it. Someone must have hidden his body, then later dumped it into the river. God knows who, God knows why. No, his body was there in the river the whole time, entangled in submerged branches. No, his body bore no signs of having spent days in the water. His belly was flat. He didn't drown. His wounds showed that he had been struck in the face with an ax. No, the ax belonged to an earlier stage of the altercation and it wasn't used on Bedo. He was hit with a rock. No, with four rocks. No, one rock, a regrettable accident that caused him to drown. No. He was murdered.

The incident occurred on a Sunday afternoon. No, it was Sunday night. Darkness. Broad daylight. One man did it. Two men, their names are the following. They are not natives of Perinet, they come from the city of Fianarantsoa, where they might or might not have been implicated in other grievous mischief, for which they might or might not have gotten off lightly. For killing Bedo, they have been sentenced to five years. No, the trial is yet to come. The two men are in jail in Ambatondrazaka. They are trying to blame it on each other. The prosecutor will be too lenient. They might receive life sentences.

The argument began at a party. Loud music and liquor. The liquor was a form of Malagasy moonshine known as *tokagasy*. Ordinarily it should be as clear as vodka; this batch was cloudy, so Bedo refused to drink. It might have been poisonous. No, he simply chose not to get drunk. No, he refused because he was haughty. He got into a fistfight with one of the hosts, who was outraged by Bedo's snub. Bedo won. No, Bedo was down on the ground, and his friend Solo brandished an ax to rescue him. Bedo and Solo then left the party. Near the railroad bridge, they were set upon. Bedo was hit—with the ax, no, with the rocks. He fell or was pushed or escaped into the water. He rose three times, the canonical number, then disappeared.

Why was he murdered? Because, in becoming the darling of foreign tourists, in earning so much fast easy money as a guide, he had provoked envy. Malagasy culture does not tolerate upstarts. Bedo spent his money on a punk haircut, a radio, flashy clothes. He had gotten abrasive and highfalutin. Some people hated him for it. No, he was just a young man groping toward maturity. Basically he was unassuming and sweet. He had great integrity. He was killed, perhaps, because his father and other officials from Eaux et Forêts had been trying to stop illegal cutting of hardwood from the Analamazaotra reserve. Illegal destruction of indri habitat. No. No, it was a meaningless village fight. He died because of alcohol and sudden anger, just like unfortunate souls in American cities every Saturday night.

According to Alison Jolly: "He was murdered for jealousy, and in a way it was we who were responsible." She means the foreign tourists and scientists who paid handsomely for his skills, disturbing the equilibrium of his world. This view has a certain logic.

By a contrary view, equally plausible, his death was a chance event.

At the time of my postmortem inquiries, Bedo lies buried in Perinet. But only temporarily. Later, I'm told, his body will be moved to the family's ancestral tomb in the north. If his family follows the Malagasy custom of *famadihana*, turning of the bones, that tomb will be reopened in a few years and Bedo's remains (among others) will be joyously rewrapped. Meanwhile, some of the scientists and conservation officials who knew him have begun talking about a different sort of commemoration: a scholarship fund, in Bedo's honor, for promising young Malagasy naturalists.

Frans Lanting mentions still another sort. "We called him *Babakoto*," says Lanting, "and as long as there are these animals screaming in the forest, he'll be memorialized." But a point remains uncertain: How long is that?

૭૩

WORLD

IN PIECES

૭૩

159

ISLAND BIOGEOGRAPHY is no longer an offshore enterprise. It has come to the mainlands. It's everywhere. The problem of habitat fragmentation, and of the animal and plant populations left marooned within the various fragments under circumstances that are untenable for the long term, has begun showing up all over the land surface of the planet. The familiar questions recur. How large? How many? How long can they survive?

How many mountain gorillas inhabit the forested slopes of the Virunga volcanoes, along the shared borders of Zaire, Uganda, and Rwanda? How many tigers live in the Sariska reserve of northwestern India? How many individuals of the Javan rhino are protected within Ujung Kulon National Park? This many dozen gorillas, this many dozen tigers, this many dozen rhino. The numbers change marginally, year to year, situation to situation, while the recurrent questions vary also in their particulars, remaining fundamentally the same. How many left? How large is their island of habitat? Do they suffer small-population risks? What are the chances that this species, this sub-species, this population will survive for another twenty years, or another hundred?

How many grizzly bears occupy the North Cascades ecosystem, a discrete patch of montane forest along the northern border of the state of Washington? Not enough. How many European brown bears are there in Italy's Abruzzo National Park? Not enough. How many Florida panthers in Big Cypress Swamp? Not enough. How many Asiatic lions in the Forest of Gir? Not enough. How many indri at Analamazaotra? Not enough. And so on. The world is in pieces.

The problem is more ubiquitous than carbon monoxide. It exists wherever *Homo sapiens* has colonized and partitioned a landscape. Although it's an old trend, it has recently become acute. Critical thresholds are being reached and passed.

Still, if we can trust Michael Soulé's judgment, even now there are no hopeless cases, only people without hope and expensive cases. The case of the Mauritius kestrel, *Falco punctatus*, seems to validate that judgment.

160

In the mid-1970s the Mauritius kestrel looked doomed. The main cause of its decline was habitat loss, although other factors such as pesticide poisoning, egg predation by monkeys, and "varmint" shooting (some Mauritians, suspecting the species of preying on chickens, called it damningly *mangeur des poules*) were probably also involved. The decline had occurred gradually, over centuries, but during the 1950s and 1960s the kestrel's state of jeopardy became acute. Its population fell so low—well below fifty, Soulé and Franklin's tentative threshold for the avoidance of short-term genetic deterioration—that there seemed no chance it could recover. It dove toward extinction. One ornithologist, reporting in 1971, could account for just four surviving birds.

In 1973 the International Council for Bird Preservation, a conservation group with headquarters in Britain, launched a small effort to rescue the species, and the first systematic census was performed. The ICBP biologist counted eight kestrels, with a possible ninth. One pair disappeared soon afterward, probably having been shot. The loss left just two breeding pairs, plus two or three nonbreeding singletons. The ICBP man trapped a pair to serve as founders for a captive-breeding program, but his captive female promptly died. He trapped another female to replace her. By the start of the breeding season in 1974, then, two birds lived in captivity and four others were known in the wild. Maybe a few additional singletons survived elsewhere on the island, and maybe not. If the total world population of *F. punctatus* exceeded six individuals, there was no evidence.

One of the wild pairs succeeded in hatching and rearing three young. But a few months later Cyclone Gervaise struck Mauritius and tore through the habitat like God's fury. Limbs broke, trees toppled, animals hunkered. The wild kestrel pairs eventually reappeared, but they had been dislocated from their old territories and fallen weeks behind their usual breeding schedule. The captive-breeding program, meanwhile, was going nowhere.

This twilight of the kestrels continued for several more years, with the wild population increasing or decreasing slightly from one season to the next, always vulnerable to the possibility that a stroke of bad luck—a stochastic perturbation, in Mark Shaffer's terminology—might destroy it completely. The ICBP project yielded no tangible results, aside from its regular censuses that showed a species on the verge of oblivion. During 1978, again, only one pair reared young in the wild. Carl Jones, the sarcastic Welshman with the sheepdog haircut, came to Mauritius in 1979. He had been sent out by the ICBP. His assigned mission was to take over the project just long enough to shut it down. The case of the kestrel appeared hopeless.

And hopeless cases waste money. And money is a limiting factor, of course, in the design and implementation of conservation efforts. There just isn't enough cash to deliver a program of individualized rescue to every endangered species of bird, reptile, mammal, insect, and plant. About the time Jones got his assignment, a respected international conservationist named Norman Myers published a sobering book on extinctions, *The Sinking Ark*, in which he suggested the need for a triage strategy toward jeopardized species. As practiced by battlefield doctors, Myers reminded his readers, triage involves sorting the casualties into three categories and then essentially ignoring the third category (those wounded so badly they probably can't be helped) and the second category (those wounded slightly, who can probably survive without help) while attending to the first category (those for whom the provision or lack of medical help will be mortally decisive). Conservationists need to make the same sort of hard decisions, said Myers; they should focus their resources on those cases that show promise of recovery. "This would mean that certain species would simply disappear because we pulled the carpet out from under them," he wrote. Then he snatched an example out of the air: "We might abandon the Mauritius kestrel to its all-but-inevitable fate, and utilize the funds to proffer stronger support for any of the hundreds of threatened bird species that are more likely to survive." The ICBP, having already spent money on the kestrel and gotten no results, was inclined to agree.

Carl Jones, gangly and brusque, has never quite forgiven either Myers or the ICBP. He takes acidulous pleasure in repeating the "all-but-inevitable" quote.

Jones has earned the right to strong opinions on this subject. Those opinions extend not just to what should or shouldn't be done for *F. punctatus* but also to the causes of its decline. "The reason why the kestrel disappeared was quite simple," he says. And then, typical of Jones, he describes something complicated: habitat fragmentation and its biological consequences.

The forest loss on Mauritius occurred in a relatively brief span of time. The early Dutch sailors and settlers, who inflicted such harm on the native Mauritian wildlife, had made little direct impact on the woodlands, since they traveled by sail and their demands for timber were modest. But the French were another matter. During the late eighteenth and early nineteenth century, French colonists on Mauritius cleared away much of the lowland forest for agriculture. They planted vast fields of sugar cane. They cut firewood to fuel their steam-powered sugar mills. The heaviest labor was done not by the French themselves but by African slaves. Slavery on Mauritius was finally abolished in 1835, and soon afterward the colonists began importing Indian workers to replace the slaves, with the net result of a huge increase in the island's population. More people cut more trees for more fuel and shelter. Roads were built, and then also railroad lines, on which ran woodburning steam engines. The island was carved up, its dry lowland forest and its palm savanna supplanted by cane and other crops, by road networks, by dense human settlements. "They chopped down the forest," says Jones, "and they cut down all the corridors between the main breeding areas, which are on the mountain chains." In the mountains themselves, some forest remained intact. But it was only a small amount, confined to areas such as the Moka Range in the northwest and the Bambous Mountains in the southeast, mostly along steep slopes and high volcanic brows.

"Sometime towards the beginning of this century," says Jones, "the populations of kestrels became isolated in three mountain chains." He mentions the Moka and the Bambous, as well as the Black River Gorges, where most of his own kestrel work has been done. For this particular bird, Jones explains, cutting the corridors of forest between those breeding areas had dire consequences. Why? Because *F. punctatus* is reluctant to disperse across wide stretches of landscape, and even more reluctant if the wide stretches are naked of forest. Although it belongs to the falcon family, in its physical capacities the Mauritius kestrel is more like a hawk. Its wings are short and stubby, adapted for quick and erratic flight through the trees (where it picks geckos and other small prey off branches), not for great speed in the open, like most falcons. And it's philopatric, in the jargon of ecology—that is, home-loving, disinclined to make adventurous explorations of new territory. Put a captive-bred individual into the wild, Jones says, and it will seldom move more than a few kilometers from where you've released it. Animal species endemic to small islands are generally unadventurous—the adventurous genes have long since been buried at sea—but you wouldn't expect to see that so markedly in a falcon. The kestrel, an acrobatic predator, is nevertheless timid about making dispersal flights between those insularized remnants of habitat.

Pesticides were another problem. Between 1948 and 1973, the Mauritius landscape was dosed abundantly with DDT and another nasty compound known as BHC, toward the goal of eliminating malaria. DDT was also used for agriculture, at least until 1970. Birds of prey are especially susceptible to DDT poisoning, since it accumulates at the top of the food chain and reduces the hatchability of eggs. Around the time of the anti-malaria campaign, both the Bambous and the Moka populations of kestrels died out. They had been too small to absorb the latest losses and too remote to be rescued by immigrants. The Black River Gorges, with almost no human population or agriculture, were spared from spraying. That's where the last handful of kestrels survived.

But even at Black River, even without any further cutting and burning and plowing, the kestrels were losing habitat. Exotic plant species had been introduced onto the island, and some of those plants (especially strawberry guava and a species of privet) were causing as much harm as the alien mammals. Strawberry guava is an aggressive species that competes well against native Mauritian plants. It had invaded the upland forests and grown so dense in certain areas that the native vegetation couldn't regenerate. Privet elbowed its way onto the steeper slopes. To make things worse, pigs and monkeys fed on the guava fruit and spread the seeds widely, while deer (also exotic, introduced for sport hunting) browsed down the seedlings of native species. So the two sets of alien species, plant and animal, helped each other take hold on the island. Guava and privet made incursions on the uplands and the slopes of the Black River Gorges. Their presence changed not only the species composition of the forest but also its physical structure. They formed a tangled scrub, closing those understory gaps through which kestrels had customarily flown. As the habitat deteriorated, the food-gathering difficulties may have caused some adult kestrels to starve, and may have inhibited reproduction among others. A nesting pair of kestrels had to find food for themselves and deliver weeks' worth of surplus to their hungry offspring. No surplus, no offspring.

Breeding success was also affected by exotic predators. In the Black River area, the kestrels commonly nested in small natural cavities pocking the high basalt cliffs. A nest consisted of three eggs, or occasionally four, laid in a scrape on the floor of the cavity. The female would incubate the eggs—and defend them, insofar as she could—while the male hunted for food. Not all those cliff cavities were perfectly secure. The basalt cliffs at Black River rise steeply for hundreds of feet above the forest, but the rock is irregular and some cavities can be reached along narrow ledges by a clambering animal. Before the invasion by humans and their entourage of pests, when the island contained no agile mammals, the difference between cavities was moot. Under modern circumstances, a cavity accessible to rock-climbing predators can be a disastrously bad choice for nesting kestrels. Monkeys and rats are both capable of eating a clutch of eggs, and the mongoose *Herpestes auropunctatus* (another alien species, brought from the Asian mainland) may eat not just eggs but also hatchlings and possibly even a careless adult.

Still another problem for the bird was genetic impoverishment. Although there are no data by which to measure inbreeding during the 1970s and earlier, several assumptions can be made. Reduced to such low numbers, and held low for consecutive generations, the kestrel population must have lost much of its genetic variation through genetic drift. Losing variation, it would also have lost adaptability. Potentially useful alleles would have disappeared from the line of inheritance. The flamingo pink socks, among other items, would have been omitted from posterity's suitcase. And with mating options so limited, inbreeding was inescapable. Whether the degree of inbreeding produced inbreeding depression is a separate issue—a tricky one, to which I'll return. For now it's enough to say that the kestrel would certainly have felt the burden of its genetic load, however heavy or light that load may have been.

So: With reduced habitat, fragmented habitat, degraded habitat, DDT, cyclones, paranoid poultry owners, exotic predators, and genetic impoverishment all standing against it, *F. punctatus* was in a world of hurt. Norman Myers could hardly have picked a better example of hopelessness. By 1979, in the blunt language of Carl Jones, the Mauritius kestrel was "a bag of bones."

161

THINGS CHANGED. The new man, this Jones, was resourceful and godawful stubborn. He had scant patience for human diplomacy and no gift for truckling obedience, but at birds he was good. His approach was more aggressive, more interventional, less cautious, and he isn't too shy to say so.

For instance: "I was the first person *this century* to visit a Mauritius kestrel nest." He climbed up into a cliff cavity and made off with a handful of eggs. Incubating the eggs in captivity might improve their chances of hatching and survival, Jones hoped. Anyway, it would protect them from monkeys and mongooses and rats. And maybe a wild female, deprived of her eggs, would immediately lay another clutch to replace them. This egg stealing was a risky venture. Jones knew that it might conceivably do more harm than good. On the other hand, judging from experiences with other kestrels, there seemed a reasonable hope of success. And since the species appeared to be fading inexorably under present circumstances, why not try?

"Were you trained as a bird person?" I ask him. What I mean is, did he follow a straight track toward ornithology, or did he do his academic work as an ecological generalist?

Trained? he wonders. No, not that he can remember. "I've been a bird person all my life."

He grew up in rural western Wales, Jones tells me, where his great boyhood passion was wildlife. "My parents' backyard was full of all sorts of wildlife. I had badgers and foxes and polecats, ferrets and rabbits, falcons and hawks and owls. All these different animals. I was a very bad student, and I used to play truant. My father never knew, but I used to play truant to go out and look after my animals." He did a bit of falconry. Cared nothing about school, he says, only wildlife. "Fiddling around with my animals. I was very successful in breeding a number of species. You have to realize that this was at a time when people *weren't* breeding birds of prey in captivity. And I was breeding kestrels, and European buzzards. In fact, I was one of the first people *ever* to breed European buzzards. And I bred owls. As a schoolboy. My schoolteachers used to say to me, they used to say, 'Jones, for goodness sake, stop messing around with all those animals, and do some *real* work. Study your books. And perhaps you'll then *get* somewhere in life.' And I told them, I said, 'Sorry, but all I want to do is

wander around the world and go to tropical islands and study birds.'
My headmaster said to me, he said, 'Jones, don't be ridiculous. To go
and do things like that you have to be either intelligent or rich. And
you're neither.' "

Jones recounts it all zestfully, doing the headmaster ("And you're
nigh-thuh") in a snotty Oxford voice.

A minor but persistent theme in Carl Jones's life has been Showing
the Sods They Were Wrong. Voices of authority—the headmaster,
Norman Myers, the ICBP—have variously told him that he couldn't,
he wasn't, he mustn't. His response has been to demonstrate that he
could indeed, he was indeed, and furthermore he bloody well *would*. If
he had been less headstrong as a young man, he might have surrendered to his own all-but-inevitable fate and settled for a thwarted
existence in Wales, teaching school, or selling widgets, or working as
a mine engineer and keeping budgies as a hobby. He refused. He
doesn't believe much in fate. What he believes in, it seems, is the glo-
riously improbable gambit. Improbable gambits carry their own existential rewards but also, by God, they sometimes deliver results. Jones
has a favorite expression, which he shouts to the heavens—to no one
and everyone—at moments of ecstatic satisfaction with the texture of
life as he finds it on this remote, troubled, biologically spectacular island: "Eat yer heart out!"

His fascination with the Mauritius kestrel began long ago, when he
was an undergraduate in England. One day he heard a lecture by an
American named Tom Cade, a raptor expert who had founded the
Peregrine Fund. Cade talked about certain scientific methods of
endangered-species work that were new at the time. Some of these
methods, pertaining especially to birds, had been used by the Peregrine Fund in the course of its work on the peregrine falcon: captive
breeding, artificial incubation, captive rearing, fostering chicks of one
species under adults of a less-endangered species, and the "hacking"
technique from traditional falconry, whereby a newly fledged bird is
gradually weaned from human-supplied food and learns to hunt in
the wild. To Jones, the notion of using these methods for the rescue
of endangered bird species sounded ingenious and thrilling. It meant
that much of what he'd done as a boy had some larger potential value.
In the course of his talk, Cade also mentioned a creature called *Falco
punctatus*, the Mauritius kestrel. The world's rarest bird, on the
threshold of extinction, Cade said. But maybe even *it* could be saved.

"I just felt so excited about this," Jones recalls. "I said to myself,

I've got to go to America and see what these people are doing, see how they've managed to breed all these birds, find out what's really going on." In 1976 he went. Besides visiting the Peregrine Fund people, he met the American biologist who had started the ICBP project on Mauritius. "I promised myself that one day I'd go to Mauritius to see the kestrel," Jones says. "I never thought I'd ever *work* with it." A year later, he heard that the project was having troubles—might soon be terminated, in fact—but that meanwhile the ICBP had need of a new project officer. It was a dead-end position at a meager salary with excellent prospects for frustration and gloom. Jones campaigned for it and got it. "So I was sent out here, ostensibly for one, maybe two years. And my original brief was to close down the project." Instead, congenitally insubordinate, he found ways of making it work.

He learned that the species *could* be successfully bred in captivity, if the breeder was skillful and if the birds' aviary and food supply were uncontaminated by DDT. He made an important behavioral discovery: that wild kestrels in the Black River Gorges could be conditioned to accept supplemental food. He began supplementing their diet with raw meat. The extra food might allow some pairs to increase their reproductive rates, he figured. Increased reproductive potential might show itself as a second clutch of eggs, if Jones stole the first clutch for hatching and rearing in captivity. And lo, that's what happened. He also learned that by "pulling" eggs from a nest—taking them one at a time, day by day, as they were laid—he could induce a female to lay as many as eight eggs in a clutch instead of the usual three or four. He experimented with both artificial incubation (in electric machines) and foster incubation by a captive European kestrel. All these techniques had been developed elsewhere by Peregrine Fund people or others; Jones and his eventual collaborators in Mauritius adapted them to *F. punctatus*. The hatching and rearing success rates among his captive kestrels were low in the first years, but they improved. He learned that artificial insemination could be used to increase breeding success. He learned that hatchlings would do rather well on a diet of minced quail. In the early phase he had just a few captive-reared birds to work with. He kept some of them in the aviaries as breeding stock. He saw evidence of inbreeding depression (specifically, a tendency toward terminal oviduct disorders) among the offspring of his breeders, but the problem affected only a few individuals, and Jones worked around it. He chose his newer breeders from among the healthy birds while allowing the birds with disorders to die out.

Other captive-reared birds were released into the wild by means of hacking. Jones deduced that supplemental feeding of a wild pair in marginal habitat could make the difference as to whether they produced offspring or not. He switched to white mice as his supplemental offering and established that wild kestrels would accept the dead mice yet still continue hunting for themselves. He discovered that a habituated kestrel would even come to his call. With this discovery, an amazing little stunt became a regular feeding routine. Jones would appear below a kestrel's roosting site and signal to the bird that he was carrying a mouse. Then he would toss the mouse straight up, alley-oop, and the kestrel would dive off its perch, swooping down to intercept the meal ten feet above Jones's head.

He learned that eggs from a European kestrel could be substituted under an inexperienced young female in the wild, giving her practice at incubation while her own eggs were being pampered at the aviary. He learned that all manner of such switcheroos could be perpetrated, each with its own little measure of advantage for the Mauritius kestrel population.

Jones didn't labor alone. As time passed and the project showed signs of promise, he was joined by some colleagues and assistants. Richard Lewis, an old chum from Britain, brought his own expertise to the kestrel after a long stint of fieldwork on the monkey-eating eagle of the Philippines. Willard Heck, a wizard hatching-and-rearing specialist employed by the Peregrine Fund, visited Mauritius each year for the breeding season and played godmother to the eggs in captivity. Wendy Strahm's plant-ecology work shed light on the habitat situation. Grad students from the United States and Britain helped with radio tracking of kestrels, among other chores. Jones positioned the project under a new aegis, the Mauritius Wildlife Fund, and drummed up additional outside support. The Peregrine Fund played a crucial role, contributing money as well as the services of Willard Heck. The Jersey Wildlife Preservation Trust became the other major sponsor. Headquartered on the British island of Jersey, the JWPT was finely attuned both to the rescue of rare species and to the special problems of insularity. Its founder and guiding spirit was the author Gerald Durrell, who by all accounts was himself just raffish enough to appreciate the temperament of Carl Jones. As far as Durrell was concerned, a surly and highhanded Welshman with a wide streak of irony could be eminently worth backing, so long as he produced results. Not everyone shared that view.

"One organization, which shall be nameless," Jones says, "told us that they wouldn't work with us unless they had complete control of the worker they supported. *They* would tell him what to do. *They* would define the mission. Of course we said, 'Sod that, mate, we're not having any of it.' "

Which nameless organization? I ask.

"The ICBP."

162

As the Land Rover heads north from the village of Black River, then east toward the mountains, jouncing along on a lumpy road, Richard Lewis cradles the thermos on his lap. It's a wide-mouth thermos with a plastic handle, and inside are two small eggs, packed in a matrix of cotton balls by the careful, knowing hands of Willard Heck. These eggs, laid by a European kestrel in residence at the aviaries, are physically similar to the eggs of a Mauritius kestrel, but more expendable. If they survive the road, if they survive the climb, Lewis and Jones will entrust them to a wild Mauritius pair who have nested in a cliff cavity at Trois Mamelles.

The site is postcard scenic, and this afternoon we've got photographer's light. The peaks known as Trois Mamelles are three pointy volcanic cones, lushly forested along the lower slopes, stark and basaltic above, with a profile that evidently evoked udders (not necessarily bovine) to the Frenchman who named them. Earlier in the century, kestrels cruised these slopes and nested above in the cliffs, but they died out during the bad times and were absent for decades—until recently, when Jones and his crew released a captive-bred male and the male managed to lure in a singleton female. The female, probably an offspring of other released birds, is presumably inexperienced at incubating and rearing, so the foster eggs are meant to give her practice. Her original two-egg clutch has already been snatched and put in the care of Heck, who can't personally *lay* kestrel eggs but is vastly experienced at hatching them. When the Trois Mamelles female has proven her competence with European eggs, she will be allowed to hatch some of her own.

Jones, Lewis, and I hike up a steep trail to the base of one cliff, a high face of crumbling purple-black rock. Finding small horizontal ledges along the cliff face, enough for toeholds and fingergrips, we climb, traverse, and continue climbing. Before long we've got a modest bit of altitude and exposure. The green apron of forest sweeps downward below us, flattens, breaks up into scrub vegetation, and then gives way to a broad vista of cane. Miles to the east is a town. The view is grand but mostly we ignore it. Twenty feet below the nest cavity we stop, having come to a dead end with our nub-and-ledge clambering.

The cliff here is nearly vertical. There's no easy route of further ascent, and a hasty impression would be that there's no route at all. This inaccessibility is exactly what makes the nest site a good one. We cling to the face and gawk upward. I think: *Where?* These guys are not jaunty young rock climbers with helmets and ropes, after all, they're a couple of pragmatic ornithologists carrying eggs. Lewis will proceed alone, since the next pitch is severe and the nesting kestrels will be distressed at one intruder, let alone more. That's fine by me; I'm close enough for journalism. Lewis makes the ascent hand over hand along a thick vine growing flush to the cliff.

The two kestrels, which have been spooked from their nest at our approach, now return. They position themselves in midair not far behind Lewis's neck. They are handsome little hawk-shaped birds, tails splayed, wings open, their mottled vanilla breast plumage darkened in silhouette against the afternoon sky. They hover on thermals, steady as a pair of Oriental kites tethered in the breeze—and the object to which they are tethered is Richard Lewis. They watch every move as he struggles up the vine toward their nest. If kestrels can glower, they glower. Sometimes a kestrel who sees her eggs being filched will defend them forcibly, Jones tells me. She'll make a nice dive and slam her talons into the silly bloke's scalp. Jones knows this because on past occasions he has been the silly bloke, though today the role has fallen to Lewis. Jones himself is feeling quite larky.

"Just *look* at that aerial display. Both of them, in formation. What a bit of flying." He tilts his face toward the sky, which is full of creamy tropical clouds. "Eat yer heart out!" he hollers, at all the absent and imagined procurators of dreary conventionalism and rational triage, as Lewis chuffs and scrambles to pull himself into the cavity.

The thermos, with its eggs, is gingerly hoisted up on a length of twine. Lewis sets the foster eggs onto the nest scrape and retrieves two dummy eggs that were left there, on the previous visit, as place holders. The dummy eggs are made of glass; not even Willard Heck could hatch them. Their purpose was to keep the female interested during the few days since her own clutch was taken. Mauritius eggs, dummy eggs, European eggs, Mauritius eggs again: It's an ornithological version of three-card monte, contrived by Jones and his accomplices to trick a few Mauritius kestrels into producing exceptional numbers of offspring.

The kestrels dislike these intrusions by lummoxy two-legged creatures, but they don't seem to be permanently put off by any remnant physical evidence of nest tampering, and they generally resume their doting attention to whatever eggs have been left in position. Are they stupid? To decide that, we would have to agree on an ornithologically appropriate definition of stupidity, which wouldn't be easy. But there's no question that *F. punctatus*, like so many other species endemic to islands, is ecologically naïve. These birds don't show much instinctive fear of mammals, human or otherwise. They tolerate a measure of disruptive hoodwinkery that would be unacceptable to most mainland falcons. This unwariness makes the egg-switching a little easier. The other reason that Jones's methods have succeeded, I suspect, is that he and his colleagues are very damned skillful.

Back in the Land Rover, we retrace our route and then turn off the pavement again, thumping down another bad road toward another set of mountains. Lewis wants to check on a new pair, which have nested in an unfamiliar cavity at a site known as Yemen. A week ago, that female laid one egg and then stopped, leaving him to wonder what happened.

We drive between cane fields, through a zone of scrubby savanna, then out across a clearing that shows the grazed-down, grubbed-up signs of heavy traffic by deer and pigs. The forest here has been cut, in order to open more area as grassland and thereby boost the population of those pestiferous deer, which provide recreation for a few wealthy Mauritian hunters. At the clearing edges, we see some of the shooting stations—log towers—that make the sport genteel and convenient, if not very sporting. "It's criminal, what's going on in places like this," says Lewis. "They're *still* cutting down Mauritius's heritage." I'm reminded that the official national emblem of Mauritius shows two animals flanking a crest: a dodo and a stag. One is a memory, the other is an imported ecological blight that has been transmogrified into a fantasy called *la chasse*. Up ahead, a dozen gray macaques gallop across the road, fleet and furtive, reminding me further that the Mauritian landscape is haunted by ecological strangers.

We pass through a gate. The dirt road leads into a narrow canyon that the cane farmers and the deer ranchers have left alone—so far. The steep wooded slopes give way, above, to more basalt cliffs. "This is some of the best forest left in Mauritius," says Jones. "Up there, near the plateau, where you see that lighter green leaking down? That's junk." By junk he means the invasive exotic plants, especially guava and privet. He waves his hand from one spot to another, like an implacable God dispensing damnation and blessing. "But over here, and down there, that's indigenous. The real stuff. Nice, innit?"

A foot trail leads upward through a tunnel of forest. Half a mile on, we cross a high chicken-wire fence designed to keep the deer where they belong. Then we come to the tree-shaded base of a cliff. This one is uncompromisingly vertical, even a little undercut, and I can't see just how far it rises above the treetops. Someone has stowed a long wooden ladder nearby, but Lewis and Jones ignore that, presumably because it would only cover the first fifteen or twenty feet anyway. Instead, they start to climb another cliff-hugging vine, like the one Lewis used at Trois Mamelles. Hand over hand, finding leverage for their feet on the rough rock, Lewis first and then Jones behind him, they move upward into the canopy. No one has briefed me on this drill. It seems as unwise and unreal as climbing a giant beanstalk through the clouds. But I follow.

When I'm thirty feet up, a tree branch flicks off my glasses, which drop to the ground. I could go down and retrieve them, sure, that would be sensible, but I'd fall too far behind the cheerful maniacs and I might be tempted to stay down. Fortunately, or maybe not, I've got another pair of glasses in my shoulder bag.

"Do you trust this vine?" I call up to Jones. Gangly but tall, he must weigh two hundred pounds, and from this angle I can appreciate the size of his feet.

"Not greatly."

We ratchet our way upward, slowly, on the cliff face. It isn't Half Dome but it's more perilous than the average birdwatching stroll. We rise out above the valley. As we move beyond the treetops, I give myself an explicit mental reminder: Fall from here and you don't go home. I focus on the vine. Finally, Jones and I catch up with Lewis on a narrow rock shelf, like a window ledge ten stories above Lexington Avenue. Lewis greets us with bad news. The nest cavity is empty. The dummy egg he left here earlier is cold. No sign of the female.

"She may be just off in those trees," Lewis says. "Waiting. That part doesn't concern me." What concerns him is that she seems to have renounced the nest, letting the glass egg go cold and not laying any more eggs herself.

"One possibility is that I disrupted her when I did the first switch," Lewis says.

Jones doesn't think so. He has full confidence in Lewis's technique, and all experience with the species suggests that a gentle egg-switch would not have routed her. But he's puzzled. Every clutch counts, for the Mauritius kestrel, and there must be some reason why this nesting failed. What happened? Jones and Lewis cogitate for a few minutes as we stand on the ledge like a trio of suicidal stockbrokers. I gaze out at the panorama—the forested canyon below us, the deer ranch beyond, and the cane plantation beyond that, all spreading westward for five miles to the crescent of beach and then the great turquoise plane of the Indian Ocean. The sun has gone low into a bank of wet western clouds. The light is sepia, the cane fields are chartreuse. It's a magnificent, kestrel-eye view. But where are the kestrels?

Lewis edges off to scout for a clue. Jones and I follow on a tightrope traverse. Somewhere far below, feral pigs are probably eating my shattered glasses. I devote close attention to the placement of each foot. Then the ledge opens out into a small balcony and I notice a wood-and-wire contraption, a box trap of the falling-door type, which evidently was hauled up and set during an earlier visit. The trap is empty.

From farther along comes Lewis's voice, sounding portentous. "Oh. Well then. Look what we have." Jones and I find him standing over another trap, inside which is a mongoose.

It's a slender brown animal resembling an oversized weasel. When Jones bends for a closer look, the mongoose throws itself into the wire, hissing, snarling, snapping its toothy snout. A weasel on steroids and too much caffeine.

"We know that mongooses take kestrels," Lewis tells me. "But this is the first time we've caught one on the bloody cliffs. It's depressing. To have one's worst fears confirmed."

Herpestes auropunctatus, the mongoose species in question, is a somewhat more recent problem than the rats and the pigs and the monkeys. It was first brought to Mauritius in the mid–nineteenth century and for decades lived as a semi-domesticated species, valued for rat catching. Those barnyard mongooses seemed reasonably innocent and useful, though they occasionally killed poultry. In 1900, because of an outbreak of plague that fanned everyone's loathing of rats, the colonial government decided to unleash mongooses into the wild. A very bad idea, some people argued at the time. The plan went ahead anyway, plague being more scary than an inchoate notion of disrupted faunal balance. Jamaica and Hawaii had already experienced bad mongoose infestations, with considerable loss of native wildlife, but Mauritius had to learn for itself. By 1905 there was a thriving wild population with a ravenous appetite for fowl and imported game birds. So the government put a bounty on mongooses, trying futilely to undo what had been done. Besides preying on chickens and partridges, the mongooses probably played a role in eliminating at least one seabird population from the island, and by now, nine decades later, they have eaten up an indeterminate number of kestrels. Jones kills and dissects them whenever he can, gathering evidence of their depredations. "I've looked at two hundred mongoose stomachs," he says. "Let me tell you about that animal sometime. Next to feral cats, it's the most successful small carnivore in the world." The mongooses constitute a diverse and far-flung group that includes about forty species, but *Herpestes auropunctatus* is the same species famed on the Indian subcontinent for doing battle with cobras. It's a ferocious hunter, a prolific breeder, and agile enough to have scaled this cliff. Smart enough, probably, to have found a safer route than the one we took.

The trapped mongoose, Lewis speculates, either ate one of the kestrel pair or drove them both out before the female produced more eggs. Too bloody bad. The question now is what to do with the little bastard. We squat on our ledge for half an hour as Lewis and Jones consider their options. Neither of them fancies trying to climb back down the vine with this trap roped to his back while the mongoose claws and nips through the wire. They could just toss it over the edge, of course, but the trap itself would be wrecked. Jones has a better idea.

Distracting the mongoose to one end of the trap, he raises the door—just a crack—behind it. When the mongoose makes an incautious move, Jones grabs its tail. He pulls it backward, pinning it against the door. The trap is still closed, the mongoose is livid, but Jones has control. With help from Lewis, he shifts the trap and himself toward the precipice. On a signal, Lewis pops open the door and Jones swings the mongoose out, overhead, and down against the brow of the cliff. Thwoppp.

Skull shattered, it's still thrashing. So Jones brains it again.

Then he holds up the limp body, like a trout for the camera. Blood leaks from its mouth. Jones seems relieved—he might have been bit, he might have fallen to his death, or worse still, he might have been bit and then fallen. He's also mildly surprised at his own adroit lethality. This is not just a revenge killing, though. It's conservation biology.

"Cor," he says. "Don't see that every day, do ya?"

163

IT's A HECTIC week for Jones. Besides coping with bird chores and mongooses and bothersome writers, he needs to throw a few things into a suitcase and then catch a plane for Europe. In the Netherlands, he'll receive an international conservation award—actually it's a noble order of some sort, Ritter of the Golden Arc—for his work on the Mauritius kestrel. Right now he's wondering out loud whether he owns any appropriate clothes. What does a bloke wear to his investiture of Ritterhood? Damned if he knows. The invitation sounds very formal. A dark suit ought to do it, Jones figures, and yes, yes in fact, he does have one of those. Black shoes, that's another question. When the Duke and Duchess of York visited Mauritius, Jones tells me, he took them up to one of the kestrel sites and showed them an alley-oop feeding, which they quite liked. In gratitude, they invited him to a tony royal dinner. He wore his dark suit and a pair of climbing boots. Maybe he'll see fit this time to invest in black pumps, or at least borrow some, but it won't greatly matter, I suspect. Presumably the Queen of Holland, like Gerald Durrell, has chosen to appreciate Jones not for his sartorial elegance but for his results. He's the mad Welshman who took a seemingly hopeless case and, in not many years, turned it into an emblem of possibility. He's the fellow who (with help from Richard Lewis and Willard Heck and others) saved the world's rarest falcon from its not-quite-inevitable fate.

The process was slow and the results, for a while, were tenuous. The first time he swiped eggs from a wild kestrel nest was in 1981. That year, there were only two mated pairs known in the wild. From one pair he took a clutch of three eggs. Two of those eggs were fertile and he succeeded in rearing one chick, while the wild pair produced a new clutch. From the second pair he also took three eggs, hatched them all, and of the three chicks he successfully reared two. Net gain from Jones's first season of intervention: three captive kestrels. It might sound insignificant, but it was a major increment of the total population. Unluckily, all three of the young captive birds were male—exactly the kind of demographic misfortune that Mark Shaffer warned about.

As time went on, Jones's luck improved along with his understanding of just how to cope with this species. The following year he took two clutches again, of which one turned out to be infertile. From the fertile clutch he reared three birds: two males and a precious female. During 1983, by supplemental feeding, he encouraged a wild pair to breed successfully in marginal habitat. This was a heartening breakthrough, given the shortage of habitat that *wasn't* marginal. The nadir had been passed; the population trend now was steadily upward. In 1985, six wild pairs fledged a total of eleven kestrel young, and the number of captive-bred birds increased too. Three years later, there were roughly forty kestrels in the wild—including a few pairs reintroduced into the Bambous Mountains—as well as twenty-one in the aviaries and another fifteen at the Peregrine Fund compound outside Boise, Idaho. The birds in Idaho were being bred as an insurance population, in case a catastrophe struck the Black River aviaries. Some of those Idaho birds would eventually be returned to Mauritius for release into the wild. By these various measures, the total number of adult birds had climbed back to almost eighty by 1988. In November of that year, in the cramped lab that functioned also as Jones's office, Willard Heck was meanwhile incubating more than a dozen eggs. The first of those eggs hatched on a Wednesday, yielding a homely pink chick—a groping and trembling little thing with matted white down and a pair of huge blue-black eyes, not yet open. I stood over Heck's shoulder as he fed it chopped quail with tweezers.

Within a few years after that, more than two hundred kestrels had gone from the Black River aviaries back into the wild, and small populations were occupying a handful of Mauritian sites (the Bambous Mountains and Trois Mamelles, among others) from which the species had earlier disappeared. The wild population in the Black River Gorges was larger than it had been in decades.

Today, the Mauritius kestrel isn't beyond all danger. Given the shortage of habitat, given the small size of the island, given human population growth, it may never be really secure. What the kestrel's story represents is just a provisional success. But in a world full of failed chances and failing hopes, even a provisional success—and especially one so improbable—has great scientific and symbolic importance. "If you can save the Mauritius kestrel," Jones tells me just before he flies off for the Netherlands with his black suit and, probably, his climbing boots, "you can save virtually anything."

164

JONES MAY be right. Take a species that's down to its last half-dozen individuals, save it from extinction despite a doom-threatening convergence of problems, and maybe that means you ought to be able to save anything. But there is an irony in the case of *F. punctatus*, making it not quite so cheeringly paradigmatic as it might seem. Jones knows this as well as anyone, though in an exuberant moment he might let himself forget. He knows that one minor but crucial circumstance distinguishes the Mauritius kestrel from the tigers of Sariska, from the lions of Gir, from the indri of Analamazaotra, from any other creatures for whom insularized habitat and small population size are recent problems. The kestrel, unlike those tigers or lions or indri, has been confined to a tiny island throughout its whole history as a species. Its population was *never* large.

Mauritius itself is simply too small to support thousands of kestrels. Even before the early colonists began cutting forest and planting cane, the island's meager size restricted the bird's abundance. According to one estimate, the maximum population of *F. punctatus* was probably about three hundred breeding pairs, plus maybe a hundred singletons. Jones's own estimate of the kestrel's territorial requirements suggests a somewhat higher maximum, but it couldn't have been *much* higher. Each pair of kestrels needs about a square kilometer of territory, and the arithmetic of available area sets a stingy limit.

Three hundred pairs and a hundred singletons—seven hundred altogether—may sound like a comfortably large total for the kestrel during that bygone era. It wasn't. Why not? First, because the singletons wouldn't have counted in the effective population, since they weren't passing their genes along to offspring. Second, because a three-hundred-pair maximum implies that during average years the number fell lower and during bad years it probably fell much lower. In the grip of a viral epidemic, the three hundred might have dropped to two hundred. If a severe drought or a disease among geckos caused a collapse of the food supply, the two hundred might have gone to one hundred. A cyclone around the same time might have reduced the number still further. We've been through this sort of scenario often enough, by now, that you know how it can work. Even if the kestrel population had subsequently recovered, breeding back up to three hundred pairs, the short-term fluctuations would have had long-term significance. By the complicated mathematics that reflect the complicated reality, temporary declines would have reduced the effective population lastingly. Ian Franklin made precisely that point, with a supporting equation, in his chapter of the Brown Book.

If the effective population of the kestrel was chronically small, the species must have been chronically inbred. Throughout earlier millennia of its history, close relatives must have mated with each other rather frequently. There is no empirical proof of this chronic inbreeding, but it can safely be assumed. Now the irony: Chronic inbreeding has a biological advantage, at least as compared to an abrupt episode of *acute* inbreeding. Chronic inbreeding prevents the accumulation of a heavy genetic load. It allows potentially lethal mutations to be combed out of the population gradually.

When inbreeding is chronic, only a few offspring in a generation are likely to turn up homozygous for any harmful recessive allele. Those unlucky offspring, fated for barrenness or premature death, disappear without leaving offspring of their own. The harmful alleles are thereby purged from the gene pool, and the population's genetic load is lightened. That spreads the impact of negative mutations across time, instead of concentrating the same impact within a period of low-population crisis.

Inbreeding depression doesn't necessarily follow from inbreeding, in a population whose genetic load is light. And the kestrel's load, most likely, was light. *F. punctatus* couldn't afford to carry a large burden of harmful recessive alleles, since there was never a large population of heterozygous individuals in which to store those recessives innocuously. The harmful recessive alleles that mutation did produce, and the homozygous individuals in which they expressed themselves, were continually being eliminated as centuries passed.

So the kestrel was prepared, by its long history of natural rarity, to endure an acute phase of inbreeding without suffering acutely from inbreeding depression. Other species, from the bigger world, have no such advantage.

165

WHILE Jones coaxed the kestrel back from oblivion, using skills borrowed from falconry and from his own misspent youth, the science of conservation biology was maturing. The Brown Book, the Green Book, and the Gray Book had been published. The Soulé-Salwasser workshop on population viability had occurred, and the Blue Book that derived from it was being edited. As the theoretical work progressed, so did the collective self-consciousness and the interchange between theory and practice. "People were just starting to read some of this stuff," Mike Gilpin says, "and starting to use it in management."

The concept of minimum viable population had even come to be known by its acronym, MVP. Those letters carried a subliminal hint, at least to some minds, that the concept itself represented a "most valuable player" in the new scientific game. Gilpin himself still wasn't happy with the MVP concept. No, he felt, it really wasn't so all-fired valuable. It was merely a first poke at the problem, and there was need for something better. Michael Soulé shared that view. Reducing the notion of viability to a single MVP number was too neat, too simplistic. It didn't do justice to the interaction of variables that could affect a small population. The 50/500 rule of Soulé and Franklin had attracted too much literalistic credulity. Some scientists and wildlife managers were inclined to take it as an absolute, not as a cautionary guideline reflecting only the genetic concerns. Gilpin recalls that Soulé had received copies of management recommendations, from the far side of the world, concerning an endangered parrot that was down to twenty-five individuals. The unnerving thrust of those recommendations, in Gilpin's paraphrase, was that "it's below both of Soulé's magic numbers, so let's give up on it, let's spend our money elsewhere." Twenty-five parrots were less than fifty, therefore vulnerable to inbreeding depression in the short term, therefore (by this faulty logic) beyond hope of rescue.

Aware of Norman Myers's fatalistic advice in *The Sinking Ark*, and of how Carl Jones and his co-workers have successfully flouted it, Gilpin adds, "I guess the Mauritius kestrel would be a more interesting example of the same point."

Gilpin and Soulé discussed all this. They agreed that the MVP concept, implying a single minimum number, was inadequate. Granted, such a generalized rule of thumb did seem temptingly useful for re-source managers and easy for the public to grasp. But the early MVP work had taken generalization too far. Gilpin and Soulé were dissatis-fied. They realized that the risk factors affecting any species would in-teract in a complex, multidimensional, context-specific way. There would be feedback and synergy. A population of some specified size might be viable under one set of environmental circumstances but not viable under another set; or viable for one species but not for another species. Fifty Mauritius kestrels would face a different gauntlet of un-certainties than fifty whooping cranes. Reptiles would be different from insects, carnivorous mammals would be different from herbi-vores. Big-island species would be different from small-island species.

"So we decided that what we needed was a process of analysis that would evaluate all of this," Gilpin says. They outlined such a process and called it *population viability analysis*, or PVA. As imagined by Gilpin and Soulé, the PVA process would be to the MVP concept what a complete physical checkup is to a cholesterol reading.

Soulé had meanwhile organized another big meeting. The Second International Conference on Conservation Biology took place in Ann Arbor in May 1985. The phrase "conservation biology" had by then become more widely familiar, more intellectually respectable, though it wasn't yet established in the mainstream scientific vernacular. Many geneticists, ecologists, population biologists, biogeographers, zoo veterinarians, conservation planners, land managers, and other pro-fessionals were beginning to think of themselves as allied in a single discipline—a movement, even—and the movement was gathering momentum. Dan Simberloff and Jared Diamond both attended the 1985 meeting. Tom Lovejoy, Hal Salwasser, Paul Ehrlich, and John Terborgh came. Norman Myers presented a paper. Mike Gilpin also talked, describing his and Soulé's idea of population viability analysis. As usual for Gilpin, he spoke quickly, informally, without a manu-script.

The essence of PVA, Gilpin explained, was to consider the poten-tial net impact of a number of processes functioning interactively, all to the detriment of a small population. Each of these processes could be seen as a feedback loop supplying input to others. One sort of problem would exacerbate another. Demographic distress would af-fect genetics. Genetic distress would affect adaptability. Loss of adaptability might affect the spatial distribution of the population, possibly dividing it into still smaller fragments. Fragmentation would affect demography and genetics. Whatever went around would come around. These interactive loops could be thought of as vortices. Having crossed a threshold of jeopardy in any single dimension—demographic, genetic, spatial, evolutionary—a small population could be swept further downward within the vortex, by interactive effects, in a swirling descent toward extinction. A population viability analysis would attempt to define where those thresholds lay for a given population within its ecological context, and to predict the consequences of a threshold's being crossed.

Besides its greater complexity, the PVA process as Gilpin and Soulé proposed it would differ from the old MVP concept in one other way: It would focus on actual situations, not on an idealized standard. It would assess the viability of real populations of animals or plants as they exist in the real world. It would look at empirical circumstances and offer hope, discouragement, or warning, rather than looking at theoretical principles and offering goals. The MVP approach had asked: How many individuals are necessary to ensure such-and-such a likelihood that a population of some species will survive for such-and-such a length of time? The PVA process would address the differing realities of particular species and their contexts. It would ask a different sort of question: Given that only *this* many individuals of *this* species now exist under *these* circumstances of habitat, behavioral character, ecological relations, genetic diversity, and spatial arrangement, what are the chances that the population will remain viable over time? In plainer language: Can these suckers survive, or are they toast?

As he delivered his talk at the Second International Conference, Gilpin showed diagrams on an overhead projector. Each one was a tangle of circles, arrows, boxes, and dotted lines illustrating a vortical route to extinction. *This* arrow punctured *that* circle, out of which came another arrow, which stabbed a box, out of which came still another arrow, which turned two sharp corners and punctured another circle—all indicating in abstraction the truth that one damn thing leads to another.

The diagrams had their own little history. Soulé, whose scientific brainstorming comes in spatial and graphic terms, had used "tons of paper," in his words, to sketch out the vortex idea. He recalls driving to Colorado for a season of summer fieldwork, "and while I was driving I was making these diagrams. These circles and these arrows. I do my best thinking sometimes when I'm driving long distances and get away from all the input." He and Gilpin together had refined both the PVA concept and the diagrams, using more paper still, in time for Gilpin to stand up before the biologists convened in Ann Arbor. Gilpin created the final versions on his Macintosh. That was easier for him than tracing them out with a ruler and the base of a coffee mug. And it gave them a reassuring exactitude.

There were four vortices, each one depicted in a diagram: the demographic vortex, the inbreeding vortex, the fragmentation vortex, the adaptation vortex. If a species fell into the ambit of one vortex, Gilpin explained, it was thereupon more liable to be swept into the others. But each species would differ in its vulnerability to each of the vortices. A prolific and very mobile species of insect, for instance, would be less vulnerable than other creatures to the fragmentation vortex, because it could multiply quickly, travel easily, and refill the gaps between fragmented populations. A huge-bodied and long-lived mammal would be less vulnerable to the demographic vortex, because its population size would tend to hold stable from one year to the next. And a chronically inbred bird (like the Mauritius kestrel) would be less vulnerable to the inbreeding vortex, because it had already reduced its genetic load. Inherent differences between types of species, as well as contingent differences in circumstances, would play out as differences of susceptibility or resistance. The task of population viability analysis, as Gilpin and Soulé defined it, would be to estimate how vulnerable a particular population might be to each of those vortices and to all four combined, based on the population's own unique traits and circumstances. There would be no magic number marking the viability threshold for a given species. There couldn't be—biological reality was too multifarious. Instead there would be a systematic process that yielded particularized diagnoses.

The presentation was well received. "I remember Hal Salwasser running up and saying, 'Yes, yes, yes, that's exactly what I've been thinking we needed,'" says Gilpin. And Salwasser wasn't the only wildlife manager who took notice.

So Gilpin and Soulé wrote a manuscript, which appeared in the volume that emerged from the 1985 conference. Soulé again did the overall editing, and this Yellow Book (less colorfully known as *Conservation Biology: The Science of Scarcity and Diversity*) joined the rainbow.

The Gilpin-and-Soulé paper, chapter two of the Yellow Book, included those vortical diagrams. The circles, the boxes, and the dotted lines were drawn crisply in black and white. The arrows pointed every which way.

166

ONE DIRECTION they pointed was toward the Atlantic forest of southeastern Brazil, where the largest of all New World monkeys, *Brachyteles arachnoides*, commonly known as the muriqui, survives in just two dozen or so tiny populations within isolated fragments of habitat.

The Brazilian Atlantic forest is not so famously celebrated as the Amazon or the rainforests of Madagascar and New Guinea, but it ranks in the same class as a reservoir of biological diversity. Or it did until recently. It's quite unlike the Amazon, from which it stands separated by the high plains of central Brazil, and among other points of dissimilarity there's this: Most of the Atlantic forest is already gone.

Five centuries ago, it covered a half-million square miles of mountain and river-valley terrain along the Brazilian coast. It was a long diagonal swath of woodland, extending from the headwaters of the Rio Uruguay in the south, through the region that now includes São Paulo and Rio de Janeiro, to a point hundreds of miles north near the easternmost bulge of the continent. The topography of this forested zone was ruggedly vertical, at least in contrast to the flatness of the Amazon basin. Its soil characteristics, its cycles of temperature and rainfall, and its history of vegetational changes were also distinct. It supported great biological richness—by one estimate, as much as seven percent of all the world's plant and animal species. Many of its birds, its reptiles, and its plants were endemic species, occurring nowhere else in South America, nowhere else in the world. Today the Brazilian Atlantic forest still harbors an impressive assortment of endemics, including about fifty species of mammals, of which a remarkably large fraction are primates. The brown howler monkey is found only here. So is the golden-headed lion tamarin, a bizarre little beast with bright orange fur and the mane of an African lion. So are the masked titi, the black lion tamarin, the golden lion tamarin, and four species of marmoset. So is the muriqui. But many of those primates are now endangered. Their ecosystem has been cut to pieces.

Arriving on this shore in the early sixteenth century, the Portuguese found an unspoiled tropical forest. They also found that coastal Brazil (like Mauritius under the French) was a fine place for growing sugar cane. Settlements took hold and expanded. Big areas of landscape were converted to cane. Toward the end of the century,

gold and diamond strikes in the interior lured some of the more adventurous settlers inland. Mining towns grew, a few to the size of
cities. Then, as the gold boom died out around 1750, those cities and
towns turned into centers of agriculture. Coffee became an important
cash crop during the nineteenth century. Cattle grazed on newly
opened meadows. Iron ore was found, the indigenous people were
routed or killed, and in the 1860s a railroad was built, which encouraged still more production and export of coffee and cane. In other
words, it was the same sad scenario that is all too familiar throughout
the annals of discovery, colonization, and empire. In 1910 a steel mill
was founded on the bank of a river in the state of Minas Gerais, about
two hundred miles north of Rio de Janeiro. Steel production required
large quantities of charcoal, supplied by cutting and roasting more of
the forest. As the human population increased, there was also more
timber cut for construction and more firewood for cooking. Highways, hydroelectric dams, and dozens of additional steel mills were
built. The southeastern coast became Brazil's most densely settled
and industrialized region, with half of the country's population
packed into just a tenth of its geographical territory. Urban sprawl,
agricultural expansion, and the demands for firewood and timber all
took their shares of Atlantic forest until, by 1980, less than five percent of the original ecosystem remained. Most of it was in small strips
and patches on hilltops, temporarily protected by inconvenience.
Those strips and patches were surrounded by coffee plantations and
pastures, cane fields, roads and towns, the relentless encirclement of
humanity.

The big monkey, *Brachyteles arachnoides*, suffered badly throughout
this period from hunting as well as from habitat loss, because each animal represented such a temptingly sizable piece of meat. An adult
muriqui can weigh thirty pounds. Dangling by its lanky arms from a
tree limb, partly supported by its prehensile tail, belly hanging out
like a melon, it looks almost as large and disreputable as a human. It
feeds on a varied diet of fruit, leaves, and flowers, and its body size allows for subsistence on high-volume meals of the less nutritious leaf
matter during seasons when neither flowers nor fruit are available. Its
fur is a light golden gray, contrasting handsomely with the bare skin
of its face, which is as dark as creosote. The word *muriqui* comes from
the Tupi Indian language, but among Portuguese-speaking Brazilians
the species is best known as *o mono carvoeiro*—translated loosely, the
charcoal monkey. Although that name alludes to the soot-colored

face, it could apply also to the fact that *B. arachnoides* has been figuratively incinerated—as its habitat has been literally incinerated—by the fires of civilization.

In better days, before the Portuguese landed, there may have been four hundred thousand muriquis. That's a rough but informed estimate, based on territorial spacing and available habitat. During recent centuries, while the forest was reduced in area and chopped into fragments, the muriqui population dropped like a diving duck. In the late 1960s a Brazilian conservationist surveyed its entire former range and found only three thousand individuals.

Another estimate, published in 1972 by one of the country's leading primatologists, put the total at just two thousand. The number was disconcertingly low, but the spatial arrangement made things worse. Fragmentation compounded rarity. The two thousand muriquis didn't constitute a single population. Instead, they were marooned in lesser clusters—three dozen here, four dozen there, five dozen somewhere else. Each cluster contained just a modest minority of the whole, and each (because muriquis, like indri, are disinclined to set foot on the ground and highly unlikely to cross open areas) was cut off from the others. Few of those populations, if any, were large enough to be considered viable. Each faced the four kinds of uncertainty described by Mark Shaffer. Each faced the vortical routes to doom represented in the diagrams of Gilpin and Soulé.

Among other troubles, inbreeding depression was likely to be serious, because the muriqui (unlike the Mauritius kestrel) had fallen quickly to rarity from its previous status of great abundance, leaving it untempered by chronic inbreeding. It was now an insularized creature that wasn't ready, genetically, for insular life.

The total number of muriquis continued dropping. By 1987 *B. arachnoides* was one of the world's rarest and most desperately threatened primates. A report published that year, based on collaborative work by a team of Brazilian and American researchers, could account for only 386 muriquis. This was a conservative estimate, the researchers explained, and they hoped that at least a few more animals were out there—either hiding shyly in the areas that had been surveyed or else still undiscovered in other forest remnants. But no one could assume it was so. These 386 muriquis were the sole surviving representatives of the species, as far as anyone knew. They lived on eleven separate patches of forest scattered sparsely across the region. In the published report was a map showing southeastern Brazil and, in outline, the former expanse of muriqui range throughout the Atlantic forest. The eleven habitat patches looked like a lonely little archipelago protruding above the ghost of Atlantis. Another report has recently added fourteen other small populations, comprising another three hundred individuals, to the full population estimate.

Some of the habitat patches are public parks or reserves. Some are private parcels of forest. One of the private parcels is on a coffee plantation called Fazenda Montes Claros, east of the town of Caratinga in Minas Gerais. The forest at Montes Claros measures 880 hectares, almost precisely the same size as Madagascar's Analamazaotra reserve. It supports about eighty muriquis, one of the largest concentrations anywhere. It is circumscribed by coffee groves and pasture. It has been preserved for the past fifty years, despite commercial enticements and human pressures on all sides, by the benevolent whim of a coffee grower named Senhor Feliciano Miguel Abdalla.

The muriqui population at Montes Claros has been more closely studied than any other. But even at Montes Claros, the research record is a short one, spanning just two decades. It was only in 1976 that Brazilian scientists discovered this population, though presumably Senhor Feliciano had been quietly doting on them for years. In 1977, a Japanese primatologist spent two weeks in the area gathering data. Also around that time, the U.S. branch of the World Wildlife Fund (in the person of Russ Mittermeier, the influential primate conservationist who supported Pat Wright's early work in Madagascar) began, cooperatively with Brazilian conservationists, to encourage long-term research on Atlantic forest primates. One of the first targets of concern was the muriqui population at Montes Claros. In 1982 a young American woman named Karen Strier visited there.

Strier was a grad student in biological anthropology at Harvard, scouting a possible research topic for her doctorate. She stayed for a couple of months and became happily intrigued with the muriqui. The next year she returned for a longer spell of fieldwork. She learned Portuguese, learned her way around Montes Claros, learned to ignore the danger of poisonous snakes, and learned how to move through the forest quickly without losing track of the animals or causing them fright. She collected thousands of hours of field observations, went home to the United States, began writing her dissertation on *B. arachnoides*, and then, before she could finish the task, found herself drawn back again to Montes Claros; there was more to learn. The young adult females whom Strier had gotten familiar with were

BIOLOGICAL
DYNAMICS
OF FOREST
FRAGMENTS
PROJECT

Manaus
(Barra)

Rio Uaupés

Rio Negro

Rio Amazon

Rio Madeira

Rio Tapajós

Santarém

Belém
(Pará)

Rio Tocantins

FAZENDA
MONTES
CLAROS

Rio de
Janeiro

Rio Uruguay

SOUTH
AMERICA

0   150   300

MILES

now carrying infants of their own, and she needed to know how motherhood would affect their ecological needs and their social relationships.

There would always be more to learn. Eventually Strier finished the doctorate and took a teaching job in Wisconsin, but the muriqui had become central to her scientific life. She returned again to Montes Claros, and again, and has continued returning ever since. She won the trust of the monkeys and of Senhor Feliciano. She trained Brazilian students in the methods of biological field research. She helped develop a research station at Montes Claros. She was in it for the long haul. She still spends a sizable portion of each year at Montes Claros. Her work there on *B. arachnoides* is the only prolonged and still ongoing study of the species. If Karen Strier isn't the world's foremost authority on the muriqui, it would be impossible to say who is.

She's a private person and she works hard. She resents interruptions that reduce her productivity. Her annual field season in Brazil is short, leaving no time to squander. I learn this much, unambiguously, in the course of a telephone call. But yes, all right, she tells me, she'll tolerate my presence at Montes Claros. I can drop in for a glimpse of her project, so long as the responsible Brazilian official approves my visit.

Write to such-and-such a fellow for permission, she says. Come down in October. Don't dream of staying more than two or three days. Take the morning bus east out of Caratinga on a little dirt road toward Ipanema, and ask the driver to drop you at the Montes Claros gate. Make sure it's the right bus, since this isn't the godawful famous Ipanema but another one, just a little country town, far off the beaten paths between Caratinga and other regional centers, and you could easily be steered wrong. If you don't speak any Portuguese, better learn some. And the bus driver, by the way, he might know Montes Claros under another name—he might call the place Feliciano. Just a farm crossroads. Then walk. It's less than two miles to the station lab. Wait there at the lab until I return from the day's work, Strier says. Do not, repeat *not*, wander into the forest and get lost. Evidently she has some prior experience with journalists. She gives me one other bit of instruction, and this one sounds almost friendly. When you go through the city of Rio, she says, be careful.

167

MORE PRECISELY, she says, "Be careful at the Rodoviaria."
The Rodoviaria is Rio de Janeiro's central bus station, an ugly concrete maze of platforms and tunnels and elevators and loading bays,
teeming with people, most but not all of whom are innocent folk going about their own business. It's the departure point for the
overnight coach to Caratinga, Strier has explained. It's also an excellent place for incautious Brazilians and numbskull gringos to get
themselves robbed by deft *ladrãos*. A *ladrão* is a thief. A *carteirista* is a
pickpocket. *Entorpecido* means numb and, although as I step off the
plane in Rio I can't yet tell you the word for skull, I have been trying
as instructed to learn a little Portuguese.

It's not my first visit to Brazil, but it's my first to the big city. Manaus, up in the central Amazon, is by comparison just a drowsy provin-
cial outpost. Rio is Rio, a great urban stewpot full of crumbling
colonial dignity, the manic spirit of Carnival, new wealth, old poverty,
new poverty in its most desperate late-modern forms, and a milieu of
sybaritic international beach-lizard tourism. The place offers its own
distinct set of wonders and perils, and those wonders and perils, like
the language itself, lie mostly beyond my ken. But the schedule of my
trip up to Montes Claros allows me a few days of gawking in Rio. So,
within just hours of having arrived, I do several things that are smart
and several that aren't. For instance, I divide my money and my important papers and my credit cards into three separate wallets, one of
which can be worn as a pouch under my shirt. This is smart. Then I
leave the hotel, wearing the pouch wallet and carrying the two others.
This is dumb dumb dumb. I take a cab to the Rodoviaria.

Buy your bus ticket in advance, Karen Strier has coached me. I figure I'll do it promptly, and then I can get away from that place and relax. I've hit town on Saturday and the Rodoviaria is acrawl with
humanity. I button my hip pocket, wary of *carteiristas*. I avoid the
thickest clots of people, working my way timidly along the fringes.
My pulse shifts to a higher gear. *Quero um bilhete para Caratinga*, I re-
hearse in my head. One ticket to Caratinga. Dozens of bus companies
operate out of the Rodoviaria, it's like a trade fair of mass-transit options, but finally I locate the right one. *Quero um bilhete para
Caratinga*, I say. *Amanha, por favor*. Tomorrow. This much is by rote.
The ticket agent looks at me blankly from inside his window and rat-

tles off a question. Duh, I say. Then, rather than getting entangled in indecipherable details (he may have asked if I want a window seat, he may have asked if I wouldn't prefer a cattle truck to Buenos Aires), I push some cruzados at him. He gives me a ticket. Caratinga. *Obrigado,* I say, ten million thanks. I glance around. So far, so good. I rebutton my pocket. It's all quite foolish and paranoid, I can see now, since the bus station *ladrões* have better things to interest them than me. Relieved, I take a taxi back to the majestic old section of central Rio, near my hotel, and go off on a celebratory stroll in the bright Saturday sunlight. This part of the city is full of graceful granite buildings from the boom years of the nineteenth century, cathedrals of culture and finance interspersed with snazzy shops and movie marquees and other touches of metropolitan glitz. The very sidewalks are elegant: cobbled mosaics of black-and-white stone in floral patterns. The outdoor cafés are doing a good business. Passing an equestrian statue, I wander out through a grassy park. Three guys appear from nowhere and knock me down.

Having never before been mugged, I behave badly, forgetting in the heat of the moment that it's poor form to shout for help and highly inadvisable to struggle. The three muggers are skinny young street thugs. One of them snatches a watch off my wrist, one digs for my hip-pocket wallet, and with the third I conduct a fierce little tug of war over the strap of my shoulder bag. Meanwhile I scuffle and holler—very ingloriously, like a rat pinned to the ground with a barbecue fork. Since we're only fifty yards from a busy streetcorner, my hollering makes them panicky, angry, hasty. The more jangled they get, the louder I holler, losing sight of the obvious fact that to induce professional frustration in a mugger is to invite knife or gun injury. But I'm lucky: These twerps don't seem to be armed. My end of the strap breaks and I say goodbye to the bag. They settle for it and run.

Picking myself up off the ground, I discover that I'm unhurt. No blood, no punctures, not even a scratch. I find my glasses in the dirt, including the missing lens, and get them jiggered back together. The pouch wallet holding my passport is still with me, dangling loose inside my shirt on a broken cord. From my hip pocket I've lost the button but not the other wallet. The muggers have gotten only the watch, the bag, the third wallet (containing my plane tickets, some cash, and a fallback American Express card), a Portuguese-English dictionary, and the notebook in which I recorded Karen Strier's directions for reaching Montes Claros by bus. The tickets can be replaced and the card can be canceled. The directions are more precious, but fortunately there's a duplicate in my memory.

Several taxi drivers have witnessed my little drama and pulled their cabs to the curb nearby, honking and shouting. Their attentions, I realize now, must have helped scare away the three punks. With miraculous promptness, cops appear. It seems that mugging is a crime taken seriously by the Rio police, and the mugging of foreigners—what with trade deficits and tourism revenues being as high as they are, I suppose—may perhaps be taken more seriously still. I'm a major law-enforcement event within minutes. There's a noisy convergence of sirens, screaming tires, unmarked cars, amid which a dozen officers materialize, half of them immediately fanning out on foot, chasing my muggers through the neighborhood. Which way, which way? Back toward the equestrian statue? No no, through the shrubbery here, down this drive, over that fence. Where's the victim? *He's* the victim? Ask him how many, ask him what they look like, which way they ran. He can't say? What, he doesn't have the vocabulary of a three-year-old or he was too witless to notice? Both? Oh Christ, not another American.

Most of these cops are jaunty young plainclothes men in colorful sport shirts and pastel pants who have evidently imprinted their professional self-image from reruns of *Miami Vice*. To me at the moment they seem princely dudes, hip but righteous, and I'm grateful for their solicitude aside from the fact that they keep waving their pistols and shotguns around like fly whisks and dinning me with questions I can't understand. *Não falo bem portugues*, I tell them—I don't speak this language too well—a howling understatement and another of my carefully practiced phrases. Possibly I've practiced that one *too* carefully because, pronounced with what I fancy is a near flawless accent, it only confuses matters further. *Não falo bem portugues*, I repeat, and they perk. They chatter at me reassuringly. Yes yes, no problem, no need to apologize for your grammar, say the young cops (or that's what I guess they're saying, anyway), and they plunge ahead with my debriefing. Many important questions. Quick, now. A streamlined effort to sort out the facts. *Quack quack quack gobble gobble hubba dubba dubba*, is what I hear, and deep in the heat core of my chest I begin wishing that all the good folk on all the islands and fragmented mainlands of the world spoke a single language. Any single language would do, though English would be preferable.

Then suddenly I find myself in a supercharged Volkswagen with a handful of these cops, their pistols are still out, they don't seem to own holsters, we're roaring along the clogged streets and banked viaducts of downtown Rio, dodging between lanes to the accompaniment of what I momentarily take for gunshots. Apparently we've been diverted from the trivial business of my mugging into a serious chase, or something, and if the traffic doesn't kill us, the drug lords or the terrorists when we catch up with them undoubtedly will. This is it, says the voice of my gloom. Now I'll never see *Brachyteles arachnoides*. I should have stayed home and simply read Strier's dissertation.

The voice is mistaken, though, and my luck holds. The Volkswagen slams to a halt at the Third Delegado police station. The gunshots turn out to have been backfires. Must be that I'm still a bit jumpy.

We march up a concrete stairway to the fourth floor, where a clerkish officer sits at a desk behind an antique typewriter in a large room otherwise bare of furniture, waiting to record my statement of victimhood. Amazingly, one of the three perpetrators is here too, having managed somehow to get himself caught, an unheard-of gaffe in the field of mugging. He wears a bloody nose—from resisting arrest, maybe, or because the cops saw fit to rough him up on general principles—and no shirt. His shoulders are narrow. The predaceous mien that I witnessed earlier has been whomped out of him, and now he is miserable and scared and repentant. He gives me a simpering look, as though hoping for a softhearted gringo sympathizer against police brutality. Nothing doing, pal. I avoid his glance. Then they handcuff him to the leg of a table.

After all the excitement of the last few minutes, time itself now slows to a bureaucratic trudge, and I spend the next six hours stuck here on the fourth floor, trying to communicate with the clerkish officer at the typewriter, then with a certain Inspector Mota, an amiable little man with the face of a parish priest, and finally with a more sternly commanding personage in razor-crease trousers whom I take to be Mota's boss, all the while waiting for a translator who never appears but who eventually phones in his services. The translator's name is Rolf. Rolf explains to me that I can forget about my stolen bag, nothing has been recovered, nothing will be, there's only the one mugger caught empty-handed. The good news from Rolf is that if I cooperate patiently with Inspector Mota I can probably escape the nuisance of being involved in a trial. Rolf himself is an academic of some kind, a friend of Mota's, intervening here as a personal favor, and no, he has no intention of spending Saturday evening down at the cop station himself. Thanks, Rolf, so long. But I can't blame him. The Third Delegado station is a bleak place, with a lot of empty gray rooms, almost no chairs or desks, men coming and going amid few traces of institutional permanence, nothing on the walls, elevators that don't work, trash and filth in the stairwells. It resembles a movie set built for *Blade Runner*. Only the *chefe* in the creased trousers, looking out over the top of his reading glasses, seems fastidious and authoritative. He oversees the interrogation and wants to know, with passing interest, how the suspect came by that bloody nose. He wants to know how long I've been in Rio. And what exactly have I lost? And where am I staying? And can I confirm, he hopes, that this seedy bare-chested character chained to the table is indeed one of the *ladrões* who jumped me? Several other questions on the mind of the *chefe* must also find answers, somehow. *Não falo bem portugues*, I recite. Yes, I rather knew that, he says with his narrowed eyes. Inspector Mota collates the details into an incident report, dictated to the clerkish officer, who eventually finishes typing. Okay, sign this, they tell me. Three pages, single-spaced, multiple carbons, who knows what it says. Into the blank I write, "I do not read, speak, or understand Portuguese," which provokes a flurry of displeasure—no no no, *assinatura*, just a signature not a manifesto, you jackass—and then, okay, I sign my name. By now it's almost midnight. They take the mugger away. He's looking at five years, I'm told. *Cinco anos. Cinco*, no kidding? that's a lot of *anos*, I say. Probably I should feel terrible for the young thug, on grounds of socioeconomic extenuation, but in the weakness of the moment my liberal knee fails to jerk and *cinco anos* sounds fine. Mota pantomimes jail bars with his fingers, smiling cheerily. He bids me a cordial goodbye. Yes, praise the Lord it's true, I'm free to go. I can pick up my copy of the incident report here at the station on Tuesday. *Aqui, terça-feira*, is what he says. Huh? *Terça* what? Oh, Tuesday, yes. I think: Why on earth would I want to come back here to pick up a copy? Is it mandatory? Will I be checked at the airport for my incident report when I try to fly home? One cop offers me a ride to my hotel but, remembering the Volkswagen, I slide away while he's distracted and instead catch a cab.

Next day I resume my gawker's tour of Rio, finding security in the crowds, avoiding equestrian statues. I buy a new Portuguese-English dictionary. One day behind schedule, having had no problems whatsoever at the Rodoviaria, I board the overnight bus to Caratinga.

168

"This is quite unusual," Karen Strier tells me. "That you're getting to see two copulations. In the first hour of watching the monkeys." I say nothing, though I'd like to believe I enjoy watching monkeys mate as much as the next person does.

Truth is that I'm more interested in the gloomy details of their population status than in the passing ecstasies of their sex lives. But I came here to Montes Claros for a look at the muriqui in its element, not just to collect facts, and on that basis the day does seem to be off to a good start. We have hiked out into the forest before daylight, after a breakfast of bread and coffee back at the candlelit shack that functions as Strier's bunkhouse and laboratory. The howler monkeys were howling in the distance as we walked, the cicadas began screeching at dawn, the air was warm but not muggy, and the muriquis allowed us to locate them quickly. The forest is dry, with a sparse canopy that admits shafts of sunlight. The terrain is hilly but the footing isn't bad. The female muriqui in the tree above us has been lustily eager to mate. If reproductive behavior has any strong relevance to the question of viability for the Montes Claros muriquis and those other small populations—which it does—then the timing of my visit is excellent and the peepshow is highly germane.

This female muriqui is named Cher. That's how Strier chooses to identify her, anyway. How the muriqui might choose to identify herself is anyone's guess.

Some primatologists prefer the neutrality of numbers, but Strier, favoring personal names for her study animals (as Eleonore Setz does for her golden-faced sakis up in the Amazon), has labeled other muriquis in this forest with the names of her mother (Arlene), her sister (Robin), her grandmother (Bess), and her dissertation advisor back at Harvard (Dr. Irven DeVore—though, mercifully, the monkey is simply called Irv). There's also a Brigitte, as in Bardot, and a Beatrice, I suppose as in Dante. It can get lonely, doing fieldwork in a tropical forest. Cher, as in Cher, is a robust voluptuary with a tan-colored face who, when she was young, took to spending much of her time with a male that Strier had named Sonny, as in "I Got You, Babe."

Cher is accompanied by her daughter, Catarina, a rambunctious juvenile not quite three years old. Catarina in her short life has never before seen her mother in mating condition, and she finds this copulation business fascinating. She has been climbing around on all sides of Cher in the interest of getting a close look. That's fine with Cher. Her hormones are flowing like spring runoff and she doesn't care who knows or who watches. A handful of males are waiting their turns for a go with her. They chortle and grunt to one another in mild guttural voices. Occasionally one droops an arm over the shoulders of another, or two of them trade a brotherly hug. Conviviality prevails. There is no combat among the male suitors, neither real nor symbolic: no bonking of heads, no baring of canine teeth, no puffing-up and fist-thumping of chests. What strikes me most forcefully this morning about muriqui society is that, notwithstanding the imperatives of survival and reproduction, it seems far more pacific than, say, Rio.

Strier has made several important discoveries during her decade of study at Montes Claros, and that's one: Social relations within a muriqui group, including the potentially fractious matter of reproductive competition among males, are remarkably gentle. Female pragmatism sets the tone. Male bonding gives the group cohesion. Those bonds among males are reaffirmed by way of good-natured embraces, buddy-buddy familiarity at the physical level, and are tempered during collective emergencies when males cooperate to defend the group against enemies. A muriqui is a big agile animal, capable of impressively belligerent behavior. But horny young muriqui bucks just don't fight with one another over sexual access to females, in the manner of male elk or male humans or so many other testosterone-addled mammals. Muriquis seem to have worked out a more subtle and peaceable method.

What's the method? If the prevailing hypothesis (which Strier herself favors) is correct, it involves a strategy that biologists call sperm competition. Male muriquis are endowed with huge and highly productive testes. When a male ejaculates, he does it in bounteous volume, like a shaving-cream canister with the button held down. Those big testes and the overflow quantities of seminal fluid suggest a nonviolent standard for judging the fitness of one male against the fitness of another: Whose sperm is better? That question subdivides into others. Whose sperm is more abundant? Whose sperm swims faster? Whose is more thoroughly fertile? Whose excess ejaculate clogs up the female most effectively, congealing into a nice pulpy plug that impedes penetration by other males? The muriqui strategy for maximizing heritable fitness seems to be founded on this standard.

Each female mates with many different males of her own choosing, taking them one after another as though checking off their names on her dance card. The males accept her choices quite docilely, showing no concern over exclusiveness or priority. And then, evidently, the competition for fatherhood is decided by the comparative effectiveness of the males' ejaculate plugs and by the speed and vigor of their sperm, which go scuttering off toward an unfertilized ovum along the female's gynecological racetrack. That hypothesis is premised on the assumption that a male with superior jism will be genetically superior in other traits too. The premise seems to be valid in the muriqui's case, because for a long span of time (until humans destroyed much of its habitat, a dilemma that big balls can't solve) *B. arachnoides* thrived as a species. The whole strategy makes good evolutionary sense, especially for an arboreal primate as large as this one—almost too heavy for the treetops and in danger of serious injury if it falls. Sperm competition is safer than fighting and, at least under these circumstances, more efficient. Everybody's happy. Nobody gets hurt.

"With an endangered species, it's so important to know about reproduction," says Strier. "If we're ever going to have to do any captive breeding."

Captive breeding is an option that has got to be entertained. Most of the twenty-some populations are so small that they can't possibly be considered viable. A relatively recent census suggested just how tenuous those populations are in sheer numerical terms: one site, a private forest much tinier than the forest of Montes Claros, harbored just a dozen muriquis; a different private forest held only eight; the Cunha State Reserve, in the state of São Paulo, supported no more than about sixteen. Over the short term such populations may linger, but over the long term they are probably doomed.

Another of Strier's discoveries is still more important to the issue of what should be done for the muriqui. She has learned that it's females, not males, who generally disperse from one muriqui group to another as they near adulthood. Dispersal of these adolescent individuals is crucial, because that's what allows a family-based group to avoid inbreeding and receive genetic enrichment from other groups. Adolescent dispersal is a stick that stirs the gene pool. In some other primates, such as baboons and macaques, it's the males who disperse and the females who remain bonded with their mothers and sisters. By historical accident—the fact that early primatological studies focused heavily on baboons and macaques—the impression took hold that male dispersal among primates is the norm. But the norm is a weak one, brooking many exceptions, and the muriqui turns out to be one of them. Females go; males stay. Strier's long-term research has revealed it.

"I feel that's my main contribution to this conservation question," she tells me. "At least I know now, if animals have to be moved, which *gender* we should move."

Again it's the issue of what should be done to cope with the muriqui's precarious situation. Strier would like to ignore that issue, if the world were an ideal place, and concern herself purely with ecology and behavior; but she knows that the world isn't and she can't. She mentions the "main contribution" that her basic research has made to "this conservation question" in order to forestall the accusation, I suspect, that her research has made no contribution whatsoever. It's an accusation that might be made against any biologist who studies a jeopardized species but who is more recognized, as Strier is, for the rigor of her science than for the passion of her conservation activism. As for the reason why animals might have to be moved— that's so obvious that she doesn't mention it: With twenty-some tiny populations all separated from one another on patches of insular habitat, those animals *aren't* moving on their own. There is no migration between patches, and consequently no genetic enrichment of one population by individuals from another—no stick stirring the larger gene pool. The total number of surviving muriquis (roughly seven hundred) would seem, under even the best conditions, to be perilously few. Under the current conditions of fragmented habitat, the situation is worse. The fragmentation vortex, among others, is sucking at muriqui feet.

Strier bears that in mind as we talk. She maintains a dogged optimism on behalf of the species, buttressed mainly, I think, by her personal devotion and her years of professional investment. To those of us without special buttresses, the muriqui's prospects look bleak. No population viability analysis, on the plan formalized by Gilpin and Soulé, has at this point been performed. We can only guess. A plausible guess is that *B. arachnoides* won't survive—not in the wild, anyway—for as long as a hundred years. Not unless something drastic is done to connect the populations.

What transfer of individuals does occur nowadays is merely from one family group to another within a single island of habitat. The forest at Montes Claros is large enough to support two family groups, each containing a few dozen animals. One group ranges mainly in the northern half of the forest, one ranges mainly in the south. Strier has focused her fieldwork on the southern group, leaving the northerners to themselves. But the territories aren't rigidly demarcated, the feeding circuits sometimes overlap, and the two groups occasionally come face to face for a vociferous argument over a fruit tree. The groups maintain distant and mainly hostile relations, complicated by an ambivalent sort of interdependence: They trade genes, via dispersing adolescent females. When a young female approaches sexual maturity, she transfers her membership from one group to the other, because she is no longer welcome among her closest blood relatives and there is nowhere else for her to go. (If she's lucky, that is, she transfers membership; less lucky, and she just wanders off solitarily toward death and oblivion outside the forest.) The female whom Strier calls Cher, for example, so busily mating today with males of the southern group, is an immigrant. She came from the northern group. Cher was the first young muriqui whose transfer and acceptance Strier witnessed, a half-dozen years ago.

"I've known her since she was just a teenager. Now she's had two kids. And it looks like she's working on her third." It does indeed. If Cher isn't pregnant by sundown today, she can't be faulted for not trying. "Both of her infants have survived," Strier says. The second of those infants is Catarina. "But she's not a very attentive mother. She likes to eat. She doesn't like to carry her kids. She leaves them crying. But they've lived." Cher has succeeded at procreation despite her maternal fecklessness. What she lacks in childrearing behavior she makes up for, apparently, with good genes.

Cher's success reflects a trend that Strier has detected, adding another bit of complexity to the full picture: good population growth at this study site, in counterpoint to the general decline. While *B. arachnoides* overall has remained sorely endangered, and some of the other populations have shrunk badly, the population here at Montes Claros has boomed. From forty-some animals in the early 1980s, when Strier began watching, the number rose to fifty-two by 1987 and in the years since to about eighty. That's almost a hundred percent increase within less than two muriqui generations. Fertility has been high, infant survival has been high. Both of those parameters tend to suggest that, notwithstanding its small size, the Montes Claros population isn't suffering from inbreeding depression—at least not yet, not noticeably.

Strier's current research concerns cycles of reproductive activity.

How do the manifest behavioral cycles correlate with physiological factors? she wants to know. Her chosen technique is to approach muriqui physiology from behind—by way of muriqui feces. Today her goal is to collect fecal samples from five different females, one called Louise, one called Didi, the familially named Robin and Arlene, and Cher. Fecal-sample collection is no picnic. And for a study of behavioral cycles, the collections must be made daily. Strier tracks each female in turn, following on foot while the female swings through the treetops, watching constantly for defecations, bushwhacking tirelessly up hillsides and down gullies, staying close, staying quiet, making notes, all the while hoping to get shat upon. Muriqui shit smells faintly of cinnamon, from the high dietary intake of cinnamon leaves, but you wouldn't mistake it for seasoned apple butter. Sometimes it comes down as a fine brown rain; sometimes it comes in dollops. Each dollop from one of her target females is precious to Strier, a morsel of biochemical data to be gathered joyfully into a vial. She freezes the samples and will eventually hand-carry them back to a lab in Wisconsin, where a colleague will do an assay for hormones. By this means Strier will investigate the connection between hormonal triggering and lubricious behavior like Cher's.

It might be easier to tranquilize animals by dart gun and take blood samples. But Strier has never darted a muriqui. She uses nonintrusive field methods, figuring that her very presence and the presence of her students (who follow the muriquis and gather further data during the months when Strier is in the United States) has been intrusion enough. After so many years of observing them so sedulously, she has never even touched a muriqui at Montes Claros.

She didn't come here for easy intimacies. She didn't come to hand out bananas and shake her fist at the evil poachers and pride herself that the muriquis have accepted her as a friend. She came to do science. She studies these creatures, as she loves them, from a distance. And now, having made such extraordinary efforts to leave the objects of her research unmolested, she is chary about other molestation, of even the most well-intended sort. She holds a strong opinion about whether the Montes Claros population should be subjected to drastic, intrusive conservation measures—removal of wild individuals for captive breeding, artificial arrangement of mating matches, introduction of captive-bred animals to the wild—such as Carl Jones applied to the Mauritius kestrel. Her opinion is, no.

If muriquis are needed for captive breeding, Strier hopes they will

be captured elsewhere. If animals must be moved between populations, Strier hopes that none will be taken from Montes Claros and
none will be brought here. She points to the recent population
growth, to the high birth rate and the infant survival. Most of the
Montes Claros adults, too, seem to be healthy. They carry few parasites, compared with other muriquis. She points to the apparent lack
of inbreeding depression. The Montes Claros population doesn't
seem to be broken, she argues, so let's not try to fix it.

"I don't want this population to be the experimental one," she says
fervently.

Cher finishes her current spate of matings and ambles off through
the treetops. Departing, she leaves behind one mustardy green blip of
shit, fallen conveniently onto a rotten log. Strier collects it with
tongs. She shows me the thing, like a dubious bit of sushi between
chopsticks, before dropping it into its vial. Then we begin stalking
Robin. Hours pass, we climb through the undergrowth, we sit, we
wait, we watch, we climb further through the undergrowth, more
watching, more waiting, until finally Robin lets fly, not quite on my
head but close, and then we turn our attention to Arlene.

One sample from each female, at least every other day, is what
Strier needs for her hormone analysis. On a good day she will spend
ten hours in the forest and get a grand total of three or four samples.
But many days aren't so good, and during the earlier weeks of the
fecal-sample study, as her notebook shows, Strier suffered a frustrating paucity of data. Not enough falling shit. This is the routine behind the supposed romance of tropical field biology.

Today is a good day. By noon Strier has gotten a sample from Arlene. We hike to another part of the forest and locate Louise.
Promptly, accommodatingly, Louise too unloads her bowel, and after
all the previous frustrations Strier can't resist a triumphant hoot.
"See? It's because I have them so well trained!"

Didi, the last of the targeted females, is not so accommodating. She
makes us find her and then she makes us be patient. She sprawls languidly on a branch, as though she might be finished for the day—finished moving, finished eating, finished defecating, finished doing
anything whatsoever that could interest a primatologist. While the
sun tilts low into the west, we sit beneath Didi's tree, again waiting. It
may be that we'll wait here until dark for nothing. It may be that
Didi's bowel has already been quite satisfactorily purged.

During the lull, Strier makes a reluctant but forthright admission.

She wants me to know that certain other knowledgeable primatologists disagree with her judgment concerning Montes Claros. Specifically, they disagree that the muriqui population here should be exempted from such emergency management measures as translocation and captive breeding. Strier reaffirms that this is her own view—leave them alone, yes, absolutely—but she feels compelled to acknowledge that not every authority shares it. She names several who don't.

These other primatologists have studied muriquis too. Not at Montes Claros, and not so continuously as Strier herself, but elsewhere, and for significant stretches of time. One of them is a woman whom Strier considers a friend as well as a valued colleague. They are good scientists, these other folk—conscientious, bright, experienced, cognizant of Strier's own arguments. They don't know all the details of the Montes Claros population, but they know some. They know that this population is larger and seemingly more secure than most of the others. They know primate biology in general. They know that there's a balance of jeopardies between intervening too little, in the case of a desperately rare species, and intervening too much. Nevertheless, they don't share her strong conviction that the Montes Claros muriquis should be left alone.

Strier confides that she anguished a bit, before I arrived at Montes Claros, about whether to leave me ignorant of the whole question. She was tempted to do that, because she fears that these other scientists will tell me things she herself would prefer I not hear. Tell me what? Well, for instance, that the Montes Claros population should be included immediately in an intrusive management project for the overall good of the species. The idea is abhorrent to Strier, but does that mean she should suppress it? She fought with her scientific conscience, she says, and decided not. So in a spirit of open discourse she has offered me the names of the experts who will gainsay her.

Now, far into the embers of the afternoon, Didi stirs from her siesta. She unfolds her long limbs and prepares to move. Possibly she's hungry again, or restless, or there's a male upwind with whom she fancies she might like to mate. Strier gathers herself for another chase. Less avidly, I gather myself too. Strier's got her tongs, her vials, her notebook, her binoculars, she's got the blood back into her legs, all set, here we go. Didi locks one hand on a vine, swings away from her perch, furls out her other arm to grab another vine, and as she hangs momentarily suspended between trees—like the world's largest spider in silhouette against an afternoon sky—she lets go a healthy crap.

"Oh," says Strier, almost speechless with gratitude. "Oh, Didi."

And then: "Oh, look. Here comes another." Strier watches the second gob fall—watches it studiously, all the way to the ground, with the meticulousness of a PGA caddy. She fixes the spot in her mind. This will make five fecal samples from five target females, she reminds me, one of the rare hundred percent days. "You have no idea how happy I am," she says, and I don't doubt that she's right.

She gallops to the prize. There's more than enough for a vial. "I think I'm being paid back," Strier says. For the confession about the opposing opinion, she means, though I wouldn't have guessed her for the superstitious type.

169

NOT LONG after the Second International Conference on Conservation Biology, Michael Soulé heard from a fellow named Jim Johnson, of the U.S. Fish and Wildlife Service regional headquarters in Albuquerque. Johnson was chief of the Office of Endangered Species for that region, with jurisdictional responsibilities and headaches stretching eastward into Texas. He had read some of Soulé's earlier work on population viability and, like Hal Salwasser of the Forest Service, he had been impressed by the possibility that these ideas could be useful to wildlife management. Now he was calling, as Salwasser had earlier done, to see whether Soulé might help him with a problem.

Johnson's problem was that a rare and obscure subspecies of reptile, *Nerodia harteri paucimaculata*, commonly known as the Concho water snake, stood in the pathway of what the U.S. Army Corps of Engineers and certain august Texas citizens were pleased to consider progress. Specifically, plans had been made for a seventy-million-dollar reservoir project at the confluence of the Concho River and the Colorado River (not the Grand Canyon river by that name but a much smaller Texas version), on the dry plains about halfway between Austin and Odessa. The linchpin of the project was to be an edifice called Stacy Dam, intercepting the Colorado just downstream from its juncture with the Concho, not far from the small town of Paint Rock. The construction of Stacy Dam promised several blessings: It would create a reservoir to serve as a municipal water supply, it would push federal tax dollars into the local cash registers, and it would keep earth-moving, cement-slinging crews from the Corps of Engineers feeling happily purposeful for a couple of years. On the negative side stood one small concern, so minor that it almost seemed risible: The project would be detrimental to some pissant snake.

The snake was an unprepossessing creature, not large or decorative or poisonous, with no economic clout and not much ecological charisma. It was pale mottled brown on its dorsal surface, orange-bellied underneath. The brown mottling was just similar enough to the color pattern of a prairie rattlesnake to get *Nerodia harteri paucimaculata* persecuted by hysterical snake-loathing people, and no one could say how many Concho water snakes had been killed with tire irons and irrigating shovels under the animus of mistaken identity. But such direct human-caused mortality wasn't the most grievous threat to its collective survival. The most grievous threat involved habitat—*N. h. paucimaculata* had precious little, and if Stacy Dam were built it would have still less. It was a localized subspecies of Harter's water snake, *Nerodia harteri*, about which the only notable claim to be made is that it's the only snake species whose range falls entirely within Texas. As reflected in its name, the Concho subspecies was native mainly to the Concho River. It also inhabited the Colorado, but an earlier reservoir project on the upper stretch of that river had already eliminated a major portion of its habitat, and its lingering presence elsewhere on the Colorado was sparse. Aside from those two rivers, it existed nowhere.

The snake's total population was no more than a few thousand. Because of the peculiarities of its life history, its effective population was much smaller still—maybe only a tenth of the total. It found food and security in the fast-flowing, shallow riffles of the Concho and the Colorado, and it seemed to have little use for the slow-water sections between riffles. This special requirement of habitat was probably the main factor limiting its population, since the supply of riffles was sparse. Some people felt that *N. h. paucimaculata* should have been listed as "endangered," or at least "threatened," under the provisions of the Endangered Species Act, which would have mandated rigorous protection, conceivably to the point of prohibiting the dam. But pork-barrel politics had weighed in against scientific judgment (by one account, a powerful Texas politician with good Washington leverage made menacing noises to the Fish and Wildlife Service), and until late in 1986 the Concho water snake remained unlisted. Then, holy miracle, it was indeed recognized as "threatened." Its state of jeopardy reflected its shortage of habitat. Nevertheless, it was about to lose more.

The proposed reservoir behind the proposed dam would engulf a wishbone-shaped portion of river around the confluence of the Colorado and the Concho. By drowning the riffles there, it would create a large gap at the center of the snake's distributional range. The reservoir wouldn't destroy all of the snake's habitat; farther upstream on the Concho, farther upstream on the Colorado, and farther downstream on the Colorado below the dam, three separate stretches would remain. But concern had arisen—and even the Corps of Engineers had been made to perceive this concern—that the reduction and fragmentation of habitat caused by the reservoir might decrease the snake's likelihood of long-term survival.

Decrease it how badly? officials wondered. A small decrease might be okay. Some loss of habitat, some further decline in population, might be acceptable, so long as such changes weren't *too* likely to lead to extinction. Maybe the world didn't need all three thousand, or however-dang-many, Concho water snakes. Maybe a mere five hundred of these miserable serpents would be adequate, say, to preserve the subspecies under foreseeable circumstances, and thereby to comply with the Endangered Species Act and keep the Texas herpetophiles and lunatic conservationists placated. The Corps had duly conferred with the U.S. Fish and Wildlife Service for an expert opinion. Jim Johnson was the man into whose lap this problem fell.

Johnson himself turned to Soulé, well known by then for his work on small-population jeopardies. Was Dr. Soulé available to analyze the population viability of *N. h. paucimaculata*? Would he offer a scientific judgment of how it might be affected by Stacy Dam?

Dr. Soulé was available. He recruited Mike Gilpin to collaborate. They both wanted badly to bridge the chasm between abstruse ecological theory and conservation management as practiced by government agencies, and this would be a showcase demonstration of population viability analysis. They accepted a modest one-month contract to do the job.

During the first week, Soulé flew to Texas and met with herpetologists and hydrologists at a motel in Odessa. He picked their brains about the population status of the snake, about its ecology and behavior, about the flow regimes of the two rivers, about how those regimes (especially the episodes of flood and of drought) might affect snake survival in various locations. Back in California, he and Gilpin sat down to digest and contemplate several boxes of data.

They learned that the snake was a fast-maturing and prolific creature, but short-lived, with one-year-old females producing broods of about eighteen young and most adults dying before the age of three. They learned that the snake fed on small fish, which it caught by diving in riffles. They learned that riffles, in fact, seemed to be the only sort of habitat in which young Concho water snakes could survive to maturity. The slow-water sections were inhospitable, at least partly because those sections harbored a larger species of water snake that preyed upon *N. h. paucimaculata*. Since riffles were few and far between, separated by long slow-water sections, the snake's population existed in spaced clusters arranged linearly, like beads on a rosary. Dispersal from one cluster to another—from one riffle to another, across the dangerous slow-water sections—was a precarious adventure that succeeded only rarely. Soulé and Gilpin learned too that a recent genetic study had found almost no variation throughout the population. The individual snakes were too much alike—high in homozygosity, low in variety of alleles—to be capable of much adaptive flexibility. This genetic uniformity suggested that the entire population had descended from only a small number of common ancestors.

Beyond these facts, there wasn't a great deal known. The herpetological literature offered little on *Nerodia harteri*, let alone its Concho subspecies. For lack of other sources, Soulé and Gilpin carefully scrutinized an unpublished master's thesis on the species, done almost twenty years earlier for a degree at Texas Tech.

And they had a visual portrait of the river itself in the form of videotapes. The videotapes, shot from a low-flying plane and supplied by Jim Johnson, showed mile after mile of the Concho unrolling itself ever onward, like a rough cut for some ill-advised Imax travelogue. "Rio Concho Getaway!" was not destined to be a box-office hit. As captured on video, the riverscape seemed to consist chiefly of slow-water sections winding through gentle bends and interrupted occasionally by riffles. From watching the tapes, Gilpin recalls, he got a good sense of the habitat distribution. He also got airsick.

170

"So we have a month to do this project," Gilpin says, seated before the screen of a Macintosh in his office in San Diego. The screen shows a mesmerizing pattern of stationary oval shapes and moving dots. The oval shapes quiver, the dots flit between them, as Gilpin re-counts the Concho episode.

Their first week had gone to gathering expert opinion and hard data. "So then Soulé and I, we just worked sixteen hours a day for three weeks on this thing and built lots of computer models. And it was as exciting a thing as I had ever done in science." It was exciting because this humble snake, on a couple of dribbly rivers in central Texas, represented a real-world test case for twenty-five years' worth of theory. What Gilpin and Soulé were doing, during that fevered month, was in some measure a culmination of what Robert Mac-Arthur and Ed Wilson had begun in the 1960s.

Among the most crucial inputs to their computer modeling was what Gilpin calls "the fragmentation data." These data charted two factors: habitat distribution and actual snake distribution. The data on actual snake distribution derived from two separate field censuses taken by competent observers. The habitat data comprised a single quantified portrait (more useful and less nauseating than the video-tapes) of the distribution of hospitable riffles along the two rivers.

As revealed by the habitat data, the riffles were unevenly spaced. Adopting a quantitative convention that had been used in the censuses, Soulé and Gilpin parsed the whole river system into five-kilometer seg-ments and considered each of those segments as a unitary patch. Some segments encompassed more riffles than other segments. Translated into an analytical abstraction as portrayed visually in their model: Some habitat patches were larger than others. The lower stretch of the Con-cho River just up from the confluence (a stretch that the reservoir would obliterate) was especially rich in riffles. So the habitat patches along there, as Soulé and Gilpin represented them, were big. The up-per stretch of the Colorado River was especially poor, with not many riffles per five-kilometer segment. Those patches were represented as being smaller.

Against the habitat background stood the snake-distribution data: which habitat patches were occupied, which weren't, and what changes of status had occurred over time.

The earlier set of snake-distribution data came from 1979, when a team of herpetologists had surveyed the rivers by boat. Riffle by riffle, the herpetologists had worked their way downstream, finding snakes in some riffles, not in others. A similar survey was done four years later. Under the schema of Soulé and Gilpin, one snake caught in one riffle within a five-kilometer habitat patch was taken to indicate a colony of *N. h. paucimaculata* within that patch. The combined results of the two surveys showed that not every habitat patch supported a colony of snakes and that not every colony was permanent. Some of those colonies had winked on or off over time.

Soulé and Gilpin called the winking "turnover." Although they borrowed the word from MacArthur and Wilson, their usage was slightly different, referring to the distributional status of one subspecies, not the diversity status of one island. But it was no accident that their take on the Concho situation harked back to the equilibrium theory. That connection wasn't just a matter of language.

Each riffle was an island of habitat, within which could occur immigration and extinction. During any one year, a given riffle might be colonized by immigrant snakes riding downstream on a flood or swimming upstream for relief during low water. In a later year, the colony within that riffle might go extinct, owing to drought or disease or lack of breeding success or some other stroke of bad luck. Then the riffle might remain vacant for several more years until a new set of colonists arrived. Meanwhile other riffles would have changed status too—from vacant to occupied or vice versa. On a larger scale, any five-kilometer patch might be vacant or occupied, and its status might change with immigration or extinction. This was turnover as defined by Soulé and Gilpin. Because of the Concho water snake's short lifespan, its capacity for fast reproduction, and the likelihood of drought episodes that could make a whole patch uninhabitable, the system's turnover rate was expected to be high. That is, extinction and recolonization would occur often.

An occupied patch could send immigrants either downstream or upstream, providing new colonists for empty habitat or reinforcements for a colony on the verge of extinction. The statistical probability of extinction on any patch would be inversely related to the patch's size—that is, to the amount of good riffle habitat within the five-kilometer segment. This was the area effect, as construed by Soulé and Gilpin. If a patch was especially remote from other occupied patches, it would seldom receive any immigrants. Its probability of extinction would therefore be higher (for lack of reinforcements) than that of less remote patches, and its probability of recolonization after extinction would be lower. This was the distance effect, again as Soulé and Gilpin chose to adapt it from MacArthur and Wilson. Depending on what happened elsewhere within the system, an especially remote patch might not be recolonized for ten years, or for twenty, or ever. What happened elsewhere within the system was therefore highly consequential to each colony in each patch of habitat. The overall fate of the aggregate population depended on the individual fates of its semi-discrete parts. Extinction was effectively contagious—more likely to strike in a given habitat patch if it had already struck in most of the neighboring patches. Survival was contagious too. There was no mainland in this system. There was no huge area of habitat that would always continue supplying immigrants. That's what made the Concho water snake case so engrossing to Gilpin and Soulé. It wasn't just their first PVA; it was a chance to investigate the shape of the future, in a world without ecological mainlands.

They brought one other bit of fancy terminology into the discourse: *metapopulation*.

A metapopulation, defined loosely, is a population of populations. The word dates back to an obscure but important study of the extinction process, published in 1970 by a mathematical ecologist named Richard Levins. That study was based on an even more obscure theoretical model he had proposed a year earlier, in a paper on the biological control of pestiferous insects. Still farther back, the general concept of a metapopulation can be traced to an ecology text from the 1950s. Gilpin himself had explored its implications in some of his theoretical work during the mid-1970s. Ed Wilson too had cited Levins's model, not in any island-related work but in *Sociobiology*. The metapopulation idea is closely related to MacArthur and Wilson's equilibrium theory, since it entails the same sort of dynamic balance between the same offsetting processes, immigration and extinction. But there's a crucial difference. The equilibrium theory of island biogeography assumes the existence of a mainland. The metapopulation concept assumes the contrary: no larger whole, only fragments.

In an equilibrium-theory situation, the mainland serves as an inextinguishable source of immigrants, while the satellite islands undergo turnover. In a metapopulation, the mainland is nonexistent; only the individual habitat patches supply immigrants and reinforcements to one another, and every patch is subject to turnover. So the possibility is inherent in a metapopulation that *all* the patches might wink out successively, dropping the whole system into oblivion.

Soulé and Gilpin saw this possibility in central Texas.

They realized that the Concho water snake system didn't constitute a population, in the precise biological sense of that word, since a population is a group that interbreeds. In the typical case of a population, habitat is continuous and individuals can mix freely for mating. In the Concho case, with its highly fragmented habitat, extinction and recolonization of colonies were more likely to occur than interbreeding between colonies. There was more turnover than mixing. And there was (as I've already said but can hardly overemphasize) no mainland. To consider this system a metapopulation, as Soulé and Gilpin did, was to recognize its particular intricacy.

Using all these bits of fact and theory, they designed a state-of-the-art analysis. Gilpin, the Macintosh jockey, did most of the programming. Within a couple of weeks he and Soulé had developed a computer model that mimicked the current status and predicted the future dynamics of the system. Their model was only a complicated cartoon of a vastly more complicated reality, but it was realistic enough to be useful. It took into account the habitat fragmentation and also the environmental uncertainty (droughts and floods, derived from real flow data for the two rivers) that the snake was likely to face. It considered immigrations, extinctions, reinforcements, the area effect, the distance effect, turnover, and the vagaries of chance. The model's purpose was to portray a large assortment of possible outcomes within a vastly compressed scope of time. Soulé and Gilpin ran many consecutive simulations on the model, each run taking the system through three hundred years' worth of possible events, based on the empirical data and the theoretical assumptions built into it. Then they tallied the results. What their model told them was that the Concho water snake, as presently situated, was probably viable for the long term. Patch occupancy winked off and on, here and there, but overall the metapopulation endured.

That much might have been good news for Jim Johnson, of the Fish and Wildlife Service, if he hadn't been facing the prospect of Stacy Dam. Soulé and Gilpin, his two honcho theorists, said that the snake would most likely survive—if left alone. But the Corps of Engineers wasn't going to leave it alone.

171

I STARE AT the glowing oval shapes on the screen of Gilpin's computer. There are roughly a hundred, aligned in the form of a wishbone, representing the Concho and Colorado rivers at their confluence near

Paint Rock, Texas. Although each stands for a five-kilometer segment of river, the ovals appear in a range of sizes, reflecting greater or lesser amounts of good riffle habitat. The zooming dots travel from one oval to another like bursts of gunfire in a video game. It's the Concho model, pulled up by Gilpin so he can show me its workings.

Some of the ovals are solid, indicating occupancy by *N. h. pauci-maculata*. Some are hollow, indicating vacancy. The zooming dots signify colonization possibilities per year. When a dot scores a hit on a hollow oval, the oval switches instantly to solid, indicating a successful colonization. What I see is a metapopulation of photons and, at least for the moment, it's sustaining itself.

"So now we modeled the creation of the dam," Gilpin says, "by simply zeroing all the patches to be inundated by the reservoir, and by allowing no colonization across the reservoir." Tapping a few keys, he starts the model through another run, this time in the mode that includes Stacy Dam. "So, where we once had a connected one-dimensional system, we've now got three disconnected one-dimensional systems." Immediately a cluster of patches at the wishbone juncture goes blank. That cluster represents the reservoir.

Dots continue to zoom between the remaining large ovals within each of the three disconnected systems. Along the Concho River the dot traffic is relatively heavy. Along the other two stretches—the upper and the lower Colorado—the dots are sparse. As I watch, those ovals along the Colorado begin to fall empty at a faster rate than empties are refilled. Turnover has become attrition. Before long, the upper Colorado stretch goes totally blank. All the ovals are empty. Then the lower Colorado stretch goes blank. The dot traffic along both stretches has ceased. The connected system has been busted to pieces, two-thirds of which have proven to be unsustainable.

Gilpin says, "And quite robustly—we did a lot of simulations of this—quite robustly we found that two of the stretches of rivers went extinct."

Gilpin and Soulé were peeking into the snake's hypothetical future, generating informed predictions as to the combined effect of chance and Stacy Dam. Their model predicted that the third stretch of habitat patches, along the Concho River, would retain its snakes throughout a three-hundred-year span. Even after the reservoir had drowned a major share of habitat and had isolated the two other stretches beyond reach, the Concho still held enough riffles to support an abiding (though perilously small) metapopulation of the snake.

"So the dam would not directly produce extinction in the short run," says Gilpin, choosing his words advisedly.

By "extinction" in this context he means total extinction of *N. h. paucimaculata*. By "not directly" he implies that the Stacy Dam project *would* indirectly make the remnant metapopulation more vulnerable to extinction. At the back of his mind are those vortices, described in his talk at Ann Arbor: the looping interactions of detriment and mischance. Loss of habitat would exacerbate distributional fragmentation. Population decline would exacerbate inbreeding. Loss of adaptability would exacerbate the difficulties of immigration between habitat patches, which would further contribute to fragmentation, which would increase inbreeding, and so on. Stacy Dam might not kill *N. h. paucimaculata* outright, but it would move the snake closer to those vortices of doom. "And we honestly presented that result in our report," he says.

Their report went to Jim Johnson in November 1986. It was a twenty-page typescript, with another dozen pages of tables and figures, including a printout of the constellation of ovals. Although the dam probably wouldn't directly cause extinction, said the report, it would "lead to the rapid disappearance of the snake in the Colorado River" and would "exacerbate its vulnerability to catastrophe" along the Concho. What sort of catastrophe might that be, that could finish off the snake? A relatively small one—a toxic spill from one tanker truck on a highway bridge, for example. Even a bad sewage leak might do it. In the meantime there would be an increased chance of inbreeding depression. These problems could be mitigated, maybe, by certain engineering and management measures.

"Of course, before they even received our report," Gilpin says, "they had pretty much committed themselves to build the dam." Construction began a year later.

The reservoir is now full. Jim Johnson has moved on, but other concerned folk at the U.S. Fish and Wildlife Service continue to monitor the Concho water snake. The latest data are inconclusive. Will it live, will it die? Can its dilemma be mitigated? While those questions do have their importance, the larger significance of this case doesn't lie in the survival or extinction of one subspecies of snake. The larger significance, as Gilpin and Soulé understood, is that we have entered the era of metapopulations.

If it's not the Concho water snake, it's the muriqui. If it's not the muriqui, it's the Florida panther. If it's not the Florida panther, it's the eastern barred bandicoot in Australia, or the tiger in Asia, or the cheetah in Africa, or the indri in Madagascar, or the northern spotted owl in the Pacific Northwest, or the black-footed ferret in Wyoming, or the Bay checkerspot butterfly in California. Or it's the grizzly bear, which in the contiguous United States is now confined to a half-dozen islands of montane forest (including the North Cascades ecosystem and the greater Yellowstone ecosystem), most of them too small to accommodate a viable population of grizzlies. The pattern is widespread. All over the planet, the distributional maps of imperiled species are patchy. The patches are winking. In some instances they're winking off and on, but in many instances they're merely winking off.

During the years since his effort on the Concho water snake, Mike Gilpin has focused his professional life on the study of metapopulation dynamics. Why? Because he knows that the mainlands are going, going, gone, the world is in pieces, and reality conforms to a new model.

❦

MESSAGE

FROM ARU

❦

172

THERE'S a voice that says: *So what?*

It's not my voice, it's probably not yours, but it makes itself heard in the arenas of public opinion, querulous and smug and fortified by just a little knowledge, which as always is a dangerous thing. *So what if a bunch of species go extinct?* it says. *Extinction is a natural process. Darwin himself said so, didn't he? Extinction is the complement of evolution, making room for new species to evolve. There have always been extinctions. So why worry about these extinctions currently being caused by humanity?* And there has always been a pilot light burning in your furnace. So why worry when your house is on fire?

Biologists and paleontologists speak of a background level of extinctions throughout the history of life. That background level is the routine average rate at which species disappear. It's generally offset by the rate of speciation, the rate at which new species evolve. These two together, extinction and speciation, constitute still another form of turnover—in this case, on the global scale. Rates of extinction in the remote past can't be calculated precisely, because gaps in the fossil record prevent us from knowing what has been lost. But a cautious paleontologist named David Jablonski has made an informed guess, placing the background level at "perhaps a few species per million years for most kinds of organisms." A few mammal species, a few bird species, a few fish species lost to extinction every million years—with that rate, evolution can keep up, adding a few species to each group by speciation. Such losses, counterbalanced by gains, yield no net loss of biological diversity. Extinction at that level, the background level, is an ordinary and sustainable process.

Against that background, a small number of big events have emerged to the foreground. These cataclysms, anything but ordinary, are the mass extinctions that scientists now recognize as major punctuation marks in the history of life. Some of them are famous: the Cretaceous extinction, the Permian extinction. In such a mass extinction, compressed within a relatively brief span of years, the extinction rate far exceeds the rate of speciation, and the richness of the biosphere plummets. Niches fall vacant. Intricate networks of ecological relationships are thrown into disarray. Entire ecosystems are left raw and ragged. Millions of years pass, then, before speciation fills the gaps and brings the overall diversity back up to previous levels.

No one knows just what caused the mass extinctions of the distant past. The competing hypotheses range from gradual climate change (reflected in habitat changes that proved intolerable for many species) to a so-far-undetected Death Star that orbits mutually with our sun, exerting cosmic gravitational drag and pulling a shitstorm of killer asteroids through the vicinity of Earth every twenty-six million years. The debate over the competing hypotheses is a fascinating story but not one I'm going to pursue here. It's enough to note that mass extinctions of the first magnitude occurred at just five points in distant geological time, and that each was caused by some indeterminate set of natural factors among which humanity (not yet on the scene then) can't be implicated. The Cretaceous extinction, 65 million years ago, claimed the last of the dinosaurs; the Permian extinction, 250 million years ago, eliminated more than half the extant families of invertebrate marine creatures. Other mass extinctions struck at the end of the Ordovician period (440 million years ago), in the late Devonian period (370 million years ago), and at the end of the Triassic period (215 million years ago, give or take a few million). Additionally, a sizable roster of large-bodied animals disappeared during the later millennia of the Pleistocene epoch, only tens of thousands of years ago, and in this case humanity *may* have been partly responsible; those Pleistocene extinctions occurred about the time that humans began hunting in armed and cooperative packs. Compared to the five big events, though, the Pleistocene spasm was minor, mostly confined to mammals.

And there have been still others, lesser episodes during which the extinction rate only modestly exceeded the background level. One way of defining a mass extinction as distinct from a lesser episode, according to Jablonski, is that it entails an extinction rate double the background level among many different plant and animal groups.

By this rigorous standard, we're experiencing one now.

It started a few thousand years ago, when humans from Neolithic cultures along the fringes of the continents began venturing across the open sea in primitive boats. Colonizing remote islands such as Madagascar, New Zealand, New Caledonia, and the Hawaiian archipelago, the human invaders promptly killed off some endemic bird species. Many of these extinct birds were giant forms, flightless and ecologically naïve. You've read enough by now to imagine how it went, island after island around the world. The Neolithic wave of human invasion had accomplished its damage centuries before the first Portuguese ship landed at Mauritius. But the results were similar, and the European-inflicted insular extinctions were actually just a second phase of the larger process. The case of the dodo was only one of hundreds.

From the time of the Neolithic voyages until the present, twenty percent of the world's bird species have gone extinct. During recent centuries, the rate of extinction has increased further and the range of jeopardy has widened—from birds to animals and plants of all kinds, and from islands to continents—as humanity's impact has grown in direct correlation with the growth of human population, technological efficaciousness, and hubris. Nowadays it's not just a question of dodos and elephant birds and moas. Nowadays we're losing a little of everything.

Within a few decades, if present trends continue, we'll be losing a *lot* of everything. As we extinguish a large portion of the planet's biological diversity, we will lose also a large portion of our world's beauty, complexity, intellectual interest, spiritual depth, and ecological health. You've heard this doom song before, so I won't chant all the dreary, important verses about how sterilizing our own biosphere represents a form of suicide. But I offer you, from some familiar authorities, a bit of numerical perspective on the scope of the damage being done. Paul Ehrlich, one of the grandfathers of conservation biology, who was a respected ecologist before he gained fame as a population-crisis maven, estimates that the current extinction rate, just among birds and mammals, is roughly a hundred times the background level. Ed Wilson, based on surveys of the diversity of invertebrates within tropical forests, estimates that the current loss of rainforest species is at least a thousand times above normal. Dan Simberloff, who is nobody's idea of a careless alarmist, has published a skeptical review of the evidence, titled "Are We on the Verge of a Mass Extinction in Tropical Rain Forests?" Is the situation really so dire as the arm-wavers would have us believe? asks Simberloff. After thirteen pages of cool argument and conscientiously crunched numbers, he concludes that it is. Yes, Simberloff predicts, the current cataclysm of extinctions is indeed likely to stand among the worst half-dozen such events in the history of life on Earth.

This time around, *we're* the Death Star.

But with a difference. Our own devastating impact on the biosphere will probably be a singular event, not part of a recurrent pattern. Why? Because we probably won't survive long enough, as a species, to do it again. The richness of Earth's ecosystems might recover to previous levels within, oh, ten or twenty million years, assuming that *Homo sapiens* itself has meanwhile gone extinct too. When we ourselves do go, the sparrows and the cockroaches and the rats and the dandelions that survive us should eventually give rise to a full new inflorescence of diversity. I'll leave it to you to decide whether that represents a gloomy scenario or a cheery one.

Eons in the future, paleontologists from the planet Tralfamadore will look at the evidence and wonder what happened on Earth to cause such vast losses so suddenly at six points in time: at the end of the Ordovician, in the late Devonian, at the end of the Permian, at the end of the Triassic, at the end of the Cretaceous, and again about sixty-five million years later, in the late Quaternary, right around the time of the invention of the dugout canoe, the stone ax, the iron plow, the three-masted sailing ship, the automobile, the hamburger, the television, the bulldozer, the chain saw, and the antibiotic.

173

JUST a hundred years ago this cataclysm was unimagined. People didn't worry much about the extinction of species. The very notion that species *could* go extinct was still a new idea in the late nineteenth century, part of the same intellectual revolution that Charles Darwin had led. Extinction seemed a phenomenon of the past; it was deduced from the fossil record but wasn't observed much in the living world. As seen in the fossils, it constituted a crucial class of evidence (along with the patterns of biogeography and the observable data on domestic animal breeding) supporting the idea of evolution by natural selection. Among those few nineteenth-century thinkers who addressed the subject at all, most addressed it with scientific detachment, grounded in the evolutionary perspective on the geological scale of time. Alfred Wallace addressed it that way himself, sometimes.

But Wallace was an exceptional man, by this measure as by so many others. He was prescient enough to be concerned.

His most vivid book, *The Malay Archipelago*, was published in 1869. That was seven years after Wallace had returned to England, eleven years after he had caused Darwin conniptions, and twelve years after he had sailed away from the trading town of Dobbo, in the Aru Islands, at the end of his memorable collecting season there. He placed his chapter on Aru late in the Malay book, where it served as a nice literary climax.

The stay in Aru had begun badly, as I've already described. After all the hassles of getting to Dobbo under force of the westerly monsoon, then the further hassles of finding transport to the forest interior, he met weather too wet for decent collecting. This was a maddening setback. Also, his living conditions were harsh and he was tormented by sand flies. For a while he got few insects or birds. But as he began to despair, one of his boys brought in a bird of paradise, which seemed to repay him for everything.

In the book he reported his "transports of admiration and delight" at laying hands on this specimen. It was the scarlet species with the green spiral tail wires, known as *Cicinnurus regius* to science and goby-goby to the Aruese—the smaller of the two birds of paradise for which Aru was renowned. Wallace described his sense of privilege at the thought of "how few Europeans had ever beheld the perfect little organism I now gazed upon, and how very imperfectly it was still known in Europe." He added a modest disclaimer to the effect that the emotions elicited by such a rarity, in the mind of a yearning naturalist, "require the poetic faculty fully to express them." Insecure about his own writing, Wallace claimed no such faculty but made a try anyway:

> The remote island in which I found myself situated, in an almost unvisited sea, far from the tracks of merchant fleets and navies; the wild, luxuriant tropical forest, which stretched far away on every side; the rude, uncultured savages who gathered round me—all had their influence in determining the emotions with which I gazed upon this "thing of beauty."

Aru itself was part of the magic.

He recalled holding the bird's carcass and contemplating its lineage, throughout "the long ages of the past, during which the successive generations of this little creature had run their course—year by year being born, and living and dying amid these dark and gloomy woods, with no intelligent eye to gaze upon their loveliness; to all appearance such a wanton waste of beauty. Such ideas excite a feeling of melancholy." Mark well that part about "a wanton waste of beauty." At this point in *The Malay Archipelago* the thrust of Wallace's remarks shifted subtly from description to argument. To appreciate that shift, you need a little sense of the historical context.

The book was an act of hindsight, written from field notes after Wallace had returned to England. This was the late 1860s. Charles Darwin, his co-discoverer of natural selection and later his cherished colleague and friend, was now the world's most famous biologist; *The Origin of Species* had gone through its fourth edition and made converts all over the world. But the theory propounded by Darwin and Wallace was still controversial. Creationists, including not just hidebound Anglican bishops but also some secular thinkers, were still vehemently resisting it. One of their arguments involved excessive beauty—that is, the gaudy anatomical modifications found in iridescent-feathered humming-birds, in peacocks, in elaborate orchids, in birds of paradise, and in other extravagantly decorative creatures. The problem with such excessive beauty was that it looked anything but adaptive. Natural selection couldn't explain it, according to the anti-Darwinians. Only God's intervention could. The Duke of Argyll, a reputable intellectual hobbyist, had recently made this argument in a book titled *The Reign of Law*.

As propounded by Argyll, the reign of law in the realm of nature entailed an omnipotent, intrusive Creator with the mentality of a parlor magician. Yes, Argyll conceded, evolutionary change among species might occur. But natural selection wasn't driving it. God himself was, the tricky devil. God decreed and guided those marvelous modifications. God produced excessive beauty by manipulating the mechanisms of inheritance and evolutionary change. Why did He do it? Good question. Argyll's answer was this: as evidence of His greatness. God produced excessive beauty for the purpose of edifying humanity. Why else, why on earth, could the peacock's tail exist?

It was a hybrid sort of argument: evolution for people who still preferred special creation. Wallace had lately refuted it in a stern review of Argyll's book. Now in *The Malay Archipelago* he answered it again, obliquely, as he remembered the birds of Aru:

> It seems sad that on the one hand such exquisite creatures should live out their lives and exhibit their charms only in these wild, inhospitable regions, doomed for ages yet to come to hopeless barbarism; while on the other hand, should civilized man ever reach these distant lands, and bring moral, intellectual, and physical light into the recesses of these virgin forests, we may be sure that he will so disturb the nicely-balanced relations of organic and inorganic nature as to cause the disappearance, and finally the extinction, of these very beings whose wonderful structure

and beauty he alone is fitted to appreciate and enjoy. This consideration must surely tell us that all living things were *not* made for man.

The point of enduring interest, in this passage, is not his attitude toward Argyll's logic. The point of enduring interest is his attitude toward extinction. Wallace was a century ahead of the wave.

The extinction of species, to him, was not just a form of biological destiny; it could also be contingent and regrettable. It could be an offense against the order and fullness of the world, for which *Homo sapiens* should be held to account. If humans destroy this place and these birds, Wallace said, then humans have transgressed.

In my copy of *The Malay Archipelago*, dog-eared from use and tattooed with marginalia, that passage is neatly outlined in pencil.

174

MY COPY OF *The Malay Archipelago* is an old Dover reprint, warped from tropical travel, though the glue is still good. This book has ridden a few airplanes. If it had its own frequent-flyer account, in fact, it would be entitled to a vacation in Honolulu. I can't recall how many times it has been wrapped in a plastic bag, bound with a rubber band, stuffed into the only flat space in a pack. I've carried it around the world for a number of years, rereading various sections in various locales by the bad light of a hotel-room bulb or a headlamp.

Repeatedly I've gone back to the pencil-framed passage on page 340: Aru and its exquisite, vulnerable birds. I've worried those words the way nervous fingers worry a scab:

> . . . should civilized man ever reach these distant lands, and bring moral, intellectual, and physical light into the recesses of these virgin forests, we may be sure that he will so disturb the nicely-balanced relations of organic and inorganic nature as to cause the disappearance, and finally the extinction . . .

I've left marks there in the margin, notes to myself, paper clips on the page—all in recognition that this passage is somehow axial.

It binds yin to yang. It links the slow pageant of evolution to the speedy juggernaut of human-caused extinction. It connects Wallace's world, and his concerns, to our own.

Aru the spectacular, Aru the remote. Aru the tiny and perishable. In the course of retracing a good deal of Wallace's mental and physical journeying, I've developed a burden of curiosity as to what has become of that place and its birds. He left Dobbo on July 2, 1857. Few biologists have visited since. There's no public airport, scanty boat traffic. Set apart from the modern trade routes by hundreds of miles of gleaming Moluccan ocean, Aru is probably still one of the most inaccessible spots on Earth, barely mentioned in even the literature of island biogeography. Not many Western travelers have gone in, not much information has come out. Wallace foresaw the extinction of its bird-of-paradise species, but he didn't say when. Now I'm bothered by a question that can't be answered with computer models or erudite theory. Has it already happened, or is there still time?

175

AT THE PORT of Ambon, on the island that Wallace knew as Amboyna, in the southern Moluccas, I board a boat. It's the *Kembang Matahari*, a motorized replica of a Buginese schooner, one hundred feet long with sleek lines, a high prow, a small upper deck above vertical ladders, a pair of log masts, and heavy canvas sails for supplemental power and decoration. The captain is a smiling Indonesian named Latif. Under charter to a Dutch company, his *Matahari* is outbound for a pleasure cruise to the islands of Banda, Kei, and a handful of others scattered sparsely between Ambon and New Guinea. Its easternmost point of contact will be Aru. Then, after sailing north for a touch on the New Guinea mainland, assuming all goes as charted, it will return toward Ambon along an elliptical orbit. We'll be three weeks in transit. I've booked my passage through a fax number in Amsterdam, don't ask how many thousand guilders. It's the slow and expensive way to reach Aru. I've chosen it for a sound reason: The cheap and quick way doesn't exist.

The *Matahari* is a comfortable, trim ship, just large enough for two dozen passengers, twenty crewmen, five or six parrots, and a hold full of canned lychees and Bintang beer. The passengers, except for myself, are Dutch tourists on a lark. The crew are amiable, sea-hardy Indonesians. There are also two tour leaders—an expatriate Dutchman named Piet, a Moluccan woman named Saar—and their daily briefings are to be delivered in Dutch, utterly incomprehensible to me. By the end of the first evening I've discovered that the language rings in my ears with a pleasantly musical inscrutability, like the late Beethoven quartets. The briefings, focused on tourist attractions and beach outings, aren't crucial to my needs anyway. What's crucial is knowing when dinner is served and, in due course, which set of islands constitutes Aru. Still, I'm armed again with a pocket dictionary—actually, a brace of them this time, one for Indonesian, one for Dutch—and the multilingual Hollanders seem inclined to show me some friendly indulgence. If all else fails, I can talk with the parrots.

My cabin, behind the wheelhouse on the upper deck, is as big as the packing crate for a piano. The other paying passengers are lodged below deck in a hive of air-conditioned, double-occupancy staterooms. Although my place is tinier, more spartan, and stiflingly hot, it's also private, and therefore will do nicely. It leaves me in closer contact with the crew, the warm sea air, and the great Moluccan sky. I can hold the Beethovian chatter at a distance, watch some astonishing sunsets and some subtle dawns, reread appropriate chapters of Wallace, and wait. If I don't slide off the upper deck and fall over the rail in the blackness of some storm-thrashed midnight, as I tightrope toward the head, I'll be fine.

We sail southeastward across the Banda Sea, making good mileage by night. The first dawn comes as a pearl-shell glow off the left of our bow, backlighting tall cumulus clouds that rise like morels on the horizon. Around midday, in midocean, we circle a solitary sea rock called Suanggi. It's a steep volcanic tooth jutting up out of deep water to a forested crown, inhabited only by frigate birds and boobies. The boobies cruise in high circles, roost on the rocky cliffs, dive for fish. The frigate birds spend their time trying to steal meals from the boobies. Some of the frigates, distracted, tilt into low passes above us, no doubt expecting handouts or garbage. They know about ships. These are long-distance sea birds, capable of soaring hundreds of miles on very little effort and inclined to nest on remote islands. They offer their own sort of biogeographical evidence that already we have pushed our way out into a lonely zone of sea. I haven't seen frigate birds in such number since the Galápagos.

Sailing on, we send flying fish skimming away from our bow. We pass a small island called Run, just a hummock of green on the horizon but notable as the piece of real estate that certain cagey Dutch bargainers acquired from the English, three hundred years ago, in exchange for a less fertile island on the far side of the world: Manhattan. Run, along with other islands hereabouts, was valued in those days for its nutmeg plantations. Nutmeg was a strategic resource, precious even in small quantities, like chrome or titanium, and particularly precious if it could be monopolized. For the Dutch East India Company, getting possession of Run and its nutmeg trees was like buying Park Place if you already own Boardwalk. Cloves and black pepper were also important trade crops, but in this particular corner of the Spice Islands (as they were famed then), nutmeg was preeminent. The southern Moluccas are still marked with the visible remnants of its imperial significance: Dutch forts, Dutch ports, Dutch churches, ramshackle old drying-and-storage barns for the harvested nutmeg, plantation groves now being reconquered by wild vegetation. Not far beyond Run, we kill a few days at anchor near an old nutmeg entrepôt in the Banda Islands, glowered over by a volcano on one side of the bay and a restored Dutch fort on the other. Here I walk into the market to buy durian fruit, the notorious regional delicacy with the rich custardy taste and the disreputable smell (like a sewer, some people say, though to me it seems no worse than an unventilated gym locker in August), and a snacking supply of good local bananas. The very streetlamps of the town sprout out of kitschy concrete bases molded and painted to represent nutmegs. Later I tag along with the tour group on an excursion to an old Dutch missionary church, from the same era as the fort, and stroll among the weathered gravestones of pious Dutch imperialists with two of my shipmates. These two companionable souls—a distinguished Dutch novelist named Max and his wife, Annelies, a formidable woman with a wry sense of humor—have befriended me, providing some cheery mitigation to my outsider status on the *Matahari*. They both feel uncomfortable wandering idly through this place where their countrymen, several centuries earlier, slaughtered thousands of Bandanese natives. In the churchyard, at the fringe of the group, Annelies murmurs to me sardonically about history, national guilt, and tourism. I appreciate her attitude, though I'm preoccupied with my own mission of historical revisitation. From Banda we sail eastward, toward the Kei archipelago, which was Wallace's last port of call before Aru.

Wallace had an agreeable layover in Kei around New Year's of 1857. He hiked in the forest, collected butterflies, and noticed a few interesting birds, including some noisy red lories and a large green pigeon that fed on nutmeg. He saw no birds of paradise, because there were none to be seen. The eighty-mile sea gap between Kei and Aru is also a gap between faunal regions. Kei has its affinity with the oceanic Moluccas, Aru with New Guinea, and so only on Aru will we enter the bird-of-paradise realm.

On Kei I skip the day's programmed divertissement—a welcoming dance staged in one of the villages—because of my constitutional loathing for quasi-traditional welcoming dances performed for white people with moneybelts and cameras. It's now a week since we left Ambon. I've been brushing up on Indonesian phrases but have decided to remain ignorant of Dutch, since the mental solitude that comes with my total incomprehension of table talk compensates some for the lack of physical solitude on a crowded boat. This evidently falls as a cultural insult upon certain of my fellow passengers. They have also noted my bad attitude about group outings. The social chemistry has gone a bit sour—but never mind, I didn't book onto this journey for social chemistry. Happy to evade being danced at, I walk through the village and up onto a forested ridge, gawking at lizards and butterflies. I buy more durian, so weirdly aromatic that not even Max and Annelies care to share it, thanks much. That night, on a medium swell against brisk winds, we cross the Arafura Sea toward our outermost reach: Aru.

A half-hour before dawn. Who could sleep at this moment? I'm watching eagerly from the upper deck when we come within sight of a flashing light, to the east, marking the harbor at Dobbo.

The town has changed some since Wallace saw it in 1857. For one thing, it's now Dobo. For another, above the low rooftops stand two radio towers and a satellite dish. A few rusty cargo ships hang at anchor, and a freighter from Halmahera is docked alongside the town's single pier. But the changes seem minor. The Halmahera freighter is only a metal epigone of a big Buginese prau like the one that carried Wallace before the westerly monsoon. The satellite dish and the towers share the Dobo skyline with palm trees, and the entire town seems to cover little more than the same narrow arm of coral sand as described in *The Malay Archipelago*. Beyond the town, forest. The harbor is still graced with sail-bearing dugouts tacking neatly among the iron ships. I watch one of these tiny praus glide by, heavily laden with a cargo of thatched mats, while a man in the stern uses his paddle as a rudder and a small boy in the bow bails with a clamshell. The pier is crowded with workers, stacked boxes, sacks of rice, and a half-dozen long-haired goats. Shouldering loads, the men hike down the pier toward waterfront warehouses. There are no forklifts to make the work easier. The goats go ashore under their own power.

Here in Dobo I jump ship. With good luck, I'll rejoin the *Matahari* in several days at a small coastal village to the south. If my luck is bad, or just interesting, I'll miss the rendezvous and have to scramble my way back to the world by other means. In that event, I tell the tour leaders Piet and Saar, you can leave my laptop at such-and-such a hotel in Ambon and I'm much obliged for your trouble. A sweet-spirited physician named Kees, from among the passengers, loans me a good map of the Aru cluster, which may prove crucial for finding a route along those meandering channels. I thank him warmly and promise to return the map in two days, if I make the rendezvous.

The cruise ship has been an extravagant convenience, a cushy alternative to the prau on which Wallace hitchhiked, but it's not going to get me any closer than this. Truth is, I'm glad to escape. I say goodbye to Max, Annelies, Kees and his wife, Maryke, and a few of the other Hollanders who've been generous with their good company. Before I climb off the *Matahari*, Piet does me a small, valuable favor. He supplies me with the Indonesian word for bird of paradise: *cenderawasih*.

I scribble it down, not trusting my memory. *Cenderawasih*. I try the pronunciation—four syllables, soft *c*, roll the *r*: "chendhrrawasee." I embrace it like a mantra.

176

FOR AN HOUR I prowl the town. An hour is enough to see almost everything in Dobo: the market, all twenty stalls, rich in fish and meager in vegetables; the single lane of shops, many with Chinese proprietors, dealing mostly in hardware and dry goods; the storefront pearl traders; the dried sea cucumbers, shriveled, black as charred sausages, laid out on burlap and sorted into three size classes, approximating hot dog, bratwurst, and Polish; the open sewers; the neatly swept dirt yards; the plank houses on stilts along the tidal frontage, standing stiff-legged above briny muck; the cemetery, on a little meadow so close to sea level that it's landscaped in mangroves, beneath which the corpses must brine-pickle before they decay; the two-wheeled cart that, in lieu of municipal plumbing, delivers carboys of water from the town well; the green parrots for sale from a furtive back-alley peddler; the starved, mangy, and downtrodden dogs; the bright-faced children in school uniforms, agawk at the sight of a white man; the dried shark fins and the sago cakes offered at curbside; the boutique that sells cassowary-egg souvenirs. This is all interesting in its way but it doesn't get me into the Aru backcountry. I stroll aimlessly, hoping for a bolt of serendipity.

I need a translator. I need my loyal pal Nyoman the Balinese Tailor, but he's at home at his shirt shop in Bali. Finally, in crimped English and baby-talk Indonesian, I start asking shopkeepers whether anyone knows where I might hire a small boat.

Boat? Small boat? *Perahu kecil?* No, sorry, boat not now. Is not. *Tidak.* Oh, very sorry. No no no. But maybe tomorrow is boat, or next week, maybe too. You come back, ya? Maybe then is boat. Small boat, big boat. Very good. No problem. Ya, you want to buy pearl-shell ashtray?

Shuffling from doorway to doorway, I remember Wallace spending a grim month in Dobo while he struggled to get a boat and a crew, thwarted by bad weather and a wave of alarm about pirates. It begins to look like I might be stuck here for some time myself. Before the second hour has passed, though, fool's luck brings me into a warehouse near the pier, where a balding half-Papuan man overhears my voice and climbs down from a bookkeeper's loft to introduce himself as Mr. Gaite. He spells the name for me and then adds, cheeringly, "as in *guide*." He's aglow at the prospect, rare in Dobo, of practicing his English.

A boat? Ah, a boat may be possible, yes, says Mr. Gaite. He knows a man. We will see. He will help me inquire for a boat, yes. Where am I from, he asks, Dutch? No, actually, I'm not from Dutch, I'm from America. Ah, Amerrika, Amerrika, Mr. Gaite has heard of it, yes, he's delighted to make my acquaintance. Amerrika, it is far away, yes? It is that, I say. Mr. Gaite has dropped his day's work like a rancid sea cucumber in order to concern himself with my needs. Ordinarily such solicitude would make me queasy with suspicion, but his face is so open and round and sweet that I'm willing to see where this might go.

Then I speak the magical word: *cenderawasih.*

I want a boat that can cross to the main Aru cluster and take me up one of the channels, I explain, so that I can go into the forest and see cenderawasih. Is that possible?

Cenderawasih! says Mr. Gaite. He arches his eyebrows into the bald zone. He confides to me that he himself *possesses* a cenderawasih, right here in Dobo. Would I like to see it? I say that I'd very much like to see it—and with that we're out of the warehouse and striding along the streets of Dobo, two men on an urgent midmorning mission.

His first name is Joel, as in the Bible, Mr. Gaite tells me. He is a Christian, he volunteers. Aru, it seems, is one more place where the missionaries have long since gotten everyone sorted into churches and mosques, and Mr. Gaite's statement sounds more a cultural avowal than a testimony of faith. I tell him my name—David, also as in the Bible, I note—and we shake hands for the second or third time. As I'm straining to recall the biblical Joel, he presses to know whether I too am a Christian. Yes yes, I fib, shamelessly grateful for any door-opening value to be milked from my fine old Christian name.

Mr. Gaite leads me up a back street, then down a path between houses, then along a plank catwalk to a shack on stilts over the tidal muck. We enter a shadowy room lit only by narrow shafts of sun. The room is bare of all furnishings except for a lurid color print of the cru-cified Christ and, in the far corner, a cage. The cage is a rough coop, slapped together from chicken wire and old boards. Draped over it is a mat. When the mat comes away, a cenderawasih dodges nervously from the floor to a perch and back down. It leaps up to clutch the chicken wire, then again to the perch, again down.

I kneel to see this thing better. It's a resplendent animal suffering the panic of captivity—a male of *Paradisaea apoda*, the greater bird of paradise, in full breeding plumage. The head is a deep satiny yellow; at the throat there's a patch of iridescent green; the breast, the back, and the wings are auburn; and from under the wings emerge long yel-low plumes, which fan out behind like a ceremonial cloak. It looks like a crow dressed for coronation. Not many birds on the planet are more decorative. The Duke of Argyll would certainly have called this excessive beauty, lacking faith as he did in the intricate economy of evolution. The floor of the cage is covered with birdshit and scraps of banana. The bottom half of an aluminum soft-drink can, cut jaggedly, serves as a water trough.

"So pretty," I mutter inanely.

"More pretty if you look at in nature, ya?" says Mr. Gaite. He seems suddenly ambivalent: proud of this magnificent creature but slightly embarrassed that he's dealing in contraband birds.

Ya, I agree. Ya, more pretty in nature. Are there others still living where this one came from?

177

"I'M HAPPY you drink my tea. Some peoples don't drink," says Mr. Gaite.

We have sat in the parlor of his prim little house, not far from the bird shack, beneath a portrait of him and his late wife, while his daughter has served us the glasses of sweet tea. Between sips, we have discussed cenderawasih (including the black-market value of a male in breeding plumage, like his) and birdwing butterflies (in which he has also occasionally dealt) and the condition of the Aru landscape. Ya ya, Mr. Gaite has told me, lots of good forest over there on the main cluster. But who knows what "lots" or "good" may mean to a gentleman who works in a warehouse in Dobo. How much has been cut for timber? How much has been cleared and burned for settlements, gardens, copra plantations, mining operations, and cattle? How badly has the remainder been fragmented? I've got to see for myself.

Mr. Gaite seems relieved that I haven't interrogated him about his purposes with the captive bird. He is also grateful for my questions about his wife, who died just three years ago in January. And the tea itself has been something of a test. "If not drink my tea, you are not my friend," he confides. "Thank you for you drink." With the niceties thus satisfied and a small bond established, he has led me off again through the byways of Dobo. We are going to see a man about a boat.

The man's name is Mr. Samuel. His boat is small but seaworthy, thirty feet long with a boxy cabin, just right for minor cargo chores or pearl diving. It's docked in a slip beside the Samuel house, handy as a car in the garage. Mr. Samuel is a thin, serious Moluccan man with a Chinese wife and a wait-and-see expression who speaks not a word of English, but with Mr. Gaite brokering delicately we discuss destination and price.

Some miles up Manumbai Channel in the Aru interior lies a village called Wakua, I learn. Manumbai Channel is one of those riverlike canyons of salt water that transect the main Aru cluster from west to east. We consult my borrowed map, on which Manumbai Channel shows clearly but the village of Wakua doesn't appear. Mr. Samuel's finger marks the spot: here. It's a tiny village, no more than a few dozen houses perched on the south bank of the channel, some miles inland from the west coast. Mr. Samuel knows Wakua because he was born there. Nearby, just a short distance by canoe and then an easy hike into the forest interior, stands a great tree to which cenderawasih come.

Banyak cenderawasih, ya. Many.

This renowned tree must be one of those sites, I realize, where the males congregate for displaying their plumage before an audience of mate-shopping females. Lekking behavior, biologists call it, and the display site itself is known as a lek. A tree with an accommodating shape—a sparse crown, horizontal limbs on which males can dance back and forth—might serve as a lek for many consecutive generations of *P. apoda.* Since the Aruese have been hunting these birds for centuries, it's not hard to understand why such a tree would be known in the nearest village. If I want to see cenderawasih, says Mr. Samuel, then he can take me to the village of Wakua. Someone else will take me to the tree.

Ya, this is mating season, so the birds should be assembling each day. They dance in the early morning, then again usually in the late afternoon. Ya, he confirms, the forest there is good. No cutting, no burning, not yet. Ya, Mr. Samuel can provide food and coffee. Ya, we can sleep on the boat. His two sons will serve as crew. The asked price is modest. I'm thrilled by the whole prospect, because I know that Manumbai is the same channel Wallace ascended to a village that in those days was called Wanumbai, where he based himself for six crucial weeks. The village of Wakua, as Mr. Samuel has placed it on my map, is within a twitch of where Wanumbai stood on Wallace's. Speaking through Mr. Gaite, I have just one further question. When can we leave?

By late afternoon we're under way. We cruise southeastward across a choppy sea, doused by occasional squalls, while the lowering sun breaks between cumulus mounds to put one lonely shrimp boat into silhouette on the western horizon. The shrimp boat is hauling its nets. Besides it and ourselves, there isn't much traffic. The afternoon light paints a patchy pattern of dun gray and quicksilver onto the water. Mr. Samuel sits on the forward deck, signaling course corrections with a wave of his arm but trusting his number-one son, Johnny, to hold the helm. Johnny is a handsome young man in his early twenties, with flowing black hair and Chinese-Moluccan features. Despite his mildly thuggish appearance—acne scars, a muscle shirt, a black baseball cap to the front of which is riveted a tin plate reading COOL AS ICE—he seems diffident and polite. I've told him my name but he persists in calling me "Meester." The younger brother is more slight, and not quite so cool as ice, though he seems to be more quick-witted and verbal than Johnny. I've asked his name too, but gotten no answer. The smattering of English that Younger Brother commands and the smithereens of Indonesian at my disposal account for most of the spo-

A R U   I S L A N D S

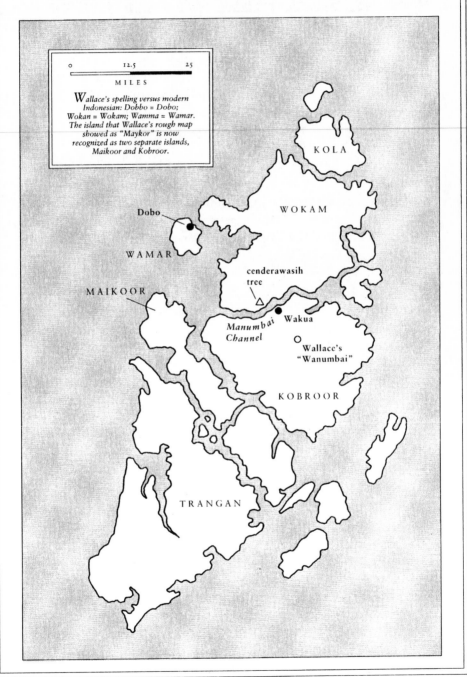

0   12.5   25

MILES

Wallace's spelling versus modern Indonesian: Dobbo = Dobo; Wokan = Wokam; Wamma = Wamar. The island that Wallace's rough map showed as "Maykor" is now recognized as two separate islands, Maikoor and Kobroor.

KOLA

WOKAM

Dobo

WAMAR

cenderawasih tree

MAIKOOR

△

Manumbai Channel

Wakua

○ Wallace's "Wanumbai"

KOBROOR

TRANGAN

ken communication that will pass between me and the Samuel family during our days together. It's not much but it's enough. Mostly I converse with them by pointing, grunting a word or two in a questioning tone, then nodding as though I understand the reply.

After two hours at sea we approach the main Aru cluster, a long shape of land rising to low green hills. From this angle it looks like a single large island. Tracking faithfully on his father's arm signals, Johnny steers toward a point along its coast. Just before sunset the western end of Manumbai Channel opens to us like an estuary. We swing about, riding in on a gentle tide.

The channel banks are thick with vegetation. In some stretches they rise sheer from the water into modest bluffs of light gray limestone, overhung with ferns and other dangling brush. Elsewhere the banks are low, fronting the channel with a wall of mangroves. The channel itself is only two hundred yards wide, and as we putter along, tracing its meanders, the optical illusion takes hold. With the banks paralleling each other at a uniform breadth, the salty strait is confusingly reconfigured to appear as a river.

We could be ascending a tributary of the Zambezi. Or the Orinoco. Or the Mekong. We could be anywhere, but most likely on a jungle watercourse that drains the interior of some continental landscape. Most likely *not* amid a tiny archipelago in the Arafura Sea. The distinction between ocean and river, the distinction between salt water and fresh, the distinction between island and mainland—they have all gone blurry.

178

FOR A SHORT distance along the channel, there's no sign of humanity. Then we pass a fish trap in the shallows. Just beyond that is a settlement. This isn't Wakua, just another small village, nameless on my map and nameless on Wallace's. It comprises thirty or forty thatched houses, a wooden dock, a few coconut and banana trees planted around the clearing, and a tin-roofed Protestant church, atop the steeple of which stands a wind-vane in the shape of a rooster. Dutch missionaries prefer the rooster icon to a cross, I've been told. Less literal, less gruesome, it's intended to symbolize the clarion cock-a-doodle of God's call. To me it suggests nothing so much as *Paradisaea apoda*, the greater bird of paradise, raising its plumes in lubricious display.

Sunset has come and gone quickly. I have barely noticed it tonight, ignoring the luminous pink and apricot tints on the high western clouds, too engrossed in staring eastward along our channel. The darkness isn't absolute but it's deep. The water catches a faint metallic glint. The tree line along each bank is a lumpy horizon of black, blacker than the starry sky. If there are fish traps or docks along this stretch, I can't see them. If there are villages, no one is burning a lantern. If there are navigational hazards, I couldn't say what or where. "*Mau makan*, meester?" Wanna eat? Ya, that sounds good, I say. Younger Brother serves up a dinner of rice and sardines on the deck. I eat, he eats, Mr. Samuel and Johnny nibble but remain focused on fine-tuning our course up the channel. We don't stop, as I expected we would at nightfall. We don't even slow down. This surprises me, since other small-boat Indonesian skippers of my limited experience—Captain Sultan in Komodoland, for instance—have been adamant about not traveling shallow waters by night. No such caution for Mr. Samuel. Three hours longer we motor upchannel in the dark, guided only by starlight and his memory. The air is cool.

When we dock at Wakua, in late evening, the whole village rouses. Twenty men, looking more Papuan than Moluccan, with broad noses and curly hair and warm wide grins, swarm onto the boat. Mr. Samuel introduces me to a fellow named Peter, an eminent hunter. Peter knows where to find the cenderawasih tree, I'm told. Mr. Samuel and Peter conduct a lengthy conference, of which I catch almost nothing except the name of the bird. Peter seems amenable in principle to the notion of my visit. Then Mr. Samuel turns back to me, and there's a confusing interlude during which he, and then he and Johnny, and then he and Johnny and Younger Brother all pooling their patience and their vocabularies, discreetly and painstakingly attempt to explain something or other, no doubt in perfectly lucid Indonesian, and I try doltishly to figure out what they're saying. My first guesses are wrong. "*Besok?*" Tomorrow we go? Ya ya, tomorrow—but that's not it. "*Besok pagi?*" Tomorrow morning? Ya ya, tomorrow morning, sure—although I haven't yet taken their meaning. "*Pagi-pagi?*" Ya ya ya, *early* tomorrow morning—but that's still beside the point. I try, "*Tidur sekarang?*" This hopeful stab means: Now we all go to bed and get some rest? No. Negatory. The discomfited shaking of heads indicates that we do not go to bed until this other thing has been settled.

Mr. Samuel is hemming and hawing. The men of Wakua are embarrassed by my obtuseness. Finally I get it. I grope for a word in my phrasebook. I say: "*Bayar* Peter?" Should I *pay* Peter? Lamentably crude but, finally, correct—and they forgive me the indelicacy. Ya ya ya ya ya, everyone says gaily. Johnny's pitted face, in particular, lights with relief; he seems glad not to have to be ashamed of me. Fine, yes of course, I will happily recompense Peter for his expert guiding services and for the hospitality of his village in allowing me to enter their territory. The asked price, again, is reasonable. A deal is sealed. And *now* we can all go to bed, ya?

No, not quite yet. I realize that one further concern needs attending. Clumsily I say: "Peter, *mengerti saya mau lihat cenderawasih, tidak memburu cenderawasih, ya?*" Peter does understand, doesn't he, that I want to *see* birds of paradise but I don't want to *hunt* them? Ya ya ya, I'm assured. "*Tidak memburu*," no hunting. He understands.

Johnny is awake again at three A.M. to start yams boiling for breakfast. At four, Mr. Samuel cranks the engine. We cast off from the Wakua dock and, having made a wide turn, motor back down the channel for a couple of miles, with Peter aboard and his dugout canoe towed on a painter behind us. At a spot along the north bank marked by no visible clues—none visible to me in the dark, anyway—we tie up at a mangrove. The sky is now clear and the Milky Way is a bright smear. I sit on the deck, wrapped in a sarong for warmth, drinking sweet coffee, waiting. As daylight leaks in, I watch a small bat working low over the water. For breakfast there's smoked wild pig, rubbery but tasty, and Johnny's bland yams. Then it's time. Peter has his bow and a few iron-tipped arrows, I notice, but presumably he carries those any time he steps into the forest, on the chance that he might see a pig. He and Johnny and Younger Brother and I all climb down into Peter's dugout.

Like most dugout canoes, this one is shallow and light, with almost no freeboard—carefully crafted so as to be astonishingly uncomfortable and unstable for people with stiff joints and pale skin. Foreseeing that much, I have double-wrapped my binoculars in plastic. If I go overboard, I'll try like a maniac to tread water while holding the notebook-and-binoculars packet above my head. At least there are no piranha in the channels of Aru. Peter paddles us up a little mangrove-choked branch, a secondary channel no bigger than a creek, while Younger Brother, in the front, chops limbs out of our way with a machete. Johnny sits just behind me, counterbalancing valiantly. Within a half-mile, the mangroves are too thick and the little channel becomes impassable. We beach the canoe and start hiking.

At first there's no trail. We bushwhack up a hillside toward a ridge. The terrain is limestone, padded with only a thin layer of leaf litter and virtually no soil, so that even on this steep slope we have traction. Peter is barefoot. I break into a sweat as I climb. I savor the landscape. Yes, it's good mature tropical rainforest—as good as Mr. Gaite and Mr. Samuel have promised—graced with epiphytes in the limb crotches, lewd fungi on the deadfalls, dangling lianas, a thick canopy, and a fair number of great trees with buttress roots like the blades on a giant ship's propeller. It has known the machete and maybe the ax, passingly, but it looks to be innocent of the chain saw. Even before we have reached the lekking tree on the second ridge, therefore, my question about Aru has been answered. The sad, dire things that have happened elsewhere, in so many parts of the world—biological imperialism, massive habitat destruction, fragmentation, inbreeding depression, loss of adaptability, decline of wild populations to unviable population levels, ecosystem decay, trophic cascades, extinction, extinction, extinction—haven't yet happened here. Probably they soon will. Meanwhile, though, there's still time. If time is hope, there's still hope.

From some distance above us comes a chorus of squawking. It sounds like a truckload of hysterical geese. "Meester," Johnny alerts me. Here in the depth of the forest, he's still wearing his COOL AS ICE cap. But now even he seems thrilled. *"Sudah, suara cenderawasih!"* Already, he says, it's the song of the cenderawasih.

GLOSSARY

adaptive radiation. The evolutionary splitting of a single lineage into a number of distinct species. Sometimes it's loosely defined so as to be synonymous with *archipelago speciation*. More strictly defined (and as I use it in this book), it differs from archipelago speciation by including the factors of sympatry and competition. In adaptive radiation, a number of closely related species evolve from a single lineage; then their geographical separation (if any) is breached and they find themselves sharing a region of habitat; under circumstances of sympatry, they continue diverging from one another owing to interspecific competition.

allele. One of several possible forms of a given gene.

allopatric speciation. Speciation that occurs under circumstances of geographical separation. In contrast to *sympatric speciation*.

archipelago speciation. Speciation that occurs while a number of populations are geographically isolated on insular patches of habitat within a cluster of such patches. The most obvious sort of context for this process, but not the only sort, is an archipelago of oceanic islands. The result of archipelago speciation is a pattern in which each patch of habitat harbors a single endemic species from the related group. Archipelago speciation is simpler than adaptive radiation, in that it doesn't necessarily entail the subsequent colonizations from one patch to another that would result in sympatry and competition among the species.

area effect. This shows itself, according to MacArthur and Wilson (1967), in the fact that small islands tend to harbor fewer species than large islands. Small islands host fewer immigrations and more extinctions, with the area of each island a determining factor. A related concept is *distance effect*.

autecology. A style of ecological research emphasizing efforts to understand individual organisms or species and their relations with their respective environments. In contrast to *synecology*.

background level of extinctions. The average rate of species extinctions over long stretches of time. Generally it's quite low, relative to the heightened rate during a mass-extinction event like the one at the end of the Permian period, the one at the end of the Cretaceous, or the one that's occurring currently.

biogeography. Come on, you've read the book.

boreal. Native to northern forests.

canonical distribution. Frank Preston's (1962) term for a particular pattern of relative species abundance within a biological community—a pattern that includes some species that are very rare, many species that are fairly abundant, and just a few species that are very abundant. On a coordinate graph with logarithmic calibration, that pattern forms a bell curve. Preston called it canonical because it seemed to be mandated by some sort of natural law.

census population size. The total number of individuals in a population, as determined by counting or estimation. In contrast to *effective population size*.

continental island. An island that lies in the relatively shallow water of a continental shelf, subject therefore to being reconnected with the mainland by a relatively small lowering of sea level. In contrast to *oceanic island*.

deterministic factors. For our purposes, any human activities that can cause the decline of species or populations—such as hunting, trapping, destroying habitat, or applying pesticides. Although deterministic factors are susceptible to prediction and control, they can nevertheless drive either large or small populations to the brink of extinction. Compare with *stochastic factors*.

disharmony. A discrepancy of species representation, either positive or negative, between one region of habitat and another—in particular, between an island and its neighboring mainland.

distance effect. This shows itself, according to MacArthur and Wilson (1967), in the fact that remote islands tend to harbor fewer species than islands closer to a mainland or to other islands. Remote islands see fewer immigrations, with distance a determining factor. A related concept is *area effect*.

ecological naïveté. My own term for the inherent shortage of defensive adaptations often seen in creatures native to islands. It results from their having adapted to ecological circumstances in which predation and competition are minimal. Not to be confused with tameness—as in Karl Angermeyer's iguanas, among whom that tameness has resulted from ecological naïveté plus handouts of rice.

ecosystem decay. Tom Lovejoy's term for the process of species loss from severely fragmented ecosystems.

effective population size. A number derived mathematically, reflecting not just the raw head-count of some population but also the patterns of breeding participation, gene flow, and loss of genetic variation. In contrast to *census population size*.

endemic. Indigenous to a given region. More precisely, an endemic species is one that evolved in and has remained restricted to a particular place. A related concept is *relict*.

extinction vortices. Gilpin and Soulé's (1986) term for what they described as "feedback loops of biological and environmental interactions," which can be detrimental to a small population, sucking it still closer to extinction.

founder principle. When an isolated population is founded by a very small number of pioneering individuals, the new population may turn out to be quite different from the base population from which the pioneers came. Why? One reason is that the sample of genes carried by the few founders may have been less diverse than, or otherwise unrepresentative of, the gene pool of the base population. Ernst Mayr (1942) noted the significance of this fact and labeled it the founder principle.

genetic drift. A type of change in the representation of alleles within a population, resulting from sheer chance, not from evolutionary adaptation. It can be especially significant in small populations.

genetic load. The burden of potentially harmful recessive alleles that a population carries. Being recessive, those alleles remain unexpressed so long as they don't occur homozygously.

genetic revolutions. Ernst Mayr's (1954) theory about drastic spates of evolutionary change that can occur in small populations when they become geographically isolated. The changes, Mayr argued, reflect adaptation not just to a new physical and biological environment but also to a "genetic environment" that has taken new form. Mayr's "genetic environment" is the matrix of interdependent effects linking the alleles of a given gene pool with each other. His theory of genetic revolutions subsumes his founder principle, proposed earlier, but goes beyond it.

hectare. A measure of area equal to a square with 100-meter sides. That converts to 2.47 acres.

heterozygous. Having two different alleles at a given gene locus.

homozygous. Having two identical alleles at a given gene locus.

impoverishment. A relative paucity of species resident in some place. Islands tend to be impoverished relative to continental samples of similar size and with similar physical conditions.

inbreeding depression. The manifest symptoms of too much inbreeding. As harmful recessive alleles begin to occur homozygously among inbred individuals, those alleles are expressed in ways that reduce evolutionary fitness.

interspecific competition. Competition between species.

isolate. As a noun, an isolated parcel of habitat. In contrast to *sample*.

land-bridge island. An island that was formerly connected by dry land to a nearby mainland, until rising water level submerged its connection.

lek. Among certain animal species, a gathering place where males competitively display their studly merits to an audience of discerning females.

local extinction. Extinction of a population in a particular place, though the species to which that population belonged survives elsewhere.

logarithm. A mathematical thing. Never mind.

lognormal. This too is mathematicians' talk. Frankly, I only read their lips.

matinha. Portuguese for "cute little forest."

meiosis. The type of cell division that forms sexual gametes—sperm cells and egg cells. Meiosis is also known as reduction division, because it involves reduction, by half, of the genetic material passed along to each cell. It stands in contrast to *mitosis*, the type of cell division that forms other body cells, involving the full duplication of the genetic material for each cell. Normal nonscientific folk, possibly including you and certainly including me, have trouble remembering which is which, thanks to mandatory memorization of these two concepts in high school biology. My own mnemonic system depends on the fact that the word *meiosis* has been "reduced," from *mitosis*, to the extent of not containing the letter *t*. Of course there's still potential confusion in the fact that *mitosis* has been "reduced" of the letter *e*, but I get past that using brute memory.

metapopulation. A population of populations. That is, a number of populations of one species, occupying a constellation of insular habitat patches, with each patch subject to the chance of local extinction or (if extinct) the chance of recolonization from other patches. Crucial fact: In such a system of insular patches, there is no mainland.

minimum viable population. The smallest population size that seems likely to allow for the population's long-term survival. It's a tricky concept, encompassing societal choices and expectations as well as biological measurements and predictions. For instance, how do we choose to define "likely" and "long-term"?

natural selection. The particular theory of evolutionary change proposed in 1858 by Alfred Russel Wallace. Oh, and by Charles Darwin.

oceanic island. An island that rises from ocean depths, not on a continental shelf and not subject to being reconnected with the mainland by a relatively small lowering of sea level. In contrast to *continental island*.

panmixia. Random mating, and therefore unimpeded mixing of genes, throughout a population.

phyletic evolution. The evolutionary changes that continue to occur between species over long stretches of time—giving rise to the larger differences recognizable among broader taxonomic groups, such as genera, families, and orders. In contrast to *speciation*.

population. A discrete group of localized and interbreeding creatures.

population viability. Is the population numerous enough to maintain itself over the long term, despite fluctuations caused by various forms of bad luck?

population viability analysis. A methodical effort, as first outlined by Gilpin and Soulé (1986), to analyze whether a particular population in a particular set of circumstances is viable.

rare. A species can be rare in two ways. It can be broadly distributed, but at low overall density; or it can be restricted to a small region, or just a few isolated patches, of habitat. In the second sort of case, even if it's relatively abundant in a certain tiny area, it will still be absolutely rare.

relict. Left behind in a given region after extinction elsewhere. A related concept is *endemic*.

reproductive isolation. Two populations are reproductively isolated from each other if they don't interbreed, because of some genetic, behavioral, or ecological form of incompatibility. A related but distinct concept is *geographical isolation*, which is self-explanatory.

Rodoviaria. A bus station in Rio de Janeiro that's not the only place in town, as it turns out, where a person might get mugged.

$S = cA^z$. The species-area equation, almost as familiar to an island biogeographer as $E = mc^2$ is to a physicist. The letter S stands for number of species; the letter A represents area; and, as expert sources explain the equation, z and c are "fitted constants" that vary from one situation to another. If the equation is illustrated by plotting real data from some group of islands on a coordinate graph, using logarithmic calibration on both axes, z is the slope of the curve. Are you comfortable with that explanation? Me neither. For instance, whenever I see the words "fitted constants," I think "fudge factors." The simple essence, for our purposes, is that the number of resident species tends to be higher on large islands, and on large samples, than on small ones. The mathematical details involve consideration of the questions "How *much* higher?" and "Under what circumstances?"

sample. As a noun, a demarcated parcel of habitat that is representative of, and not ecologically isolated from, a larger area of similar, contiguous habitat. In contrast to *isolate*.

slope. In a species-area curve, the extremeness of the correlation between size of area and number of resident species.

special creation. A theory, popular during the early Victorian era, stating that biological diversity derives not from evolution but from many individual acts of creation performed (some of them, rather recently) by a frequently intervening God. This theory portrayed the deity like a retired engineer with a woodworking shop in his basement—always restless, always looking for little projects.

speciation. The process of species formation, in which one species becomes divided into two or more species.

species-area curve. A line on a coordinate graph depicting the correlation—among a group of different-sized parcels of habitat—between the area of each parcel and the number of resident species. The parcels can be either samples or isolates. Every group of parcels will show its own particular curve.

species-area relationship. Whereas species-area curves are particularized, this is a sweeping generalization: Larger areas tend to support more species.

stochastic factors. For our purposes, any random or seemingly random factors that can cause the decline of species or populations—such as drought, disease, lightning-caused fire, volcanic eruption, natural climate change, or aberrations in the gender ratio of offspring. Stochastic factors are effectively accidental. They can push small populations over the brink of extinction. Compare with *deterministic factors*.

subfossil. A remnant bone or bone fragment; different from a fossil in that it's not petrified.

sympatric speciation. Speciation that occurs in a single region—that is, despite a lack of geographical separation between the newly distinct species. In contrast to *allopatric speciation*.

sympatry. The condition of two or more species inhabiting the same region.

synecology. A style of ecological research emphasizing efforts to understand whole communities or ecosystems as integral entities. In contrast to *autecology*.

trophic cascades. Falling ecological dominoes, with each domino a species and each fall an extinction.

turnover. Changes that occur in the roster of resident species for a given place, as some species are lost by extinction while others are added by immigration.

variety. A group of creatures within a species that all share some form of genetic variation distinguishing them—though not unbreachably—from other members of the species.

walckvögel. Dodo.

AUTHOR'S NOTE

THIS IS a book of diagnosis, not of prescription. Some readers, I hope, will want more. I foresee a further question arising in their minds: What can we do about it?

The question is simple, but unfortunately there's no simple answer. If a solution exists to this complicated set of problems—landscape fragmentation, insularization of species, small-population jeopardies, extinction, ecosystem decay—that solution can't be sketched on a thumbnail. Certain mitigating measures offer some promise of helping us avert the future we're presently headed toward, which is a future of soul-withering biological loneliness. But to do justice to the range of those measures, in their scientific and political complexity, would take another book. All I can do here is point toward several useful publications that address various aspects of the "What can we do about it?" question from various angles.

Michael Soulé, in collaboration with Dave Foreman (an affable and clear-headed fellow who happens to have been one of the founders of Earth First!), has lately become involved in something called the Wildlands Project. This audacious initiative is dedicated to rethinking the traditional fragment-by-fragment approach to landscape and species conservation in North America, and to replacing it with an approach emphasizing connectivity. In lieu of our insularized parks, refuges, private reserves, and wilderness areas, the Wildlands Project envisions regional networks of natural landscape, in which core areas would be linked to each other by multiple-use buffer zones and landscape corridors. Major fragments of landscape such as the greater Yellowstone ecosystem (centered on the northwestern corner of Wyoming, with overlap into Montana and Idaho) and the northern Continental Divide ecosystem (including Glacier National Park and the Bob Marshall Wilderness, almost two hundred miles from Yellowstone in the northwestern corner of Montana) would be linked to each other, and to similar fragments in Idaho, as a reticular whole. Another network would connect the remaining fragments of natural landscape throughout southern Appalachia. Still another would encompass the Adirondack region. If the landscape corridors allowed passage of dispersing animals from one core area to another, as intended, the networks might go some way toward promoting population viability. Scientifically, this vision is firmly anchored in the works of MacArthur and Wilson, Jared Diamond, Mike Gilpin, William Newmark, Mark Shaffer, Soulé himself, and others. The political barriers to its realization are huge but maybe not impassable. A preliminary level of detail is offered in a hefty pamphlet published as a special issue of the journal *Wild Earth*, with essays by both Soulé and Foreman. That pamphlet and other information are available through the Wildlands Project, P.O. Box 13768, Albuquerque, NM 87192.

Another approach, very different but not contradictory, has lately begun emerging under the label *community-based conservation*. The first premise of this approach is that the conservation of species and ecosystems must be achieved not just within parks, refuges, and other types of statutorily protected area, but also on the human-occupied rural landscapes that lie outside the boundaries of statutory protection. According to one tally, there are now about eight thousand protected areas worldwide, constituting roughly four percent of the planet's land surface. To concern ourselves only with that four percent is to kiss off the other ninety-six, much of which still supports considerable biological diversity. The second premise of this approach is that human needs can't be ignored. Hunger, poor health, cultural legacy, birth rate, parental devotion, fear, aspiration toward a marginally higher level of security or comfort, and all the other factors that drive humans to cause severe impact on their own landscapes—these are realities that must be addressed by conservationists as well as by development practitioners. For conservation efforts to succeed within human-occupied landscapes, local people must be the proprietors and the managers of those efforts, sharing directly in the tangible benefits. That's the essence of community-based conservation. Among the most articulate and experienced of its proponents is Dr. David Western, a Kenyan ecologist who has worked closely for decades with the Maasai people in the vicinity of Amboseli National Park. A thorough introduction to the subject is *Natural Connections: Perspectives in Community-Based Conservation*, which contains papers originally prepared for a small international workshop on community-based conservation held at Airlie House, in rural Virginia, during the week of October 18, 1993. *Natural Connections*, edited by David Western and R. Michael Wright with assistance from Shirley C. Strum, was published by Island Press. Additionally, a concise overview of the subject is presented in a pamphlet, "The View from Airlie," published by and available from the Liz Claiborne and Art Ortenberg Foundation (650 Fifth Avenue, New York, NY 10019), which convened and sponsored the Airlie event.

These are works of prescription, not of diagnosis. They can lead you to still others. *Conservation for the Twenty-first Century*, also edited by David Western, this time in collaboration with Mary Pearl, and *Essentials of Conservation Biology*, by Richard B. Primack, are authoritative books that describe positive measures to be taken. Michael Soulé's Yellow Book, *Conservation Biology: The Science of Scarcity and Diversity*, includes chapters on restoration ecology and reserve design. And the journal *Conservation Biology* tracks these subjects continually. A well-informed person will have no trouble finding avenues of constructive action.

If you really want to do something, a healthy first step might be to prepare yourself for the cold shock of sacrifice. The number of children you produce, the number of miles you drive, and your yearning for a home in the country (wilderness getaways for Thoreau wanna-bes represent a severe cause of habitat loss and fragmentation in affluent temperate-zone nations) all have their impacts on jeopardized populations of other species and on the cohesiveness of ecosystems.

To despair of the entire situation is another reasonable alternative. But the unsatisfactory thing about despair, in my view, is that besides being fruitless it's far less exciting than hope, however slim.

∽

ACKNOWLEDGMENTS

In 1982 a graphic artist named Kris Ellingsen went to graduate school in evolutionary biology and began coming home with the name Robert MacArthur on her lips. MacArthur, who had been dead then just a decade, hovered like an intellectual archangel over the department of which she had become part. I was (and am) lucky enough to be married to Ms. Ellingsen, and those early dinner-table conversations with her about the man whom Edward O. Wilson has called "the most important ecologist of his generation" were the real beginning of this book. During that period, while I sat in a solitary study writing magazine essays about peculiar animals, she was bringing the revolution in theoretical ecology home from school.

Throughout later years, Kris has remained my first and most trusted scientific consultant and literary sounding board. For all her varied contributions to *The Song of the Dodo* (including the maps), my gratitude is best expressed privately—on a daily basis, but especially on wedding anniversaries.

Two other ferociously bright and patient women have also seen this project along its way, providing crucial support, advocacy, and guidance: Renée Wayne Golden and Maria Guarnaschelli. To call them the book's godmothers would be only half enough; they are also its obstetricians.

The John Simon Guggenheim Memorial Foundation favored me with a fellowship devoted to this project during the year 1988–1989, making possible (among other things) my first trip to Madagascar. For the Foundation's trusting generosity, I'm very grateful. I'm also grateful to Ann Gudenkauf, Thad Holt, Alan Lightman, Barry Lopez, and David Roe for helping make that first year of work possible.

Tom Lovejoy, Ed Wilson, and Mike Gilpin provided important early encouragement. Special thanks are due also to scientific and literary friends who, near the other end of the process, read the entire manuscript and gave me valuable reactions, criticisms, and suggestions: Tim Cahill, Greg Cliburn, Mike Gilpin again, and Bob Holmes. Two other people have my gratitude for support and encouragement beyond description, crucial to this book and every other one I've written: Mary and Will Quammen.

Many busy but generous individuals read portions of the manuscript and offered corrections on matters of scientific, historical, and cultural fact. Exuberant thanks on this account to Bob Beck, Rob Bierregaard, Ted Case, Jared Diamond, Owen Griffiths, Willard Heck, Carl Jones, Tom Lovejoy, Russ Mittermeier, William Newmark, Gordon Rodda, Ilse Schreiner, Eleonore Setz, Mark Shaffer, Daniel Simberloff, Ted Smith, Michael Soulé, P. J. Stephenson, Karen Strier, Ed Wilson, and Pat Wright. Although the criticisms and suggestions of these expert readers were vastly important in bringing me back on line where I had wobbled off, any lingering mistakes are obviously my own responsibility, not theirs. Gloria Thiede transcribed many hours of taped interviews full of abstruse technical terminology and indecipherable scientific names, saving me weeks of frustration. Chuck West performed an equally precious service, mastering a body of rules and numbers far more arcane than island biogeography in order to keep me aright with the IRS. Jenny Dossin of Scribner not only designed the published book but allowed me the privilege of advice and consent. Katya Rice gave her fine eye to the copyediting. Susan Moldow, Roz Lippel, Tysie Whitman, Nan Graham, Sharon Dynak, John Fontana, Maria C. Hald, and many others of the Scribner crew, as well as Mike Jacobs, Larry Norton, Marie Florio, and their colleagues at the umbrella entity, Simon & Schuster, have been valued collaborators. Walton Ford's jacket art captured beautifully the lost, wondrous bird that dances in my mind's eye too. For the glossary, I have drawn especially on the guidance of Art (1993), Brown and Gibson (1983), MacArthur and Wilson (1967), Mayr (1991), Shafer (1990), Steen (1971), and Wilson (1992).

I'm also deeply indebted to the scientists and fieldworkers (some but not all of whom I've already mentioned) who welcomed me into their offices, their field camps, their homes, in some cases their lives; who shared their publications, their food, occasionally even their data; who indulged me to explore their work and their thoughts. Again alphabetically: Bob Anderson, David Baum, Bob Beck, Peter Becker, Rob Bierregaard, Todd Bishop, Doug Bolger, Ted Case, Jodie

Carter, Paul Conry, James Cook, Peggy Cymerys, Patrick Daniels, Jared Diamond, Hugues Ducenne, Tom Fritts, R. Gajeelee, Mike Gilpin, R. H. Green, Owen and Mary Ann Griffiths, Sue Haig, David Hau, Willard Heck, Jon Herron, Roger Hutchings, Jim Johnson, Alison Jolly, Carl Jones, Michael Jones, Valerie Kapos, Dick Knight, Linda Laack, Olivier Langrand, Richard Lewis, Tom Lovejoy, Dave Mattson, Mike McCoid, Andy Mack, Debra Wright Mack, David Maehr, Adina Merenlender, David Meyers, Don Nafus, William Newmark, Martin Nicoll, Deborah Oberdorff, Sheila O'Connor, Mark Pidgeon, Ray Radtkey, Chris Ray, Gordon Rodda, Jungle Jim Ryan, Hal Salwasser, Urs Schroff, Ilse Schreiner, Herbert Schubart, Chris Servheen, Eleonore Setz, Mark Shaffer, Dan Simberloff, Andrew Smith, M. Noor Soetawijaya, Michael Soulé, Patricia Spyer, P. J. Stephenson, Wendy Strahm, Karen Strier, Michael Tewes, Bas van Balen, Bas van Helvoort, Eduardo Veado, David Western, Gary Wiles, Ed Wilson, Summer Wilson, Michael Winslett, Marina Wong, and last again but never least, Pat Wright.

And there were the other good souls whose various acts of trust and generosity helped me along the pathways of travel and research. In addition to those whose contributions are described in the text: Philip and Sue Adlard, Caroline Alexander, Joseph Andrianjaka, Loretta Barrett, Dick Bergsma, Peter Brussard, Victor Bullen, Larry Burke, George Butler, Mark Bryant, Kerry Byrd, Matthew Childs, Liz Claiborne, Doug Crichton, Larry Dew, Andrew Dobson, Uta Henschell, Michael H. Jackson, Mark Jaffe, Amy Kemmerer, Frans Lanting, Lewis Lapham, James D. Lazell, Jr., Les Line, Rod Mast, Bob Mecoy, Pascale Moehrle, Jim Murtaugh, Nick Nichols, Grace O'Connor, Art Ortenberg, Mary Pearl, John Peck, Hallie and Charles Rabenarivo, Emil Rajeriarison, Loret Rasabo, Solomon Razafindratandra, Alison Richard, Bill Roberson, Yvonne Rush, George Schatz, Gary Soucie, Rudy Specht, Effendy Sumardja, Ian Thornton, Simon Tonge, Makarim Wibisono, Todd Wilkinson, Amanda Wright, Steve Zack, and Charles Zerner. And to still other folk who helped me materially and whose names have momentarily and regrettably leaked from my porous memory: thanks and sorry.

The young Malagasy man identified here as Bedo, and known more formally as Joseph Rabeson, falls in a special category; I can't begin to thank him for what he gave to this story because, besides giving the song of the indri, he gave his life. I'm deeply obliged to his family members—his father, Jaosolo Besoa, his brother, Maurice, his sister, Eugenie—for entrusting me with some personal memories and some glimpses of their anger and grief.

Also in a special category is my faithful friend, interpreter, and traveling companion for Indonesian journeys, I. Nyoman Wirata. *Terima kasih*.

SOURCE NOTES

25 *"During the few days which I stayed . . ."* Wallace (1869), p. 155 in the 1962 edition.
25 *"On crossing over to Lombock . . ."* Ibid.
26 *"these islands differ . . ."* Wallace (1880), p. 4 in the 1975 edition.
28 *"I was quite alone . . ."* Wallace (1905), Vol. I, p. 354.
29 *"Having always been interested . . ."* Ibid., pp. 354–355.
29 Journal of Researches into . . . Although the title said "1832 to 1836," the *Beagle* in fact left England on December 27, 1831.
30 *"only yellow and very beautiful . . ."* Quoted in George (1980), p. 82.
31 *"a sort of waterboar . . ."* Ibid., p. 85.
31 *"one of the slowest beastes . . ."* Ibid., p. 84.
31 *"the snout very long . . ."* Ibid., p. 95.
31 *"covered with scales . . ."* Ibid., p. 98.
31 *"birds which they bring over dead . . ."* Ibid., p. 92.
32 *"To compare it to any European animal . . ."* Quoted in Younger (1988), p. 48.
33 *"About 1638, as I walked London streets . . ."* Quoted in Fuller (1988), p. 117.
35 *"If we therefore enquire . . ."* Quoted in Browne (1983), p. 18.
35 *"a kind of living museum . . ."* Ibid.
36 *"proper soil . . . proper climate . . ."* Ibid., p. 19.
38 *"a tiresome, rather mean-spirited . . ."* Moorehead (1987), p. 67.
38 *"miscellaneous low-powered . . ."* J. C. Beaglehole quoted in Moorehead (1987), p. 67.
39 *"Islands only produce a greater or less number . . ."* Forster (1778), p. 169. Also quoted (slightly misquoted) in Browne (1983), p. 35.
40 *"simple and obvious"* Henry Walter Bates; November 19, 1856. Quoted in Marchant (1916), p. 52 in the 1975 edition.
41 *"Every species has come into existence . . ."* Wallace (1855), 1891 edition, p. 6. For reasons known only to himself, Wallace wrote "both in space and time" in his first statement of the law (p. 6 of the 1891 edition) and then "both in time and space" when he echoed it, otherwise verbatim, at the end of the paper (p. 18).
43 Experts on tenrec taxonomy . . . Eisenberg and Gould (1970), pp. 28–30.
43 *"climbers and ricochettors . . ."* Ibid., p. 28.
47 *"Wallace's paper: Laws of Geograph. Distrib . . ."* McKinney (1972), p. 118.
47 *"Wallace, Index Book 1 . . ."* November 28, 1855; in Leonard G. Wilson (1970), p. 3.

48 *"a very ancient island having obtained . . ."* Wallace (1855); in Wallace (1891), p. 9.
48 *"On the other hand, no example . . ."* Ibid.
49 *"The origin of the organic world . . ."* April 29, 1856; in Leonard G. Wilson (1970), p. 60.
54 *"Frogs are not found . . ."* April 13, 1856; ibid., p. 53.
54 *"Darwin finds frogs' spawn . . ."* Ibid.
54 *"I never was in a country where . . ."* February 1854; ibid., p. xxxix.
54 *"It seems to me that many species . . ."* November 17, 1855; ibid., p. xli.
55 *"But I must not run on . . ."* Ibid.
55 *"Such phenomena as are exhibited . . ."* Wallace (1855); in Wallace (1891), pp. 8–9.
56 *"The natural history of these islands . . ."* Darwin (1845), p. 363 in the 1980 edition.
56 *"a little world within itself . . ."* Ibid.
56 *"we seem to be brought somewhat near . . ."* Ibid.
56 *"The Galapagos are a volcanic group . . ."* Wallace (1855); in Wallace (1891), p. 9.
56 *"They must have been first peopled . . ."* Ibid.
57 *"By your letter and even still more . . ."* May 1, 1857; in *The Correspondence of Charles Darwin*, Vol. 6, p. 387.
57 *"This summer will mark the 20th year . . ."* Ibid.
58 *"I have never heard how long . . ."* Ibid.
59 *"I also learnt from him . . ."* Wallace (1905), Vol. I, p. 237.
60 *"This new pursuit . . ."* Ibid.
60 *"The primary causes which have led . . ."* Quoted in Browne (1983), p. 167.
61 *"I have rather a more favourable opinion . . ."* December 28, 1847; quoted in Wallace (1905), Vol. I, p. 254.
62 *"fever-heat of expectation . . ."* Letter to the Mechanics' Institution, Neath, 1849; quoted in Wallace (1905), Vol. I, p. 270.
62 *"On my first walk into the forest . . ."* Ibid.
63 *"We hired one of the heavy iron boats . . ."* October 23, 1848; quoted in Brooks (1984), pp. 21–22.
63 *"Our collections were chiefly made . . ."* Ibid, p. 22.
64 *"Two beautiful Consignments . . ."* Ibid., p. 20.
64 *"I have got thus far . . ."* Ibid., p. 22.
64 *"and in these, Lepidoptera . . ."* Ibid., pp. 22–23.
64 *"Pray write whenever you can . . ."* Ibid., p. 23.
65 *"The Tapajoz here is clear . . ."* Ibid.
65 *"The more I see . . ."* Ibid.
66 *"This encounter pleased me much . . ."* Wallace (1853), p. 166 in the 1889 edition.

67   *"inclined to think it is a mere white variety . . ."*
Ibid., p. 249.

68   "On April 1st we passed . . ." Ibid., p. 251.

68   *"some little fear of the passage . . ."* Ibid., p. 218.

68   *"one of my most valuable . . ."* Ibid., p. 256.

68   *"which, in a small canoe . . ."* Ibid.

69   *"I'm afraid the ship's on fire . . ."* Ibid., p. 271.

70   *"the foremast stood . . ."* Ibid., pp. 273–274.

70   *"During the night I saw several meteors . . ."*
Ibid., p. 275.

70   *"We also saw a flock . . ."* Ibid.

71   *"It was now, when the danger . . ."* Ibid., 277.

71   *"And now everything was gone . . ."* Ibid., p.
278.

71   *"Fifty times since I left Para . . ."* Wallace
(1905), Vol. I, p. 309.

74   *"I have never heard of monkeys swimming . . ."*
Wallace (1853), p. 328 in the 1889 edition.

74   *"sloth monkeys"* Wallace (1854), p. 451.

74   *"During my residence . . ."* Ibid., p. 454.

82   *"During my constant attendance . . ."* Wallace
(1905), Vol. I, p. 326.

82   *"From my first arrival in the East . . ."* Ibid., p.
385.

83   *"the Babirusa stands completely isolated . . ."*
Wallace (1869), p. 213 in the 1962 edition.

83   *"I trembled with excitement . . ."* Ibid., p. 167.

83   *"Whole families and genera are altogether . . ."*
December 1, 1856; quoted in Brooks (1984),
p. 140.

84   *"More important than all these . . ."* Wallace
(1869), p. 309 in the 1962 edition.

84   *"These islands are quite out . . ."* Ibid.

85   *"The trade to these islands . . ."* Ibid.

85   *Cicinnurus regius, a small scarlet species . . .*
Wallace called it *Paradisea regia* in his narrative account of the Aru visit in *The Malay
Archipelago*, but later in the book (in a descriptive chapter on birds of paradise) mentioned its alternate name, which is now
standard: *Cicinnurus regius.* Wallace (1869),
pp. 339 and 426 in the 1962 edition.

85   *"He who has made it . . ."* Ibid., pp. 309–310.

88   *"just wide enough to contain three rows . . ."*
Ibid., p. 327.

89   *"as contented as if I had obtained . . ."* Ibid., p.
328.

89   *"could hardly believe I had really succeeded . . ."*
Ibid.

89   *"it is quite another thing to capture . . ."* Ibid.,
pp. 328–329.

91   *"It soon became apparent . . ."* Wallace (1857),
p. 474.

91   *"Not a man will stir . . ."* Wallace (1869), p.
334 in the 1962 edition.

91   *"an enormous amount of talk . . ."* Ibid., p. 337.

91   *"which repaid me for months . . ."* Ibid., pp.
338–339.

91   *"The greater part of its plumage . . ."* Ibid., p.
339.

92   *"These two ornaments . . ."* Ibid.

92   *"a great prize"* Ibid., p. 341.

92   *"penetrating to every part . . ."* Ibid., p. 353.

93   *"the little village of Wanumbai . . ."* Ibid., p.
347.

94   *"Instead of rats and mice . . ."* Ibid., pp.
356–357.

94   *"After a month's incessant punishment . . ."*
Ibid., p. 353.

94   *"When I crawled down to the river-side . . ."*
Ibid.

94   *"kept prisoner . . . are to be met with . . ."* Ibid.

95   *"I fully intended . . ."* Ibid., p. 360.

95   *"There was a damp stagnation . . ."* Ibid., p.
365.

96   *"The Chinamen killed . . ."* Ibid., p. 367.

96   *"I had made the acquaintance . . ."* Ibid., p 369.

98   *"Such an identity occurs . . ."* Wallace (1857), p.
478.

98   *"they seem portions of true rivers"* Ibid., p. 479.

98   *"by supposing them to have been . . ."* Ibid.

99   *"The general explanation given . . ."* Ibid., p. 480.

100   *"In neither is there any marked dry season . . ."*
Ibid., p. 481.

100   *"If kangaroos are especially adapted . . ."* Ibid.

100   *"We can hardly help concluding . . ."* Ibid.

100   *"In a former Number . . ."* Ibid., p. 482.

101   *"new species have been gradually introduced . . ."*
Ibid.

101   *"The species which at the time . . ."* Ibid.

102   *"extremely glad to hear . . ."* December 22,
1857; in *The Correspondence of Charles Darwin*,
Vol. 6, p. 514.

102   *"You say that you have been somewhat surprised
. . ."* Ibid.

102   *"I have not yet seen your paper . . ."* Ibid.

103   *"The number of species which migrated . . ."*
April 16, 1856; quoted in Leonard G. Wilson
(1970), p. xliv.

103   *"Mr. Wallace['s] introduction . . ."* Ibid., p. xlv.

104   *"With respect to your suggestion . . ."* May 3,
1856; in *The Correspondence of Charles Darwin*,
Vol. 6, p. 100.

104   *"a good talk with Lyell . . ."* May 9, 1856; ibid.,
p. 106.

105   *"This I have begun to do . . ."* Darwin to
William Darwin Fox, June 8, 1856; ibid., pp.
135–136.

105   *"gazing, lost in admiration"* Wallace (1869), p.
328 in the 1962 edition.

106   *"most illogical"* Wallace (1858a); reprinted in
Brooks (1984), p. 152.

107   *"that fact is one of the strongest arguments . . ."*
Ibid.

108   *"This is the hardest story . . ."* Quoted in Brackman (1980), p. 348.

108   *"mutual nobility"* Eiseley (1961), p. 292.

108   *"a monument to the natural generosity . . ."*
Quoted in Beddall (1968), p. 301.

108   *"Never was there a more beautiful example . . ."*
Quoted in Marchant (1916), p. 99 in the 1975
edition.

109   *"a few words of personal statement . . ."* Wallace
(1891), p. 20 in the 1969 edition.

109   *"the question of how changes of species . . ."* Ibid.

109   *"no satisfactory conclusion . . ."* Ibid.

109   *"These checks—war, disease . . ."* Ibid.

110   *"has to-day sent me the enclosed . . ."* In *The Correspondence of Charles Darwin*, Vol. 7, p. 107.
Also, the complete letter is reprinted in
Brooks (1984), p. 262. Although the conventional dating of the letter is June 18, 1858,
the letter itself is headed only "Down [Dar-
win's home] Bromley Kent/ 18th," and
Brooks argues that it was "probably" written
on May 18, 1858.

111   *"Please return me the M.S. . . ."* In *The Correspondence of Charles Darwin*, Vol. 7, p. 107.

111 *"Your words have come true . . ."* Ibid.

112 *"I cared very little whether . . ."* Darwin (1969), p. 124.

112 *"I had absolutely nothing whatever . . ."* January 25, 1859; in *The Correspondence of Charles Darwin*, Vol. 7, p. 240.

112 *"I should be extremely glad now . . ."* June 25, 1858; ibid., p. 117. Also, the complete letter is reprinted in Brooks (1984), pp. 263–264. Brooks indicates that the word "now" had been double-underlined, and that the underlinings of both "extremely" and "now" were omitted from the letter as published in the "authoritative" *Life and Letters of Charles Darwin*, as edited (1887) by Darwin's son Francis. In the more recent (and more reliable?) edition from Cambridge University Press, *The Correspondence of Charles Darwin*, Vol. 7, edited by Burkhardt and Smith, the emphases on those two words do appear.

113 *"Somebody cleaned up . . ."* Quoted in Brackman (1980), p. 348.

114 *"Wallace heartily agreed to"* Himmelfarb (1968), p. 243.

120 *"The fame of the islands was founded . . ."* Moorehead (1971), p. 187.

123 *"so great that they will creepe . . ."* Quoted in Stoddart and Peake (1979), p. 148.

123 *"odious food"* Ibid.

124 *"The Land Turtles are also . . ."* Ibid.

124 *"soupe de tortue . . ."* Ibid., p. 150.

125 *"lande turtles of so huge a bignes . . ."* Ibid.

126 *"it would be possible to live for years . . ."* Fryer (1910), p. 258.

126 *"No plan will effectively prevent . . ."* Quoted in Stoddart and Peake (1979), p. 156.

128 *"It was not until the period . . ."* Mayr (1982), p. 525.

128 *"the laws of life and health"* Desmond and Moore (1991), p. 29.

129 *"The individuals of a genus strike out . . ."* Quoted in Mayr (1982), p. 411.

129 *"Finally, these varieties become constant . . ."* Ibid.

131 *"The formation of a genuine variety . . ."* Ibid., p. 563.

131 *"the formation of a new race . . ."* Ibid.

131 *"With respect to original creation . . ."* Quoted in Sulloway (1979), p. 32.

132 *"I remember well, long ago . . ."* Ibid., p. 31.

132 *"Most Wretched Rubbish"* Ibid., p. 50.

133 *"Species are groups of actually or potentially . . ."* Mayr (1940), quoted in Mayr (1942), p. 120 in the 1964 edition.

136 *"Geographic isolation alone cannot . . ."* Mayr (1942), p. 226 in the 1964 edition.

136 *"It was one of the major theses . . ."* Mayr (1982), p. 565.

136 *"Since 1942 the importance . . ."* Ibid.

140 *the experts offer lists in this vein . . .* In particular I'm alluding to Mark Williamson, whose book *Island Populations* discusses most of the items I include on the Insular Menu. *Island Populations* is an excellent overview of the subject from a biologist's perspective, and I've followed its guidance and its terminology on many points.

144 *"To my surprise I found . . ."* Darwin (1859), p. 354 in the 1979 edition.

145 *"floating islands"* Quoted in Carlquist (1965), p. 17.

145 *"These mats of vegetation . . ."* Ibid.

145 *"similar rafts with trees . . ."* Wallace (1880), p. 74 in the 1975 edition.

145 *"twisted round the trunk of a cedar tree . . ."* Ibid., p. 75.

146 *"It was curious and interesting . . ."* Quoted in Simkin and Fiske (1983), p. 152.

148 *one account puts it again . . .* The two differing accounts are MacArthur and Wilson (1967), p. 45, and Thornton (1984), p. 219, both of them citing original work by K. W. Dammermann.

149 *"Child of Krakatau"* Thornton (1984), p. 223.

153 *"on finding ground beneath her feet . . ."* Quoted in Johnson (1980), p. 396.

154 *"Full grown elephants swim perhaps better . . ."* G. P. Sanderson, quoted in Johnson (1978), p. 218.

154 *"Its feat of riding out the storm . . ."* Quoted in Johnson (1980), p. 398.

155 *"Of the 38 extinct mammalian . . ."* Johnson (1978), p. 213.

156 *"Because an elephant . . ."* Audley-Charles and Hooijer (1973), p. 197.

161 *"Much of the notoriety . . ."* Auffenberg (1981), p. 20.

162 *or anyway it was "reputed" . . .* Ibid.

162 *"in this case we had been deceived . . ."* G. Piazzini, quoted in Auffenberg (1981), p. 22.

162 *"That size is related to predation . . ."* Auffenberg (1981), p. 288.

163 *"the evolution of body size . . ."* Ibid.

163 *"Thus the evolution of large size . . ."* Ibid., p. 289.

168 *"A clue to the evolution of oras . . ."* Diamond (1987), p. 832.

170 *"The vine stopped the youngster . . ."* Auffenberg (1981), p. 320.

175 *"are especially susceptible of reaching the limit . . ."* Foster (1964), p. 235.

175 *"Insular adaptations are likely attained . . ."* Ibid.

177 *"To do science is to search for repeated patterns . . ."* MacArthur (1972), p. 1 in the 1984 edition.

178 Foster's table: Foster (1964), p. 234.

178 Case's equations: Case (1978), pp. 6–7.

194 *"First, not all described species . . ."* Fuller (1988), p. 15.

195 *"a very ancient type of bird . . ."* Wallace (1876), Vol. II, pp. 370–371.

199 *"I think there can be little doubt . . ."* Darwin (1859), p. 175 in the 1979 edition.

200 *"the wingless condition . . ."* Ibid., p. 177.

200 *"For during thousands of successive generations . . ."* Ibid.

202 *"Darwin's explanation was too simple . . ."* Darlington (1943), p. 39.

204 *"Limitation of area often limits . . ."* Ibid., p. 42.

205 *"incurably unsuspicious of man"* Darlington (1957), p. 499.

205 *"tame, stupid, and extremely inquisitive"* Fryer (1910), p. 260.

205 *"extreme tameness"* In the first edition (1839) of the journal he noted only "the tameness of the birds" (p. 475), but in the revised edition of 1845 he added the emphatic "extreme"; p. 383 in the 1980 edition.

205 *"There is not one which will not approach . . ."* Darwin (1839), p. 475.

205 *"A gun is here almost superfluous . . ."* Ibid.
205 *"I often tried . . ."* Ibid., p. 476.
205 *"Turtle-doves were so tame . . ."* Quoted in Darwin (1839), p. 476.
206 *"It is a hideous-looking creature . . ."* Darwin (1839), p. 466.
206 *"minced sea-weed . . . I do not recollect . . ."* Ibid., p. 467.
207 *"there is in this respect . . ."* Ibid., p. 468.
207 *"It swam near the bottom . . ."* Ibid.
207 *"I several times caught this same lizard . . ."* Ibid.
207 *"Perhaps this singular piece . . ."* Ibid.
215 *"the different islands to a considerable extent . . ."* Darwin (1845), p. 379 in the 1980 edition. This point wasn't made so explicitly in the 1839 edition. Between 1839 and 1845, Darwin's comprehension of his Galápagos experience changed significantly—as I'll discuss further in connection with the finches—and that change was reflected, somewhat subtly, in the 1845 revision of his *Beagle* journal. Most of my quotations from the journal, from here on, will derive from the 1845 edition. There were some further (small) revisions for an 1860 edition, which was the last Darwin worked on himself. Darwin's various revisions, as well as his 1845 preface and his 1860 postscript, were incorporated into an edition published by J. M. Dent and Sons as *The Voyage of the "Beagle,"* in 1959, and reprinted in paperback in 1980. That's the version, referred to by me as the 1980 edition, that I've used as the source of the 1845 text.
215 *"My attention was first called to this fact . . ."* Darwin (1845), p. 379 in the 1980 edition.
215 *"a better taste when cooked"* Ibid.
215 *"I did not for some time pay sufficient attention . . ."* Ibid.
216 *"Unfortunately most of the specimens . . ."* Ibid., p. 380.
216 *"My attention was first thoroughly aroused . . ."* Ibid., pp. 379–380.
217 *"Adaptive radiation is the diversification . . ."* Brown and Gibson (1983), pp. 177, 179.
218 *The fourteen subspecies of Galápagos tortoise arose . . .* Older sources, such as Hendrickson (1966) and Thornton (1971), put the number of subspecies at fifteen; but Michael H. Jackson, both in Jackson (1985) and in a recent conversation in my backyard, has argued persuasively that the supposed subspecies on the tiny island of Rábida was illusory (based on one individual animal, probably translocated there by humans) and that fourteen is most likely the correct number.
223 *"In an English garden, finches . . ."* Lack (1947), p. v. in the 1968 edition.
224 *"Just over a hundred years ago . . ."* Ibid.
225 *"These facts origin (especially latter) . . ."* Quoted in Sulloway (1982), p. 23.
226 *"a most singular group of finches . . ."* Darwin (1845), p. 364 in the 1980 edition.
226 *"Seeing this gradation and diversity . . ."* Ibid., p. 365.
226 *"involves the claim that the different forms . . ."* Sulloway (1982), p. 38.
227 *"holds that Darwin's observations . . ."* Ibid.

228 *"Had been greatly struck . . ."* Quoted in Sulloway (1982), pp. 22–23.
229 *"As it turns out, Darwin made . . ."* Sulloway (1982), p. 39.
231 *"the finest example known . . ."* Pratt et al. (1987), p. 295.
232 *"Of all the groups of organisms . . ."* Williamson (1981), p. 168.
238 *"I had just about given up hope . . ."* Wright (1988b), pp. 57–58.
243 *"In fact, the ring of wet and dry forest . . ."* Jolly, Albignac, and Petter (1984), p. 184.
246 *"? ? ? . . ."* Martin (1972), p. 308.
246 *"the situation in Madagascar . . ."* Ibid., p. 316.
246 *"the basis for subsequent specialization . . ."* Ibid., p. 317.
246 *"With the lemurs, it may well be the case . . ."* Ibid.
248 *"We do not know how golden bamboo lemurs . . ."* Glander et al. (1989), p. 122.
252 *"The potentiality for rapid divergent evolution . . ."* Mayr (1942), p. 236 in the 1964 edition.
253 *"a great puzzle"* Mayr (1976), p. 188.
253 *"genetic revolutions . . . perhaps the most original theory . . ."* Ibid.
253 *"genetic environment"* Mayr (1954), reprinted as "Changes of Environment and Speciation," in Mayr (1976), p. 195.
254 *"Indeed, it may have the character . . ."* Ibid., p. 201.
254 *"The physical and biotic environments . . ."* Ibid., p. 199.
255 *"I visualize all major evolutionary novelties . . ."* Ibid., p. 208.
257 *"The species of all kinds which inhabit . . ."* Darwin (1859), p. 379 in the 1979 edition.
262 *"When we held one by the leg . . ."* Quoted in Hachisuka (1953), p. 77.
262 *"The call was never clearly described . . ."* Fuller (1988), p. 120.
265 *"Grey Parrots are also common there . . ."* Quoted in Strickland and Melville (1848), p. 123.
265 *"disgusting bird"* Hachisuka (1953), p. 64.
265 *"even long boiling would scarcely . . ."* Quoted in Greenway (1967), p. 121.
265 *"is reputed more for wonder . . ."* Quoted in Fuller (1988), p. 122.
266 *"the whole crew made an ample meal . . ."* Quoted in Strickland and Melville (1848), p. 15.
266 *"and all that remained over was salted"* Ibid.
266 *"provided with sticks, nets, muskets . . ."* Ibid.
266 *"to be swollen"* Hachisuka (1953), p. 37.
266 *"round heavy lump . . . arse"* Ibid., p. 36.
266 *"sluggard"* H. E. Strickland, in Strickland and Melville (1848), p. 16.
267 *"foolish . . . simple"* Ibid.
267 *"doo-doo"* Hachisuka (1953), pp. 37, 77.
267 *"They are very serene or majestic . . ."* Ibid., p. 51.
267 *"only Turtle-doves and other birds . . ."* Quoted in Strickland and Melville (1848), p. 123.
267 *"lived on Tortoises, Dodos, Pigeons . . ."* Ibid., p. 125.
267 *"In colour they are grey . . ."* Ibid.
268 *"Dodos were hunted mercilessly . . ."* Fuller (1988), p. 122.
269 *"great destruction"* Quoted in Hachisuka (1953), p. 65.

270 *"had the pleasure to see more than 4,000 monkeys . . ."* Ibid.

273 *"Amongst other birds were those . . ."* Quoted in Cheke (1987), p. 38.

277 *"the remaining unknown part . . ."* Quoted in Hughes (1988), p. 47.

278 *"No natives showed themselves . . ."* Hughes (1988), p. 47.

278 *"footprints not ill-resembling . . ."* Quoted in Steven Smith (1981), p. 19.

281 *"Am engaged all the morn . . . ."* Quoted in Guiler (1985), p. 1.

281 *"an animal of truly singular . . ."* Quoted in Steven Smith (1981), p. 19.

281 *"a species perfectly distinct . . ."* Ibid., p. 20.

282 *"there is an animal of the panther kind . . ."* Ibid.

283 *"Where sheep run in large flocks . . ."* Ibid.

283 *"5 shillings for every male Hyena . . ."* Ibid., p. 21.

283 *"When 20 Hyenas have been destroyed . . ."* Ibid., pp. 21–22.

283 *"tiger man"* Guiler (1985), p. 95.

284 *"extremely rare . . . in a very few years . . ."* Quoted in Steven Smith (1981), p. 23.

284 *"When the comparatively small island . . ."* Ibid.

285 *"This final decline, which was very rapid . . ."* Guiler (1985), p. 26.

285 *"All of these factors combined . . ."* Ibid., p. 28.

286 *"to obtain another tiger . . ."* Quoted in Steven Smith (1981), p. 32.

292 *"a particularly clean study . . ."* Diamond (1984b), p. 213.

292 *"Species varied greatly in their resistance . . ."* Ibid., p. 214.

292 *"Carnivores were more susceptible . . ."* Ibid., p. 216.

292 *"Large carnivores were more susceptible . . ."* Ibid.

293 *"Habitat specialists were more susceptible . . ."* Ibid.

293 *"risk of extinction decreased . . ."* Ibid.

296 *"Recent Alleged Sightings of the Thylacine . . ."* Rounsevell and Smith (1982).

297 *"I had gone to sleep . . ."* Quoted in Mooney (1984), p. 177.

304 *"we felt we deserved better luck . . ."* Guiler (1985), p. 158.

305 *"a quite unsuspected abundance"* Ibid., p. 168.

305 *"probably extinct"* Steven Smith (1981), p. 103.

308 *"beyond number or imagination"* Quoted in Schorger (1955) p. 5.

308 *"The Birds most common are wild Pigeons . . ."* Ibid., p. 9.

308 *"Here in the fall, large flocks . . ."* Ibid., p. 6.

308 *"clouds of doves . . ."* Ibid., p. 12.

309 *"Almost every conceivable cause . . ."* Schorger (1955), p. 211.

310 *"Day and night the horrible business . . ."* Etta S. Wilson (1934), p. 166.

311 *"In my youth I was identified with . . ."* Ibid., p. 160.

311 *"After awhile there will be . . ."* Ibid., p. 167.

311 *"There'll be Pigeons as long as . . ."* Ibid., p. 168.

311 *"The passenger pigeon needs no protection . . ."* Quoted in Schorger (1955), p. 225.

311 *"persecution was unremitting until . . ."* Schorger (1955), p. 217.

312 *"The puzzling aspect of the passenger pigeon's demise . . ."* Halliday (1980), p. 157.

312 *"social factors, namely colony size . . ."* Ibid.

316 *"There are no mosquitoes here"* Quoted in Culliney (1988), p. 271.

316 *"Dr. Judd was called upon . . ."* Quoted in Warner (1968), p. 104.

318 *"chief who eats the island"* Culliney (1988), p. 323.

318 *"fly flaps"* Ibid., p. 256.

318 *"grievous afflictions"* Quoted in Culliney (1988), p. 258.

318 *"One may spend hours in them . . ."* Quoted in Warner (1968), p. 102.

319 *"The ohia blossoms as freely . . ."* Ibid.

319 *"The cause of the extinction of the ou . . ."* Ibid., p. 103.

320 *"Oahu can make the melancholy boast . . ."* Ibid.

320 *"brought bird diseases . . ."* Quoted in Culliney (1988), p. 259.

320 *"Any protracted visit to the lowland . . ."* Warner (1968), p. 115.

321 *"We've never seen anything like this . . ."* Quoted in Jaffe (1985).

325 *"Philippine rat snake"* Jaffe (1985).

342 *"Since species abundances depend . . ."* Diamond (1984a), p. 845.

342 *"All five species of the endemic . . ."* Ibid.

346 *"Although Barro Colorado has proven to be . . ."* Terborgh (1974), p. 718.

347 *"excessive densities"* Terborgh and Winter (1980), p. 131.

348 *"However, by 1973, only 13 old . . ."* Temple (1977), p. 885.

348 *"no germination of Calvaria seeds . . ."* Ibid.

348 *"The temporal coincidence . . ."* Ibid.

348 *"In response to intense exploitation . . ."* Ibid.

349 *"The germination and distribution . . ."* Vaughan and Wiehe (1941), p. 137.

350 *"These may well have been the first . . ."* Temple (1977), p. 886.

351 challenged by Anthony S. Cheke . . . Cheke (1987), in A. W. Diamond (1987), pp. 17–19. The tree species *Calvaria major* is now more authoritatively denominated *Sideroxylon grandiflorum*, which is the scientific name that Cheke used. The colloquial name in Mauritius is *tambalacoque*, as used in Owadally (1979).

353 *"With no defences but cunning . . ."* Turnbull (1948), p. 1.

353 *"The Tasmanian aborigines are an extinct people."* Plomley (1977), p. 1.

357 *"to open an intercourse . . ."* Quoted in Ryan (1981), p. 75.

357 *"civilized people"* Ryan (1981), p. 73.

357 *"They were now British subjects . . ."* Ibid.

358 *"The mutilated aborigines were left to drown . . ."* Turnbull (1948), p. 100.

358 *"timid, gentle people . . ."* Ryan (1981), p. 85.

359 *"prevent further atrocities . . . a conciliatory line . . ."* Quoted in Ryan, p. 90.

359 *"They already complain . . ."* Ibid., p. 93.

360 *"black catching"* Ryan (1981), p. 102.

361 *"For three weeks they beat . . ."* Ibid., p. 112.

363 *"The children have witnessed . . ."* Quoted in Ryan, p. 141.

363 *"We should fly . . ."* Ibid., p. 142.

364 *"extirpationist"* Ryan (1981), p. 151.

365 *"forcible removal . . ."* Ibid., p. 153.

365 *"They were shy and sullen . . ."* Quoted in Ryan, p. 165.

366   *"Scarcely had he spoken when . . ."* Ryan (1981),
       pp. 165–166.
366   *"Robinson would in future carry firearms . . ."*
       Ibid., p. 166.
366   *"The entire Aboriginal population . . ."* Quoted
       in Ryan, p. 170.
367   *"protector"* Ryan (1981), p. 192.
368   *"Their tactics had all the marks . . ."* Ibid., p.
       197.
368   *"At the age of fifty he represented . . ."* Ibid., p.
       197.
368   *"no blackfellows to live in . . ."* Quoted in Ryan,
       p. 190.
368   *"the race in a very short period . . ."* Ibid.
370   *"a fine young man, plenty beard . . ."* Ibid., p.
       210.
371   *"king of the Tasmanians"* Ibid., p. 214.
371   *"a poor, drunken piece of flotsam"* Turnbull
       (1948), p. 234.
371   *"I am the last man of my race . . ."* Quoted in
       Ryan (1981), p. 214.
372   *"masses of blood and fat . . ."* Quoted in Turn-
       bull (1948), p. 235.
372   *"Dr. Stokell had a tobacco pouch made . . ."* Ryan
       (1981), p. 217.
372   *"Don't let them cut me up. . . . Bury me . . ."*
       Quoted in Turnbull (1948), p. 236.
373   *"a staunch Methodist . . ."* Julia Clark (1986),
       p. 38.
374   *"Lo! with might runs the man . . ."* Quoted in
       Clark, p. 38.
374   *"The married woman hunts the kangaroo . . ."*
       Quoted in Turnbull (1948), p. 154.
375   *"half-castes"* E.g., as quoted in Ryan (1981), p.
       230.
375   *"the Islanders"* Ibid., p. 251.
375   *"It is still much easier . . ."* Ryan (1981), p. 255.
376   *"The Tasmanian aborigines are an extinct peo-
       ple."* Plomley (1977), p. 1.
377   *"Many causes contributed to the extinction . . ."*
       Ibid., p. 60.
378   *"the Tasmanians were probably never . . ."* Ibid.
380   *"Except for island cases . . ."* Soulé (1983), p.
       124.
381   *"Scores of birds, mammals, and flowering . . ."*
       Ibid., p. 111.
381   *"Now the disease appears to be spreading . . ."*
       Ibid.
385   *"Islands only produce a greater or less number . . ."*
       Forster (1778), p. 169.
387   *"Limitation of area often limits . . ."* Darlington
       (1943), p. 42.
387   *"division of area by 10 divides number of species
       . . ."* Ibid.
388   *"division of area by ten divides the amphibian . . ."*
       Darlington (1957), p. 483 in the 1982 edition.
390   *"If what we have said is correct . . ."* Preston
       (1962), Part II, p. 427.
390   *"a giant in the industry . . ."* Mann and Plum-
       mer (1995), p. 53.
390   *"Preston Laboratories . . ."* E.g., Preston
       (1962), Part I, p. 185.
390   *"American Glass Research . . ."* Ibid., p. 211.
391   *"the Arrhenius equation"* Ibid., p. 200.
391   *"This equation is not given by Arrhenius . . ."*
       Ibid.
393   *"The explanation is that . . ."* Preston (1962),
       Part II, p. 411.

393   *"But on the mainland a small area . . ."* Ibid.
393   *"This is a matter that does not seem . . ."* Ibid.
411   *"I had use of only my left eye . . ."* Wilson
       (1985), p. 466. Wilson has published three
       autobiographical works, each very different
       from the others in form and intent, that
       touch either passingly or substantially on the
       subjects at issue here: *Biophilia* (1984), "In the
       Queendom of the Ants" (1985), and *Natural-
       ist* (1994). I have chosen to quote from the
       first two; the third, *Naturalist*, is a wonderful
       book that appeared several years after my
       visit with Wilson at his lab. Although *Natu-
       ralist* covers some of the same episodes we
       discussed that day—including his last conver-
       sation with Robert MacArthur—for purposes
       here I prefer the informal immediacy of what
       Ed Wilson told me in person.
411   *"I am the last to spot a hawk . . ."* Ibid.
411   *"In a sense, the ants gave . . ."* Ibid., p. 465.
411   *"Get a global view . . ."* Ibid., p. 470.
412   *"It just didn't seem enough . . ."* Ibid., p. 474.
412   *"much of the whole range . . ."* Ibid., p. 475.
412   *"In 1962 Robert H. MacArthur and I . . ."* Wil-
       son (1984), pp. 67–68.
413   *"He was medium tall and thin . . ."* Ibid., p. 68.
413   *"By general agreement MacArthur was . . ."*
       Ibid.
430   *"Dan didn't seem to care for that joke . . ."* But in
       retrospect Dan Simberloff is a very good
       sport. Privately he has confided to me about
       the foot paddles: "They almost killed me! I
       jumped overboard wearing them in the Mi-
       ami Oolite ooze of upper Florida Bay, and
       promptly sank to my waist in mud. I was un-
       able to budge, because of the shoes, and Ed
       and Bill Bossert had to pull me back into the
       boat. One shoe remained forever buried in
       Florida Bay."
430   *History doesn't record Dan Simberloff's shoe size
       . . .* Until now. It's 10-1/2 D, Dan reports.
432   *"He may ask why . . ."* MacArthur (1972), p.
       239 in the 1984 edition.
432   *"is to search for repeated patterns . . ."* Ibid., p. 1.
433   *"showed me how interesting . . ."* Ibid., p. xii.
434   *"Someday I'll pull that out and read it . . ."* He
       did. A heartfelt account of the phone conver-
       sation with MacArthur appears in Wilson's
       Naturalist, written after my intrusion into
       that corner of his private memories.
436   *"will be well worth examination"* Quoted in
       MacArthur and Wilson (1967), p. 3.
436   *"Insularity is moreover a universal feature . . ."*
       MacArthur and Wilson (1967), p. 3.
436   *"Consider, for example, the insular nature . . ."*
       Ibid., pp. 3–4.
439   *"Apparently the present rate of immigration . . ."*
       Brown (1971), p. 477.
440   *"The same principles apply . . ."* MacArthur and
       Wilson (1967), p. 4.
442   *"is being destroyed at a rate such that . . ."* Dia-
       mond (1972b), p. 3203.
443   *"The governments of some tropical countries . . ."*
       Ibid.
443   *"If what we have said is correct . . ."* Preston
       (1962), Part II, p. 427.
444   *"revolutionized"* Simberloff (1974a), p. 163.
444   *"Island biogeography has changed . . ."* Ibid., p. 178.

444 *"scientific revolution"* Diamond (1975), p. 131.
446 *"In cases where one large area . . ."* May (1975), p. 177.
447 *"the methods and the way of thinking . . ."* Terborgh (1975), p. 369.
460 *"In sum, the broad generalizations . . ."* Simberloff and Abele (1976a), p. 286.
460 *"Their reasoning from their assumptions . . ."* Diamond (1976), pp. 1027–1028.
461 *"A refuge system that contained . . ."* Ibid., p. 1028.
461 *"could be detrimental . . ."* Terborgh (1976), p. 1029.
462 *"We regret being cast . . ."* Simberloff and Abele (1976b), p. 1032.
486 *"Because of the similarity . . ."* Zimmerman and Bierregaard (1986), p. 134.
486 *"The inescapable conclusion . . ."* Ibid.
488 *The gray wolf . . . had been present . . . but was now gone.* In early 1995 the wolf returned, by way of a carefully designed program of reintroduction directed by the U.S. Fish and Wildlife Service. Whether a wolf population endures in Yellowstone will depend, in forthcoming months and years, as much on politics and law enforcement as on the factor of insularity.
489 *The red fox was missing . . .* William Newmark's original work—the *Nature* article as derived from his dissertation—has been both criticized and defended in the years since. Recently he has produced a new paper on the same subject, titled "Extinction of Mammal Populations in Western North American National Parks," published in the journal *Conservation Biology* in 1995. The new paper represents a revision of the details of the earlier work but by no means a retraction of its central findings: that dozens of mammal populations have gone extinct from western North American national parks, and that the extinction rate correlates inversely with park area. In mentioning some of the extinctions that Newmark reported in his earlier work, I've included only examples that Newmark also reports, unchanged, in his new paper.
491 *"solely because the parks . . ."* Gleick (1987).
491 *"Several predictions follow . . ."* Newmark (1986), p. 20.
492 *"a mammalian faunal collapse"* Ibid., p. 28.
492 *"If what we have said is correct . . ."* Preston (1962), Part II, p. 427. I wouldn't keep quoting this statement if it hadn't been so presciently and far-reachingly significant.
502 *"grandfather"* Shoumatoff (1988), p. 75.
502 *"little father"* Petter and Peyriéras (1974), p. 41.
502 *"the ancestor"* Harcourt (1990), p. 197.
502 *"may not be cleared by fire . . ."* Quoted in Oxby (1985), p. 49.
506 *"a large membranous laryngeal sac"* Pollock (1986a), p. 226.
507 *"Forests inhabited by* Indri *resound . . ."* Ibid.
507 *"The song consists of a series of calls . . ."* Ibid., pp. 230, 232.
512 *"adequately satisfy the ecological requirements"* Main and Yadav (1971), p. 123.

513 *"When these data are extrapolated . . ."* Ibid., p. 130.
513 *"minimal populations"* Allee et al. (1949), p. 403.
513 *"viable populations"* Moore (1962), p. 369.
513 *"The minimum viable population (defined as . . .)"* Main and Yadav (1971), p. 132.
516 *"systematic pressures"* and *"stochastic perturbations"* Shaffer (1978), p. 8.
516 *"In general, there are four sources . . ."* Ibid.
519 *"A minimum viable population for any given species . . ."* Ibid., p. 12.
525 *"There will be a minimum population size . . ."* Franklin (1980), p. 140.
527 *"coming to a screeching halt"* Soulé (1980), p. 166.
527 *"Their rule of thumb . . ."* Ibid., p. 160.
528 *"the basic rule"* Ibid., p. 161.
528 *"This century will see the end . . ."* Ibid., p. 168.
530 *"Conservation biology is a mission-oriented discipline . . ."* Soulé and Wilcox (1980), p. 1.
530 *"But academic snobbery is no longer . . ."* Ibid., p. 2.
530 *"There is no escaping the conclusion . . ."* Ibid., p. 8.
534 *"protect and improve the forest"* Quoted in Salwasser, Mealey, and Johnson (1984), p. 422.
534 *"provide for diversity of plant and animal communities"* Ibid.
534 *"Fish and wildlife habitat shall be managed . . ."* Ibid.
537 *"The moral of this tale . . ."* Soulé (1987), p. 4.
538 *"The problem that we address in this book . . ."* Ibid., pp. 1–2.
538 *"There are no hopeless cases . . ."* Ibid., p. 181.
540 *"The* SLOSS *(single large or several small) debate . . ."* Soulé and Simberloff (1986), p. 19.
541 *Soulé and Simberloff emphasized keystone species . . .* As thinking on all these subjects has continued to evolve, the concept of keystone species has also, more recently, been criticized—by Soulé himself, among others. See Mills, Soulé, and Doak (1993).
542 *"First, the smaller the population . . ."* Shaffer (1987), 70.
542 *"Does a 95% probability of persistence . . ."* Ibid., p. 84.
542 *"the size and number of current nature reserves . . ."* Ibid.
551 *"This would mean that certain species . . ."* Myers (1980), p. 43.
551 *"We might abandon the Mauritius kestrel . . ."* Ibid.
601 *"lead to the rapid disappearance . . ."* Soulé and Gilpin (1986), p. 1.
605 *"perhaps a few species per million years . . ."* Jablonski (1986), p. 44.
609 *"transports of admiration and delight"* Wallace (1869), p. 339 in the 1962 edition.
609 *"how few Europeans had ever beheld . . ."* Ibid.
609 *"require the poetic faculty . . ."* Ibid.
609 *"The remote island in which I found myself . . ."* Ibid., pp. 339–340.
609 *"the long ages of the past . . ."* Ibid., p. 340.
610 *"It seems sad that on the one hand . . ."* Ibid.

BIBLIOGRAPHY

The editions cited here are those to which I've had access and from which, if at all, I've quoted. Those are not necessarily, in each case, the first edition or the one considered scientifically standard. But I've made several exceptions to that policy. The works of Alfred Russel Wallace and Charles Darwin, in particular, are cited by their dates of first publication; the more recent and more available editions are mentioned parenthetically. In a select few other cases—such as Mayr (1942), Lack (1947), Darlington (1957), and MacArthur (1972)—I have likewise supplied information about both original editions and reprints. The rationale for this inconsistency is that (1) the historical timing of certain publications, especially Wallace's and Darwin's, is important to the story I'm telling, and, on the other hand, (2) if I supplied the same sort of publishing history for all the books mentioned, this bibliography would be considerably longer and more boring than it is, which is already long and boring enough.

Abbott, Ian. 1983. "The Meaning of Z in Species Area Regressions and the Study of Species Turnover in Island Biogeography." *Oikos*, Vol. 41.

Ackerly, David D., Judy M. Rankin-De-Merona, and William A. Rodrigues. 1990. "Tree Densities and Sex Ratios in Breeding Populations of Dioecious Central Amazonian Myristicaceae." *Journal of Tropical Ecology*, Vol. 6.

Adams, Douglas, and Mark Carwardine. 1990. *Last Chance to See*. New York: Ballantine Books.

Adler, Gregory H., Mark L. Wilson, and Michael J. Derosa. 1986. "Influence of Island Area and Isolation on Population Characteristics of *Peromyscus leucopus*." *Journal of Mammalogy*, Vol. 67, No. 2.

Alexander, R. McN. 1983a. "Allometry of the Leg Bones of Moas (Dinornithes) and Other Birds." *Journal of Zoology*, Vol. 200.

———. 1983b. "On the Massive Legs of a Moa (*Pachyornis elephantopus*, Dinornithes)." *Journal of Zoology*, Vol. 201.

Allan, H. H. 1936. "Indigene Versus Alien in the New Zealand Plant World." *Ecology*, Vol. 17, No. 2.

Allee, W. C., Alfred E. Emerson, Orlando Park, Thomas Park, and Karl P. Schmidt. 1949. *Principles of Animal Ecology*. Philadelphia: W. B. Saunders.

Amadon, Dean. 1947a. "Ecology and the Evolution of Some Hawaiian Birds." *Evolution*, Vol. 1.

———. 1947b. "An Estimated Weight of the Largest Known Bird." *Condor*, Vol. 49.

Anderson, Tim J. C., Andrew J. Berry, J. Nevil Amos, and James M. Cook. 1988. "Spool-and-Line Tracking of the New Guinea Spiny Bandicoot, *Echymipera kalubu* (Marsupialia, Peramelidae)." *Journal of Mammalogy*, Vol. 69, No. 1.

Andrewartha, H. G., and L. C. Birch. 1954. *The Distribution and Abundance of Animals*. Chicago: University of Chicago Press.

Andriamampianina, Joseph. 1984. "Nature Reserves and Nature Conservation in Madagascar." In Jolly et al. (1984).

Angerbjörn, Anders. 1985. "The Evolution of Body Size in Mammals on Islands: Some Comments." *American Naturalist*, Vol. 125, No. 2.

Archbold, Richard. 1930. "Bevato: A Camp in Madagascar. Experiences of a Collector in an Island Forest." *Natural History*, Vol. 30, November–December 1930.

Archer, M., ed. 1982. *Carnivorous Marsupials*. Sydney: Royal Zoological Society of New South Wales.

Armbruster, Peter, and Russell Lande. 1993. "A Population Viability Analysis for African Elephant (*Loxodonta africana*): How Big Should Reserves Be?" *Conservation Biology*, Vol. 7, No. 3.

Arnold, E. N. 1979. "Indian Ocean Giant Tortoises: Their Systematics and Island Adaptations." *Philosophical Transactions of the Royal Society of London*, Series B, Vol. 286.

Arredondo, Oscar. 1976. "The Great Predatory Birds of the Pleistocene of Cuba." Translated and amended by Storrs L. Olson. *Smithsonian Contributions to Paleobiology*, No. 27.

Arrhenius, Olof. 1921. "Species and Area." *Journal of Ecology*, Vol. 9, No. 1.

Art, Henry W., general ed. 1993. *The Dictionary of Ecology and Environmental Science*. New York: Henry Holt.

Ashby, Gene. 1987. *Pohnpei, an Island Argosy*. Kolonia, Pohnpei: Rainy Day Press.

Audley-Charles, M. G., and D. A. Hooijer. 1973. "Relation of Pleistocene Migrations of Pygmy Stegodonts to Island Arc Tectonics in Eastern Indonesia." *Nature*, Vol. 241, January 19, 1973.

Auffenberg, Walter. 1980. "The Herpetofauna of Komodo, with Notes on Adjacent Areas." *Bulletin of the Florida State Museum, Biological Sciences*, Vol. 25, No. 2.

———. 1981. *The Behavioral Ecology of the Komodo Monitor*. Gainesville: University Presses of Florida.

Avise, John C. 1990. "Flocks of African Fishes." *Nature*, Vol. 347, October 11, 1990.

Ayala, Francisco J. 1982. *Population and Evolutionary Genetics: A Primer*. Menlo Park, Calif.: Benjamin/Cummings.

Ayres, José Márcio. 1990. "Brazilian Nutcracker Suite." *Wildlife Conservation*, November/December 1990.

Baskin, Yvonne. 1992. "Africa's Troubled Waters." *BioScience*, Vol. 42, No. 7.

Battistini, R. 1972. "Madagascar Relief and Main Types of Landscape." In Battistini and Richard-Vindard (1972).

Battistini, R., and G. Richard-Vindard, eds. 1972. *Biogeography and Ecology in Madagascar*. The Hague: Dr. W. Junk.

Battistini, R., and P. Verin. 1967. "Ecologic Changes in Protohistoric Madagascar." In Martin and Wright (1967).

———. 1972. "Man and the Environment in Madagascar." In Battistini and Richard-Vindard (1972).

Bauer, Aaron M., and Jens V. Vindum. 1990. "A Checklist and Key to the Herpetofauna of New Caledonia, with Remarks on Biogeography." *Proceedings of the California Academy of Sciences*, Vol. 47, No. 2.

Beardmore, John A. 1983. "Extinction, Survival, and Genetic Variation." In Schonewald-Cox et al. (1983).

Becker, Peter. 1975. "Island Colonization by Carnivorous and Herbivorous Coleoptera." *Journal of Animal Ecology*, Vol. 44, No. 3.

Beddall, Barbara G. 1968. "Wallace, Darwin, and the Theory of Natural Selection: A Study in the Development of Ideas and Attitudes." *Journal of the History of Biology*, Vol. 1, No. 2.

———. 1972. "Wallace, Darwin, and Edward Blyth: Further Notes on the Development of Evolution Theory." *Journal of the History of Biology*, Vol. 5, No. 1.

Beehler, Bruce. 1982. "Ecological Structuring of Forest Bird Communities in New Guinea." In Gressitt (1982).

———. 1983. "Frugivory and Polygamy in Birds of Paradise." *Auk*, Vol. 100.

———. 1985. "Conservation of New Guinea Rainforest Birds." In Diamond and Lovejoy (1985).

———. 1987. "Birds of Paradise and Mating System Theory—Predictions and Observations." *Emu*, Vol. 87.

———. 1988. "Lek Behavior of the Raggiana Bird of Paradise." *National Geographic Research*, Vol. 4.

———. 1989. "The Birds of Paradise." *Scientific American*, December 1989.

Beehler, Bruce M., Thane K. Pratt, and Dale A. Zimmerman. 1986. *Birds of New Guinea*. Princeton, N.J.: Princeton University Press.

Bellwood, Peter. 1985. *Pre-History of the Indo-Malaysian Archipelago*. Sydney: Academic Press.

Bengston, Sven-Axel, and Dorete Bloch. 1983. "Island Land Bird Population Densities in Relation to Island Size and Habitat Quality on the Faroe Islands." *Oikos*, Vol. 41.

Beresford, Quentin, and Garry Bailey. 1981. *Search for the Tasmanian Tiger*. Hobart, Tasmania, Aus.: Blubber Head Press.

Bergman, Charles. 1990. *Wild Echoes: Encounters with the Most Endangered Animals in North America*. New York: McGraw-Hill.

Berlin, Isaiah. 1957. *The Hedgehog and the Fox*. New York: Mentor/New American Library.

Berry, A. J., T. J. C. Anderson, J. N. Amos, and J. M. Cook. 1987. "Spool-and-Line Tracking of Giant Rats in New Guinea." *Journal of Zoology*, London, Vol. 213.

Berry, R. J. 1983. "Diversity and Differentiation: The Importance of Island Biology for General Theory." *Oikos*, Vol. 41.

Bibby, Cyril. 1972. *Scientist Extraordinary: The Life and Scientific Work of Thomas Henry Huxley 1825–1895*. New York: St. Martin's Press.

Bierregaard, Richard O., Jr., and Thomas E. Lovejoy. 1989. "Effects of Forest Fragmentation on Amazonian Understory Bird Communities." *ACTA Amazonica*, Vol. 19.

Blanc, Charles P. 1972. "Les Reptiles de Madagascar et des Iles Voisines." In Battistini and Richard-Vindard (1972).

Blanchard, Bonnie M., and Richard R. Knight. 1991. "Movements of Yellowstone Grizzly Bears." *Biological Conservation*, Vol. 58.

Boecklen, William J., and Nicholas J. Gotelli. 1984. "Island Biogeographic Theory and Conservation Practice: Species-Area or Specious-Area Relationships?" *Biological Conservation*, Vol. 29.

Boecklin, William J., and Daniel Simberloff. 1986. "Area-Based Extinction Models in Conservation." In Elliott (1986).

Boersma, P. Dee. 1983. "An Ecological Study of the Galapagos Marine Iguana." In Bowman et al. (1983).

Bourn, David. 1977. "Reproductive Study of Giant Tortoises on Aldabra." *Journal of Zoology*, London, Vol. 182.

Bowler, Peter J. 1989. *Evolution: The History of an Idea*. Berkeley: University of California Press.

———. 1990. *Charles Darwin: The Man and His Influence*. Oxford: Basil Blackwell.

Bowman, Robert I., ed. 1966. *The Galápagos: Proceedings of the Symposia of the Galápagos International Scientific Project*. Berkeley: University of California Press.

Bowman, Robert I., Margaret Berson, and Alan E. Leviton, eds. 1983. *Patterns of Evolution in Galápagos Organisms*. San Francisco: American Association for the Advancement of Science.

Boyce, M. S., ed. 1988. *Evolution of Life Histories: Theory and Patterns from Mammals*. New Haven, Conn.: Yale University Press.

Brackman, Arnold C. 1980. *A Delicate Arrangement: The Strange Case of Charles Darwin and Alfred Russel Wallace*. New York: Times Books.

Brent, Peter. 1981. *Charles Darwin: A Man of Enlarged Curiosity*. New York: W. W. Norton.

Brock, Robert. 1987. "Paradise Gone Awry." *Washington*, Winter 1987.

Brodkorb, Pierce. 1963. "A Giant Flightless Bird from the Pleistocene of Florida." *Auk*, Vol. 80, No. 2.

Brooks, John Langdon. 1984. *Just Before the Origin: Alfred Russel Wallace's Theory of Evolution*. New York: Columbia University Press.

Brown, James H. 1971. "Mammals on Mountaintops: Nonequilibrium Insular Biogeography." *American Naturalist*, Vol. 105, No. 945.

———. 1978. "The Theory of Insular Biogeography and the Distribution of Boreal Birds and Mammals." *Great Basin Naturalist Memoirs*, Vol. 2.

———. 1981. "Two Decades of Homage to Santa Rosalia: Toward a General Theory of Diversity." *American Zoologist*, Vol. 21.

———. 1986. "Two Decades of Interaction Between the MacArthur-Wilson Model and the Complexities of Mammalian Distributions." *Biological Journal of the Linnean Society*, Vol. 28.

Brown, James H., and Arthur C. Gibson. 1983. *Biogeography*. St. Louis: C. V. Mosby.

Brown, James H., and Astrid Kodric-Brown. 1977. "Turnover Rates in Insular Biogeography: Effect of Immigration on Extinction." *Ecology*, Vol. 58.

Brown, James H., and Mark V. Lomolino. 1989. "Independent Discovery of the Equilibrium Theory of Island Biogeography." *Ecology*, Vol. 70.

Brown, W. L., Jr., and E. O. Wilson. 1956. "Character Displacement." *Systematic Zoology*, Vol. 5.

Browne, Janet. 1983. *The Secular Ark: Studies in the History of Biogeography*. New Haven, Conn.: Yale University Press.

Brownlee, Shannon. 1987. "These Are Real Swinging Primates." *Discover*, Vol. 8, April 1987.

Brussard, Peter F. 1985. "The Current Status of Conservation Biology." *Bulletin of the Ecological Society of America*, Vol. 66.

———. 1987. "Meeting Report." *Conservation Biology*, Vol. 1, No. 3.

———. 1988. "Conservation Biology." *Evolution*, Vol. 42, No. 4.

Bryden, W. 1974. "Aborigines." In Williams (1974).

Buckley, R. 1982. "The Habitat-Unit Model of Island Biogeography." *Journal of Biogeography*, Vol. 9.

Bunge, Frederica M., ed. 1983. *Indian Ocean: Five Island Countries*. Area Handbook Series. Washington, D.C.: U.S. Government Printing Office.

Burbidge, Andrew. 1989. *Australian and New Zealand Islands: Nature Conservation Values and Management*. Western Australia: Department of Conservation and Land Management.

Burden, Douglas. 1927. "The Quest for the Dragon of Komodo." *Natural History*, Vol. 27, No. 1, January–February 1927.

Burghardt, Gordon M., and A. Stanley Rand, eds. 1982. *Iguanas of the World: Their Behavior, Ecology, and Conservation*. Park Ridge, N.J.: Noyes.

Burke, Russell L. 1990. "Conservation of the World's Rarest Tortoise." *Conservation Biology*, Vol. 4, No. 2.

Burney, David A., and Ross D. E. MacPhee. 1988. "Mysterious Island: What Killed Madagascar's Large Native Animals?" *Natural History*, Vol. 97, July 1988.

Burton, John A., with the assistance of Vivien G. Burton. 1987. Illustrated by Bruce Pearson. *The Collins Guide to the Rare Mammals of the World*. Lexington, Mass.: Stephen Greene Press.

Buettner-Janusch, John, Ian Tattersall, and Robert W. Sussman. 1975. "History of Study of the Malagasy Lemurs, with Notes on Major Museum Collections." In Tattersall and Sussman (1975).

Calder, J. E. 1875. *Some Account of the Wars, Extirpation, Habits and Etc. of the Native Tribes of Tasmania*. Hobart Town, Tasmania, Aus.: Henn.

Campbell, Douglas Houghton. 1909. "The New Flora of Krakatau." *American Naturalist*, Vol. 43, No. 512.

Carlquist, Sherwin. 1965. *Island Life: A Natural History of the Islands of the World*. Garden City, N.Y.: Natural History Press.

———. 1966. "The Biota of Long-Distance Dispersal, Part 1: Principles of Dispersal and Evolution." *Quarterly Review of Biology*, Vol. 41, No. 3.

———. 1974. *Island Biology*. New York: Columbia University Press.

———. 1982. "The First Arrivals." *Natural History*, Vol. 91, December 1982.

Carpenter, Stephen F., James F. Kitchell, and James R. Hodgson.1985. "Cascading Trophic Interactions and Lake Productivity." *BioScience*, Vol. 35, No. 10.

Carson, Hampton L. 1982. "Hawaii: Showcase of Evolution." *Natural History*, Vol. 91, December 1982.

Carson, Hampton L., and Alan R. Templeton. 1984. "Genetic Revolutions in Relation to Speciation Phenomena: The Founding of New Populations." *Annual Review of Ecology and Systematics*, Vol. 15.

Case, Ted J. 1975. "Species Numbers, Density Compensation, and Colonizing Ability of Lizards on Islands in the Gulf of California." *Ecology*, Vol. 56.

———. 1976. "Body Size Differences Between Populations of the Chuckwalla, *Sauromalus obesus*." *Ecology*, Vol. 57.

———. 1978. "A General Explanation for Insular Body Size Trends in Terrestrial Vertebrates." *Ecology*, Vol. 59, No. 1.

———. 1982. "Ecology and Evolution of the Insular Gigantic Chuckwallas, *Sauromalus hispidus* and *Sauromalus varius*." In Burghardt and Rand (1982).

———. 1983. "Niche Overlap and the Assembly of Island Lizard Communities." *Oikos*, Vol. 41.

———. 1990. "Invasion Resistance Arises in Strongly Interacting Species-Rich Model Competition Communities." *Proceedings of the National Academy of Sciences*, Vol. 87.

Case, Ted J., Douglas T. Bolger, and Adam D. Richman. 1992. "Reptilian Extinctions: The Last Ten Thousand Years." In *Conservation Biology*, P. L. Fiedler and S. K. Jain, eds. New York: Chapman and Hall.

Case, Ted J., and Martin L. Cody, eds. 1983. *Island Biogeography in the Sea of Cortez*. Berkeley: University of California Press.

———. 1987. "Testing Theories of Island Biogeography." *American Scientist*, Vol. 75.

Case, Ted J., Michael E. Gilpin, and Jared M. Diamond. 1979. "Overexploitation, Interference Competition, and Excess Density Compensation in Insular Faunas." *American Naturalist*, Vol. 113, No. 6.

Caufield, Catherine. 1984. *In the Rainforest*. Chicago: University of Chicago Press.

Chambers, Robert. 1844. *Vestiges of the Natural History of Creation*. London: John Churchill. (Facsimile edition: Humanities Press, New York, 1969.)

Chauvet, B. 1972. "The Forests of Madagascar." In Battistini and Richard-Vindard (1972).

Cheke, A. S. 1987. "An Ecological History of the Mascarene Islands, with Particular Reference to Extinctions and Introductions of Land Vertebrates." In A. W. Diamond (1987).

Cheke, Anthony S., Tony Gardner, Carl G. Jones, A. Wahab Owadally, and France Staub. 1984. "Did the Dodo Do It?" Response to Temple. *Animal Kingdom*, February/March 1984.

Christian, Keith A. 1982. "Reproductive Behavior of Galápagos Land Iguanas, *Conolophus pallidus*, on Isla Santa Fe, Galápagos." In Burghardt and Rand (1982).

Clark, Julia. 1986. *The Aboriginal People of Tasmania*. Hobart, Tasmania, Aus.: Tasmanian Museum and Art Gallery.

Clark, Ronald W. 1984. *The Survival of Charles Darwin*. New York: Random House.

Clark, Tim W. 1989. *Conservation Biology of the Black-Footed Ferret (Mustela nigripes)*. Special Scientific Report No. 3. Philadelphia: Wildlife Preservation Trust International.

Clark, Tim W., Gary N. Backhouse, and Robert C. Lacy. 1990. "The Population Viability Assessment Workshop: A Tool for Threatened Species Management." *Endangered Species Update*, Vol. 8, No. 2.

Clark, Tim W., and John H. Seebeck, eds. 1990. *Management and Conservation of Small Populations*. Chicago: Chicago Zoological Society.

Cliff, Andrew, and Peter Haggett. 1984. "Island Epidemics." *Scientific American*, Vol. 250, No. 5.

Clutton-Brock, T. H., ed. 1977. *Primate Ecology: Studies of Feeding and Ranging Behaviour in Lemurs, Monkeys and Apes*. London: Academic Press.

Cody, Martin L. 1977. "Convergent Evolution, Habitat Area, Species Diversity, and Distribution." *National Geographic Society Research Reports*, Vol. 18.

Cody, Martin L., and Jared M. Diamond, eds. 1975. *Ecology and Evolution of Communities*. Cambridge, Mass.: Belknap Press of Harvard University Press.

Coe, Malcolm, and Ian R. Swingland. 1984. "Giant Tortoises of the Seychelles." In Stoddart (1984).

Coimbra-Filho, Adelmar F., and Russell A. Mittermeier, eds. 1981. *Ecology and Behavior of Neotropical Primates*, Vol. 1. Rio de Janeiro: Academia Brasileira de Ciencias.

Cole, Blaine J. 1981. "Colonizing Abilities, Island Size, and the Number of Species on Archipelagoes." *American Naturalist*, Vol. 117, No. 5.

Coleman, Bernard D., Michael A. Mares, Michael R. Willig, and Ying-Hen Hsieh. 1982. "Randomness, Area and Species Richness." *Ecology*, Vol. 63, No. 4.

Colinvaux, Paul A. 1989. "The Past and Future Amazon." *Scientific American*, Vol. 260, No. 5.

Colinvaux, Paul A., and Eileen K. Schofield. 1976. "Historical Ecology in the Galápagos Islands, Part 1: A Holocene Pollen Record from El Junco Lake, Isla San Cristobal." *Journal of Ecology*, Vol. 64.

Conniff, Richard. 1986. "Inventorying Life in a 'Biotic Frontier' Before It Disappears." *Smithsonian*, September 1986.

Connor, Edward F., and Earl D. McCoy. 1979. "The Statistics and Biology of the Species-Area Relationship." *American Naturalist*, Vol. 113, No. 6.

Connor, Edward F., and Daniel Simberloff. 1979. "The Assembly of Species Communities: Chance or Competition?" *Ecology*, Vol. 60, No. 6.

———. 1983. "Interspecific Competition and Species Co-Occurrence Patterns on Islands: Null Models and the Evaluation of Evidence." *Oikos*, Vol. 41.

Conrad, Peter. 1989. *Behind the Mountain: Return to Tasmania*. New York: Poseidon Press.

Conry, Paul J. 1988. "High Nest Predation by Brown Tree Snakes on Guam." *Condor*, Vol. 90.

Cox, C. Barry, and Peter D. Moore. 1988. *Biogeography: An Ecological and Evolutionary Approach*. Oxford: Blackwell Scientific Publications.

Cracraft, Joel. 1974. "Phylogeny and Evolution of the Ratite Birds." *Ibis*, Vol. 116, No. 4.

Cranbrook, the Earl of. 1981. "The Vertebrate Faunas." In Whitmore (1981).

———. 1987. *Riches of the Wild: Land Mammals of South-East Asia*. Singapore: Oxford University Press.

Cree, Alison, and Charles Daugherty. 1990. "Tuatara Sheds Its Fossil Image." *New Scientist*, Vol. 128, October 20, 1990.

Crome, F. H. J. 1976. "Some Observations on the Biology of the Cassowary in Northern Queensland." *Emu*, Vol. 76.

Crome, F. H. J., and L. A. Moore. 1988. "The Cassowary's Casque." *Emu*, Vol. 88.

Crosby, Alfred W. 1986. *Ecological Imperialism: The Biological Expansion of Europe, 900–1900*. Cambridge: Cambridge University Press.

Crowell, Kenneth L. 1983. "Islands—Insight or Artifact?: Population Dynamics and Habitat Utilization in Insular Rodents." *Oikos*, Vol. 41.

Culliney, John L. 1988. *Islands in a Far Sea: Nature and Man in Hawaii*. San Francisco: Sierra Club Books.

Culver, David C. 1970. "Analysis of Simple Cave Communities, Part 1: Caves as Islands." *Evolution*, Vol. 24.

Curtis, John T. 1956. "The Modification of Mid-latitude Grasslands and Forests by Man." In *Man's Role in Changing the Face of the Earth*, ed. William L. Thomas. Chicago: University of Chicago Press.

Cushing, John, Marla Daily, Elmer Noble, V. Louise Roth, and Adrian Wenner. 1984. "Fossil Mammoths from Santa Cruz Island, California." *Quaternary Research*, Vol. 21.

Dammerman, K. W. 1922. "The Fauna of Krakatau, Verlaten Island and Sebesy." *Treubia*, Vol. 111, No. 1.

Darlington, Philip J., Jr. 1943. "Carabidae of Mountains and Islands: Data on the Evolution of Isolated Faunas, and on Atrophy of Wings." *Ecological Monographs*, Vol. 13, No. 1.

———. 1957. *Zoogeography: The Geographical Distribution of Animals*. New York: John Wiley and Sons. (Reprint edition: Robert E. Krieger, Malabar, Fla., 1982.)

———. 1965. *Biogeography of the Southern End of the World*. Cambridge, Mass.: Harvard University Press.

———. 1971. "Carabidae on Tropical Islands, Especially the West Indies." In Stern (1971).

Darwin, Charles. 1839. *Journal of Researches into the Geology and Natural History of the Various Countries Visited by H.M.S. Beagle, Under the Command of Captain Fitzroy, R.N. from 1832 to 1836*. London: Henry Colburn. (Facsimile edition: Hafner, New York, 1952. The revised edition of 1845, further revised for an 1860 edition, has also been reprinted as *The Voyage of the "Beagle."* London: J. M. Dent and Sons, 1980.)

———. 1858. "Extract from an Unpublished Work on Species, by C. Darwin, Esq., Consisting of a Portion of a Chapter Entitled, 'On the Variation of Organic Beings in a State of Nature; on the Natural Means of Selection; on the Comparison of Domestic Races and True Species.' " *Journal of the Proceedings of the Linnean Society (Zoology)*, Vol. 3, August 20, 1858. (Read to the Society, along with Wallace's contribution, on July 1, 1858.)

———. 1859. *On the Origin of Species by Means of Natural Selection, or the Preservation of Favoured Races in the Struggle for Life*. London: J. Murray. (Reprint edition: Avenel Books, New York, 1979.)

———. 1969. *The Autobiography of Charles Darwin*. (Edited, from a posthumous manuscript, by Nora Barlow.) New York: W. W. Norton.

———. 1990, 1991. *The Correspondence of Charles Darwin*, Vol. 6, 1856–1857; Vol. 7, 1858–1859. Edited by Frederick Burkhardt and Sydney Smith. Cambridge: Cambridge University Press.

Darwin, Francis, ed. 1892. *Charles Darwin: His Life Told in an Autobiographical Chapter, and in a Selected Series of His Published Letters*. London: John Murray.

———. 1909. *The Foundations of the Origin of Species*. New York: NYU Press, 1987.

Davies, David. 1973. *The Last of the Tasmanians*. Sydney: Shakespeare Head Press.

Dawson, William R., George A. Bartholomew, and Albert F. Bennett. 1977. "A Reappraisal of the Aquatic Specializations of the Galápagos Marine Iguana (*Amblyrhynchus cristatus*)." *Evolution*, Vol. 31.

Day, David. 1989. *The Encyclopedia of Vanished Species*. Hong Kong: McLaren.

De Beer, Sir Gavin. 1965. *Charles Darwin: A Scientific Biography*. New York: Anchor Books.

De Vos, Antoon, Richard H. Manville, and Richard G. Van Gelder. 1956. "Introduced Mammals and Their Influence on Native Biota." *Zoologica*, Vol. 41, No. 19.

De Vries, Tj. 1984. "The Giant Tortoises: A Natural History Disturbed by Man." In Perry (1984).

Desmond, Adrian, and James Moore. 1991. *Darwin*. New York: Warner Books.

Dewar, Robert E. 1984. "Extinctions in Madagascar: The Loss of the Subfossil Fauna." In Martin and Klein (1984).

Dewsbury, D. A., ed. 1985. *Leaders in the Study of Animal Behavior: Autobiographical Perspectives*. Lewisburg, Pa.: Bucknell University Press.

Diamond, A. W. 1981a. "The Continuum of Insularity: The Relevance of Equilibrium Theory to the Conservation of Ecological Islands." *African Journal of Ecology*, Vol. 19.

———. 1981b. "Reserves as Oceanic Islands: Lessons for Conserving Some East African Montane Forests." *African Journal of Ecology*, Vol. 19.

Diamond, A. W., ed., with the assistance of A. S. Cheke and Sir H. F. I. Elliott. 1987. *Studies of Mascarene Island Birds*. Cambridge: Cambridge University Press.

Diamond, A. W., and T. E. Lovejoy, eds. 1985. *Conservation of Tropical Forest Birds*. Technical Publication No. 4. Cambridge: International Council for Bird Preservation.

Diamond, Jared M. 1966. "Zoological Classification System of a Primitive People." *Science*, Vol. 151, March 4, 1966.

———. 1969. "Avifaunal Equilibria and Species Turnover Rates on the Channel Islands of California." *Proceedings of the National Academy of Sciences*, Vol. 64.

———. 1970. "Ecological Consequences of Island Colonization by Southwest Pacific Birds, Part 1: Types of Niche Shifts." *Proceedings of the National Academy of Sciences*, Vol. 67, No. 2.

———. 1971. "Comparison of Faunal Equilibrium Turnover Rates on a Tropical Island and a Temperate Island." *Proceedings of the National Academy of Sciences*, Vol. 68, No. 11.

———. 1972a. *Avifauna of the Eastern Highlands of New Guinea*. Cambridge, Mass.: Nuttall Ornithological Club.

———. 1972b. "Biogeographic Kinetics: Estimation of Relaxation Times for Avifaunas of Southwest Pacific Islands." *Proceedings of the National Academy of Sciences*, Vol. 69, No. 11.

———. 1973. "Distributional Ecology of New Guinea Birds." *Science*, Vol. 179, February 23, 1973.

———. 1974a. "Colonization of Exploded Volcanic Islands by Birds: The Supertramp Strategy." *Science*, Vol. 184, May 17, 1974.

———. 1974b. "Relaxation and Differential Extinction on Land-Bridge Islands: Applications to Natural Preserves." *Proceedings of the 16th International Ornithological Congress*.

———. 1975a. "The Island Dilemma: Lessons of Modern Biogeographic Studies for the Design of Natural Reserves." *Biological Conservation*, Vol. 7.

———. 1975b. "Assembly of Species Communities." In Cody and Diamond (1975).

———. 1976. "Island Biogeography and Conservation: Strategy and Limitations." First in a grouped set of responses to Simberloff and Abele (1976). *Science*, Vol. 193, September 10, 1976.

———. 1977. "Continental and Insular Speciation in Pacific Land Birds." *Systematic Zoology*, Vol. 26, No. 3.

———. 1978. "Niche Shifts and the Rediscovery of Interspecific Competition." *American Scientist*, Vol. 66.

———. 1981. "Birds of Paradise and the Theory of Sexual Selection." *Nature*, Vol. 293, September 24, 1981.

———. 1984a. "Historic Extinctions: A Rosetta Stone for Understanding Prehistoric Extinctions." In Martin and Klein (1984).

———. 1984b. " 'Normal' Extinctions of Isolated Populations." In Nitecki (1984).

———. 1984c. "Possible Effects of Unrestricted Pesticide Use on Tropical Birds." *Nature*, Vol. 310, August 1984.

———. 1986. "Biology of Birds of Paradise and Bowerbirds." *Annual Review of Ecology and Systematics*, Vol. 17.

———. 1987. "Did Komodo Dragons Evolve to Eat Pygmy Elephants?" *Nature*, Vol. 326, April 30, 1987.

———. 1988. "Founding Fathers and Mothers." *Natural History*, June 1988.

———. 1989. "This-Fellow Frog, Name Belong-Him Dakwo." *Natural History*, April 1989.

———. 1991. "World of the Living Dead." *Natural History*, September 1991.

Diamond, Jared M., K. David Bishop, and S. van Balen. 1987. "Bird Survival in an Isolated Javan Woodland: Island or Mirror?" *Conservation Biology*, Vol. 1, No. 2.

Diamond, Jared M., and Michael E. Gilpin. 1982. "Examination of the 'Null' Model of Connor and Simberloff for Species Co-Occurrences on Islands." *Oecologia*, Vol. 52.

———. 1983. "Biogeographic Umbilici and the Origin of the Philippine Avifauna." *Oikos*, Vol. 41.

Diamond, Jared M., Michael E. Gilpin, and Ernst Mayr. 1976. "Species-Distance Relation for Birds of the Solomon Archipelago, and the Paradox of the Great Speciators." *Proceedings of the National Academy of Sciences*, Vol. 73, No. 6.

Diamond, Jared M., and Robert E. May. 1976. "Island Biogeography and the Design of Natural Reserves." In May (1976).

Diamond, Jared M., and C. Richard Veitch. 1981. "Extinctions and Introductions in the New Zealand Avifauna: Cause and Effect?" *Science*, Vol. 211, January 30, 1981.

Dickerson, Roy E. 1928. *Distribution of Life in the Philippines*. Manila: Bureau of Printing.

Dobzhansky, Theodosius, and Olga Pavlovsky. 1957. "An Experimental Study of Interactions Between Genetic Drift and Natural Selection." *Evolution*, Vol. 11.

Dorst, Jean. 1972. "The Evolution and Affinities of the Birds of Madagascar." In Battistini and Richard-Vindard (1972).

Duffey, E., and A. S. Watt, eds. 1971. *The Scientific Management of Animal and Plant Communities for Conservation*. Oxford: Blackwell Scientific Publications.

Durrell, Gerald. 1977. *Golden Bats and Pink Pigeons*. New York: Touchstone/Simon & Schuster.

Dworski, Susan. 1987. "Mystery and Magic in Madagascar." *Islands*, September 10, 1987.

East, R. 1981. "Species-Area Curves and Populations of Large Mammals in African Savanna Reserves." *Biological Conservation*, Vol. 21.

Edgar, R. 1983. "Spotted-tailed Quoll." In Strahan (1983).

Ehrenfeld, David W. 1970. *Biological Conservation*. New York: Holt, Rinehart and Winston.

———. 1986. "Life in the Next Millennium: Who Will Be Left in the Earth's Community?" In Kaufman and Mallory (1986).

———. 1987. "History of the Society for Conservation Biology: How and Why We Got Here." *Conservation Biology*, Vol. 1, No. 1.

Ehrlich, Paul R. 1986. "Extinction: What Is Happening Now and What Needs to Be Done." In Elliott (1986).

Ehrlich, Paul, and Anne Ehrlich. 1981. *Extinction: The Causes and Consequences of the Disappearance of Species*. New York: Random House.

Ehrlich, Paul R., Dennis D. Murphy, and Bruce A. Wilcox. 1988. "Islands in the Desert." *Natural History*, October 1988.

Ehrlich, Paul R., and Edward O. Wilson. 1991. "Biodiversity Studies: Science and Policy." *Science*, Vol. 253, August 16, 1991.

Eiseley, Loren C. 1959. "Alfred Russel Wallace." *Scientific American*, Vol. 200, No. 2.

———. 1961. *Darwin's Century: Evolution and the Men Who Discovered It*. Garden City, N.Y.: Anchor Books.

Eisenberg, John F. 1981. *The Mammalian Radiations: An Analysis of Trends in Evolution, Adaptation, and Behavior*. Chicago: University of Chicago Press.

Eisenberg, John F., and Edwin Gould. 1970. *The Tenrecs: A Study in Mammalian Behaviour and Evolution*. Smithsonian Contributions to Zoology, No. 27. Washington, D.C.: Smithsonian Institution Press.

———. 1984. "The Insectivores." In Jolly et al. (1984).

Eldredge, Niles. 1985. *Time Frames: The Rethinking of Darwinian Evolution and the Theory of Punctuated Equilibria*. New York: Touchstone/Simon & Schuster.

Elliott, David K., ed. 1986. *Dynamics of Extinction*. New York: John Wiley and Sons.

Elton, Charles S. 1972. *The Ecology of Invasions by Animals and Plants*. London: Chapman and Hall.

Emmons, Louise H. 1984. "Geographic Variation in Densities and Diversities of Non-flying Mammals in Amazonia." *Biotropica*, Vol. 16, No. 3.

Endler, John A. 1982. "Pleistocene Forest Refuges: Fact or Fancy?" In Prance (1982).

Engbring, John, and Thomas H. Fritts. 1988. "Demise of an Insular Avifauna: The Brown Tree Snake on Guam." *Transactions of the Western Section of the Wildlife Society*, Vol. 24.

Erwin, Terry L. 1991. "An Evolutionary Basis for Conservation Strategies." *Science*, Vol. 253, August 16, 1991.

Faeth, Stanley H., and Thomas C. Kane. 1978. "Urban Biogeography." *Oecologia*, Vol. 32.

Fahrig, Lenore, and Gray Merriam. 1994. "Conservation of Fragmented Populations." *Conservation Biology*, Vol. 8, No. 1.

Ferry, Camille, Jacques Blondel, and Bernard Frochot. 1974. "Plant Successional Stage and Avifaunal Structure on an Island." *Proceedings of the 16th International Ornithological Congress*.

Fichman, Martin. 1977. "Wallace: Zoogeography and the Problem of Land Bridges." *Journal of the History of Biology*, Vol. 10, No. 1.

———. 1981. *Alfred Russel Wallace*. Boston: Twayne.

Fleet, Harriet. 1986. *The Concise Natural History of New Zealand*. Auckland, New Zealand: Heinemann.

Flessa, Karl W. 1975. "Area, Continental Drift and Mammalian Diversity." *Paleobiology*, Vol. 1.

Fonseca, Gustavo A. B. da. 1985. "The Vanishing Brazilian Atlantic Forest." *Biological Conservation*, Vol. 34.

Fooden, Jack. 1972. "Breakup of Pangaea and Isolation of Relict Mammals in Australia, South America, and Madagascar." *Science*, Vol. 175, February 25, 1972.

Forster, John Reinold [Johann Reinhold]. 1778. *Observations Made During a Voyage Round the World, on Physical Geography, Natural History, and Ethic Philosophy*. London: G. Robinson.

Fossey, Dian. 1983. *Gorillas in the Mist*. Boston: Houghton Mifflin.

Foster, J. Bristol. 1964. "Evolution of Mammals on Islands." *Nature*, Vol. 202, April 18, 1964.

Frankel, O. H., and Michael E. Soulé. 1989. *Conservation and Evolution*. (The Green Book; first published 1981.) Cambridge: Cambridge University Press.

Franklin, Ian Robert. 1980. "Evolutionary Change in Small Populations." In Soulé and Wilcox (1980).

Fretwell, Stephen D. 1975. "The Impact of Robert MacArthur on Ecology." *Annual Review of Ecology and Systematics*, Vol. 6.

Fritts, Thomas H. 1983. "Morphometrics of Galapagos Tortoises: Evolutionary Implications." In Bowman et al. (1983).

———. 1988. "The Brown Tree Snake, *Boiga irregularis*, A Threat to Pacific Islands." Biological Report 88(31). Washington, D.C.: Fish and Wildlife Service, U.S. Department of the Interior.

Fritts, Thomas H., and Michael J. McCoid. 1991. "Predation by the Brown Tree Snake *Boiga irregularis* on Poultry and Other Domesticated Animals in Guam." *Snake*, Vol. 23.

Fritts, Thomas H., Michael J. McCoid, and Robert L. Haddock. 1990. "Risks to Infants on Guam from Bites of the Brown Tree Snake (*Boiga irregularis*)." *American Journal of Tropical Medical Hygiene*, Vol. 42.

———. 1994. "Symptoms and Circumstances Associated with Bites by the Brown Tree Snake (Colubridae: *Boiga irregularis*) on Guam." *Journal of Herpetology*, Vol. 28, No. 1.

Fritts, Thomas H., Norman J. Scott, Jr., and Julie A. Savidge. 1987. "Activity of the Arboreal Brown Tree Snake (*Boiga irregularis*) on Guam as Determined by Electrical Outages." *Snake*, Vol. 19.

Fryer, Geoffrey, and T. D. Iles. 1972. *The Cichlid Fishes of the Great Lakes of Africa: Their Biology and Evolution*. Neptune, N.J.: T.F.H. Publications.

Fryer, J. C. F. 1910. "The South-west Indian Ocean." *Geographical Journal*, Vol. 37, No. 3.

Fuller, Errol. 1988. *Extinct Birds*. New York: Facts on File.

Futuyma, Douglas J. 1987. "On the Role of Species in Anagenesis." *American Naturalist*, Vol. 130, No. 3.

Ganzhorn, Jörg U. 1988. "Food Partitioning Among Malagasy Primates." *Oecologia*, Vol. 75.

———. 1989. "Primate Species Separation in Relation to Secondary Plant Chemicals." *Human Evolution*, Vol. 4, No. 2.

Ganzhorn, Jörg U., and Joseph Rabeson. 1986. "Sightings of Aye-ayes in the Eastern Rainforest of Madagascar." *Primate Conservation*, No. 7.

Gaymer, R. 1968. "The Indian Ocean Giant Tortoise *Testudo gigantea* on Aldabra." *Journal of Zoology*, London, Vol. 154.

Gehlbach, Frederick R. 1981. *Mountain Islands and Desert Seas: A Natural History of the U.S.-Mexican Borderlands*. College Station: Texas A&M University Press.

Gentry, Alwyn H., ed. 1990. *Four Neotropical Rainforests*. New Haven, Conn.: Yale University Press.

George, Wilma. 1964. *Biologist Philosopher: A Study of the Life and Writings of Alfred Russel Wallace*. London: Abelard-Schuman.

———. 1980. "Sources and Background to Discoveries of New Animals in the Sixteenth and Seventeenth Centuries." *History of Science*, Vol. 18, Part 2, No. 40.

———. 1987. "Complex Origins." In Whitmore (1987).

Ghiselin, Jon. 1981. "Applied Ecology." In Kormondy and McCormick (1981).

Ghiselin, Michael T. 1984. *The Triumph of the Darwinian Method*. Chicago: University of Chicago Press.

Gibbs, James P., and John Faaborg. 1990. "Estimating the Viability of Ovenbird and Kentucky Warbler Populations in Forest Fragments." *Conservation Biology*, Vol. 4, No. 2.

Gilbert, Bil. 1981. "Nasty Little Devil." *Sports Illustrated*, October 5, 1981.

Gilbert, F. S. 1980. "The Equilibrium Theory of Island Biogeography: Fact or Fiction?" *Journal of Biogeography*, Vol. 7.

Gill, Don, and Penelope Bonnett. 1973. *Nature in the Urban Landscape: A Study of City Ecosystems*. Baltimore: York Press.

Gilpin, Michael E. 1980. "The Role of Stepping-Stone Islands." *Theoretical Population Biology*, Vol. 17.

———. 1981. "Peninsular Diversity Patterns." *American Naturalist*, Vol. 118.

———. 1989. "Population Viability Analysis." *Endangered Species Update*, Vol. 6, No. 10.

———. 1991. "The Genetic Effective Size of a Metapopulation." In Gilpin and Hanski (1991).

Gilpin, Michael E., and Ted J. Case. 1976. "Multiple Domains of Attraction in Competition Communities." *Nature*, Vol. 261, May 6, 1976.

Gilpin, Michael E., and Jared M. Diamond. 1976. "Calculation of Immigration and Extinction Curves from the Species-Area-Distance Relation." *Proceedings of the National Academy of Sciences*, Vol. 73, No. 11.

———. 1980. "Subdivision of Nature Reserves and the Maintenance of Species Diversity." *Nature*, Vol. 285, June 1980.

———. 1982. "Factors Contributing to Non-Randomness in Species Co-Occurrences on Islands." *Oecologia*, Vol. 52.

Gilpin, Michael, and Ilkka Hanski, eds. 1991. *Metapopulation Dynamics: Empirical and Theoretical Investigations*. Reprinted from the *Biological Journal of the Linnean Society*, Vol. 42, Nos. 1 and 2. London: Academic Press.

Gilpin, Michael E., and Michael E. Soulé. 1986. "Minimum Viable Populations: Processes of Species Extinction." In Soulé (1986a).

Ginzburg, Lev R., Scott Ferson, and H. Resit Akçakaya. 1990. "Reconstructibility of Density Dependence and the Conservative Assessment of Extinction Risks." *Conservation Biology*, Vol. 4, No. 1.

Glander, Kenneth E., Patricia C. Wright, David S. Seigler, Voara Randrianasolo, and Bodovololona Randrianasolo. 1989. "Consumption of Cyanogenic Bamboo by a Newly Discovered Species of Bamboo Lemur." *American Journal of Primatology*, Vol. 19.

Glasby, G. P. 1986. "Modification of the Environment in New Zealand." *Ambio*, Vol. 15, No. 5.

Gleason, Henry Allan. 1922. "On the Relation Between Species and Area." *Ecology*, Vol. 3.

Gleick, James. 1987. "Species Vanishing from Many Parks." *New York Times*, February 3, 1987.

Godfrey, Laurie, and Martine Vuillaume-Randriamanantena. 1986. "*Hapalemur simus*: Endangered Lemur Once Widespread." *Primate Conservation*, No. 7.

Godsell, J. 1983. "Eastern Quoll." In Strahan (1983).

Golley, Frank B., and Ernesto Medina, eds. 1975. *Tropical Ecological Systems: Trends in Terrestrial and Aquatic Research*. New York: Springer-Verlag.

Goodman, Daniel. 1975. "The Theory of Diversity-Stability Relationships in Ecology." *Quarterly Review of Biology*, Vol. 50, No. 3.

———. 1987. "The Demography of Chance Extinction." In Soulé (1987).

Gordon, Kenneth R. "Insular Evolutionary Body Size Trends in *Ursus*." *Journal of Mammalogy*, Vol. 67, No. 2.

Gorman, M. L. 1979. *Island Ecology*. London: Chapman and Hall.

Gottfried, Bradley M. 1979. "Small Mammal Populations in Woodlot Islands." *American Midland Naturalist*, Vol. 102, No. 1.

Gould, Edwin, and John F. Eisenberg. 1966. "Notes on the Biology of the Tenrecidae." *Journal of Mammalogy*, Vol. 47, No. 4.

Gould, Margaret S., and Ian R. Swingland. 1980. "The Tortoise and the Goat: Interactions on Aldabra Island." *Biological Conservation*, Vol. 17.

Gould, Stephen Jay. 1979. "An Allometric Interpretation of Species-Area Curves: The Meaning of the Coefficient." *American Naturalist*, Vol. 114, No. 3.

———. 1980. *The Panda's Thumb: More Reflections in Natural History*. New York: W. W. Norton.

———. 1985. "Darwin at Sea—and the Virtues of Port." In *The Flamingo's Smile*. New York: W. W. Norton.

———. 1991. "Unenchanted Evening." *Natural History*, September 1991.

Gradwohl, Judith, and Russell Greenberg. 1988. *Saving the Tropical Forests*. London: Earthscan.

Grant, Peter R. 1974. "Population Variation on Islands." *Proceedings of the 16th International Ornithological Congress*.

———. 1986. *Ecology and Evolution of Darwin's Finches*. Princeton, N.J.: Princeton University Press.

Graves, Gary R., and Nicolas J. Gotelli. 1983. "Neotropical Land-Bridge Avifaunas: New Approaches to Null Hypotheses in Biogeography." *Oikos*, Vol. 41.

Green, Glen M., and Robert W. Sussman. 1990. "Deforestation History of the Eastern Rain Forests of Madagascar from Satellite Images." *Science*, Vol. 248, April 13, 1990.

Green, R. H. 1974. "Mammals." In Williams (1974).

Greenway, James C., Jr. 1967. *Extinct and Vanishing Birds of the World*. New York: Dover.

Gressitt, J. L., ed. 1982. *Biogeography and Ecology of New Guinea*. The Hague: Dr. W. Junk.

Griveaud, P., and R. Albignac. 1972. "The Problems of Nature Conservation in Madagascar." In Battistini and Richard-Vindard (1972).

Grue, Christian E. 1985. "Pesticides and the Decline of Guam's Native Birds." *Nature*, Vol. 316, July 25, 1985.

Grumbine, Edward. 1990a. "Protecting Biological Diversity Through the Greater Ecosystem Concept." *Natural Areas Journal*, Vol. 10, No. 3.

———. 1990b. "Viable Populations, Reserve Size, and Federal Lands Management: A Critique." *Conservation Biology*, Vol. 4, No. 2.

Grzimek, Bernhard et al., eds. 1974. *Grzimek's Animal Life Encyclopedia*. New York: Van Nostrand Reinhold.

Guiler, Eric R. 1970. "Observations on the Tasmanian Devil, *Sarcophilus harrisi* (Marsupialia: Dasyuridae)." *Australian Journal of Zoology*, Vol. 18.

———. 1985. *Thylacine: The Tragedy of the Tasmanian Tiger*. Melbourne: Oxford University Press.

Gulick, Addison. 1932. "Biological Peculiarities of Oceanic Islands." *Quarterly Review of Biology*, Vol. 7, No. 4.

Hachisuka, Masauji. 1953. *The Dodo and Kindred Birds: Or, The Extinct Birds of the Mascarene Islands*. London: H. F. & G. Witherby.

Haffer, Jürgen. 1969. "Speciation in Amazonian Forest Birds." *Science*, Vol. 165, July 11, 1969.
––––––. 1982. "General Aspects of the Refuge Theory." In Prance (1982).
Haig, Susan M., Jonathan D. Ballou, and Scott R. Derrickson. 1990. "Management Options for Preserving Genetic Diversity: Reintroduction of Guam Rails to the Wild." *Conservation Biology*, Vol. 4, No. 3.
Hairston, Nelson G., Frederick E. Smith, and Lawrence B. Slobodkin. 1960. "Community Structure, Population Control, and Competition." *American Naturalist*, Vol. 94, No. 879.
Halliday, T. R. 1980. "The Extinction of the Passenger Pigeon *Ectopistes migratorius* and Its Relevance to Contemporary Conservation." *Biological Conservation*, Vol. 17.
Hamilton, Terrell H. 1968. "Biogeography and Ecology in a New Setting." A review of MacArthur and Wilson (1967). *Science*, Vol. 159, January 5, 1968.
Hamilton, Terrell H., and Ira Rubinoff. 1963. "Isolation, Endemism, and Multiplication of Species in the Darwin Finches." *Evolution*, Vol. 17.
Hamilton, Terrell H., Ira Rubinoff, Robert H. Barth, Jr., and Guy L. Bush. 1963. "Species Abundance: Natural Regulation of Insular Variation." *Science*, Vol. 142, December 20, 1963.
Hanna, Willard A. 1976. *Bali Profile: People, Events, Circumstances (1001–1976).* New York: American Universities Field Staff.
Hanski, Ilkka, and Michael Gilpin. 1991. "Metapopulation Dynamics: Brief History and Conceptual Domain." In Gilpin and Hanski (1991).
Harcourt, A. H., and D. Fossey. 1981. "The Virunga Gorillas: Decline of an 'Island' Population." *African Journal of Ecology*, Vol. 19.
Harcourt, Caroline, with assistance from Jane Thornback. 1990. *Lemurs of Madagascar and the Comoros: The IUCN Red Data Book.* Gland, Switz.: International Union for the Conservation of Nature and Natural Resources.
Harris, Larry D. 1984. *The Fragmented Forest: Island Biogeography Theory and the Preservation of Biotic Diversity.* Chicago: University of Chicago Press.
Harris, Tom. 1990. "Losing Paradise." *Amicus Journal*, Summer 1990.
Harvey, Paul H., and A. H. Harcourt. 1984. "Sperm Competition, Testes Size, and Breeding Systems in Primates." In *Sperm Competition and the Evolution of Animal Mating Systems*, ed. Robert H. Smith. Orlando, Fla.: Academic Press.
Heaney, Lawrence R. 1984. "Mammalian Species Richness on Islands on the Sunda Shelf, Southeast Asia." *Oecologia*, Vol. 61.
Heaney, Lawrence R., Paul D. Heideman, Eric A. Rickart, Ruth B. Utzurrum, and J. S. H. Klompen. 1989. "Elevational Zonation of Mammals in the Central Philippines." *Journal of Tropical Ecology*, Vol. 5.
Heaney, L. R., and B. D. Patterson, eds. 1986. *Island Biogeography of Mammals.* Reprinted from the *Biological Journal of the Linnean Society*, Vol. 28, Nos. 1 and 2. London: Academic Press.
Hecht, Max K. 1975. "The Morphology and Relationships of the Largest Known Terrestrial Lizard, *Megalania prisca* Owen, from the Pleistocene of Australia." *Proceedings of the Royal Society of Victoria.*
Heideman, P. D., and L. R. Heaney. 1989. "Population Biology and Estimates of Abundance of Fruit Bats (Pteropodidae) in Philippine Submontane Rainforest." *Journal of Zoology*, Vol. 218.
Heideman, Paul D., Lawrence R. Heaney, Rebecca L. Thomas, and Keith R. Erickson. 1987. "Patterns of Faunal Diversity and Species Abundance of Non-Volant Small Mammals on Negros Island, Philippines." *Journal of Mammalogy*, Vol. 68, No. 4.
Hendrickson, John D. 1966. "The Galápagos Tortoises, *Geochelone* Fitzinger 1835 (*Testudo* Linnaeus 1758 in Part)." In Bowman (1966).
Herter, K. 1974. "The Insectivores." In *Grzimek's Animal Life Encyclopedia*, ed. Bernhard Grzimek et al. New York: Van Nostrand Reinhold. Vol. 10.
Hickman, John. 1985. *The Enchanted Islands: The Galapagos Discovered.* Oswestry, Shropshire: Anthony Nelson.
Higgs, A. J. 1981. "Island Biogeography Theory and Nature Reserve Design." *Journal of Biogeography*, Vol. 8.
Higgs, A. J., and M. B. Usher. 1980. "Should Nature Reserves Be Large or Small?" *Nature*, Vol. 285, June 19, 1980.
Hill, Arthur W. 1941. "The Genus *Calvaria*, with an Account of the Stony Endocarp and Germination of the Seed and Description of a New Species." *Annals of Botany*, Vol. 5, No. 20.
Hill, M. G., and D. McC. Newbery. 1982. "An Analysis of the Origins and Affinities of the Coccid Fauna (Coccoidea; Homoptera) of Western Indian Ocean Islands, with Special Reference to Aldabra Atoll." *Journal of Biogeography*, Vol. 9.
Himmelfarb, Gertrude. 1968. *Darwin and the Darwinian Revolution.* New York: W. W. Norton.
Hoage, R. J., ed. 1985. *Animal Extinctions: What Everyone Should Know.* Washington, D.C.: Smithsonian Institution Press.
Hobson, Edmund S. 1965. "Observations on Diving in the Galápagos Marine Iguana, *Amblyrhynchus cristatus* (Bell)." *Copeia*, No. 2.
Hodge, M. J. S. 1972. "The Universal Gestation of Nature: Chambers' *Vestiges* and *Explanations*." *Journal of the History of Biology*, Vol. 5, No. 1.
Hoeck, H. N. 1984. "Introduced Fauna." In Perry (1984).
Hoffman, Eric. 1989. "Madagascar in Crisis." *Animals*, Vol. 122, No. 3, May/June 1989.
Holmes, Derek, and Stephen Nash. 1989. *The Birds of Java and Bali.* Singapore: Oxford University Press.
Honegger, René E. 1980–81. "List of Amphibians and Reptiles Either Known or Thought to Have Become Extinct Since 1600." *Biological Conservation*, Vol. 19.
Hoogerwerf, A. 1953. "Notes on the Vertebrate Fauna of the Krakatau Islands, with Special Reference to the Birds." *Treubia*, Vol. 22.

Hooijer, Dirk Albert. 1951. "Pygmy Elephant and Giant Tortoise." *Scientific Monthly*, Vol. 72.
———. 1967. "Indo-Australian Insular Elephants." *Genetica*, Vol. 38.
———. 1970. "Pleistocene South-east Asiatic Pygmy Stegodonts." *Nature*, Vol. 225, January 31, 1970.
———. 1975. "Quaternary Mammals West and East of Wallace's Line." *Netherlands Journal of Zoology*, Vol. 25, No. 1.
Hooker, Sir J. D. 1896. "Lecture on Insular Floras." Delivered before the British Association for the Advancement of Science at Nottingham, August 27, 1866. London: L. Reeve.
Hooper, M. D. 1971. "The Size and Surroundings of Nature Reserves." In *The Scientific Management of Animal and Plant Communities for Conservation*, ed. E. Duffey and A. S. Watt. Oxford: Blackwell Scientific Publications.
Hope, J. H. 1974. "The Biogeography of the Mammals of the Islands of Bass Strait." In Williams (1974).
Hughes, Robert. 1988. *The Fatal Shore*. New York: Vintage Books.
Hull, David L. 1988. *Science as a Process: An Evolutionary Account of the Social and Conceptual Development of Science*. Chicago: University of Chicago Press.
Humphreys, W. F., and D. J. Kitchener. 1982. "The Effect of Habitat Utilization on Species-Area Curves: Implications for Optimal Reserve Area." *Journal of Biogeography*, Vol. 9.
Hutchinson, G. Evelyn. 1954. "Marginalia." *American Scientist*, Vol. 42, No. 2.
Huxley, Julian, ed. 1936. *T. H. Huxley's Diary of the Voyage of H.M.S. Rattlesnake*. Garden City, N.Y.: Doubleday, Doran. (Reprint edition: Kraus, New York, 1972.)
Iker, Sam. 1982. "Islands of Life." *Mosaic*, September/October 1982.
Jablonski, David. 1986a. "Mass Extinctions: New Answers, New Questions." In Kaufman and Mallory (1986).
———. 1986b. "Causes and Consequences of Mass Extinctions: A Comparative Approach." In Elliott (1986).
———. 1991. "Extinctions: A Paleontological Perspective." *Science*, Vol. 253, August 16, 1991.
Jackson, M. H. 1985. *Galápagos: A Natural History Guide*. Calgary, Alberta, Can.: University of Calgary Press.
Jackson, Peter S. Wyse, Quentin C. B. Cronk, and John A. N. Parnell. 1988. "Notes on the Regeneration of Two Rare Mauritian Endemic Trees." *Tropical Ecology*, Vol. 29.
Jaffe, Mark. 1985. "Cracking an Ecological Murder Mystery." *Philadelphia Inquirer*, June 4, 1985.
———. 1994. *And No Birds Sing: The Story of an Ecological Disaster in a Tropical Paradise*. New York: Simon & Schuster.
Janzen, Daniel H. 1968. "Host Plants as Islands in Evolutionary and Contemporary Time." *American Naturalist*, Vol. 102, November/December 1968.
———. 1974. "The Deflowering of Central America." *Natural History*, April 1974.
———. 1975. "Host Plants as Islands, Part 2: Competition in Evolutionary and Contemporary Time." *American Naturalist*, Vol. 107, No. 958.
———. 1983. "No Park Is an Island: Increase in Interference from Outside as Park Size Decreases." *Oikos*, Vol. 41.
Jenkins, Alan C. 1978. *The Naturalists: Pioneers of Natural History*. New York: Mayflower Books.
Jenkins, J. Mark. 1979. "Natural History of the Guam Rail." *Condor*, Vol. 81.
———. 1983. *The Native Forest Birds of Guam*. Ornithological Monographs No. 31. Washington, D.C.: American Ornithologists' Union.
Jenkins, Robert E. 1989. "Long-Term Conservation and Preserve Complexes." *Nature Conservancy*, Vol. 39, No. 1, January/February 1989.
Johns, Andrew. 1986. "Notes on the Ecology and Current Status of the Buffy Saki (*Pithecia albicans*)." *Primate Conservation*, No. 7.
Johnson, Donald Lee. 1978. "The Origin of Island Mammoths and the Quaternary Land Bridge History of the Northern Channel Islands, California." *Quaternary Research*, Vol. 10.
———. 1980. "Problems in the Land Vertebrate Zoogeography of Certain Islands and the Swimming Powers of Elephants." *Journal of Biogeography*, Vol. 7.
Johnson, Michael P., and Peter H. Raven. 1973. "Species Number and Endemism: The Galápagos Archipelago Revisited." *Science*, Vol. 179, March 2, 1973.
Jolly, Alison. 1966. *Lemur Behavior: A Madagascar Field Study*. Chicago: University of Chicago Press.
———. 1980. *A World Like Our Own: Man and Nature in Madagascar*. New Haven, Conn.: Yale University Press.
Jolly, Alison, Roland Albignac, and Jean-Jacques Petter. 1984. "The Lemurs." In Jolly et al. (1984).
Jolly, Alison, Philippe Oberlé, and Roland Albignac, eds. 1984. *Key Environments: Madagascar*. Oxford: Pergamon Press.
Jones, Carl G., Willard Heck, Richard E. Lewis, Yousoof Mungroo, Glenn Slade, and Tom Cade. 1995. "The Restoration of the Mauritius Kestrel *Falco punctatus* Population." *Ibis*, Vol. 137.
Jones, H. Lee, and Jared M. Diamond. 1976. "Short-Time-Base Studies of Turnover in Breeding Bird Populations on the California Channel Islands." *Condor*, Vol. 78.
Jordan, David S. 1905. "The Origin of Species Through Isolation." *Science*, Vol. 22, November 3, 1905.
Juvik, J. O., A. J. Andrianarivo, and C. P. Blanc. 1980–81. "The Ecology and Status of *Geochelone yniphora*: A Critically Endangered Tortoise in Northwestern Madagascar." *Biological Conservation*, Vol. 19.
Kaneshiro, Kenneth Y., and Alan T. Ohta. 1982. "The Flies Fan Out." *Natural History*, Vol. 91, December 1982.
Kapos, Valerie. 1989. "Effects of Isolation on the Water Status of Forest Patches in the Brazilian Amazon." *Journal of Tropical Ecology*, Vol. 5.
Karr, James R. 1982a. "Avian Extinction on Barro Colorado Island, Panama: A Reassessment." *American Naturalist*, Vol. 119, No. 2.

————. 1982b. "Population Variability and Extinction in the Avifauna of a Tropical Land Bridge Island." *Ecology*, Vol. 63, No. 6.

Kaufman, Les, and Kenneth Mallory, eds. 1986. *The Last Extinction*. Cambridge, Mass.: MIT Press.

Keast, Allen. 1968a. "Evolution of Mammals on Southern Continents, Part 1, Introduction: The Southern Continents as Backgrounds for Mammalian Evolution." *Quarterly Review of Biology*, Vol. 43, No. 3.

————. 1968b. "Evolution of Mammals on Southern Continents, Part 4, Australian Mammals: Zoogeography and Evolution." *Quarterly Review of Biology*, Vol. 43, No. 4.

————. 1971. "Adaptive Evolution and Shifts in Niche Occupation in Island Birds." In Stern (1971).

————. 1974. "Ecological Opportunities and Adaptive Evolution on Islands, with Special Reference to Evolution in the Isolated Forest Outliers of Southern Australia." *Proceedings of the 16th International Ornithological Congress*.

Keiter, Robert B., and Mark S. Boyce, eds. 1991. *The Greater Yellowstone Ecosystem: Redefining America's Wilderness Heritage*. New Haven, Conn.: Yale University Press.

Kensley, Brian, ed. 1988. *Results of Recent Research on Aldabra Atoll, Indian Ocean*. Collected papers as *Bulletin of the Biological Society of Washington*, No. 8.

Key, Robert E., ed. 1968. *A Science Teacher's Handbook to Guam*. Agana, Guam: Guam Science Teachers Association.

Kilburn, Paul D. 1966. "Analysis of the Species-Area Relation." *Ecology*, Vol. 47.

King, Laura B., Faith Campbell, David Edelson, and Susan Miller. 1989. "Extinction in Paradise: Protecting Our Hawaiian Species." Washington, D.C.: Natural Resources Defense Council.

Kingdon, Jonathan. 1989. *Island Africa: The Evolution of Africa's Rare Animals and Plants*. Princeton, N.J.: Princeton University Press.

Kingsland, Sharon E. 1988. *Modeling Nature: Episodes in the History of Population Ecology*. Chicago: University of Chicago Press.

Kirch, Patrick V. 1982. "Transported Landscapes." *Natural History*, Vol. 91, December 1982.

————. 1985. *Feathered Gods and Fishhooks: An Introduction to Hawaiian Archaeology and Prehistory*. Honolulu: University of Hawaii Press.

Kirkpatrick, J. B. 1983. "An Iterative Method for Establishing Priorities for the Selection of Nature Reserves: An Example from Tasmania." *Biological Conservation*, Vol. 25.

Kirkpatrick, J. B., and Sue Backhouse. 1985. *Native Trees of Tasmania*. Hobart, Tasmania, Aus.: Sue Backhouse.

Kitchener, Andrew. 1993. "Justice at Last for the Dodo." *New Scientist*, Vol. 139. August 28, 1993.

Kitchener, D. J. 1982. "Predictors of Vertebrate Species Richness in Nature Reserves in the Western Australian Wheatbelt." *Australian Wildlife Research*, Vol. 9.

Kitchener, D. J., A. Chapman, and B. G. Muir. 1980. "The Conservation Value for Mammals of Reserves in the Western Australian Wheatbelt." *Biological Conservation*, Vol. 18.

Knight, Bill, and Chris Schofield. 1985. "Sulawesi: An Island Expedition." *New Scientist*, Vol. 105, January 10, 1985.

Knight, Richard R., and L. L. Eberhardt. 1985. "Population Dynamics of Yellowstone Grizzly Bears." *Ecology*, Vol. 66, No. 2.

Kolar, K. 1974. "Tree Shrews and Prosimians." In *Grzimek's Animal Life Encyclopedia*, ed. Bernhard Grzimek et al. New York: Van Nostrand Reinhold. Vol. 10.

Kolata, Gina Bari. 1974. "Theoretical Ecology: Beginnings of a Predictive Science." *Science*, Vol. 183, February 1, 1974.

Kormondy, Edward J., and J. Frank McCormick, eds. 1981. *Handbook of Contemporary Developments in World Ecology*. Westport, Conn.: Greenwood Press.

Kottack, Conrad Phillip. 1980. *The Past in the Present: History, Ecology, and Cultural Variation in Highland Madagascar*. Ann Arbor: University of Michigan Press.

Kottack, Conrad Phillip, Jean-Aimé Rakotoarisoa, Aidan Southall, and Pierre Vérin, eds. 1986. *Madagascar: Society and History*. Durham, N.C.: Carolina Academic Press.

Kottler, Malcolm Jay. 1974. "Alfred Russel Wallace, the Origin of Man, and Spiritualism." *Isis*, Vol. 65, No. 227.

Krantz, Grover S. 1970. "Human Activities and Megafaunal Extinctions." *American Scientist*, Vol. 58.

Kricher, John C. 1989. *A Neotropical Companion: An Introduction to the Animals, Plants, and Ecosystems of the New World Tropics*. Princeton, N.J.: Princeton University Press.

Kuris, Armand M., Andrew R. Blaustein, and José Javier Alió. 1980. "Hosts as Islands." *American Naturalist*, Vol. 116, No. 4.

Lacava, Jerry, and Jeff Hughes. 1984. "Determining Minimum Viable Population Levels." *Wildlife Society Bulletin*, Vol. 12, No. 4.

Lack, David. 1947. *Darwin's Finches: An Essay on the General Biological Theory of Evolution*. Cambridge: Cambridge University Press. (Reprint edition: Peter Smith, Gloucester, Mass., 1968.)

————. 1969. "Subspecies and Sympatry in Darwin's Finches." *Evolution*, Vol. 23.

————. 1971a. *Ecological Isolation in Birds*. Cambridge, Mass.: Harvard University Press.

————. 1971b. "Island Birds." In Stern (1971).

————. 1973. "The Numbers of Species of Hummingbirds in the West Indies." *Evolution*, Vol. 27.

————. 1976. *Island Biology, Illustrated by the Land Birds of Jamaica*. Berkeley: University of California Press.

Lake, P. S. 1974. "Conservation." In Williams (1974).

Lamberson, Roland H., Robert McKelvey, Barry R. Noon, and Curtis Voss. 1992. "A Dynamic Analysis of Northern Spotted Owl Viability in a Fragmented Forest Landscape." *Conservation Biology*, Vol. 6, No. 4.

Lamont, Byron B., and Peter G. L. Klinkhamer. 1993. "Population Size and Viability." *Nature*, Vol. 362, March 18, 1993.

Lande, Russell. 1987. "Extinction Thresholds in Demographic Models of Territorial Populations." *American Naturalist*, Vol. 130, No. 4.
———. 1988. "Genetics and Demography in Biological Conservation." *Science*, Vol. 241, September 16, 1988.
Langrand, Olivier. 1990. *Guide to the Birds of Madagascar*. Translated by Willem Daniels. New Haven, Conn.: Yale University Press.
Lanting, Frans. 1990. *A World Out of Time: Madagascar*. New York: Aperture.
Lawlor, Timothy E. 1982. "The Evolution of Body Size in Mammals: Evidence from Insular Populations in Mexico." *American Naturalist*, Vol. 119, No. 1.
———. 1983. "The Peninsular Effect on Mammalian Species Diversity in Baja California." *American Naturalist*, Vol. 121, No. 3.
Lazell, James D., Jr. 1984. "A Population of Devils *Sarcophilus harrisii* in North-Western Tasmania." *Records of the Queen Victoria Museum*, No. 84.
———. 1987. "Beyond the Wallace Line." *Explorers Journal*, Vol. 65, No. 2.
———. 1989. *Wildlife of the Florida Keys: A Natural History*. Washington, D.C.: Island Press.
———. 1989. "Pushy Wildlife." *National Parks*, Vol. 63, Nos. 9–10.
Le Bourdiec, Paul. 1972. "Accelerated Erosion and Soil Degradation." In Battistini and Richard-Vindard (1972).
LeCroy, Mary. 1981. "The Genus *Paradisaea*—Display and Evolution." *American Museum Novitates*, No. 2714.
LeCroy, Mary, Alfred Kulupi, and W. S. Peckover. 1980. "Goldie's Bird of Paradise: Display, Natural History and Traditional Relationships of People to the Bird." *Wilson Bulletin*, Vol. 92, No. 3.
Leigh, Egbert Giles, Jr. 1981. "The Average Lifetime of a Population in a Varying Environment." *Journal of Theoretical Biology*, Vol. 90.
Leigh, Egbert G., Jr., Stanley A. Rand, and Donald M. Windsor, eds. 1982. *The Ecology of a Tropical Forest: Seasonal Rhythms and Long-term Changes*. Washington, D.C.: Smithsonian Institution Press.
Levins, Richard. 1970. "Extinction." *Some Mathematical Questions in Biology*, Vol. 2.
Lewin, Roger. 1984. "Parks: How Big Is Big Enough?" *Science*, Vol. 225, August 10, 1984.
Lincoln, G. A. 1975. "Bird Counts Either Side of Wallace's Line." *Journal of Zoology*, London, Vol. 177.
Livezey, Bradley C. 1993. "An Ecomorphological Review of the Dodo (*Raphus cucullatus*) and Solitaire (*Pezophaps solitaria*), Flightless Columbiformes of the Mascarene Islands." *Journal of Zoology*, London, Vol. 230.
Lomolino, Mark V. 1984. "Mammalian Island Biogeography: Effects of Area, Isolation, and Vagility." *Oecologia*, Vol. 61.
———. 1985. "Body Size of Mammals on Islands: The Island Rule Reexamined." *American Naturalist*, Vol. 125.
Lovejoy, Thomas E. 1979a. "The Epoch of Biotic Impoverishment." *Great Basin Naturalist Memoirs*, No. 3.
———. 1979b. "Refugia, Refuges and Minimum Critical Size: Problems in the Conservation of the Neotropical Herpetofauna." In *The South American Herpetofauna: Its Origin, Evolution, and Dispersal*, ed. William E. Duellman. Museum of Natural History Monograph 7. Lawrence: University of Kansas.
———. 1980. "Discontinuous Wilderness: Minimum Areas for Conservation." *Parks*, Vol. 5, No. 2.
———. 1982a. "Designing Refugia for Tomorrow." In Prance (1982).
———. 1982b. "Hope for a Beleaguered Paradise." *Garden*, February 1982.
———. 1986. "Damage to Tropical Forests, or Why Were There So Many Kinds of Animals?" *Science*, Vol. 234, October 10, 1986.
Lovejoy, T. E., R. O. Bierregaard, Jr., A. B. Rylands, J. R. Malcolm, C. E. Quintela, L. H. Harper, K. S. Brown, Jr., A. H. Powell, G. V. N. Powell, H. O. R. Schubart, and M. B. Hays. 1986. "Edge and Other Effects of Isolation on Amazon Forest Fragments." In Soulé (1986).
Lovejoy, Thomas E., and David C. Oren. 1981. "The Minimum Critical Size of Ecosystems." In *Forest Island Dynamics in Man-Dominated Landscapes*, ed. Robert L. Burgess and David M. Sharpe. Vol. 41 of *Ecological Studies*. New York: Springer-Verlag.
Lovejoy, Thomas E., Judy M. Rankin, R. O. Bierregaard, Jr., Keith S. Brown, Jr., Louise H. Emmons, and Martha E. Van der Voort. 1984. "Ecosystem Decay of Amazon Forest Remnants." In Nitecki (1984).
Lovejoy, Thomas E., and Eneas Salati. 1983. "Precipitating Change in Amazonia." In *The Dilemma of Amazonian Development*, ed. Emilio F. Moran. Boulder, Colo.: Westview Press.
Lovejoy, Thomas E., and Herbert O. R. Schubart. 1979. "The Ecology of Amazonian Development." In *Land, People and Planning in Contemporary Amazonia*, ed. Francoise Barbira-Scazzocchio. Proceedings of the Conference on the Development of Amazonia in Seven Countries, Centre of Latin American Studies, University of Cambridge.
Lucas, Frederic A. 1922. "Historic Tortoises and Other Aged Animals." *Natural History*, Vol. 22, May–June 1922.
Lutz, Dick, and J. Marie Lutz. 1991. *Komodo: The Living Dragon*. Salem, Ore.: Dimi Press.
MacArthur, Robert. 1955. "Fluctuations of Animal Populations, and a Measure of Community Stability." *Ecology*, Vol. 36.
———. 1957. "On the Relative Abundance of Bird Species." *Proceedings of the National Academy of Sciences*, Vol. 43.
———. 1962. "Growth and Regulation of Animal Populations." A review of Slobodkin (1961). *Ecology*, Vol. 43.
———. 1969. "The Ecologist's Telescope." *Ecology*, Vol. 50, No. 3.

———. 1972a. *Geographical Ecology: Patterns in the Distribution of Species.* New York: Harper and Row. (Reprint edition: Princeton University Press, Princeton, N.J., 1984.)

———. 1972b. "Coexistence of Species." In *Challenging Biological Problems: Directions Toward Their Solution,* ed. John A. Behnke. New York: Oxford University Press.

MacArthur, Robert H., and Joseph H. Connell. 1966. *The Biology of Populations.* New York: John Wiley and Sons.

MacArthur, Robert H., Jared M. Diamond, and James R. Karr. 1972. "Density Compensation in Island Faunas." *Ecology,* Vol. 53, No. 2.

MacArthur, Robert, and Richard Levins. 1967. "The Limiting Similarity, Convergence, and Divergence of Coexisting Species." *American Naturalist,* Vol. 101, No. 921.

MacArthur, Robert, John MacArthur, Duncan MacArthur, and Alan MacArthur. 1973. "The Effect of Island Area on Population Densities." *Ecology,* Vol. 54.

MacArthur, Robert H., and Edward O. Wilson. 1963. "An Equilibrium Theory of Insular Zoogeography." *Evolution,* Vol. 17, No. 4.

———. 1967. *The Theory of Island Biogeography.* Princeton, N.J.: Princeton University Press.

McCoid, Michael James. 1991. "Brown Tree Snake (*Boiga irregularis*) on Guam: A Worst Case Scenario of an Introduced Predator." *Micronesica* Supplement 3.

McCoy, Michael. 1980. *Reptiles of the Solomon Islands.* Handbook No. 7. Wau, Papua New Guinea: Wau Ecology Institute.

McDougal, Charles. 1977. *The Face of the Tiger.* London: Rivington Books and André Deutsch.

MacFarland, Craig G., José Villa, and Basilio Toro. 1974. "The Galápagos Giant Tortoises (*Geochelone elephantopus*): Part 1, Status of the Surviving Populations." *Biological Conservation,* Vol. 6, No. 2.

Machlis, Gary E., and David L. Tichnell. 1985. *The State of the World's Parks: An International Assessment for Resource Management, Policy, and Research.* Boulder, Colo.: Westview Press.

McIntosh, Robert P. 1980. "The Background and Some Current Problems of Theoretical Ecology." *Synthese,* Vol. 43.

———. 1986. *The Background of Ecology: Concept and Theory.* Cambridge: Cambridge University Press.

McKibben, Bill. 1989. *The End of Nature.* New York: Random House.

McKinney, H. Lewis. 1972. *Wallace and Natural Selection.* New Haven, Conn.: Yale University Press.

MacKinnon, John. 1990. *Field Guide to the Birds of Java and Bali.* Yogyakarta, Indonesia: Gadjah Mada University Press.

MacKinnon, Kathy. 1992. *Nature's Treasurehouse: The Wildlife of Indonesia.* Jakarta: Penerbit PT Gramedia Pustaka Utama.

McLellan, Charles H., Andrew P. Dobson, David S. Wilcove, and James F. Lynch. 1985. "Effects of Forest Fragmentation on New- and Old-World Bird Communities: Empirical Observations and Theoretical Implications." In *Wildlife 2000: Modeling Habitat Relationships of Terrestrial Vertebrates,* ed. Jared Verner, Michael L. Morrison, and C. John Ralph. Madison: University of Wisconsin Press.

Maddox, John. 1993. "Bringing the Extinct Dodo Back to Life." *Nature,* Vol. 365, September 22, 1993.

Maehr, David S. 1984. "Distribution of Black Bears in Eastern North America." *Proceedings of the 7th Eastern Workshop on Black Bear Research and Management,* Vol. 7.

———. 1990. "Tracking Florida's Panthers." *Defenders,* September/October 1990.

Maehr, David S., Robert C. Belden, E. Darrell Land, and Laurie Wilkins. 1990. "Food Habits of Panthers in Southwest Florida." *Journal of Wildlife Management,* Vol. 54, No. 3.

Maehr, David S., E. Darrell Land, and Jayde C. Roof. 1991. "Florida Panthers." *National Geographic Research and Exploration,* Vol. 7, No. 4.

Maehr, David S., E. Darrell Land, Jayde C. Roof, and J. Walter McCown. 1989. "Early Maternal Behavior in the Florida Panther (*Felis concolor coryi*)." *American Midland Naturalist,* Vol. 122, No. 1.

Maehr, David S., James N. Layne, E. Darrell Land, J. Walter McCown, and Jayde Roof. 1988. "Long Distance Movements of a Florida Black Bear." *Florida Field Naturalist,* Vol. 16, No. 1.

Maehr, David S., Jayde C. Roof, E. Darrell Land, and J. Walter McCown. 1989. "First Reproduction of a Panther (*Felis concolor coryi*) in Southwestern Florida, U.S.A." *Mammalia,* Vol. 53, No. 1.

Maehr, David S., Jayde C. Roof, E. Darrell Land, J. Walter McCown, Robert C. Belden, and William B. Frankenberger. 1989. "Fates of Wild Hogs Released into Occupied Florida Panther Home Ranges." *Florida Field Naturalist,* Vol. 17, No. 2.

Mahe, J. 1972. "The Malagasy Subfossils." In Battistini and Richard-Vindard (1972).

Main, A. R., and M. Yadav. 1971. "Conservation of Macropods in Reserves in Western Australia." *Biological Conservation,* Vol. 3, No. 2.

Malcolm, Jay R. 1988. "Small Mammal Abundances in Isolated and Non-Isolated Primary Forest Reserves Near Manaus, Brazil." *ACTA Amazonica,* Vol. 18, Nos. 3–4.

Mann, Charles C., and Mark L. Plummer. 1995. *Noah's Choice: The Future of Endangered Species.* New York: Alfred A. Knopf.

Marchant, James. 1916. *Alfred Russel Wallace: Letters and Reminiscences.* New York: Harper & Brothers. (Facsimile edition: Arno Press, New York, 1975.)

Margalef, Ramón. 1968. *Perspectives in Ecological Theory.* Chicago: University of Chicago Press.

Marnham, Patrick. 1980. "Rescuing the Ark." *Harper's,* October 1980.

Martin, P. S., and H. E. Wright, Jr., eds. 1967. *Pleistocene Extinctions: The Search for a Cause.* New Haven, Conn.: Yale University Press.

Martin, Paul S., and Richard G. Klein, eds. 1984. *Quaternary Extinctions: A Prehistoric Revolution.* Tucson: University of Arizona Press.

Martin, R. D. 1972. "Adaptive Radiation and Behaviour of the Malagasy Lemurs." *Philosophical Transactions of the Royal Society of London,* Series B, Vol. 264.

Martin, R. D., G. A. Doyle, and A. C. Walker, eds. 1974. *Prosimian Biology*. Pittsburgh: University of Pittsburgh Press.

Martuscelli, Paulo, Liege Mariel Petroni, and Fábio Olmos. 1994. "Fourteen New Localities for the Muriqui *Brachyteles arachnoides*." *Neotropical Primates*, Vol 2, No. 2.

Mason, Victor. 1989. *Birds of Bali*. Illustrations by Frank Jarvis. Berkeley, Calif.: Periplus Editions.

Mattson, David J. 1990. "Human Impacts on Bear Habitat Use." *International Conference on Bear Research and Management*, Vol. 8.

Mattson, David J., Bonnie M. Blanchard, and Richard R. Knight. 1991. "Food Habits of Yellowstone Grizzly Bears, 1977–1987." *Canadian Journal of Zoology*, Vol. 69.

———. 1992. "Yellowstone Grizzly Bear Mortality, Human Habituation, and Whitebark Pine Seed Crops." *Journal of Wildlife Management*, Vol. 56, No. 3.

Mattson, David J., Colin M. Gillian, Scott A. Benson, and Richard R. Knight. 1991. "Bear Feeding Activity at Alpine Insect Aggregation Sites in the Yellowstone Ecosystem." *Canadian Journal of Zoology*, Vol. 69.

Mattson, David J., and Richard R. Knight. 1991. "Application of Cumulative Effects Analysis to the Yellowstone Grizzly Bear Population." Interagency Grizzly Bear Study Team Report, 1991C.

Mattson, David J., and Matthew M. Reid. 1991. "Conservation of the Yellowstone Grizzly Bear." *Conservation Biology*, Vol. 5, No. 3.

May, Robert M. 1975. "Island Biogeography and the Design of Wildlife Preserves." *Nature*, Vol. 254, March 20, 1975.

———, ed. 1976. *Theoretical Ecology*. Sunderland, Mass.: Sinauer Associates.

———. 1981. "The Role of Theory in Ecology." *American Zoology*, Vol. 21.

———. 1988. "How Many Species Are There on Earth?" *Science*, Vol. 241, September 16, 1988.

Mayr, Ernst. 1932. "A Tenderfoot Explorer in New Guinea." *Natural History*, January/February 1932.

———. 1940. "Speciation Phenomena in Birds." *American Naturalist*, Vol. 74.

———. 1942. *Systematics and the Origin of Species*. New York: Columbia University Press. (Reprint edition: Dover Publications, New York, 1964.)

———. 1944. "Wallace's Line in the Light of Recent Zoogeographic Studies." *Quarterly Review of Biology*, Vol. 19, No. 1.

———. 1947. "Ecological Factors in Speciation." *Evolution*, Vol. 1.

———. 1954. "Change of Genetic Environment and Evolution." In *Evolution as a Process*, ed. J. Huxley, A. C. Hardy, and E. B. Ford. London: Allen & Unwin. Reprinted in Mayr (1976).

———. 1959. "Isolation as an Evolutionary Factor." *Proceedings of the American Philosophical Society*, Vol. 103, No. 2.

———. 1963. *Animal Species and Evolution*. Cambridge, Mass.: Belknap Press of Harvard University Press.

———. 1965a. "The Nature of Colonizations of Birds." In *The Genetics of Colonizing Species*, ed. H. G. Baker and G. Ledyard Stebbins. New York: Academic Press.

———. 1965b. "Avifauna: Turnover on Islands." *Science*, Vol. 150, December 17, 1965.

———. 1976. *Evolution and the Diversity of Life: Selected Essays*. Cambridge, Mass.: Belknap Press of Harvard University Press.

———. 1982. *The Growth of Biological Thought: Diversity, Evolution, and Inheritance*. Cambridge, Mass.: Belknap Press of Harvard University Press.

———. 1991. *One Long Argument: Charles Darwin and the Genesis of Modern Evolutionary Thought*. Cambridge, Mass.: Harvard University Press.

Mech, David L. 1966. *The Wolves of Isle Royale*. Fauna Series 7. Washington, D.C.: U.S. Department of the Interior.

Meier, Bernhard, Roland Albignac, André Peyriéras, Yves Rumpler, and Patricia Wright. 1987. "A New Species of *Hapalemur* (Primates) from South East Madagascar." *Folia Primatologica*, Vol. 48.

Meier, Bernhard, and Yves Rumpler. 1987. "Preliminary Survey of *Hapalemur simus* and of a New Species of *Hapalemur* in Eastern Betsileo, Madagascar." *Primate Conservation*, No. 8.

Menges, Eric S. 1990. "Population Viability Analysis for an Endangered Plant." *Conservation Biology*, Vol. 4, No. 1.

Merton, L. F. H., D. M. Bourn, and R. J. Hnatiuk. 1976. "Giant Tortoise and Vegetation Interactions on Aldabra Atoll: Part I, Inland." *Biological Conservation*, Vol. 9.

Miller, Alden H. 1966. "Animal Evolution on Islands." In Bowman (1966).

Miller, Christine. 1990. "Zen and the Art of Planet Maintenance." *Science Notes*, Vol. 16, No. 2.

Miller, Ronald I., Susan P. Bratton, and Peter S. White. 1987. "A Regional Strategy for Reserve Design and Placement Based on an Analysis of Rare and Endangered Species' Distribution Patterns." *Biological Conservation*, Vol. 39.

Miller, Ronald I., and Larry D. Harris. 1977. "Isolation and Extirpations in Wildlife Reserves." *Biological Conservation*, Vol. 12.

———. 1979. "Predicting Species Changes in Isolated Wildlife Preserves." In *Proceedings of the First Conference on Scientific Research in the National Parks*, ed. Robert M. Linn. U.S. Department of the Interior, National Park Service Transactions and Proceedings Series, Vol. 1, No. 5.

Mills, L. Scott, Michael E. Soulé, and Daniel F. Doak. 1993. "The Keystone-Species Concept in Ecology and Conservation." *BioScience*, Vol. 43, No. 4.

Mitchell, Rodger. 1974. "Scaling in Ecology." *Science*, Vol. 184, June 14, 1974.

Mittermeier, Russell A. 1987. "Monkey in Peril." *National Geographic*, Vol. 171, March 1987.

———. 1988. "Strange and Wonderful Madagascar." *International Wildlife*, Vol. 18, No. 4, July/August 1988.

Mittermeier, Russell A., Anthony B. Rylands, Adelmar Coimbra-Filho, and Gustavo A. B. Fonseca, eds. 1988. *Ecology and Behavior of Neotropical Primates*, Vol. 2. Washington, D.C.: World Wildlife Fund.

Mooney, Nick. 1984. "Tasmanian Tiger Sighting Casts Marsupial in New Light." *Australian Natural History*, Vol. 21, No. 5, Winter 1984.

Moore, N. W. 1962. "The Heaths of Dorset and Their Conservation." *Journal of Ecology*, Vol. 50.

Moore, Philip H., and Patrick D. McMakin. 1979. *Plants of Guam*. Mangilao, Guam: Cooperative Extension Service, University of Guam.

Moorehead, Alan. 1971. *Darwin and the Beagle*. Harmondsworth, Middlesex, Eng.: Penguin Books.

———. 1987. *The Fatal Impact: The Invasion of the South Pacific 1767–1840*. Sydney: Mead & Beckett.

Moors, P. J. 1985. *Conservation of Island Birds*. Technical Publication No. 3. Cambridge: International Council for Bird Preservation.

Morowitz, Harold J. 1991. "Balancing Species Preservation and Economic Considerations." *Science*, Vol. 253, August 16, 1991.

Mueller-Dombois, Dieter. 1975. "Some Aspects of Island Ecological Analysis." In Golley and Medina (1975).

Murphy, Dennis D., Kathy E. Freas, and Stuart B. Weiss. 1990. "An Environment-Metapopulation Approach to Population Viability Analysis for a Threatened Invertebrate." *Conservation Biology*, Vol. 4, No. 1.

Murphy, Dervla. 1989. *Muddling Through in Madagascar*. Woodstock, N.Y.: Overlook Press.

Myers, Norman. 1980. *The Sinking Ark: A New Look at the Problem of Disappearing Species*. Oxford: Pergamon Press.

Nafus, Donald, and Ilse Schreiner. 1989. "Biological Control Activities in the Mariana Islands from 1911 to 1988." *Micronesica*, Vol. 22.

Nelson, Gareth, and Norman I. Platnick. 1980. "A Vicariance Approach to Historical Biogeography." *Bio-Science*, Vol. 30, No. 5.

New, T. R., M. B. Bush, I. W. B. Thornton, and H. K. Sudarman. 1988. "The Butterfly Fauna of the Krakatau Islands After a Century of Colonization." *Philosophical Transactions of the Royal Society of London*, Vol. 322.

New, T. R., and I. W. B. Thornton. 1988. "A Pre-Vegetation Population of Crickets Subsisting on Allochthonous Aeolian Debris on Anak Krakatau." *Philosophical Transactions of the Royal Society of London*, Vol. 322.

Newmark, William D. 1985. "Legal and Biotic Boundaries of Western North American National Parks: A Problem of Congruence." *Biological Conservation*, Vol. 33.

———. 1986. "Mammalian Richness, Colonization, and Extinction in Western North American National Parks." Ph.D. dissertation, University of Michigan.

———. 1987. "A Land-Bridge Island Perspective on Mammalian Extinctions in Western North American Parks." *Nature*, Vol. 325, January 29, 1987.

———. 1995. "Extinction of Mammal Populations in Western North American National Parks." *Conservation Biology*, Vol. 9, No. 3.

Nilsson, Sven G., and Ingvar N. Nilsson. 1983. "Are Estimated Species Turnover Rates on Islands Largely Sampling Errors?" *American Naturalist*, Vol. 121.

Nitecki, Matthew H., ed. 1984. *Extinctions*. Chicago: University of Chicago Press.

Norse, Elliott A., Kenneth L. Rosenbaum, David S. Wilcove, Bruce A. Wilcox, William H. Romme, David W. Johnston, and Martha L. Stout. 1986. *Conserving Biological Diversity in Our National Forests*. Washington, D.C.: Wilderness Society.

Norton, Bryan G., ed. 1986. *The Preservation of Species: The Value of Biological Diversity*. Princeton, N.J.: Princeton University Press.

Noss, Reed F. 1991. "Ecosystem Restoration: An Example for Florida." *Wild Earth*, Spring 1991.

O'Brien, Stephen J., Melody E. Roelke, Naoya Yuhki, Karen W. Richards, Warren E. Johnson, William L. Franklin, Allen E. Anderson, Oron L. Bass, Jr., Robert C. Belden, and Janice S. Martenson. 1990. "Genetic Introgression Within the Florida Panther *Felis concolor coryi*." *National Geographic Research*, Vol. 6, No. 4.

O'Connor, Sheila Margaret. 1987. "The Effect of Human Impact on Vegetation and the Consequences to Primates in Two Riverine Forests, Southern Madagascar." Ph.D. dissertation, Cambridge University.

O'Hanlon, Redmond. 1987. *Into the Heart of Borneo*. New York: Vintage Books.

Ola, Per, and Emily d'Aulaire. 1986. "Lessons from a Ravaged Jungle." *International Wildlife*, September/October 1986.

Olson, Sherry H. 1984. "The Robe of the Ancestors." *Journal of Forest History*, Vol. 28, No. 4.

Olson, Storrs L. 1975. "Paleornithology of St. Helena Island, South Atlantic Ocean." *Smithsonian Contributions to Paleobiology*, No. 23. Washington, D.C.: Smithsonian Institution Press.

Olson, Storrs L., and Helen F. James. 1982. "Fossil Birds from the Hawaiian Islands: Evidence for Wholesale Extinction by Man Before Western Contact." *Science*, Vol. 217, August 13, 1982.

Oren, David C. 1982. "Testing the Refuge Model for South America." In Prance (1982).

Otte, Daniel, and John A. Endler, eds. 1989. *Speciation and Its Consequences*. Sunderland, Mass.: Sinauer Associates.

Owadally, A. W. 1979. "The Dodo and the Tambalacoque Tree." *Science*, Vol. 203, March 30, 1979.

———. 1981. "Mauritius." In Kormondy and McCormick (1981).

Oxby, Clare. 1985. "Forest Farmers: The Transformation of Land Use and Society in Eastern Madagascar." *Unasylva*, Vol. 37, No. 149.

Page, Jake. 1988. "Clear-cutting the Tropical Rain Forest in a Bold Attempt to Salvage It." *Smithsonian*, April 1988.

Pain, Stephanie. 1990. "Last Days of the Old Night Bird." *New Scientist*, Vol. 126, No. 1721, June 16, 1990.

Parsons, P. A. 1982. "Adaptive Strategies of Colonizing Animal Species." *Biological Reviews*, Vol. 57.

Perry, R., ed. 1984. *Key Environments: Galapagos*. Oxford: Pergamon Press.

Peters, Robert L., and Thomas E. Lovejoy, eds. 1992. *Global Warming and Biological Diversity*. New Haven, Conn.: Yale University Press.

Peterson, Dale. 1989. *The Deluge and the Ark: A Journey into Primate Worlds*. Boston: Houghton Mifflin.
Peterson, Rolf Olin. 1977. *Wolf Ecology and Prey Relationships on Isle Royale*. Scientific Monograph Series, No. 11. Washington, D.C.: National Park Service.
Petter, J. J. 1972. "Order of Primates: Sub-Order of Lemurs." In Battistini and Richard-Vindard (1972).
Petter, J. J. and A. Peyriéras. 1974. "A Study of Population Density and Home Ranges of *Indri indri* in Madagascar." In Martin et al. (1974).
———. 1975. "Preliminary Notes on the Behavior and Ecology of *Hapalemur griseus*." In Tattersall and Sussman (1975).
Picton, Harold D. 1979. "The Application of Insular Biogeographic Theory to the Conservation of Large Mammals in the Northern Rocky Mountains." *Biological Conservation*, Vol. 15.
Pickett, S. T. A., and John N. Thompson. 1978. "Patch Dynamics and the Design of Nature Reserves." *Biological Conservation*, Vol. 13.
Pielou, E. C. 1979. *Biogeography*. New York: John Wiley and Sons.
Pimm, Stuart L. 1987. "The Snake That Ate Guam." *Trends in Ecology and Evolution*, Vol. 2, No. 10.
Pimm, Stuart L., H. Lee Jones, and Jared Diamond. 1988. "On the Risk of Extinction." *American Naturalist*, Vol. 132, No. 6.
Plomley, N. J. B., ed. 1966. *Friendly Mission: The Tasmanian Journals and Papers of George Augustus Robinson 1829–1834*. Hobart: Tasmanian Historical Research Association.
———. 1977. *The Tasmanian Aborigines*. Launceston, Tasmania: Published by the author.
Pollock, Jonathan I. 1975. "Field Observations on *Indri indri*: A Preliminary Report." In Tattersall and Sussman (1975).
———. 1977. "The Ecology and Sociology of Feeding in *Indri indri*." In Clutton-Brock (1977).
———. 1979. "Female Dominance in *Indri indri*." *Folia Primatologica*, Vol. 31.
———. 1986a. "The Song of the Indris (*Indri indri*; Primates: Lemuroidea): Natural History, Form, and Function." *International Journal of Primatology*, Vol. 7, No. 3.
———. 1986b. "Primates and Conservation Priorities in Madagascar." *Oryx*, Vol. 20, No. 4.
———. 1986c. "A Note on the Ecology and Behavior of *Hapalemur griseus*." *Primate Conservation*, No. 7.
———. 1986d. "Towards a Conservation Policy for Madagascar's Eastern Rain Forests." *Primate Conservation*, No. 7.
Powell, A. Harriett. 1987. "Population Dynamics of Male Euglossine Bees in Amazonian Forest Fragments." *Biotropica*, Vol. 19, No. 2.
Power, Dennis M. 1972. "Numbers of Bird Species on the California Islands." *Evolution*, Vol. 26.
Prance, Ghillean T., ed. 1982. *Biological Diversification in the Tropics*. New York: Columbia University Press.
Prance, Ghillean T., and Thomas E. Lovejoy, eds. 1985. *Key Environments: Amazonia*. Oxford: Pergamon Press.
Pratt, H. Douglas, Phillip L. Bruner, and Delwyn G. Berrett. 1987. *A Field Guide to the Birds of Hawaii and the Tropical Pacific*. Princeton, N.J.: Princeton University Press.
Pratt, Thane K. 1983. "Diet of the Dwarf Cassowary *Casuarius bennetti picticollis* at Wau, Papua New Guinea." *Emu*, Vol. 82.
———. 1984. "Examples of Tropical Frugivores Defending Fruit-Bearing Plants." *Condor*, Vol. 86.
Preston, Frank W. 1948. "The Commonness, and Rarity, of Species." *Ecology*, Vol. 29, No. 3.
———. 1960. "Time and Space and the Variation of Species." *Ecology*, Vol. 41, No. 4.
———. 1962a. "The Canonical Distribution of Commonness and Rarity: Part I." *Ecology*, Vol. 43, No. 2.
———. 1962b. "The Canonical Distribution of Commonness and Rarity: Part II." *Ecology*, Vol. 43, No. 3.
———. 1968. "On Modeling Islands." *Ecology*, Vol. 49, No. 3.
Preston-Mafham, Ken. 1991. *Madagascar: A Natural History*. Oxford: Facts on File.
Primack, Richard B. 1993. *Essentials of Conservation Biology*. Sunderland, Mass.: Sinauer Associates.
Proctor-Gray, Elizabeth. 1990. "Kangaroos up a Tree." *Natural History*, January 1990.
Provine, William B. 1986. *Sewall Wright and Evolutionary Biology*. Chicago: University of Chicago Press.
———. 1987. *The Origins of Theoretical Population Genetics*. Chicago: University of Chicago Press.
Pruett-Jones, M. A., and S. G. Pruett-Jones. 1985. "Food Caching in the Tropical Frugivore, MacGregor's Bowerbird (*Amblyornis macgregoriae*)." *Auk*, Vol. 102.
Pruett-Jones, Stephen G., and Melinda A. Pruett-Jones. 1986. "Altitudinal Distribution and Seasonal Activity Patterns of Birds of Paradise." *National Geographic Research*, Vol. 2, No. 1.
———. 1988. "A Promiscuous Mating System in the Blue Bird of Paradise *Paradisaea rudolphi*." *Ibis*, Vol. 130.
Quintela, Carlos E. 1990. "An SOS for Brazil's Beleaguered Atlantic Forest." *Nature Conservancy*, March/April 1990.
Radtkey, R. R., S. C. Donnellan, R. N. Fisher, C. Moritz, K. A. Hanley, and T. J. Case. 1995. "When Species Collide: The Origin and Spread of an Asexual Species of Gecko." *Proceedings of the Royal Society of London*, Series B, Vol. 259.
Ralph, C. John. 1982. "Birds of the Forest." *Natural History*, Vol. 91, December 1982.
Raup, David M. 1991. *Extinction: Bad Genes or Bad Luck?* New York: W. W. Norton.
Raup, D. M., and J. J. Sepkoski. 1984. "Periodicity of Extinctions in the Geologic Past." *Proceedings of the National Academy of Sciences*, Vol. 81.
Raven, H. C. 1935. "Analysis of the Mammalian Fauna of the Indo-Australian Region." *Bulletin of the American Museum of Natural History*, Vol. 68.
Raxworthy, C. J. 1988. "Reptiles, Rainforest and Conservation in Madagascar." *Biological Conservation*, Vol. 43.

Ray, Chris, Michael Gilpin, and Andrew T. Smith. 1991. "The Effect of Conspecific Attraction on Metapopulation Dynamics." In Gilpin and Hanski (1991).

Reed, J. Michael, Phillip D. Doerr, and Jeffrey R. Walters. 1986. "Determining Minimum Population Sizes for Birds and Mammals." *Wildlife Society Bulletin*, Vol. 14.

Reed, Timothy M. 1983. "The Role of Species-Area Relationships in Reserve Choice: A British Example." *Biological Conservation*, Vol. 25.

Regan, Timothy W., and David S. Maehr. 1990. "Melanistic Bobcats in Florida." *Florida Field Naturalist*, Vol. 18, No. 4.

Rey, Jorge R., and Donald R. Strong, Jr. 1983. "Immigration and Extinction of Salt Marsh Arthropods on Islands: An Experimental Study." *Oikos*, Vol. 41.

Reyment, Richard A. 1983. "Palaeontological Aspects of Island Biogeography: Colonization and Evolution of Mammals on Mediterranean Islands." *Oikos*, Vol. 41.

Richard, Alison F., and Robert W. Sussman. 1975. "Future of the Malagasy Lemurs: Conservation or Extinction?" In Tattersall and Sussman (1975).

Richman, Adam D., Ted J. Case, and Terry D. Schwaner. 1988. "Natural and Unnatural Extinction Rates of Reptiles on Islands." *American Naturalist*, Vol. 131, No. 5.

Richter-Dyn, Nira, and Narendra S. Goel. 1972. "On the Extinction of a Colonizing Species." *Theoretical Population Biology*, Vol. 3.

Rick, C. M., and R. I. Bowman. 1961. "Galápagos Tomatoes and Tortoises." *Evolution*, Vol. 15.

Rickard, John. 1988. *Australia: A Cultural History*. London: Longman.

Ricklefs, Robert E. 1973. *Ecology*. Portland, Ore.: Chiron Press.

Ricklefs, Robert E., and George W. Cox. 1972. "Taxon Cycles in the West Indian Avifauna." *American Naturalist*, Vol. 106, No. 948.

Ridpath, M. G., and R. E. Moreau. 1966. "The Birds of Tasmania: Ecology and Evolution." *Ibis*, Vol. 108.

Ripley, S. Dillon. 1977. *Rails of the World: A Monograph of the Family Rallidae*. Boston: David R. Godine.

Ripley, S. Dillon, and Bruce M. Beehler. 1989. "Ornitho-Geographic Affinities of the Andaman and Nicobar Islands." *Journal of Biogeography*, Vol. 16.

Robson, Lloyd. 1989. *A Short History of Tasmania*. Melbourne: Oxford University Press.

Rodda, G. H., and T. H. Fritts. 1992. "The Impact of the Introduction of *Boiga irregularis* on Guam's Lizards." *Journal of Herpetology*, Vol. 26.

Rodda, Gordon H., Thomas H. Fritts, and James D. Reichel. 1991. "The Distributional Patterns of Reptiles and Amphibians in the Mariana Islands." *Micronesica*, Vol. 24.

Roof, Jayde C., and David S. Maehr. 1988. "Sign Surveys for Florida Panthers on Peripheral Areas of Their Known Range." *Florida Field Naturalist*, Vol. 16, No. 4.

Rose, R. W. 1986. "The Habitat, Distribution and Conservation Status of the Tasmanian Bettong, *Bettongia gaimardi* (Desmarest)." *Australian Wildlife Research*, Vol. 13.

Roth, V. Louise. 1984. "How Elephants Grow: Heterochrony and the Calibration of Developmental Stages in Some Living and Fossil Species." *Journal of Vertebrate Paleontology*, Vol. 4, No. 1.

———. 1990. "Insular Dwarf Elephants: A Case Study in Body Mass Estimation and Ecological Inference." In *Body Size in Mammalian Paleobiology: Estimation and Biological Implications*, ed. John Damuth and Bruce J. MacFadden. Cambridge: Cambridge University Press.

———. 1993. "Dwarfism and Variability in the Santa Rosa Island Mammoth (*Mammuthus exilis*): An Interspecific Comparison of Limb-bone Sizes and Shapes in Elephants." In *Third California Islands Symposium*, ed. F. G. Hochberg. Santa Barbara, Calif.: Santa Barbara Museum of Natural History.

Roth, V. Louise, and Maryrose S. Klein. 1986. "Maternal Effects on Body Size of Large Insular *Peromyscus maniculatus*: Evidence from Embryo Transfer Experiments." *Journal of Mammalogy*, Vol. 67, No. 1.

Rounsevell, D. E., and S. J. Smith. 1982. "Recent Alleged Sightings of the Thylacine (Marsupialia, Thylacinidae) in Tasmania." In Archer (1982).

Ruud, Jorgen. 1960. *Taboo: A Study of Malagasy Customs and Beliefs*. Oslo: Oslo University Press.

Ryan, James M., G. Ken Creighton, and Louise H. Emmons. 1993. "Activity Patterns of Two Species of *Nesomys* (Muridae: Nesomyinae) in a Madagascar Rain Forest." *Journal of Tropical Ecology*, Vol. 9.

Ryan, Lyndall. 1981. *The Aboriginal Tasmanians*. St. Lucia, Queensland, Aus.: University of Queensland Press.

Rylands, Anthony, and Russell Mittermeier. 1983. "Parks, Reserves, and Primate Conservation in Brazilian Amazonia." *Oryx*, Vol. 17, No. 2.

Sabath, Michael D., Laura E. Sabath, and Allen M. Moore. 1974. "Web, Reproduction and Commensals of the Semisocial Spider *Cyrtophora moluccensis* (Araneae: Araneidae) on Guam, Mariana Islands." *Micronesica*, Vol. 10, No. 1.

Salomonsen, Finn. 1974. "The Main Problems Concerning Avian Evolution of Islands." *Proceedings of the 16th International Ornithological Congress*.

Salwasser, Hal, Stephen P. Mealey, and Kathy Johnson. 1984. "Wildlife Population Viability: A Question of Risk." In *Transactions of the Forty-ninth North American Wildlife and Natural Resources Conference*, ed. Kenneth Sabol. Washington, D.C.: Wildlife Management Institute.

Sattaur, Omar. 1989. "The Shrinking Gene Pool." *New Scientist*, Vol. 123, July 29, 1989.

Sauer, Jonathan D. 1969. "Oceanic Islands and Biogeographical Theory: A Review." *Geographical Review*, Vol. 59.

Savidge, Julie A. 1984. "Guam: Paradise Lost for Wildlife." *Biological Conservation*, Vol. 30.

———. 1985. "Pesticides and the Decline of Guam's Native Birds." *Nature*, Vol. 316, July 25, 1985.

———. 1987. "Extinction of an Island Forest Avifauna by an Introduced Snake." *Ecology*, Vol. 68.

Savidge, J. A., L. Sileo, and L. M. Siegfried. 1992. "Was Disease Involved in the Decimation of Guam's Avifauna?" *Journal of Wildlife Diseases*, Vol. 28, No. 2.

Schoener, Amy. 1974. "Experimental Zoogeography: Colonization of Marine Mini-Islands." *American Naturalist*, Vol. 108, No. 964.

Schoener, Thomas W. 1974. "The Species-Area Relation Within Archipelagos: Models and Evidence from Island Land Birds." *Proceedings of the 16th International Ornithological Congress.*

——. 1983. "Rate of Species Turnover Decreases from Lower to Higher Organisms: A Review of the Data." *Oikos*, Vol. 41.

Schoener, Thomas W., and David A. Spiller. 1987. "Effect of Lizards on Spider Populations: Manipulative Reconstruction of a Natural Experiment." *Science*, Vol. 236, May 22, 1987.

Schonewald-Cox, Christine M., Steven M. Chambers, Bruce MacBryde, and W. Lawrence Thomas, eds. 1983. *Genetics and Conservation.* (The Gray Book.) Menlo Park, Calif.: Benjamin/Cummings.

Schopf, Thomas J. 1974. "Permo-Triassic Extinctions: Relation to Sea-Floor Spreading." *Journal of Geology*, Vol. 82, No. 2.

Schorger, A. W. 1955. *The Passenger Pigeon: Its Natural History and Extinction.* Madison: University of Wisconsin Press.

Schreiner, Ilse, and Donald Nafus. 1986. "Accidental Introductions of Insect Pests to Guam, 1945–1985." *Proceedings, Hawaiian Entomological Society*, Vol. 27.

Schwarzkopf, Lin, and Anthony B. Rylands. 1989. "Primate Species Richness in Relation to Habitat Structure in Amazonian Rainforest Fragments." *Biological Conservation*, Vol. 48.

Sclater, Philip Lutley. 1858a. "On the General Geographical Distribution of the Members of the Class Aves." *Journal of the Proceedings of the Linnean Society (Zoology)*, Vol. 2.

——. 1858b. "On the Zoology of New Guinea." *Journal of the Proceedings of the Linnean Society (Zoology)*, Vol. 2.

Segerstrale, Ullica. 1986. "Colleagues in Conflict: An 'In Vivo' Analysis of the Sociobiology Controversy." *Biology and Philosophy*, Vol. 1.

Servheen, Christopher. 1993. "Grizzly Bear Recovery Plan." Missoula, Mont.: U.S. Fish and Wildlife Service.

Shafer, Craig L. 1990. *Nature Reserves: Island Theory and Conservation Practice.* Washington, D.C.: Smithsonian Institution Press.

Shaffer, Mark L. 1978. "Determining Minimum Viable Population Sizes: A Case Study of the Grizzly Bear (*Ursus arctos* L.)" Ph.D. dissertation, Duke University.

——. 1981. "Minimum Population Sizes for Species Conservation." *BioScience*, Vol. 31, No. 2.

——. 1987. "Minimum Viable Populations: Coping with Uncertainty." In Soulé (1987).

——. 1990. "Population Viability Analysis." *Conservation Biology*, Vol. 4, No. 1.

Shaffer, Mark L., and Fred B. Samson. 1985. "Population Size and Extinction: A Note on Determining Critical Population Sizes." *American Naturalist*, Vol. 125, No. 1.

Shelford, Robert W. 1985. *A Naturalist in Borneo.* Singapore: Oxford University Press.

Shelton, Napier. 1975. *The Life of Isle Royale.* Washington, D.C.: National Park Service, U.S. Department of the Interior.

Shoumatoff, Alex. 1988. "Look at That." *New Yorker*, March 7, 1988.

Simberloff, Daniel S. 1969a. "Taxonomic Diversity of Island Biotas." *Evolution*, Vol. 24.

——. 1969b. "Experimental Zoogeography of Islands: A Model for Insular Colonization." *Ecology*, Vol. 50, No. 2.

——. 1974a. "Equilibrium Theory of Island Biogeography and Ecology." *Annual Review of Ecology and Systematics*, Vol. 5.

——. 1974b. "Permo-Triassic Extinctions: Effects of Area on Biotic Equilibrium." *Journal of Geology*, Vol. 82.

——. 1976a. "Experimental Zoogeography of Island: Effects of Island Size." *Ecology*, Vol. 57, No. 4.

——. 1976b. "Species Turnover and Equilibrium Island Biogeography." *Science*, Vol. 194, November 5, 1976.

——. 1978a. "Using Island Biogeographic Distributions to Determine if Colonization Is Stochastic." *American Naturalist*, Vol. 112, No. 986.

——. 1978b. "Ecological Aspects of Extinction." *Atala*, Vol. 6, Nos. 1–2.

——. 1980. "A Succession of Paradigms in Ecology: Essentialism to Materialism and Probabilism." *Synthese*, Vol. 43.

——. 1982a. "Big Advantages of Small Refuges." *Natural History*, Vol. 91, April 1982.

——. 1982b. Review of Williamson (1981). *Journal of Biogeography*, Vol. 9.

——. 1983a. "Competition Theory, Hypothesis-Testing, and Other Community Ecological Buzzwords." *American Naturalist*, Vol. 122, No. 5.

——. 1983b. "When Is an Island Community in Equilibrium?" *Science*, Vol. 220, June 17, 1983.

——. 1983c. "Biogeography: The Unification and Maturation of a Science." In *Perspectives in Ornithology*, ed. Alan H. Brush and George A. Clark, Jr. Cambridge: Cambridge University Press.

——. 1986. "Are We on the Verge of a Mass Extinction in Tropical Rainforests?" In Elliott (1986).

——. 1988. "The Contribution of Population and Community Biology to Conservation Science." *Annual Review of Ecology and Systematics*, Vol. 19.

Simberloff, Daniel S., and Lawrence G. Abele. 1976a. "Island Biogeography Theory and Conservation Practice." *Science*, Vol. 191, January 23, 1976.

——. 1976b. Untitled rebuttal to Diamond (1976a), Terborgh (1976), and Whitcomb et al. (1976). *Science*, Vol. 193, September 10, 1976.

————. 1982. "Refuge Design and Island Biogeographic Theory: Effects of Fragmentation." *American Naturalist*, Vol. 120, No. 1.

Simberloff, Daniel S., and Nicholas Gotelli. 1984. "Effects of Insularisation on Plant Species Richness in the Prairie-Forest Ecotone." *Biological Conservation*, Vol. 29.

Simberloff, Daniel S., and Edward O. Wilson. 1969. "Experimental Zoogeography of Islands: The Colonization of Empty Islands." *Ecology*, Vol. 50, No. 2.

————. 1970. "Experimental Zoogeography of Islands: A Two-Year Record of Colonization." *Ecology*, Vol. 51.

Simkin, Tom, and Richard S. Fiske. 1983. *Krakatau 1883: The Volcanic Eruption and Its Effects.* Washington, D.C.: Smithsonian Institution Press.

Simmons, Alan H. 1988. "Extinct Pygmy Hippopotamus and Early Man in Cyprus." *Nature*, Vol. 333, June 9, 1988.

Simpson, Beryl B., and Jürgen Haffer. 1978. "Speciation Patterns in the Amazonian Forest Biota." *Annual Review of Ecology and Systematics*, Vol. 9.

Simpson, George Gaylord. 1943. "Turtles and the Origin of the Fauna of Latin America." *American Journal of Science*, Vol. 241, No. 7.

————. 1961. "Historical Zoogeography of Australian Mammals." *Evolution*, Vol. 15.

————. 1977. "Too Many Lines: The Limits of the Oriental and Australian Zoogeographic Regions." *Proceedings of the American Philosophical Society*, Vol. 121, No. 2.

Slobodkin, Lawrence B. 1961. *Growth and Regulation of Animal Populations.* New York: Holt, Rinehart and Winston.

————. 1966. "Organisms in Communities." Review of MacArthur and Connell (1966). *Science*, Vol. 154, November 25, 1966.

Smith, Andrew T. 1974a. "The Distribution and Dispersal of Pikas: Consequences of Insular Population Structure." *Ecology*, Vol. 55, No. 5.

————. 1974b. "The Distribution and Dispersal of Pikas: Influences of Behavior and Climate." *Ecology*, Vol. 55, No. 6.

————. 1978. "Comparative Demography of Pikas (*Ochotona*): Effect of Spatial and Temporal Age-Specific Mortality." *Ecology*, Vol. 59, No. 1.

————. 1980. "Temporal Changes in Insular Populations of the Pika (*Ochotona princeps*)." *Ecology*, Vol. 61, No. 1.

————. 1988. "Patterns of Pika (Genus *Ochotona*) Life History Variation." In *Evolution of Life Histories: Theory and Patterns from Mammals*, ed. M. S. Boyce. New Haven, Conn.: Yale University Press.

Smith, Andrew T., and Barbara L. Ivins. 1983. "Colonization in a Pika Population: Dispersal vs. Philopatry." *Behavioral Ecology and Sociobiology*, Vol. 13.

Smith, Andrew T., and Mary M. Peacock. 1990. "Conspecific Attraction and the Determination of Metapopulation Colonization Rates." *Conservation Biology*, Vol. 4, No. 3.

Smith, Charles H. 1991. *Alfred Russel Wallace: An Anthology of His Shorter Writings.* Oxford: Oxford University Press.

Smith, Robert H. 1979. "On Selection for Inbreeding in Polygynous Animals." *Heredity*, Vol. 43, No. 2.

Smith, Steven. 1981. *The Tasmanian Tiger—1980.* Technical Report 81/1. Hobart, Tasmania, Aus.: National Parks and Wildlife Service.

Sondaar, Paul Y. 1976. "Insularity and Its Effects on Mammal Evolution." In *Major Patterns in Vertebrate Evolution*, ed. Max K. Hecht, Peter C. Goody, and Bessie M. Hecht. New York: Plenum Press.

————. 1986. "The Island Sweepstakes." *Natural History*, Vol. 95, September 1986.

Soulé, Michael E. 1966. "Trends in the Insular Radiation of a Lizard." *American Naturalist*, Vol. 100, No. 910.

————. 1972. "Phenetics of Natural Populations, Part 3: Variation in Insular Populations of a Lizard." *American Naturalist*, Vol. 106, No. 950.

————. 1980. "Thresholds for Survival: Maintaining Fitness and Evolutionary Potential." In Soulé and Wilcox (1980).

————. 1983. "What Do We Really Know About Extinction?" In Schonewald-Cox et al. (1983).

————. 1985. "What Is Conservation Biology?" *BioScience*, Vol. 35.

————, ed. 1986. *Conservation Biology: The Science of Scarcity and Diversity.* (The Yellow Book.) Sunderland, Mass.: Sinauer Associates.

————, ed. 1987. *Viable Populations for Conservation.* (The Blue Book.) Cambridge: Cambridge University Press.

————. 1989. "Risk Analysis for the Concho Water Snake." *Endangered Species Update*, Vol. 6, No. 10.

————. 1991. "Conservation: Tactics for a Constant Crisis." *Science*, Vol. 253, August 16, 1991.

Soulé, Michael E., Allison C. Alberts, and Douglas T. Bolger. 1992. "The Effects of Habitat Fragmentation on Chaparral Plants and Vertebrates." *Oikos*, Vol. 63.

Soulé, Michael E., Douglas T. Bolger, Allison C. Alberts, John Wright, Marina Sorice, and Scott Hill. 1988. "Reconstructed Dynamics of Rapid Extinctions of Chaparral-Requiring Birds in Urban Habitat Islands." *Conservation Biology*, Vol. 2, No. 1.

Soulé, Michael E., and Michael E. Gilpin. 1986. "Viability Analysis for the Concho Water Snake, *Nerodia harteri paucimaculata*." Unpublished report, submitted to Jim Johnson, U.S. Fish and Wildlife Service, Albuquerque, N. Mex. November 1, 1986.

————. 1989. "A Metapopulation Approach to Population Vulnerability: The Concho Water Snake." Unpublished paper.

Soulé, Michael E., and Daniel Simberloff. 1986. "What Do Genetics and Ecology Tell Us About the Design of Nature Reserves?" *Biological Conservation*, Vol. 35.

Soulé, Michael, and Allan J. Sloan. 1966. "Biogeography and Distribution of the Reptiles and Amphibians on Islands in the Gulf of California, Mexico." *Transactions of the San Diego Society of Natural History*, Vol. 14, No. 11.

Soulé, Michael E., and Bruce A. Wilcox. 1980. *Conservation Biology: An Evolutionary-Ecological Perspective*. (The Brown Book.) Sunderland, Mass.: Sinauer Associates.

Soulé, Michael, Bruce A. Wilcox, and Claire Holtby. 1979. "Benign Neglect: A Model of Faunal Collapse in the Game Reserves of East Africa." *Biological Conservation*, Vol. 15.

Southwood, T. R. E., and C. E. J. Kennedy. 1983. "Trees as Islands." *Oikos*, Vol. 41.

Spyer, Patricia. 1993. "The Memory of Trade: Circulation, Autochthony, and the Past in the Aru Islands (Eastern Indonesia)." Ph.D. dissertation, University of Chicago.

Stanley, Steven M. 1987. *Extinction*. New York: Scientific American Books.

Steadman, David W. 1995. "Prehistoric Extinctions of Pacific Island Birds: Biodiversity Meets Zooarchaeology." *Science*, Vol. 267, February 24, 1995.

Steen, Edwin B. 1971. *Dictionary of Biology*. New York: Harper and Row.

Stern, William L., ed. 1971. *Adaptive Aspects of Insular Evolution*. Pullman: Washington State University Press.

Stock, Chester, and E. L. Furlong. 1928. "The Pleistocene Elephants of Santa Rosa Island, California." *Science*, Vol. 68, August 10, 1928.

Stoddart, D. R. 1969. "Island Life." A review of MacArthur and Wilson (1967). *Nature*, Vol. 221, February 22, 1969.

———, ed. 1984. *Biogeography and Ecology of the Seychelles Islands*. The Hague: Dr. W. Junk.

Stoddart, D. R., D. Cowx, C. Peet, and J. R. Wilson. 1982. "Tortoises and Tourists in the Western Indian Ocean: The Curieuse Experiment." *Biological Conservation*, Vol. 24.

Stoddart, D. R., and J. F. Peake. 1979. "Historical Records of Indian Ocean Giant Tortoise Populations." *Philosophical Transactions of the Royal Society of London*, Series B, Vol. 286.

Stoddart, D. R., and Serge Savy. 1983. "Aldabra: Island of Giant Tortoises." *Ambio*, Vol. 12.

Stolzenburg, William. 1991. "The Fragment Connection." *Nature Conservancy*, July/August 1991.

Stone, Charles P., and Danielle B. Stone, eds. 1989. *Conservation Biology in Hawai'i*. Honolulu: University of Hawaii Cooperative National Park Resources Studies Unit.

Stonehouse, Bernard, and Desmond Gilmore, eds. 1977. *The Biology of Marsupials*. Baltimore: University Park Press.

Strahan, Ronald. 1983a. "Dasyurids." In Strahan (1983b).

———, ed. 1983b. *The Australian Museum Complete Book of Australian Mammals*. London: Angus & Robertson.

Strickland, H. E., and A. G. Melville. 1848. *The Dodo and Its Kindred; Or the History, Affinities, and Osteology of the Dodo, Solitaire, and Other Extinct Birds of the Islands Mauritius, Rodriguez, and Bourbon*. London: Reeve, Benham, and Reeve.

Strier, Karen. 1986. "The Behavior and Ecology of the Woolly Spider Monkey, or Muriqui (*Brachyteles arachnoides* E. Geoffroy 1806)." Ph.D. dissertation, Harvard University.

———. 1987a. "The Muriquis of Brazil." *AnthroQuest*, Vol. 38.

———. 1987b. "Ranging Behavior of Woolly Spider Monkeys, or Muriquis, *Brachyteles arachnoides*." *International Journal of Primatology*, Vol. 8, No. 6.

———. 1989. "Effects of Patch Size on Feeding Associations in Muriquis (*Brachyteles arachnoides*)." *Folia Primatologica*, Vol. 52.

———. 1992. *Faces in the Forest: The Endangered Muriqui Monkeys of Brazil*. New York: Oxford University Press.

———. 1993. "Menu for a Monkey." *Natural History*, Vol. 102, No. 3, March 1993.

Strong, Donald R., Jr., Daniel Simberloff, Lawrence G. Abele, and Anne B. Thistle, eds. 1984. *Ecological Communities: Conceptual Issues and the Evidence*. Princeton, N.J.: Princeton University Press.

Stuenes, Solweig. 1989. "Taxonomy, Habits, and Relationships of the Subfossil Madagascan Hippopotami *Hippopotamus lemerlei* and *H. madagascariensis*." *Journal of Vertebrate Paleontology*, Vol. 9, No. 3.

Suchy, Willie J., Lyman L. McDonald, M. Dale Stickland, and Stanley H. Anderson. 1985. "New Estimates of Minimum Viable Population Size for Grizzly Bears of the Yellowstone Ecosystem." *Wildlife Society Bulletin*, Vol. 13.

Sullivan, Arthur L., and Mark L. Shaffer. 1975. "Biogeography of the Megazoo." *Science*, Vol. 189, July 4, 1975.

Sulloway, Frank J. 1979. "Geographic Isolation in Darwin's Thinking: The Vicissitudes of a Crucial Idea." *Studies in the History of Biology*, Vol. 3.

———. 1982, "Darwin and His Finches: The Evolution of a Legend." *Journal of the History of Biology*, Vol. 15, No. 1.

Sussman, Robert W., and Alison F. Richard. 1986. "Lemur Conservation in Madagascar: The Status of Lemurs in the South." *Primate Conservation*, No. 7.

Swingland, Ian R. 1977. "Reproductive Effort and Life History Strategy of the Aldabran Giant Tortoise." *Nature*, Vol. 269, September 29, 1977.

———. 1988. "The Ecology and Conservation of Aldabran Giant Tortoises." In Kensley (1988).

Swingland, Ian R., and Malcolm Coe. 1978. "The Natural Regulation of Giant Tortoise Populations on Aldabra Atoll: Reproduction." *Journal of Zoology*, London, Vol. 186.

Swingland, Ian R., and C. M. Lessells. 1979. "The Natural Regulation of Giant Tortoise Populations on Aldabra Atoll: Movement Polymorphism, Reproductive Success and Mortality." *Journal of Animal Ecology*, Vol. 48.

Tattersall, Ian. 1982. *The Primates of Madagascar*. New York: Columbia University Press.

Tattersall, Ian, and Robert W. Sussman. 1975. *Lemur Biology*. New York: Plenum Press.

Temple, Stanley A. 1977. "Plant-Animal Mutualism: Coevolution with Dodo Leads to Near Extinction of Plant." *Science*, Vol. 197. August 26, 1977.

———. 1979. Response to Owadally (1979). *Science*, Vol. 203, March 30, 1979.

———. 1981. "Applied Island Biogeography and the Conservation of Endangered Island Birds in the Indian Ocean." *Biological Conservation*, Vol. 20.

———. 1983. "The Dodo Haunts a Forest." *Animal Kingdom*, February/March 1983.

———. 1986. "Recovery of the Endangered Mauritius Kestrel from an Extreme Population Bottleneck." *Auk*, Vol. 103.

Templeton, Alan R. 1980. "The Theory of Speciation Via the Founder Principle." *Genetics*, Vol. 94.

Tennant, Alan. 1985. *A Field Guide to Texas Snakes*. Austin: Texas Monthly Press.

Terborgh, John. 1971. "Distribution on Environmental Gradients: Theory and a Preliminary Interpretation of Distributional Patterns in the Avifauna of the Cordillera Vilcabamba, Peru." *Ecology*, Vol. 52.

———. 1974. "Preservation of Natural Diversity: The Problem of Extinction Prone Species." *BioScience*, Vol. 24.

———. 1975. "Faunal Equilibria and the Design of Wildlife Preserves." In Golley and Medina (1975).

———. 1976. Response to Simberloff and Abele (1976). *Science*, Vol. 193, September 10, 1976.

———. 1988. "The Big Things That Run the World: A Sequel to E. O. Wilson." *Conservation Biology*, Vol. 2, No. 4.

Terborgh, John, and Blair Winter. 1983. "A Method for Siting Parks and Reserves with Special Reference to Colombia and Ecuador." *Biological Conservation*, Vol. 27.

Tewes, Michael E. 1983. "Brush Country Cats." *Texas Parks and Wildlife*, February 1983.

———. 1986. "Ecological and Behavioral Correlates of Ocelot Spatial Patterns." Ph.D. dissertation, University of Idaho.

Tewes, Michael E., and Daniel D. Everett. 1986. "Status and Distribution of the Endangered Ocelot and Jaguarundi in Texas." In *Cats of the World: Biology, Conservation, and Management*, ed. S. Douglas Miller and Daniel D. Everett. Washington, D.C.: National Wildlife Federation.

Thomas, C. D. 1990. "What Do Real Population Dynamics Tell Us About Minimum Viable Population Sizes?" *Conservation Biology*, Vol. 4, No. 3.

Thomas, D. G. 1974. "Some Problems Associated with the Avifauna." In Williams (1974).

Thompson, Steven D., and Martin E. Nicoll. 1986. "Basal Metabolic Rate and Energetics of Reproduction in Therian Mammals." *Nature*, Vol. 321, June 12, 1986.

Thornton, Ian. 1971. *Darwin's Islands: A Natural History of the Galápagos*. Garden City, N.Y.: Natural History Press.

———. 1984. "Krakatau: The Development and Repair of a Tropical Ecosystem." *Ambio*, Vol. 13, No. 4.

———. 1986a. "Krakatau." *Australian Geographic*, Vol. 1, No. 2, April/June 1986.

———, ed. 1986b. "1985 Zoological Expedition to the Krakataus: Preliminary Report." Miscellaneous Series No. 2. Bundoora, Victoria, Aus.: La Trobe University, Department of Zoology.

———, ed. 1987. "1986 Zoological Expedition to the Krakataus: Preliminary Report." Miscellaneous Series No. 3. Bundoora, Victoria, Aus.: La Trobe University, Department of Zoology.

Thornton, I. W. B., and S. R. Graves. 1988. "Colonization of the Krakataus by Bacteria and the Development of Antibiotic Resistance." *Philosophical Transactions of the Royal Society of London*, Series B, Vol. 322.

Thornton, I. W. B., and T. R. New. 1988. "Krakatau Invertebrates: The 1980s Fauna in the Context of a Century of Recolonization." *Philosophical Transactions of the Royal Society of London*, Series B, Vol. 322.

Thornton, I. W. B., T. R. New, D. A. McLaren, H. K. Sudarman, and P. J. Vaughan. 1988. "Air-Borne Arthropod Fall-Out on Anak Krakatau and a Possible Pre-Vegetation Pioneer Community." *Philosophical Transactions of the Royal Society of London*, Series B, Vol. 322.

Thornton, I. W. B., and N. J. Rosengren. 1988. "Zoological Expeditions to the Krakatau Islands, 1984 and 1985: General Introduction." *Philosophical Transactions of the Royal Society of London*, Series B, Vol. 322.

Thornton, I. W. B., R. A. Zann, P. A. Rawlinson, C. R. Tidemann, A. S. Adikerana, and A. H. T. Widjoya. 1988. "Colonization of the Krakatau Islands by Vertebrates: Equilibrium, Succession, and Possible Delayed Extinction." *Proceedings of the National Academy of Sciences*, Vol. 85.

Tichnell, David L., Gary E. Machlis, and James R. Fazio. 1983. "Threats to National Parks: A Preliminary Survey." *Parks*, Vol. 8, No. 1.

Tilson, Ronald L., and Ulysses S. Seal, eds. 1987. *Tigers of the World: The Biology, Biopolitics, Management, and Conservation of an Endangered Species*. Park Ridge, N.J.: Noyes.

Toft, Catherine A., and Thomas W. Schoener. 1983. "Abundance and Diversity of Orb Spiders on 106 Bahamian Islands: Biogeography at an Intermediate Trophic Level." *Oikos*, Vol. 41.

Tonge, Simon. 1989. "Round Island—Saved?" In *Eilanden en Natuurbescherming* (*Islands and the Protection of Nature*). Proceedings of a symposium under the auspices of the Netherlands Commission for International Nature Protection. *Mededelingen*, No. 25.

Towns, David, and Ian Atkinson. 1991. "New Zealand's Restoration Ecology." *New Scientist*, Vol. 130, April 20, 1991.

Trauers, Robert. 1968. *The Tasmanians: The Story of a Doomed Race*. Melbourne: Cassell Australia.

Trimble, Stephen. 1989. *The Sagebrush Ocean: A Natural History of the Great Basin*. Reno: University of Nevada Press.

Tryon, Rolla. 1971. "Development and Evolution of Fern Floras of Oceanic Islands." In Stern (1971).

Tudge, Colin. 1991. "Time to Save Rhinoceroses." *New Scientist*, Vol. 131, September 28, 1991.

———. 1992. "Birds of Prey Fly Again." *New Scientist*, Vol. 135, August 22, 1992.

Turnbull, Clive. 1948. *Black War: The Extermination of the Tasmanian Aborigines*. Melbourne: F. W. Cheshire.

Turner, Tom. 1990. "Saving a Honeycreeper." *Defenders*, July/August 1990.

Turrill, W. B. 1953. *Pioneer Plant Geography: The Phytogeographical Researches of Sir Joseph Dalton Hooker*. The Hague: Martinus Nijhoff.

———. 1963. *Joseph Dalton Hooker: Botanist, Explorer, and Administrator*. London: Scientific Book Club.

Udall, James R. 1991. "The Pika Hunter." *Audubon*, March 1991.

Valle, Celio. 1982. "Campaign to Save the Highly Endangered Muriqui Now Underway in Brazil." *IUCN/SSC Primate Specialist Group Newsletter*, No. 2.

Van Balen, S., E. T. Margawati, and Sudaryanti. 1986. "Birds of the Botanical Gardens of Indonesia at Bogor." *Berita Biologi*, Vol. 3, No. 4.

Van Heekeren, H. R. 1949. "Early Man and Fossil Vertebrates on the Island of Celebes." *Nature*, Vol. 163, March 26, 1949.

Van Helvoort, B. E. "An Attempt to a Population-Genetic Analysis of the American Captive Bali Starling Population (*Leucopsar rothschildi* Stresemann 1912)." Draft for comment. Cambridge: International Council for Bird Preservation.

Vankat, John L., and Jack Major. 1978. "Vegetation Changes in Sequoia National Park, California." *Journal of Biogeography*, Vol. 5.

Vaughan, R. E. 1984. "Did the Dodo Do It?" Response to Temple. *Animal Kingdom*, February/March 1984.

Vaughan, R. E., and P. O. Wiehe. 1937. "Studies on the Vegetation of Mauritius, Part 1: A Preliminary Survey of the Plant Communities." *Journal of Ecology*, Vol. 25.

———. 1941. "Studies on the Vegetation of Mauritius, Part 3: The Structure and Development of the Upland Climax Forest." *Journal of Ecology*, Vol. 29.

Veevers-Carter, W. 1979. *Land Mammals of Indonesia*. Jakarta: PT Intermasa.

Vuilleumier, Beryl Simpson. 1971. "Pleistocene Changes in the Fauna and Flora of South America." *Science*, Vol. 173, August 27, 1971.

Vuilleumier, François. 1970. "Insular Biogeography in Continental Regions, Part 1: The Northern Andes of South America." *American Naturalist*, Vol. 104, No. 938.

Walker, Alan. 1967. "Patterns of Extinction Among the Subfossil Madagascan Lemuroids." In Martin and Wright (1967).

Wallace, Alfred Russel. 1853. *A Narrative of Travels on the Amazon and Rio Negro, with an Account of the Native Tribes, and Observations on the Climate, Geology, and Natural History of the Amazon Valley*. London: Reeve. (Revised edition: Ward, Lock and Co., London, 1889.)

———. 1854. "On the Monkeys of the Amazon." (A paper read at a meeting of the Zoological Society of London on December 14, 1852.) *Annals and Magazine of Natural History*, Vol. 14, December 1854.

———. 1855. "On the Law Which Has Regulated the Introduction of New Species." *Annals and Magazine of Natural History*, Vol. 16, September 1855. Reprinted in Wallace (1891).

———. 1857. "On the Natural History of the Aru Islands." *Annals and Magazine of Natural History*, Supplement to Vol. 20, December 1857.

———. 1858a. "Note on the Theory of Permanent and Geographical Varieities." *Zoologist*, Vol. 16, January 1858.

———. 1858b. "On the Tendency of Varieties to Depart Indefinitely from the Original Type." *Journal of the Proceedings of the Linnean Society (Zoology)*, Vol. 3, August 20, 1858. (Read to the Society, along with Darwin's contribution, on July 1, 1858.)

———. 1869. *The Malay Archipelago, the Land of the Orang-Utan and the Bird of Paradise: A Narrative of Travel with Studies of Man and Nature*. London: Macmillan. (Reprint edition: Dover, New York, 1962. The Dover edition is a reprint of the last edition revised by Wallace, 1891.)

———. 1876. *The Geographical Distribution of Animals*. Vols. I and II. New York: Harper & Brothers.

———. 1880. *Island Life, or the Phenomena and Causes of Insular Faunas and Floras, Including a Revision and Attempted Solution of the Problem of Geological Climates*. London: Macmillan. (Reprint edition: AMS Press, New York, 1975. The AMS reprint is a facsimile of the revised third edition of 1911.)

———. 1889. *Darwinism: An Exposition of the Theory of Natural Selection, with Some of Its Applications*. London: Macmillan. (Revised edition: Macmillan, London, 1905.)

———. 1891. *Natural Selection and Tropical Nature: Essays on Descriptive and Theoretical Biology*. London: Macmillan. (Facsimile edition: Gregg International Publishers, Westmead, Eng., 1969. *Natural Selection* was originally published in 1870. *Tropical Nature* was originally published in 1878.)

———. 1899. *The Wonderful Century*. New York: Dodd, Mead.

———. 1905. *My Life: A Record of Events and Opinions*. Vols. I and II. London: Chapman and Hall. (Reprint edition: Gregg International Publishers, Westmead, Eng., 1969.)

Walters, Edgar T., Thomas J. Carew, and Eric R. Kandel. 1981. "Extinctions and Introductions in the New Zealand Avifauna: Cause and Effect?" *Science*, Vol. 211, January 30, 1981.

Warner, Richard E. 1968. "The Role of Introduced Diseases in the Extinction of the Endemic Hawaiian Avifauna." *Condor*, Vol. 70.

Watson, Lyall. 1987. *The Dreams of Dragons: Riddles of Natural History*. New York: William Morrow.

Watts, Dave. 1987. *Tasmanian Mammals: A Field Guide*. Hobart: Tasmanian Conservation Trust.

Waters, Tom. 1988. "Return to Krakatau." *Discover*, October 1988.

Webb, S. David. 1969. "Extinction-Origination Equilibria in Late Cenozoic Land Mammals of North America." *Evolution*, Vol. 23.

Weiner, Jonathan. 1990. *The Next One Hundred Years: Shaping the Fate of Our Living Earth*. New York: Bantam Books.

————. 1994. *The Beak of the Finch: A Story of Evolution in Our Time.* New York: Alfred A. Knopf.
Western, David. 1992. "The Biodiversity Crisis: A Challenge for Biology." *Oikos*, Vol. 63.
Western, David, and Mary C. Pearl, eds. 1989. *Conservation for the Twenty-first Century.* New York: Oxford University Press.
Western, David, and Ssemakula, James. 1981. "The Future of the Savannah Ecosystems: Ecological Islands or Faunal Enclaves?" *African Journal of Ecology*, Vol. 19.
Western, David, and R. Michael Wright, eds. Shirley C. Strum, associate ed. 1994. *Natural Connections: Perspectives in Community-Based Conservation.* Washington, D.C.: Island Press.
Whitcomb, Robert F. 1977. "Island Biogeography and 'Habitat Islands' of Eastern Forest." *American Birds*, Vol. 31, No. 1.
Whitcomb, Robert F., James F. Lynch, Paul A. Opler, and Chandler S. Robbins. 1976. Response to Simberloff and Abele (1976a). *Science*, Vol. 193, September 10, 1976.
Whitehead, Donald R., and Claris E. Jones. 1969. "Small Islands and the Equilibrium Theory of Insular Biogeography." *Evolution*, Vol. 23.
Whitfield, Philip, ed. 1988. *The Macmillan Illustrated Encyclopedia of Birds: A Visual Who's Who in the World of Birds.* New York: Macmillan.
Whitlock, Ralph. 1981. *Birds at Risk: A Comprehensive World-Survey of Threatened Species.* Bradford-on-Avon, Wiltshire, Eng.: Moonraker Press.
Whitmore, T. C., ed. 1981. *Wallace's Line and Plate Tectonics.* Oxford: Clarendon Press.
————, ed. 1987. *Biogeographical Evolution of the Malay Archipelago.* Oxford: Clarendon Press.
Whitten, Anthony J., Sengli J. Damanik, Jazanul Anwar, and Nazaruddin Hisyam. 1987. *The Ecology of Sumatra.* Yogyakarta, Indonesia: Gadjah Mada University Press.
Whitten, Anthony J., Muslimin Mustafa, and Gregory S. Henderson. 1987. *The Ecology of Sulawesi.* Yogyakarta, Indonesia: Gadjah Mada University Press.
Whitten, Tony, and Jane Whitten; photographs by Gerald Cubitt. 1992. *Wild Indonesia: The Wildlife and Scenery of the Indonesian Archipelago.* Cambridge, Mass.: MIT Press.
Wilcox, Bruce A. 1978. "Supersaturated Island Faunas: A Species-Age Relationship for Lizards on Post-Pleistocene Land-Bridge Islands." *Science*, Vol. 199, March 3, 1978.
————. 1986. "Extinction Models and Conservation." *Trends in Ecology and Evolution*, Vol. 1, No. 2.
————. 1987. Editorial. *Conservation Biology*, Vol. 1, No. 3.
Wilcox, Bruce A., and Dennis D. Murphy. 1985. "Conservation Strategy: The Effects of Fragmentation on Extinction." *American Naturalist*, Vol. 125.
Wilcox, Bruce A., Dennis D. Murphy, Paul R. Ehrlich, and George T. Austin. 1986. "Insular Biogeography of the Montane Butterfly Faunas in the Great Basin: Comparison with Birds and Mammals." *Oecologia*, Vol. 69.
Williams, C. B. 1943. "Area and Number of Species." *Nature*, Vol. 152, September 4, 1943.
Williams, W. D., ed. 1974. *Biogeography and Ecology in Tasmania.* The Hague: Dr. W. Junk.
Williams-Ellis, Amabel. 1966. *Darwin's Moon: A Biography of Alfred Russel Wallace.* London: Blackie.
Williamson, Mark. 1969. Review of MacArthur and Wilson (1967). *Journal of Animal Ecology*, Vol. 38, No. 2.
————. 1981. *Island Populations.* Oxford: Oxford University Press.
————. 1983. "The Land-Bird Community of Skokholm: Ordination and Turnover." *Oikos*, Vol. 41.
Willis, Edwin O. 1974. "Populations and Local Extinctions of Birds on Barro Colorado Island, Panama." *Ecological Monographs*, Vol. 44.
————. 1980. "Species Reduction in Remanescent Woodlots in Southern Brazil." *Proceedings of the 17th International Ornithological Congress*, Vol. II.
————. 1984. "Conservation, Subdivision of Reserves, and the Anti-Dismemberment Hypothesis." *Oikos*, Vol. 42.
Willis, Edwin O., and Eugene Eisenmann. 1979. "A Revised List of Birds of Barro Colorado Island, Panama." *Smithsonian Contributions to Zoology*, No. 291. Washington, D.C.: Smithsonian Institution Press.
Wilson, Bob. 1989. "Promoting a Preserve for Primates." *Duke Magazine*, Vol. 75, No. 3.
Wilson, Edward O. 1958. "Patchy Distributions of Ant Species in New Guinea Rain Forests." *Psyche*, Vol. 65.
————. 1959. "Adaptive Shift and Dispersal in a Tropical Ant Fauna." *Evolution*, Vol. 13.
————. 1961. "The Nature of the Taxon Cycle in the Melanesian Ant Fauna." *American Naturalist*, Vol. 95, No. 882.
————. 1969. "The Species Equilibrium." In *Diversity and Stability in Ecological Systems*, Brookhaven Symposia in Biology, No. 22.
————. 1984. *Biophilia.* Cambridge, Mass.: Harvard University Press.
————. 1985. "In The Queendom of the Ants: A Brief Autobiography." In *Leaders in the Study of Animal Behavior: Autobiographical Perspectives*, ed. D. A. Dewsbury. Lewisburg, Pa.: Bucknell University Press.
————, ed. 1988. *Biodiversity.* Washington, D.C.: National Academy Press.
————. 1992. *The Diversity of Life.* Cambridge, Mass.: Belknap Press of Harvard University Press.
————. 1994. *Naturalist.* Washington, D.C.: Island Press.
Wilson, Edward O., and Daniel S. Simberloff. 1969. "Experimental Zoogeography of Islands: Defaunation and Monitoring Techniques." *Ecology*, Vol. 50, No. 2.
Wilson, Etta S. 1934. "Personal Recollections of the Passenger Pigeon." *Auk*, Vol. 51.
Wilson, Leonard G., ed. 1970. *Sir Charles Lyell's Scientific Journals on the Species Question.* New Haven, Conn.: Yale University Press.
Witteman, Gregory J., and Robert E. Beck, Jr. 1991. "Decline and Conservation of the Guam Rail." In *Wildlife Conservation: Present Trends and Perspectives for the 21st Century*, ed. Naoki Maruyama, Boguslaw

Bobek, Yuiti Ono, Wayne Regelin, Ludek Bartos, and Philip R. Ratcliffe. Tokyo: Japan Wildlife Research Center.

Witteman, Gregory J., Robert E. Beck, Jr., Stuart L. Pimm, and Scott R. Derrickson. 1991. "The Decline and Restoration of the Guam Rail, *Rallus owstoni*." *Endangered Species Update*, Vol. 8, No. 1.

Wolf, Edward C. 1987. "On the Brink of Extinction: Conserving the Diversity of Life." Worldwatch Paper 78. Washington, D.C.: Worldwatch Institute.

Wong, Marina, and Jorge Ventocilla. 1986. *A Day on Barro Colorado Island*. Panama City: Smithsonian Tropical Research Institute.

Wool, David. 1987. "Differentiation of Island Populations: A Laboratory Model." *American Naturalist*, Vol. 129, No. 2.

World Conservation Monitoring Centre, compilers. 1990. *1990 IUCN Red List of Threatened Animals*. Gland, Switz.: International Union for Conservation of Nature and Natural Resources.

Worster, Donald. 1977. *Nature's Economy: A History of Ecological Ideas*. Cambridge: Cambridge University Press.

———. 1984. "History as Natural History: An Essay on Theory and Method." *Pacific Historical Review*, Vol. 53, No. 1.

Wrangham, Richard W. 1980. "An Ecological Model of Female-Bonded Primate Groups." *Behaviour*, Vol. 75.

Wright, Patricia. 1988a. "Lemurs' Last Stand." *Animal Kingdom*, January/February 1988.

———. 1988b. "Lemurs Lost and Found." *Natural History*, Vol. 97, No. 7, July 1988.

———. 1990. "Patterns of Paternal Care in Primates." *International Journal of Primatology*, Vol. 11, No. 2.

———. 1992. "Primate Ecology, Rainforest Conservation, and Economic Development: Building a National Park in Madagascar." *Evolutionary Anthropology*, Vol. 1, No. 1.

Wright, Patricia C., Patrick S. Daniels, David M. Meyers, Deborah J. Overdorff, and Joseph Rabeson. 1987. "A Census and Study of *Hapalemur* and *Propithecus* in Southeastern Madagascar." *Primate Conservation*, No. 8.

Wright, Robert. 1988. *Three Scientists and Their Gods: Looking for Meaning in an Age of Information*. New York: Harper & Row.

Wright, Sewall. 1938. "Size of Population and Breeding Structure in Relation to Evolution." *Science*, Vol. 87, May 13, 1938.

Wright, S. Joseph. 1980. "Density Compensation in Island Avifaunas." *Oecologia*, Vol. 45.

Wright, S. Joseph, and Stephen P. Hubbell. 1983. "Stochastic Extinction and Reserve Size: A Focal Species Approach." *Oikos*, Vol. 41.

Wright, S. Joseph, Robert Kimsey, and Claudia J. Campbell. 1984. "Mortality Rates of Insular *Anolis* Lizards: A Systematic Effect of Island Area?" *American Naturalist*, Vol. 123, No. 4.

Younger, R. M. 1988. *Kangaroo: Images Through the Ages*. Hawthorn, Victoria: Century Hutchinson Australia.

Ziegler, Alan C. 1977. "Evolution of New Guinea's Marsupial Fauna in Response to a Forested Environment." In Stonehouse and Gilmore (1977).

———. 1982. "An Ecological Check-list of New Guinea Recent Mammals." In Gressitt (1982).

Zimmerman, B. L., and R. O. Bierregaard. 1986. "Relevance of the Equilibrium Theory of Island Biogeography and Species-Area Relations to Conservation with a Case from Amazonia." *Journal of Biogeography*, Vol. 13.

Zimmerman, Elwood C. 1971. "Adaptive Radiation in Hawaii with Special Reference to Insects." In Stern (1971).

Ziswiler, Vincenz. 1967. *Extinct and Vanishing Animals: A Biology of Extinction and Survival*. Revised English edition by Fred and Pille Bunnell. New York: Springer-Verlag.